IMPORTANT INFORMATION

PLEASE RETURN THIS BOOK TO:

____

____

____

____

phone no. ________________

Examination date: ________________ Time: ________________

location: ____________________

____

____

Examination board address: ____________________

____

____

phone no.: ____________________

Application was sent on this date: ________________

Application was accepted on this date: ________________

Information received during the examination (such as booklet numbers) follows:

Fundamentals of Engineering Review

ISBN 0-9614760-0-1

Printed in Grand Rapids, Michigan

9 8 7 6 5 4 3

ENGINEER-IN-TRAINING/ FUNDAMENTALS OF ENGINEERING REVIEW

A complete review for the FE/EIT examination

Edited by: MERLE C. POTTER, PhD, PE
Professor, Michigan State University

Contributors:

R. O. Barr, PhD, PE	Electrical Engineering
D. G. Farnum, PhD	Chemistry
B. Weinberg, PhD, PE	Computer Science
F. Hatfield, PhD, PE	Civil Engineering
K. Mukerjee, PhD	Materials Science
P. N. Parks, MS	Physics
M. C. Potter, PhD, PE	Mechanical Engineering
J. A. Soper, PhD	Electrical Engineering

published by
GREAT LAKES PRESS
P.O. Box 483
Okemos, MI 48864

Table of Contents

TABLE OF CONTENTS (continued)

TABLE OF CONTENTS (continued)

TABLE OF CONTENTS (continued)

Preface

This book was written to provide a quick, thorough review for the Fundamentals of Engineering (FE) — also known as the Engineer-In Training (EIT) — Examination. Other review books are either too detailed, with too much material included, or they simplistically present sample problems with no subject matter review. We have attempted to present a review that includes the major equations and the most important ideas and concepts in each of the thirteen major subject areas covered by the exam. Undoubtedly, there will be too few questions to cover everything. By far the majority of questions will, however, be covered so that an effective review is possible.

The practice problems are designed to be as exam-like as possible. We have included theory questions as well as numerical problems, all with five-part multiple-choice answers — just like the exam. In some chapters selected problems are starred (*) for a quicker review.

To effectively prepare for the exam you should first develop a strategy. For your information, the afternoon session of the test has five required sets of questions containing 50 questions, plus another four sets, two of which are to be selected, each containing 10 questions; these are identified in our Introduction. Hence, there are a total of seven very important subject areas from the thirteen areas included in this review book. You should predetermine the two areas to be selected so you do not waste time trying to decide during the exam, and so you can focus your review on those 7 areas. Remember, all seven areas are also covered in the morning session. So, the strategy is:

1. Choose the two areas to be selected for the afternoon session.

2. Focus your review on the seven subjects of the afternoon session.

3. Quickly review the other areas you are familiar with.

4. Do not spend time reviewing material that you are unfamiliar with and which is not included in the afternoon session.

5. Become familiar with one review book and use it. Do not take a pile of books into the exam room.

6. Prepare a one-page summary of the important formulas and information in each subject area for quick reference.

Keep in mind that the objective of your review is to pass the Fundamentals of Engineering Examination — your first step toward becoming a registered engineer. The point is not to gain once again a proficiency in all those subject areas that you have not been directly involved with. So use your time wisely by reviewing in detail the afternoon subject areas (you'll be tested over those areas twice). Get a good night's sleep before the exam, and good luck!

Merle C. Potter

Okemos, Michigan
November 1, 1984

Introduction

To become registered as an engineer, a state may require that you:

1. Graduate from an ABET-accredited engineering program.

2. Pass the Fundamentals of Engineering (FE) exam — formerly the Engineer-In-Training (EIT) exam.

3. After several years of engineering experience, pass the Principles and Practice of Engineering (PE) exam.

Registration is necessary if an engineer works as a consultant, and is highly recommended in certain industries, especially when one is in — or hopes to be in — a management position.

This book presents a review of the subject areas the FE/EIT exam covers, although certain chapters are also recommended for review in preparation for the PE exam. This book contains:

- Short, succinct reviews of the theoretical aspects of the subject areas.
- Example problems with detailed solutions illustrating important concepts.
- Practice problems similar to those that could be encountered on the FE/EIT exam.
- Solutions to all the practice problems.

The FE/EIT exam is open book; textbooks, handbooks, bound reference books, and a silent calculator (it may be preprogrammed) are allowed. Unbound tables or notes, and exchanging of materials during the exam are not allowed. Blank pages are included at the end of this book so you can prepare summaries of equations and information for quick reference. Scratch paper is provided in every test booklet.

The FE/EIT exam is composed of 2 four-hour sessions separated by a one-hour break. The morning session has 140 multiple choice questions covering the subject areas listed in the following table:

Morning Session

Subject Area	*Approximate Number*
Chemistry	10
Computer Science	8
Dynamics	13
Economics	6
Electrical Theory	18
Fluids	14
Materials Science	6
Mathematics	12
Mechanics of Materials	13
Physics	5
Statics	13
Systems Theory	8
Thermodynamics	14

Each problem is worth one-half point so a maximum of 70 is possible.

The afternoon session consists of 100 multiple choice questions, but only 70 of the questions may be answered. Questions must be answered from economics, electrical theory, mathematics, and mechanics. They are arranged as follows:

Afternoon Session

Required Subjects	*Number*
Economics	10
Electrical Theory	10
Mathematics	15
Mechanics (Dynamics and Statics)	15
Other subjects (select two)	
Computer Science	10
Electrical Theory	10
Fluids	10
Mechanics of Materials	10
Thermodynamics	10

Each problem is worth one point so the two-session total is 140 points with both sessions having equal weight. A predetermined passing percentage is not established nor is the exam graded on a curve.

All questions can be worked using either English units or SI units; the answers will be the same using either set of units. The problems in this book are in SI units only, except that practice problem sets in the three chapters involving pressure or stress are also given in English units. A table of conversions is presented in this Introduction.

Answers are seldom given with more than three significant figures, and may be given with two significant figures. The closest answer should be selected.

All State Boards of Registration administer the National Council of Engineering Examiners (NCEE) uniform examination, described above. The dates of the exams include a span of three days in April and three days in October. The specific dates are selected by each State Board. Also, to be accepted to take the FE/EIT exam an applicant must apply well in advance. For information regarding specific requirements by a State Board, contact the State Board; State Board addresses can be obtained from the Executive Director of NCEE, P.O. Box 1686, Clemson, SC 29633-1686.

State Boards of Registration Phone Numbers

Alabama	(205) 832-6100
Alaska	(907) 465-2540
Arizona	(602) 255-4053
Arkansas	(501) 371-2517
California	(916) 445-5544
Colorado	(303) 866-2396
Connecticut	(203) 566-3386
Delaware	(302) 656-7311
District of Columbia	(202) 727-6038
Florida	(904) 488-9912
Georgia	(404) 656-3926
Hawaii	(808) 548-3086
Idaho	(208) 334-3860
Illinois	(217) 785-0872
Indiana	(317) 232-1840
Iowa	(515) 281-5602
Kansas	(913) 296-3053
Kentucky	(502) 564-2680
Louisiana	(504) 581-7938
Maine	(207) 289-3236
Maryland	(301) 659-6322
Massachusetts	(617) 727-3088
Michigan	(517) 373-3880
Minnesota	(612) 296-2388
Mississippi	(601) 354-7241
Missouri	(314) 751-2334
Montana	(406) 449-3737
Nebraska	(402) 471-2021
Nevada	(702) 329-1955
New Hampshire	(603) 271-2219
New Jersey	(201) 648-2660
New Mexico	(505) 827-9940
New York	(518) 474-3846
North Carolina	(919) 781-9499
North Dakota	(701) 852-1220
Ohio	(614) 466-8948
Oklahoma	(405) 521-2874
Oregon	(503) 378-4180
Pennsylvania	(717) 783-7049
Puerto Rico	(809) 725-7060
Rhode Island	(401) 277-2565
South Carolina	(803) 758-2855
South Dakota	(605) 394-2510
Tennessee	(615) 741-3221
Texas	(512) 475-3141
Utah	(801) 530-6628
Vermont	(802) 828-2363
Virginia	(804) 786-8818
Washington	(206) 753-6966
West Virginia	(304) 348-3554
Wisconsin	(608) 266-1397

The morning session consists of 140 problems that are, for the most part, unrelated. Consequently, *less than two minutes* (on the average) can be spent on each problem. This makes fast recall essential as time will not allow a book search for helpful hints. A summary sheet of important equations is most useful for quick reference.

The afternoon session is different in that, for most of the subject areas, the problems are related to a common situation; thus 35 minutes may be spent solving a more difficult problem that is divided into 10 sub-problems. This allows approximately 3½ minutes per problem so that time may be spent in some research. However, quick recall is still very helpful so that more time can be spent on the difficult problems.

If time permits only a quick review, be sure to study those subjects that are included on both the morning and afternoon sessions. Predetermine the two optional subject areas you will select; this will save time during the exam so you will not have to read over the other three groups of questions. Also, the practice problems that have been starred should be selected for a quick review.

The following tables presenting SI (Systems International) units and their conversion to SI units conclude this introduction.

Si Prefixes

Multiplication Factor	*Prefix*	*Symbol*
10^{15}	peta	P
10^{12}	tera	T
10^{9}	giga	G
10^{6}	mega	M
10^{3}	kilo	k
10^{-1}	deci	d
10^{-2}	centi	c
10^{-3}	mili	m
10^{-6}	micro	μ
10^{-9}	nano	n
10^{-12}	pico	p
10^{-15}	femto	f

SI base units

Quantity	*Name*	*Symbol*
length	meter	m
mass	kilogram	kg
time	second	s
electric current	ampere	A
temperature	kelvin	K
amount of substance	mole	mol
luminous intensity	candela	cd

SI Derived Units

Quantity	*Name*	*Symbol*	*In Terms of Other Units*
area	square meter		m^2
volume	cubic meter		m^3
velocity	meter per second		m/s
acceleration	meter per second squared		m/s^2
density	kilogram per cubic meter		kg/m^3
specific volume	cubic meter per kilogram		m^3/kg
frequency	hertz	Hz	s^{-1}
force	newton	N	$m \cdot kg/s^2$
pressure, stress	pascal	Pa	$kg/(m \cdot s^2)$
energy, work, heat	joule	J	N·m
power	watt	W	J/s
electric charge	coulomb	C	A·s
electric potential	volt	V	W/A
capacitance	farad	F	C/V
electric resistance	ohm	Ω	V/A
conductance	siemens	S	A/V
magnetic flux	weber	Wb	V·s
inductance	henry	H	Wb/A
viscosity	pascal second		Pa·s
moment (torque)	meter newton		N·m
heat flux	watt per square meter		W/m^2
entropy	joule per kelvin		J/K
specific heat	joule per kilogram-kelvin		J/(kg·K)
conductivity	watt per meter-kelvin		W/(m·K)

Conversion Factors to SI Units

English	*SI*	*SI Symbol*	*To Convert from English to SI Multiply by*
	Area		
square inch	square centimeter	cm^2	6.452
square foot	square meter	m^2	0.09290
acre	hectare	ha	0.4047
	Length		
inch	centimeter	cm	2.54
foot	meter	m	0.3048
mile	kilometer	km	1.6093
	Volume		
cubic inch	cubic centimeter	cm^3	16.387
cubic foot	cubic meter	m^3	0.02832
gallon	cubic meter	m^3	0.004546
gallon	liter	ℓ	3.785
	Mass		
pound mass	kilogram	kg	0.4536
slug	kilogram	kg	14.59
	Force		
pound	newton	N	4.448
kip (1000 lb)	newton	N	4448
	Density		
pound/cubic foot	kilogram/cubic meter	kg/m^3	16.02
pound/cubic foot	grams/liter	g/ℓ	16.02
	Work, Energy, Heat		
foot-pound	joule	J	1.356
BTU	joule	J	1055
BTU	kilowatt-hour	kWh	0.000293
therm	kilowatt-hour	kWh	29.3
quad	giga joule	GJ	1.055×10^9

Conversion Factors to SI Units (continued)

English	SI	SI Symbol	To Convert from English to SI Multiply by
	Power, Heat Rate		
horsepower	watt	W	745.7
foot pound/sec	watt	W	1.356
BTU/hour	watt	W	0.2931
BTU/hour-ft²-°F	watt/meter squared-degree celsius	W/m²·°C	5.678
tons of refrig.	kilowatts	kW	3.517
	Pressure		
pound/square inch	kilopascal	kPa	6.895
pound/square foot	kilopascal	kPa	0.04788
inches of H_20	kilopascal	kPa	0.2486
inches of Hg	kilopascal	kPa	3.374
one atmosphere	kilopascal	kPa	101.3
	Temperature		
Fahrenheit	Celsius	°C	5/9(°F-32)
Fahrenheit	kelvin	K	5/9(°F+460)
	Velocity		
foot/second	meter/second	m/s	0.3048
mile/hour	meter/second	m/s	0.4470
mile/hour	kilometer/hour	km/h	1.609
	Acceleration		
foot/second squared	meter/second squared	m/s^2	0.3048
	Torque		
pound-foot	newton-meter	N·m	1.356
pound-inch	newton-meter	N·m	0.1130
	Viscosity, Kinematic Viscosity		
pound-sec/square foot	newton-sec/square meter	$N \cdot s/m^2$	47.88
square foot/second	square meter/second	m^2/s	0.09290
	Flow Rate		
cubic foot/second	cubic meter/second	m/s	0.0004719
cubic foot/second	liter/second	ℓ/s	0.4719
	Frequency		
cycles/second	hertz	Hz	1.00

Conversion Factors

Length

1 cm	= 0.3937 in
1 m	= 3.281 ft
1 yd	= 3 ft
1 mi	= 5280 ft
1 mi	= 1760 yd
1 km	= 3281 ft

Area

1 cm^2	= 0.155 in^2
1 m^2	= 10.76 ft^2
1 ha	= 10^4 m^2
1 are	= 100 m^2
1 acre	= 4047 m^2
1 acre	= 43560 ft^2

Volume

1 ft^3	= 28.32 ℓ
1 ℓ	= 0.03531 ft^3
1 ℓ	= 0.2642 gal
1 m^3	= 264.2 gal
1 ft^3	= 7.481 gal
1 m^3	= 35.31 ft^3
1 acre-ft	= 43560 ft^2

Velocity

1 m/s	= 3.281 ft/s
1 mph	= 1.467 ft/s
1 mph	= 0.8684 knot
1 knot	= 1.688 ft/s
1 km/h	= 0.2778 m/s
1 km/h	= 0.6214 mph

Force

1 lb	= 4.448 x 10^5 dyne
1 lb	= 32.17 pdl
1 lb	= 0.4536 kg
1 N	= 10^5 dyne
1 N	= 0.2248 lb
1 kip	= 1000 lb

Mass

1 oz	= 23.35 g
1 lb	= 0.4536 kg
1 kg	= 2.205 lb
1 slug	= 14.59 kg
1 slug	= 32.17 lb

Work and Heat

1 Btu	= 778.2 ft-lb
1 Btu	= 1055 J
1 Cal	= 3.088 ft-lb
1 J	= 10^7 ergs
1 kJ	= 0.9478 ft-lb
1 Btu	= 0.2929 W · hr
1 ton	= 12000 Btu
1 kWh	= 3414 Btu
1 quad	= 10^{15} Btu
1 therm	= 10^5 Btu

Power

1 Hp	= 550 ft-lb/s
1 HP	= 33000 ft-lb/min
1 Hp	= 0.7067 Btu/s
1 Hp	= 2545 Btu/hr
1 Hp	= 745.7 W
1 W	= 3.414 Btu/hr
1 kW	= 1.341 Hp

Volume Flow Rate

1 cfm	= 7.481 gal/min
1 cfm	= 0.4719 ℓ/s
1 m^3/s	= 35.31 ft^3/s
1 m^3/s	= 2119 cfm
1 gal/min	= 0.1337 cfm

Torque

1 N · m	= 10^7 dyne · cm
1 N · m	= 0.7376 lb-ft
1 N · m	= 10197 g · cm
1 lb-ft	= 1.356 N · m

Viscosity

1 lb-s/ft^2	= 478 poise
1 poise	= 1 g/cm · s
1 N · s/m^2	= 0.02089 lb-s/ft^2

Pressure

1 atm	= 14.7 psi
1 atm	= 29.92 in Hg
1 atm	= 33.93 ft H_20
1 atm	= 1.013 bar
1 atm	= 1.033 kg/cm^2
1 atm	= 101.3 kPa
1 psi	= 2.036 in Hg
1 psi	= 6.895 kPa
1 psi	= 68950 dyne/cm^2
1 ft H_20	= 0.4331 psi

1. Mathematics

The engineer uses mathematics as a tool to help solve the problems encountered in the analysis and design of physical systems. We will review those parts of mathematics that are used fairly often by the engineer and which may appear on the exam. The topics include: algebra, trigonometry, analytic geometry, linear algebra (matrices), calculus, differential equations, and probability and statistics. The review here is intended to be brief and not exhaustive. The majority of the questions on the exam will be answerable based on the material included in this chapter. There may be a few questions, however, that will require information not covered here; to cover all possible points would not be in the spirit of this review.

1.1 Algebra

It is assumed that the reader is familiar with most of the rules and laws of algebra, as applied to both real and complex numbers. We will review some of the more important of these and illustrate several with examples. The three basic rules are:

commutative law:	$a + b = b + a$	$ab = ba$	(1.1.1)
distributive law:	$a(b+c) = ab + ac$		(1.1.2)
associative law:	$a + (b+c) = (a+b) + c$	$a(bc) = (ab)c$	(1.1.3)

1.1.1 Exponents

Laws of exponents are used in many manipulations. For positive x and y we use

$$\begin{aligned} x^{-a} &= 1/x^a \\ x^a x^b &= x^{a+b} \\ (xy)^a &= x^a y^a \\ x^{ab} &= (x^a)^b \end{aligned} \qquad (1.1.4)$$

1.1.2 Logarithms

Logarithms are actually exponents. For example, if $b^x = y$ then $x = \log_b y$; that is, the exponent x is equal to the logarithm of y to the base b. Most engineering applications involve common logs which have a base of 10, written as $\log y$, or natural logs which have a base of e ($e = 2.7183\cdots$), written as $\ln y$. If any other base is used it will be so designated, such as $\log_5 y$.

Remember, logarithms of numbers less than one are negative, the logarithm of one is zero, and logarithms of numbers greater than one are positive. The following identities are often useful when manipulating logarithms:

$$\begin{aligned} \ln x^a &= a \ln x \\ \ln(xy) &= \ln x + \ln y \\ \ln(x/y) &= \ln x - \ln y \\ \ln x &= 2.303 \log x \\ \log_b b &= 1 \\ \ln 1 &= 0 \\ \ln e^a &= a \end{aligned} \qquad (1.1.5)$$

1.1.3 The Quadratic Formula and the Binomial Theorem

We often encounter the quadratic equation $ax^2 + bx + c = 0$ when solving engineering problems. The quadratic formula provides its solution; it is

$$x = \frac{-b \pm \sqrt{b^2 - 4ac}}{2a} . \tag{1.1.6}$$

If $b^2 < 4ac$, the two roots are complex numbers. Cubic and higher order equations are most often solved by trial and error.

The binomial theorem is used to expand an algebraic expression of the form $(a + x)^n$. It is

$$(a + x)^n = a^n + n\,a^{n-1}x + \frac{n(n-1)}{2!}\,a^{n-2}\,x^2 + \cdots \tag{1.1.7}$$

If n is a positive integer, the expansion contains $(n + 1)$ terms. If it is a negative integer or a fraction, an infinite series expansion results.

1.1.4 Partial Fractions

A rational fraction $P(x)/Q(x)$, where $P(x)$ and $Q(x)$ are polynomials, can be resolved into partial fractions for the following cases.

Case 1: $Q(x)$ factors into n different linear terms,

$$Q(x) = (x-a_1)(x-a_2)\cdots(x-a_n).$$

Then

$$\frac{P(x)}{Q(x)} = \sum_{i=1}^{n} \frac{A_i}{x - a_i} . \tag{1.1.8}$$

Case 2: $Q(x)$ factors into n identical terms,

$$Q(x) = (x-a)^n.$$

Then

$$\frac{P(x)}{Q(x)} = \sum_{i=1}^{n} \frac{A_i}{(x-a)^i} . \tag{1.1.9}$$

Case 3: $Q(x)$ factors into n different quadratic terms,

$$Q(x) = (x^2+a_1\,x+b_1)(x^2+a_2\,x+b_2)\cdots(x^2+a_n\,x+b_n).$$

Then

$$\frac{P(x)}{Q(x)} = \sum_{i=1}^{n} \frac{A_i\,x+B_i}{x^2+a_i\,x+b_i} . \tag{1.1.10}$$

Case 4: $Q(x)$ factors into n identical quadratic terms,

$$Q(x) = (x^2 + ax + b)^n.$$

Then

$$\frac{P(x)}{Q(x)} = \sum_{i=1}^{n} \frac{A_i x + B_i}{(x^2 + a x + b)^i}. \tag{1.1.11}$$

Case 5: $Q(x)$ factors into a combination of the above. The partial fractions are the obvious ones from the appropriate expansions above.

EXAMPLE 1.1

The temperature at a point in a body is given by $T(t) = 100e^{-0.02t}$. At what value of t is $T = 20$?

Solution. The equation takes the form

$$20 = 100\, e^{-0.02t}$$

$$0.2 = e^{-0.02t}.$$

Take the natural logarithm of both sides and obtain

$$\ln 0.2 = \ln e^{-0.02t}.$$

Using a calculator, we find

$$-1.6094 = -0.02t.$$

$$\therefore t = 80.47.$$

EXAMPLE 1.2

Find an expansion for $(9 + x)^{1/2}$.

Solution. Using the binomial theorem Eq. 1.1.7, we have

$$(9 + x)^{1/2} = 3(1 + \frac{x}{9})^{1/2}$$

$$= 3\left[1 + \frac{1}{2}\left(\frac{x}{9}\right) + \frac{1/2(-1/2)}{2}\left(\frac{x}{9}\right)^2 + \frac{1/2(-1/2)(-3/2)}{3\cdot 2}\left(\frac{x}{9}\right)^3 + \cdots\right]$$

$$= 3 + \frac{x}{6} - \frac{x^2}{216} + \frac{x^3}{3888} + \cdots .$$

EXAMPLE 1.3

Resolve $\dfrac{x^2 + 2}{x^4 + 4x^3 + x^2}$ into partial fractions.

Solution. The denominator is factored into

$$x^4 + 4x^3 + x^2 = x^2 (x^2 + 4x + 1).$$

Using Cases 2 and 3 there results

$$\frac{x^2 + 2}{x^4 + 4x^3 + x^2} = \frac{A_1}{x} + \frac{A_2}{x^2} + \frac{A_3 x + B_3}{x^2 + 4x + 1} .$$

This can be written as

$$\frac{x^2 + 2}{x^4 + 4x^3 + x^2} = \frac{A_1 x(x^2 + 4x + 1) + A_2(x^2 + 4x + 1) + (A_3 x + B_3) x^2}{x^2(x^2 + 4x + 1)}$$

$$= \frac{(A_1 + A_3)x^3 + (4A_1 + A_2 + B_3)x^2 + (A_1 + 4A_2)x + A_2}{x^2(x^2 + 4x + 1)}$$

The numerators on both sides must be equal. Equating the coefficients of the various powers of x provides us with the four equations:

$$A_1 + A_3 = 0$$

$$4A_1 + A_2 + B_3 = 1$$

$$A_1 + 4 A_2 = 0$$

$$A_2 = 2 .$$

These are solved quite easily to give $A_2 = 2$, $A_1 = -8$, $A_3 = 8$, $B_3 = 31$.

Finally,

$$\frac{x^2 + 2}{x^4 + 4x^3 + x^2} = -\frac{8}{x} + \frac{2}{x^2} + \frac{8x + 31}{x^2 + 4x + 1} .$$

1.2 Trigonometry

The primary functions in trigonometry involve the ratios between the sides of a right triangle. Referring to the right triangle in Fig. 1.1 on the next page, the functions are defined by

$$\sin \theta = \frac{y}{r}, \qquad \cos \theta = \frac{x}{r}, \qquad \tan \theta = \frac{y}{x} . \tag{1.2.1}$$

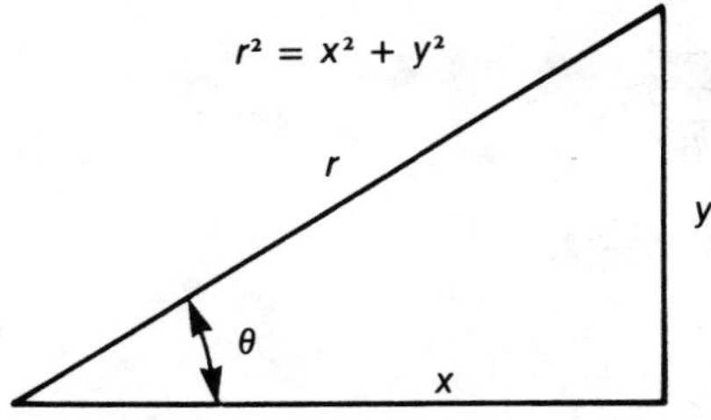

Figure 1.1 **A right triangle.**

In addition, there are three other functions that find occasional use, namely,

$$\cot\theta = \frac{x}{y}, \qquad \sec\theta = \frac{r}{x}, \qquad \csc\theta = \frac{r}{y}. \tag{1.2.2}$$

The trig functions are periodic functions with a period of 2π. Fig. 1.2 shows a plot of the three primary functions.

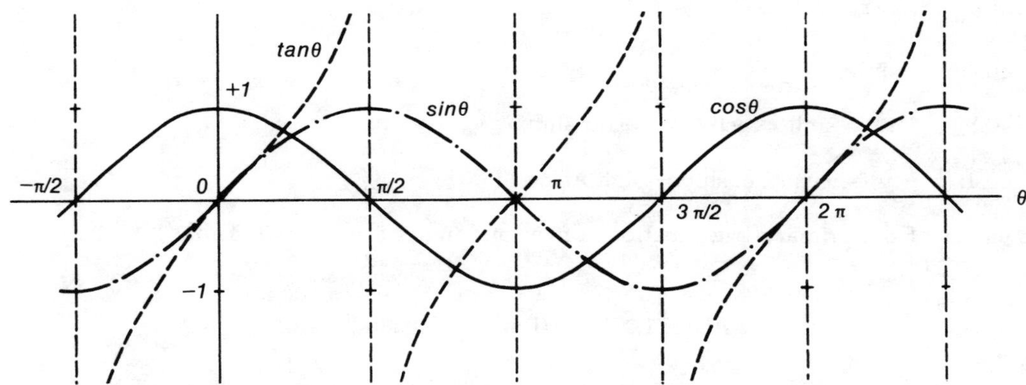

Figure 1.2. **The Trig Functions.**

In the above relationships, the angle θ is usually given in radians for mathematical equations. It is possible, however, to express the angle in degrees; if that is done it may be necessary to relate degrees to radians. This can be done remembering that there are 2π radians in 360°. Hence, we multiply radians by $(180/\pi)$ to obtain degrees, or multiply degrees by $(\pi/180)$ to obtain radians. A calculator may use either degrees or radians for an input angle.

Most problems involving trigonometry can be solved using a few fundamental identities. They are

$$\sin^2\theta + \cos^2\theta = 1 \tag{1.2.3}$$

$$\sin 2\theta = 2\sin\theta\cos\theta \tag{1.2.4}$$

$$\cos 2\theta = \cos^2\theta - \sin^2\theta \tag{1.2.5}$$

$$\sin(\alpha \pm \beta) = \sin\alpha\cos\beta \pm \sin\beta\cos\alpha \tag{1.2.6}$$

$$\cos(\alpha \pm \beta) = \cos\alpha\cos\beta \mp \sin\alpha\sin\beta. \tag{1.2.7}$$

A general triangle may be encountered, such as that shown in Fig. 1.3. For this triangle we may use the following equations:

$$\text{law of sines:} \quad \frac{\sin \alpha}{a} = \frac{\sin \beta}{b} = \frac{\sin \gamma}{c} \tag{1.2.8}$$

$$\text{law of cosines: } a^2 = b^2 + c^2 - 2bc \cos \alpha. \tag{1.2.9}$$

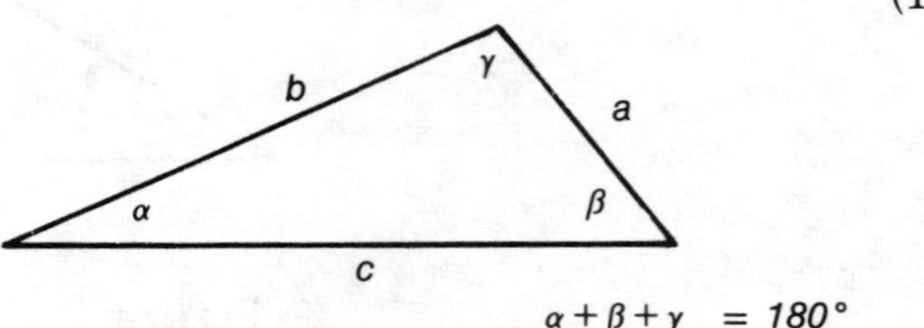

Figure 1.3. **A general triangle.**

Note that if $\alpha = 90°$, the law of cosines becomes the Pythagorean formula

$$a^2 = b^2 + c^2. \tag{1.2.10}$$

The hyperbolic trig functions also find occasional use. They are defined by

$$\sinh x = \frac{e^x - e^{-x}}{2}, \qquad \cosh x = \frac{e^x + e^{-x}}{2}, \qquad \tanh x = \frac{\sinh x}{\cosh x}. \tag{1.2.11}$$

Useful identities follow:

$$\cosh^2 x - \sinh^2 x = 1 \tag{1.2.12}$$

$$\sinh (x + y) = \sinh x \cosh y + \cosh x \sinh y \tag{1.2.13}$$

$$\cosh (x + y) = \cosh x \cosh y + \sinh x \sinh y. \tag{1.2.14}$$

The values of the primary trig functions of certain angles are listed in Table 1.1.

TABLE 1.1. Functions of Certain Angles.

	0	30°	45°	60°	90°	135°	180°	270°	360°
$\sin \theta$	0	1/2	$\sqrt{2}/2$	$\sqrt{3}/2$	1	$\sqrt{2}/2$	0	-1	0
$\cos \theta$	1	$\sqrt{3}/2$	$\sqrt{2}/2$	1/2	0	$-\sqrt{2}/2$	-1	0	1
$\tan \theta$	0	$1/\sqrt{3}$	1	$\sqrt{3}$	∞	-1	0	$-\infty$	0

EXAMPLE 1.4

Express $\cos^2\theta$ as a function of $\cos 2\theta$.

Solution. Substitute Eq. 1.2.3 into Eq. 1.2.5 and obtain

$$\cos 2\theta = \cos^2\theta - (1 - \cos^2\theta)$$
$$= 2\cos^2\theta - 1.$$

There results

$$\cos^2\theta = \frac{1}{2}(1 + \cos 2\theta).$$

EXAMPLE 1.5

If the $\sin\theta = x$, what is $\tan\theta$?

Solution. Think of $x = x/1$. Thus, the hypotenuse is of length unity and the side opposite θ is of length x. The other side is of length $\sqrt{1-x^2}$. Hence,

$$\tan\theta = \frac{x}{\sqrt{1-x^2}}\ .$$

EXAMPLE 1.6

An airplane leaves Lansing flying due southwest at 300 km/hr, and a second leaves Lansing at the same time flying due west at 500 km/hr. How far apart are the airplanes after 2 hours?

Solution. After 2 hours, the respective distances from Lansing are 600 km and 1000 km. A sketch is quite helpful. The distance d that the two airplanes are apart is found using the law of cosines. It is

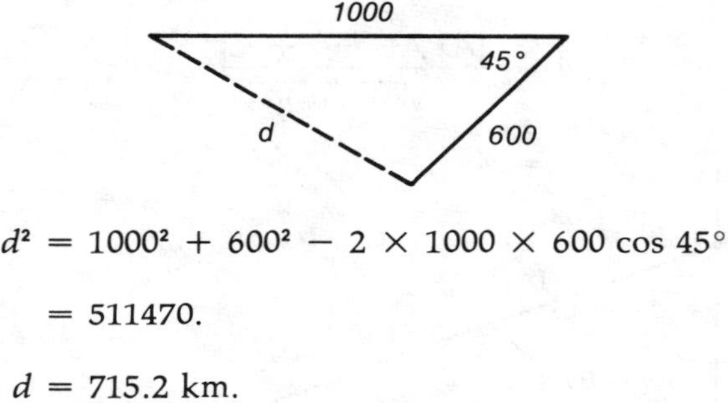

$$\begin{aligned} d^2 &= 1000^2 + 600^2 - 2 \times 1000 \times 600 \cos 45° \\ &= 511470. \\ \therefore d &= 715.2 \text{ km}. \end{aligned}$$

1.3 Geometry

A regular polygon with n sides has a vortex angle (the central angle subtended by one side) of $2\pi/n$. The included angle between two successive sides is given by $\pi(n-2)/n$.

Some common geometric shapes are displayed in Fig. 1.4.

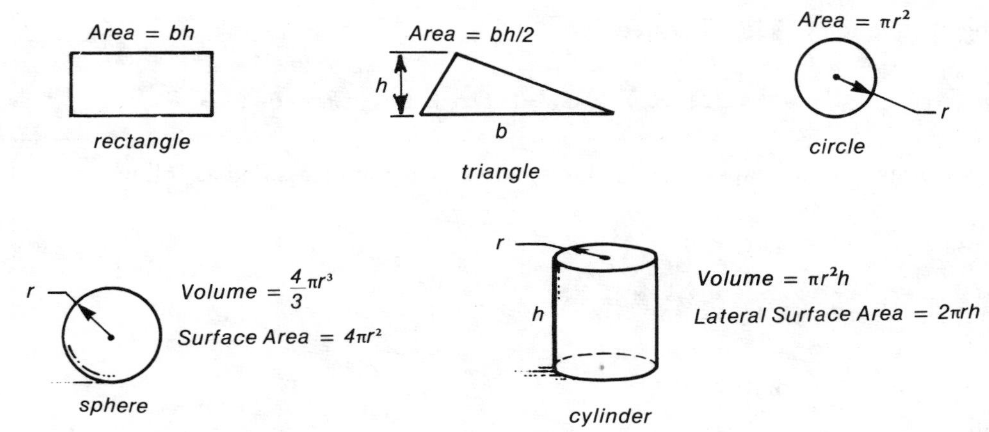

Figure 1.4. Some common geometric shapes.

The equation of a straight line can be written in the general form

$$Ax + By + C = 0. \tag{1.3.1}$$

There are three particular forms that this equation can take; they are:

Point-slope: $$y - y_1 = m(x - x_1) \tag{1.3.2}$$

Slope-intercept: $$y = mx + b \tag{1.3.3}$$

Two-intercept: $$\frac{x}{a} + \frac{y}{b} = 1. \tag{1.3.4}$$

In the above equations "a" is the x-intercept and "b" is the y-intercept.

The perpendicular distance d from the point (x_3, y_3) to the line $Ax + By + C = 0$ is given by (see Fig. 1.5)

$$d = \frac{|Ax_3 + By_3 + C|}{\sqrt{A^2 + B^2}} \tag{1.3.5}$$

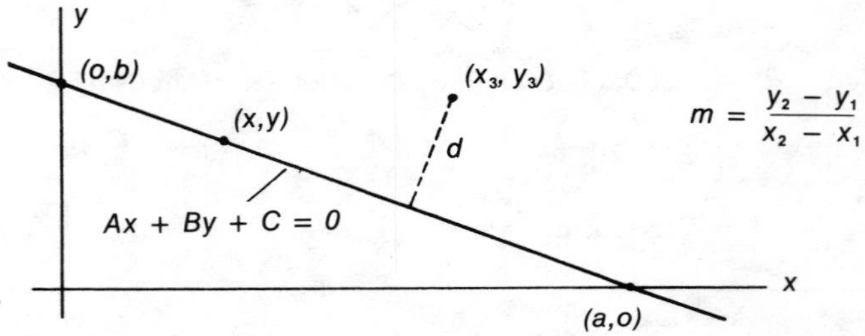

Figure 1.5 **A straight line.**

The equation of a plane surface is given as

$$Ax + By + Cz + D = 0 \tag{1.3.6}$$

The general equation of second degree

$$Ax^2 + Bxy + Cy^2 + Dx + Ey + F = 0 \tag{1.3.7}$$

represents a set of geometric shapes called *conic sections.* They are classified as follows:

$$\begin{aligned} &\text{ellipse:} && B^2 - 4AC < 0 \\ &\text{parabola:} && B^2 - 4AC = 0 \\ &\text{hyperbola:} && B^2 - 4AC > 0. \end{aligned} \tag{1.3.8}$$

If $A = B = C = 0$, the equation represents a line in the xy-plane, not a parabola. Let's consider each in detail.

Circle: The circle is a special case of an ellipse with $A = C$. Its general form can be expressed as

$$(x - a)^2 + (y - b)^2 = r^2 \tag{1.3.9}$$

where its center is at (a,b) and r is the radius.

Ellipse: The sum of the distances from the two foci to any point on an ellipse is a constant. For an ellipse centered at the origin

$$\frac{x^2}{a^2} + \frac{y^2}{b^2} = 1 \tag{1.3.10}$$

where a and b are the semimajor and semiminor axes. The foci are at $(\pm c,0)$ where $c^2 = a^2 - b^2$. See Fig. 1.6a.

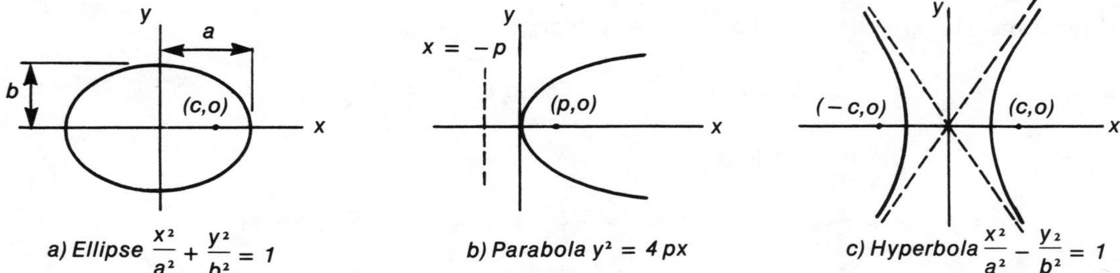

Figure 1.6 **The three conic sections.**

Parabola: The locus of points on a parabola are equidistant from the focus and a line (the directrix). If the vertex is at the origin and the parabola opens to the right, it is written as

$$y^2 = 4px \tag{1.3.11}$$

where the focus is at $(p,0)$ and the directrix is at $x = -p$. See Fig. 1.6b. For a parabola opening to the left, simply change the sign of p. For a parabola opening upward or downward, interchange x and y.

Hyperbola: The difference of the distances from the foci to any point on a hyperbola is a constant. For a hyperbola centered at the origin opening left and right, the equation can be written as

$$\frac{x^2}{a^2} - \frac{y^2}{b^2} = 1 \tag{1.3.12}$$

The lines to which the hyperbola is asymtotic are given by

$$y = \pm \frac{b}{a} x. \tag{1.3.13}$$

If the lines are perpendicular, a rectangular hyperbola results. If the asymtotes are the x and y axes, the equation can be written as

$$xy = \pm k^2. \tag{1.3.14}$$

Finally, in our review of geometry, we will present three other coordinate systems often used in engineering analysis. They are the polar (r,θ) coordinate system, the cylindrical (r,θ,z) coordinate system, and the spherical (r,θ,ϕ) coordinate system. For the plane, polar coordinate system

$$x = r \cos \theta, \qquad y = r \sin \theta. \tag{1.3.15}$$

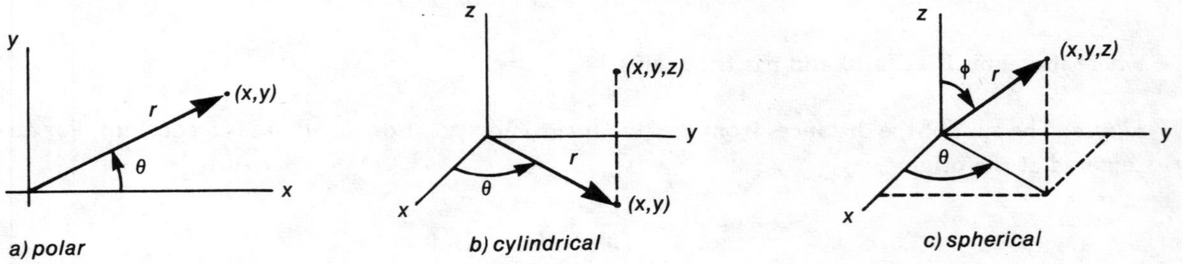

Figure 1.7. The polar, cylindrical and spherical coordinate systems.

For the three-dimensional cylindrical coordinate system

$$x = r\cos\theta, \qquad y = r\sin\theta, \qquad z = z. \tag{1.3.16}$$

And, for the spherical coordinate system

$$x = r\sin\phi\cos\theta \qquad y = r\sin\phi\sin\theta \qquad z = r\cos\phi \tag{1.3.17}$$

EXAMPLE 1.7

What conic section is represented by

$$2x^2 - 4xy + 5x = 10?$$

Solution. Comparing this with the general form Eq. 1.3.7, we see that

$$A = 2, \qquad B = -4, \qquad C = 0.$$

Thus, $B^2 - 4AC = 16$ which is greater than zero. Hence, the conic section is a hyperbola.

EXAMPLE 1.8

Write the general form of the equation of a parabola, vertex at (2,4), opening upward, with directrix at $y = 2$.

Solution. The equation of the parabola (see Eq. 1.3.11) can be written as

$$(x - x_1)^2 = 4p(y - y_1).$$

For this example, $x_1 = 2$, $y_1 = 4$, and $p = 2$ (p is the distance from the vertex to the directrix). Hence, the equation is

$$(x - 2)^2 = 4(2)\,(y - 4)$$

or, in general form,

$$x^2 - 4x - 8y + 36 = 0.$$

EXAMPLE 1.9

Express the point (3,4,5) in cylindrical coordinates, and spherical coordinates.

Solution. In cylindrical coordinates

$$r = \sqrt{x^2 + y^2}$$

$$= \sqrt{3^2 + 4^2} = 5\ ,$$

$$\theta = \tan^{-1} y/x$$

$$= \tan^{-1} 4/3 = 0.927 \text{ rad.}$$

Thus, in cylindrical coordinates, the point is located at (5,0.927,5).

In spherical coordinates

$$r = \sqrt{x^2 + y^2 + z^2}$$

$$= \sqrt{3^2 + 4^2 + 5^2} = 7.071\ ,$$

$$\phi = \cos^{-1} z/r$$

$$= \cos^{-1} 5/7.071 = 0.785 \text{ rad,}$$

$$\theta = \tan^{-1} y/x$$

$$= \tan^{-1} 4/3 = 0.927 \text{ rad.}$$

Finally, in spherical coordinates, the point is located at (7.071, 0.927, 0.785).

1.4 Complex Numbers

A complex number consists of a real part x and an imaginary part y, written as $x + iy$, where $i = \sqrt{-1}$. (In electrical engineering, however, it is common to let $j = \sqrt{-1}$ since i represents current.) In real number theory, the square root of a negative number does not exist; in complex number theory, we would write $\sqrt{-4} = \sqrt{4(-1)} = 2i$. The complex number may be plotted using the real x-axis and the imaginary y-axis, as shown in Fig. 1.8.

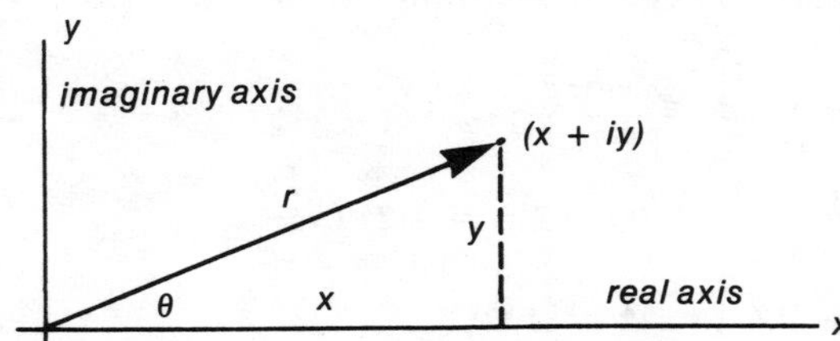

Figure 1.8. The complex number.

It is often useful to express a complex number in polar form as

$$x + iy = re^{i\theta} \tag{1.4.1}$$

where we use Euler's equation

$$e^{i\theta} = \cos\theta + i\sin\theta \tag{1.4.2}$$

to verify the relations

$$x = r\cos\theta, \quad y = r\sin\theta. \tag{1.4.3}$$

Using Euler's equation we can show that

$$\sin\theta = \frac{e^{i\theta} - e^{-i\theta}}{2i}, \quad \cos\theta = \frac{e^{i\theta} + e^{-i\theta}}{2}. \tag{1.4.4}$$

It is usually easier to find powers and roots of complex numbers using the polar form.

EXAMPLE 1.10

Divide $(3 + 4i)$ by $(4 + 3i)$.

Solution. We perform the division as follows:

$$\frac{3+4i}{4+3i} = \frac{3+4i}{4+3i}\,\frac{4-3i}{4-3i} = \frac{12+16i-9i+12}{16+9} = \frac{24+7i}{25} = 0.96 + 0.28i.$$

Note that we multiplied the numerator and the denominator by the complex conjugate of the denominator. A complex conjugate is formed simply by changing the sign of the imaginary part.

EXAMPLE 1.11

Find $(2 + 3i)^4$.

Solution. This can be done using the polar form. Hence,

$$r = \sqrt{2^2 + 3^2} = \sqrt{13}, \qquad \theta = \tan^{-1} 3/2 = 0.9828 \text{ rad}.$$

We normally express θ in radians. The complex number, in polar form, is

$$2+3i = \sqrt{13}\, e^{0.9828i}.$$

Thus,

$$(2 + 3i)^4 = (\sqrt{13})^4\, e^{4(0.9828i)}$$

$$= 169\, e^{3.9312i}$$

Converting back to rectangular form we have

$$169e^{3.9312i} = 169\,(\cos 3.9312 + i \sin 3.9312)$$

$$= 169\,(-0.7041 - 0.7101i)$$

$$= -119 - 120i.$$

EXAMPLE 1.12

Find the three one-third roots of $2 + 3i$.

Solution. We express the complex number (see Example 1.11) in polar form as

$$2 + 3i = \sqrt{13}\; e^{0.9828i}\,.$$

Since the trig functions are periodic we know that

$$\sin\theta = \sin(\theta + 2\pi) = \sin(\theta + 4\pi)$$

$$\cos\theta = \cos(\theta + 2\pi) = \cos(\theta + 4\pi).$$

Thus, in addition to the first form, we have

$$2 + 3i = \sqrt{13}\; e^{(0.9828\,+\,2\pi)i} = \sqrt{13}\; e^{(0.9828\,+\,4\pi)i}.$$

Taking the $1/3^{rd}$ root of each form, we find the three roots to be

$$(2 + 3i)^{1/3} = (\sqrt{13}\,)^{1/3}\, e^{0.3276i}$$

$$= 1.533\,(0.9468 + 0.3218i) = 1.452 + 0.4935i.$$

$$(2 + 3i)^{1/3} = (\sqrt{13}\,)^{1/3}\, e^{2.422i}$$

$$= 1.533\,(-0.7521 + 0.6591i) = -1.153 + 1.010i.$$

$$(2 + 3i)^{1/3} = (\sqrt{13}\,)^{1/3}\, e^{4.516i}$$

$$= 1.533\,(-0.1951 - 0.9808i) = -0.2991 - 1.504i.$$

If we added 6π to the angle, we would be repeating the first root so obviously this is not done If we were finding the $1/4^{th}$ root, four roots would result.

1.5 Linear Algebra

The primary objective in linear algebra is to find the solution to a set of n linear, algebraic equations for n unknowns. To do this we must learn how to manipulate a matrix, a rectangular array of quantities arranged into rows and columns.

An $m \times n$ matrix has m rows (the horizontal lines) and n columns (the vertical lines). We are primarily interested in n square matrices since we usually have the same number of equations as unknowns, such as

$$\begin{aligned} a_{11}x_1 + a_{12}x_2 + a_{13}x_3 + a_{14}x_4 &= r_1 \\ a_{21}x_1 + a_{22}x_2 + a_{23}x_3 + a_{24}x_4 &= r_2 \\ a_{31}x_1 + a_{32}x_2 + a_{33}x_3 + a_{34}x_4 &= r_3 \\ a_{41}x_1 + a_{42}x_2 + a_{43}x_3 + a_{44}x_4 &= r_4 \end{aligned} \tag{1.5.1}$$

In matrix form this can be written as

$$[a_{ij}][x_j] = [r_i] \tag{1.5.2}$$

where $[x_j]$ and $[r_i]$ are column matrices. (A row matrix is often referred to as a vector.) The coefficient matrix $[a_{ij}]$ and the column matrix $[r_i]$ are assumed to be known quantities. The solution $[x_j]$ is expressed as

$$[x_j] = [a_{ij}]^{-1}[r_i] \tag{1.5.3}$$

where $[a_{ij}]^{-1}$ is the *inverse* matrix of $[a_{ij}]$. It is defined as

$$[a_{ij}]^{-1} = \frac{[a_{ij}]^+}{|a_{ij}|} \tag{1.5.4}$$

where $[a_{ij}]^+$ is the *adjoint* matrix and $|a_{ij}|$ is the *determinant* of $[a_{ij}]$. Let us review how the determinant and the adjoint are evaluated.

In general, the determinant may be found using the *cofactor* A_{ij} of the element a_{ij}; the cofactor is defined to be $(-1)^{i+j}$ times the determinant obtained by deleting the ith row and the jth column. The determinant is then

$$|a_{ij}| = \sum_{j=1}^{n} a_{ij}\, A_{ij}, \tag{1.5.5}$$

where i is any value from 1 to n. Recall that a third-order determinant can be evaluated by writing the first two columns after the determinant and then summing the products of the elements of the diagonals using negative signs with the diagonals sloping upward.

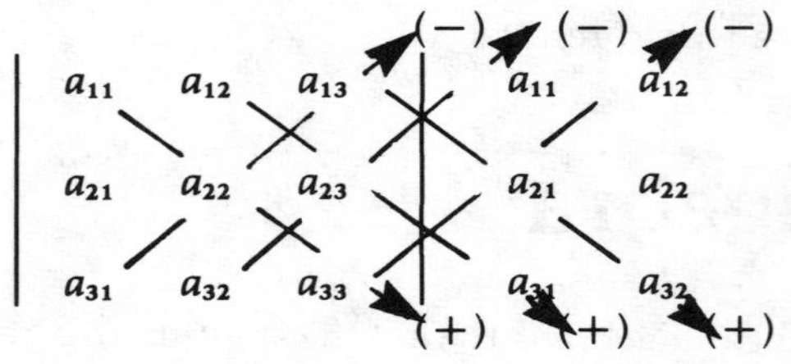

The elements of the adjoint $[a_{ij}]^+$ are the cofactors A_{ij} of the elements a_{ij}; for the matrix $[a_{ij}]$ of Eq. 1.5.1 we have

$$[a_{ij}]^+ = \begin{bmatrix} A_{11} & A_{21} & A_{31} & A_{41} \\ A_{12} & A_{22} & A_{32} & A_{42} \\ A_{13} & A_{23} & A_{33} & A_{43} \\ A_{14} & A_{24} & A_{34} & A_{44} \end{bmatrix} \tag{1.5.6}$$

Note that A_{ij} takes the position of a_{ji}.

Finally, the solution $[x_j]$ results if we multiply a square matrix by a column matrix. In general, we multiply the elements in each left-hand matrix row by the elements in each right-hand matrix column, add the products, and place the sum at the location where the row and column intersect. The following examples will illustrate.

Before we work some examples, though, we should point out that the above matrix presentation can also be presented as Cramer's rule, which states that the solution element x_i can be expressed as

$$x_i = \frac{|b_{ij}|}{|a_{ij}|} \tag{1.5.7}$$

where $|b_{ij}|$ is formed by replacing the ith column of $|a_{ij}|$ with the elements of the column matrix $[r_i]$.

Note: If the system of equations is homogeneous, i.e., r_i=0, a solution may exist if $|a_{ij}| = 0$.

EXAMPLE 1.13

Calculate the determinants of $\begin{bmatrix} 2 & -3 \\ 1 & 4 \end{bmatrix}$ and $\begin{bmatrix} 2 & 3 & 0 \\ 1 & 4 & -2 \\ 0 & 3 & 5 \end{bmatrix}$

Solution. For the first matrix we have

$$\begin{vmatrix} 2 & -3 \\ 1 & 4 \end{vmatrix} = 2 \times 4 - 1(-3) = 11.$$

The second matrix is set up as follows:

$$\begin{vmatrix} 2 & 3 & 0 \\ 1 & 4 & -2 \\ 0 & 3 & 5 \end{vmatrix} \begin{matrix} 2 & 3 \\ 1 & 4 \\ 0 & 3 \end{matrix} = 40 + 0 + 0 - 0 - (-12) - 15 = 37.$$

(−) (−) (−)

(+) (+) (+)

EXAMPLE 1.14

Expanding with cofactors, evaluate the determinant $D = \begin{vmatrix} 1 & 0 & -2 \\ -1 & 2 & 0 \\ 1 & 2 & 1 \end{vmatrix}$.

Solution. Choosing the first row ($i = 1$ in Eq. 1.5.5) we have

$$D = 1\begin{vmatrix} 2 & 0 \\ 2 & 1 \end{vmatrix}(-1)^2 + 0 + (-2)\begin{vmatrix} -1 & 2 \\ 1 & 2 \end{vmatrix}(-1)^4$$

$$= 2 + 0 - 2(-4) = 10.$$

Note: If any two columns (or rows) are a multiple of each other, the determinant is zero.

EXAMPLE 1.15

Find the adjoint of the matrix $[a_{ij}] = \begin{bmatrix} 1 & 0 & -2 \\ -1 & 2 & 0 \\ 1 & 2 & 1 \end{bmatrix}$.

Solution. The cofactor of each element of the matrix must be determined. The cofactor is found by multiplying $(-1)^{i+j}$ times the determinant formed by deleting the ith row and j^{th} column. They are found to be

$$A_{11} = 2, \quad A_{12} = 1, \quad A_{13} = -4$$

$$A_{21} = -4, \quad A_{22} = 3, \quad A_{23} = -2$$

$$A_{31} = 4, \quad A_{32} = 2, \quad A_{33} = 2$$

The adjoint is then

$$[a_{ij}]^+ = \begin{bmatrix} 2 & -4 & 4 \\ 1 & 3 & 2 \\ -4 & -2 & 2 \end{bmatrix}.$$

Note: The matrix $[A_{ji}]$ is called the *transpose* of $[A_{ij}]$.

EXAMPLE 1.16

Find the inverse of the matrix $[a_{ij}] = \begin{bmatrix} 1 & 0 & -2 \\ -1 & 2 & 0 \\ 1 & 2 & 1 \end{bmatrix}$.

Solution. The inverse is defined to be the adjoint matrix divided by the determinant $|a_{ij}|$. Hence, the inverse is (see Example 1.15)

$$[a_{ij}]^{-1} = \frac{1}{10}\begin{bmatrix} 2 & -4 & 4 \\ 1 & 3 & 2 \\ -4 & -2 & 2 \end{bmatrix} = \begin{bmatrix} 0.2 & -0.4 & 0.4 \\ 0.1 & 0.3 & 0.2 \\ -0.4 & -0.2 & 0.2 \end{bmatrix}.$$

EXAMPLE 1.17

Find the solution to

$$x_1 \qquad - 2x_3 = 2$$

$$-x_1 + 2x_2 \qquad = 0$$

$$x_1 + 2x_2 + x_3 = -4\,.$$

Solution. The solution matrix is (see Example 1.14 and 1.15)

$$[x_j] = [a_{ij}]^{-1}[r_i] = \frac{[a_{ij}]^{+}}{|a_{ij}|}[r_i]$$

$$= \frac{1}{10}\begin{bmatrix} 2 & -4 & 4 \\ 1 & 3 & 2 \\ -4 & -2 & 2 \end{bmatrix}\begin{bmatrix} 2 \\ 0 \\ -4 \end{bmatrix}.$$

The two matrices are multiplied row by column as follows:

$$2\cdot 2 + (-4)\cdot 0 + 4\cdot(-4) = -12$$

$$1\cdot 2 + 3\cdot 0 + 2\cdot(-4) = -6$$

$$-4\cdot 2 - 2\cdot 0 - 2\cdot(-4) = -16$$

The solution matrix is then

$$[x_i] = \frac{1}{10}\begin{bmatrix} -12 \\ -6 \\ -16 \end{bmatrix} = \begin{bmatrix} -1.2 \\ -0.6 \\ -1.6 \end{bmatrix}.$$

In component form, the solution is

$$x_1 = -1.2\,, \qquad x_2 = -0.6\,, \qquad x_3 = -1.6.$$

EXAMPLE 1.18

Use Cramer's rule and solve

$$\begin{aligned} x_1 \qquad\quad - 2x_3 &= 2 \\ -x_1 + 2x_2 \qquad\quad &= 0 \\ x_1 + 2x_2 + x_3 &= -4 \end{aligned}$$

Solution. The solution is found by evaluating the ratios

$$x_1 = \frac{\begin{vmatrix} 2 & 0 & -2 \\ 0 & 2 & 0 \\ -4 & 2 & 1 \end{vmatrix}}{D} = \frac{-12}{10} = -1.2$$

$$x_2 = \frac{\begin{vmatrix} 1 & 2 & -2 \\ -1 & 0 & 0 \\ 1 & -4 & 1 \end{vmatrix}}{D} = \frac{-6}{10} = -0.6$$

$$x_3 = \frac{\begin{vmatrix} 1 & 0 & 2 \\ -1 & 2 & 0 \\ 1 & 2 & -4 \end{vmatrix}}{D} = \frac{-16}{10} = -1.6$$

where

$$D = |a_{ij}| = \begin{vmatrix} 1 & 0 & -2 \\ -1 & 2 & 0 \\ 1 & 2 & 1 \end{vmatrix} = 10.$$

Note that the numerator is the determinant formed by replacing the ith column with right-hand side elements r_i when solving for x_i.

1.6 Calculus

1.6.1 Differentiation

The slope of a curve $y = f(x)$ is the ratio of the change in y to the change in x as the change in x becomes infinitesimally small; this is written as

$$\frac{dy}{dx} = \lim_{\Delta x \to 0} \frac{\Delta y}{\Delta x} . \qquad (1.6.1)$$

This may be written using abbreviated notation as

$$\frac{dy}{dx} = Dy = y' = \dot{y} . \qquad (1.6.2)$$

The second derivative is written as

$$D^2y = y'' = \ddot{y} , \qquad (1.6.3)$$

and is defined by

$$\frac{d^2y}{dx^2} = \lim_{\Delta x \to 0} \frac{\Delta y'}{\Delta x} \qquad (1.6.4)$$

Some derivative formulas, where f and g are functions of x and k is a constant, are given below.

$$\frac{dk}{dx} = 0$$

$$\frac{d(k\,x^n)}{dx} = k\,n\,x^{n-1} .$$

$$\frac{d}{dx}(f + g) = f' + g' .$$

$$\frac{df^n}{dx} = n\,f^{n-1}\,f' .$$

$$\frac{d}{dx}(fg) = fg' + gf' . \qquad (1.6.5)$$

$$\frac{d}{dx}(\ell nx) = \frac{1}{x} .$$

$$\frac{d}{dx}(e^{kx}) = ke^{kx}$$

$$\frac{d}{dx}(\sin x) = \cos x$$

$$\frac{d}{dx}(\cos x) = -\sin x$$

If a function f depends on more than one variable, partial derivatives are used. If $z = f(x,y)$, then $\partial z/\partial x$ is the derivative of z with respect to x holding y constant. It would represent the slope of a line tangent to the surface in a plane of constant y.

1.6.2 Maxima and Minima

Derivatives are used to locate points of inflection, maxima, and minima. The following are used:

$f'(x) = 0$ at a maximum or a minimum.
$f''(x) = 0$ at an inflection point.
$f''(x) > 0$ at a minimum.
$f''(x) < 0$ at a maximum.

An inflection point always exists between a maximum and a minimum.

1.6.3 L'Hospital's Rule

Differentiation is also useful in establishing the limit of $f(x)/g(x)$ as $x \rightarrow x_0$ if $f(x_0)$ and $g(x_0)$ are both zero or $\pm\infty$. L'Hospital's rule is used in such cases; it is

$$\lim_{x \rightarrow x_0} \frac{f(x)}{g(x)} = \lim_{x \rightarrow x_0} \frac{f'(x)}{g'(x)}. \tag{1.6.6}$$

1.6.4 Taylor's Series

Derivatives are used to expand a continuous function as a power series around $x = a$; using Taylor's series we have

$$f(x) = f(a) + (x - a) f'(a) + (x - a)^2 f''(a)/2! + \cdots \tag{1.6.7}$$

This series is often used to express a function as a polynomial near a point $x = a$ providing the series can be truncated after a few terms. Using Taylor's series we can show that (expanding about $a = 0$)

$$\begin{aligned} \sin x &= x - x^3/3! + x^5/5! - \cdots \\ \cos x &= 1 - x^2/2! + x^4/4! - \cdots \\ \ell n(1 + x) &= x - x^2/2 + x^3/3 - \cdots \\ \frac{1}{1 + x} &= 1 + x + x^2 + \cdots \\ e^x &= 1 + x + x^2/2! + x^3/3! + \cdots. \end{aligned} \tag{1.6.8}$$

1.6.5 Integration

The inverse of differentiation is the process called integration. If a curve is given by $y = f(x)$, then the area under the curve from $x = a$ to $x = b$ is given by

$$A = \int_a^b y \, dx. \tag{1.6.9}$$

The length of the curve between the two points is expressed as

$$L = \int_a^b (1 + y'^2)^{1/2}\, dx. \tag{1.6.10}$$

Volumes of various objects are also found by an appropriate integration.

If the integral has limits, it is a *definite integral*; if it does not have limits, it is an *indefinite integral* and a constant is always added. Some common indefinite integrals follow:

$$\int dx = x + C.$$

$$\int cydx = c \int ydx$$

$$\int x^n dx = \frac{x^{n+1}}{n+1} + C \qquad n \neq -1$$

$$\int x^{-1} dx = \ell n\, x + C$$

$$\int e^{ax} dx = \frac{1}{a} e^{ax} + C \tag{1.6.11}$$

$$\int \sin x\, dx = -\cos x + C$$

$$\int \cos x\, dx = \sin x + C$$

$$\int \cos^2 x\, dx = \frac{x}{2} + \frac{1}{4} \sin 2x + C$$

$$\int u\, dv = uv - \int v\, du.$$

This last integral is often referred to as "integration by parts." If the integrand (the coefficient of the differential) is not one of the above, then, in the last integral, $\int v\,du$ may in fact, be integrable. An example will illustrate.

EXAMPLE 1.19

Find the slope of $y = x^2 + \sin x$ at $x = 0.5$.

Solution. The derivative is

$$y'(x) = 2x + \cos x.$$

At $x = 0.5$ the slope is

$$y'(0.5) = 2 \cdot 0.5 + \cos 0.5 = 1.878.$$

EXAMPLE 1.20

Find $\frac{d}{dx}(\tan x)$.

Solution. Writing $\tan x = \sin x / \cos x = f(x) \cdot g(x)$ we find

$$\frac{d}{dx}(\tan x) = \frac{1}{\cos x}\frac{d}{dx}(\sin x) + \sin x \frac{d}{dx}(\cos x)^{-1}$$

$$= \frac{\cos x}{\cos x} + \frac{\sin^2 x}{\cos^2 x} = 1 + \tan^2 x$$

$$= \frac{\cos^2 x + \sin^2 x}{\cos^2 x} = \frac{1}{\cos^2 x} = \sec^2 x.$$

Either expression is acceptable although the latter is usually used.

EXAMPLE 1.21

Locate the maximum and minimum points of the function $y(x) = x^3 - 12x - 9$ and evaluate y at those points.

Solution. The derivative is

$$y'(x) = 3x^2 - 12.$$

The points at which $y'(x) = 0$ are at

$$x = 2, -2.$$

At these two points the extrema are

$$y_{min} = (2)^3 - 12 \cdot 2 - 9 = -25.$$

$$y_{max} = (-2)^3 - 12(-2) - 9 = 7.$$

Let us check the second derivative. At the two points we have

$$y''(2) = 6 \cdot 2 = 12.$$

$$y''(-2) = 6 \cdot (-2) = -12.$$

Obviously, the point $x = 2$ is a minimum since its second derivative is positive there.

EXAMPLE 1.22

Find the limit as $x \to 0$ of $\sin x / x$.

Solution. If we let $x = 0$ we are faced with the ratio of 0/0, an undefined quantity. Hence, we use L'Hospital's rule and differentiate both numerator and denominator to obtain

$$\lim_{x \to 0} \frac{\sin x}{x} = \lim_{x \to 0} \frac{\cos x}{1}.$$

Now, we let $x = 0$ and find

$$\lim_{x \to 0} \frac{\sin x}{x} = \frac{1}{1} = 1.$$

EXAMPLE 1.23

Verify that $\sin x = x - x^3/3! + x^5/5! - \cdots$.

Solution. We expand in a Taylor's series about $x = 0$; that is,

$$f(x) = f(0) + x\,f'(0) + x^2 f''(0)/2! + \cdots .$$

Letting $f(x) = \sin x$ so that $f' = \cos x$, $f'' = -\sin x$, etc., there results

$$\sin x = 0 + x\,(1) + x^2\,(0/2!) + x^3(-1)/3! + \cdots$$

$$= x - x^3/3! + x^5/5! - \cdots .$$

EXAMPLE 1.24

Find the shaded area shown.

Solution. We can find this area by using either a horizontal strip or a vertical strip. We will use both. First, for a horizontal strip:

$$A = \int_0^2 x dy$$

$$= \int_0^2 y^2 dy = \left.\frac{y^3}{3}\right|_0^2 = 8/3.$$

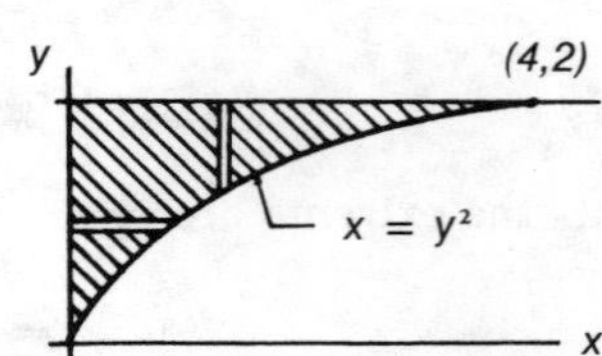

Using a vertical strip we have

$$A = \int_0^4 (2 - y)\,dx$$

$$= \int_0^4 (2 - x^{1/2})\,dx = \left[2x - \frac{2}{3}x^{3/2}\right]_0^4 = 8 - \frac{2}{3}\,8 = 8/3.$$

Either technique is acceptable; the first appears to be the simpler one.

EXAMPLE 1.25

Find the volume enclosed by rotating the shaded area of Example 1.24 about the y-axis.

Solution. If we rotate the horizontal strip about the y-axis we will obtain a disc with volume

$$dV = \pi x^2\,dy.$$

This can be integrated to give the volume; it is

$$V = \int_0^2 \pi x^2 dy$$

$$= \pi \int_0^2 y^4\,dy = \left.\frac{\pi y^5}{5}\right|_0^2 = \frac{32\pi}{5}.$$

Now, let us rotate the vertical strip about the y-axis; it will form a cylinder with volume

$$dV = 2\pi x(2 - y)\,dx.$$

This can be integrated to yield

$$V = \int_0^4 2\pi x(2 - y)\,dx$$

$$= \int_0^4 2\pi x(2 - x^{1/2})\,dx = 2\pi\left[x^2 - \frac{2x^{5/2}}{5}\right]_0^4 = \frac{32\pi}{5}.$$

Again, the horizontal strip is simpler.

EXAMPLE 1.26

Find $\int x\,e^x\,dx$.

Solution. This is not one of our tabulated integrals, so let's attempt the last integral of (1.6.11). Define

$$u = x, \qquad dv = e^x\,dx.$$

Then,

$$du = dx, \qquad v = \int e^x dx = e^x$$

and we find that

$$\int x\,e^x\,dx = x\,e^x - \int e^x dx$$

$$= x\,e^x - e^x + C = (x - 1)e^x + C.$$

1.7 Differential Equations

A differential equation is *linear* if no term contains the dependent variable to a power other than one (terms that do not contain the dependent variable are not considered in the test of linearity). For example,

$$y'' + 2xy' - y\sin x = 3x^2 \tag{1.7.1}$$

is a linear differential equation; the dependent variable is y and the independent variable is x. If a term contained y'^2, or $y^{1/2}$, or $\sin y$ the equation would be nonlinear.

A differential equation is *homogeneous* if all of its terms contain the dependent variable. Eq. 1.7.1 is nonhomogeneous because of the term $3x^2$.

The *order* of a differential equation is established by its highest order derivative. Eq. 1.7.1 is a second order differential equation.

The general solution of a differential equation involves a number of arbitrary constants equal to the order of the equation. If conditions are specified, the arbitrary constants may be calculated.

1.7.1 First Order

A first order differential equation is *separable* if it can be expressed as

$$M(x)\,dx + N(y)\,dy = 0. \tag{1.7.2}$$

The solution follows by integrating each of the terms.

If $M = M(x,y)$ and $N = N(x,y)$, the solution $F(x,y) = C$ can be found if Eq. 1.7.2 is *exact*, that is $\partial M/\partial y = \partial N/\partial x$; then $M = \partial F/\partial x$ and $N = \partial F/\partial y$. Note that Eq. 1.7.2 is, in general, nonlinear.

The linear, first order differential equation

$$y' + h(x)\,y = g(x) \tag{1.7.3}$$

has the solution

$$y(x) = \frac{1}{u} \int u\,g(x)\,dx + \frac{C}{u} \tag{1.7.4}$$

where

$$u(x) = e^{\int h(x)\,dx}\quad. \tag{1.7.5}$$

The function $u(x)$ is called an integrating factor.

1.7.2 Second Order, Linear, Homogeneous, with Constant Coefficients

The general form of a second order, linear, homogeneous differential equation with constant coefficients is

$$y'' + Ay' + By = 0\,. \tag{1.7.6}$$

To find a solution we must first solve the characteristic equation

$$m^2 + Am + B = 0. \tag{1.7.7}$$

If $m_1 \neq m_2$ and both are real, the general solution is

$$y(x) = c_1\,e^{m_1 x} + c_2\,e^{m_2 x}\,. \tag{1.7.8}$$

If $m_1 = m_2$, the general solution is

$$y(x) = c_1\,e^{m_1 x} + c_2\,xe^{m_2 x}. \tag{1.7.9}$$

Finally, if $m_1 = a + ib$ and $m_2 = a - ib$, the general solution is

$$y(x) = (c_1 \sin bx + c_2 \cos bx)\,e^{ax}. \tag{1.7.10}$$

1.7.3 Linear, Nonhomogeneous, with Constant Coefficients

If Eq. 1.7.6 were nonhomogeneous, it would be written as

$$y'' + Ay' + By = g(x). \tag{1.7.11}$$

The general solution is found by finding the solution $y_h(x)$ to the homogeneous equation (simply let the right-hand side be zero and solve the equation as in Section 1.7.2) and adding to it a particular solution $y_p(x)$ found by using Table 1.2.

TABLE 1.2 Particular Solutions

g(x)	$y_p(x)$	Provisions
a	C	
$ax + b$	$Cx + D$	
$ax^2 + bx + c$	$Cx^2 + Dx + E$	
e^{ax}	Ce^{ax}	if m_1 or $m_2 \neq a$
	Cxe^{ax}	if m_1 or $m_2 = a$
$b \sin ax$	$C \sin ax + D \cos ax$	if $m_{1,2} \neq \pm ai$
	$Cx \sin ax + Dx \cos ax$	if $m_{1,2} = \pm ai$
$b \cos ax$	(same as above)	

EXAMPLE 1.27

Find the general solution to

$$y' + y = xe^{-x}.$$

Solution. The integrating factor is

$$u(x) = e^{\int dx} = e^x.$$

The solution is then

$$y(x) = \frac{1}{u} \int ug(x)\, dx + \frac{C}{u}$$

$$= e^{-x} \int e^x (xe^{-x})\, dx + Ce^{-x}$$

$$= e^{-x} (x^2/2) + Ce^{-x} = e^{-x} (x^2/2 + C).$$

EXAMPLE 1.28

Find the solution to

$$y' + 2y = 4x, \qquad y(0) = 2.$$

Solution. We will use the method outlined in Section 1.7.2. The characteristic equation of the homogeneous equation is

$$m + 2 = 0. \qquad \therefore m = -2.$$

The homogeneous solution is then

$$y_h(x) = c_1 e^{-2x}.$$

The particular solution is assumed to be of the form

$$y_p(x) = Ax + B.$$

Substituting this into the original differential equation gives

$$A + 2(Ax + B) = 4x. \qquad \therefore A = 2, \quad B = -1.$$

The solution is then

$$\begin{aligned} y(x) &= y_h(x) + y_p(x) \\ &= c_1 e^{-2x} + 2x - 1. \end{aligned}$$

Using the given condition

$$y(0) = 2 = c_1 - 1. \qquad \therefore c_1 = 3.$$

Finally,

$$y(x) = 3e^{-2x} + 2x - 1.$$

EXAMPLE 1.29

A simple spring-mass system is represented by

$$M\ddot{y} + C\dot{y} + Ky = F(t)$$

where the mass M, the damping coefficient C, the spring constant K, and the forcing function $F(t)$ have the appropriate units. Find the general solution if $M = 2$, $C = 0$, $K = 50$, and $F(t) = 0$.

Solution. The differential equation simplifies to

$$2\ddot{y} + 50y = 0.$$

The characteristic equation is then

$$2m^2 + 50 = 0.$$

$$\therefore m_1 = 5i, \; m_2 = -5i.$$

The solution is then (see Eq. 1.7.10)

$$y(t) = c_1 \sin 5t + c_2 \cos 5t.$$

Note that we have used dots to represent the time derivative, $dy/dt = \dot{y}$. Also, note that the coefficient of t in $\sin 5t$ divided by 2π is the frequency in hertz, i.e., $f = 5/2\pi$ Hz.

EXAMPLE 1.30

In the equation of Ex. 1.29, let $M = 2$, $C = 12$, $K = 50$, and $F(t) = 60 \sin 5t$. Find the general solution.

Solution. The differential equation is

$$2\ddot{y} + 12\,\dot{y} + 50\,y = 60 \sin 5t\,.$$

The characteristic equation of the homogeneous differential equation is found to be

$$2m^2 + 12m + 50 = 0.$$

$$\therefore m_1 = -3 + 4i, \qquad m_2 = -3 - 4i.$$

The homogeneous solution is (see Eq. 1.7.10)

$$y_h(t) = e^{-3t}(c_1 \sin 4t + c_2 \cos 4t).$$

The particular solution is found by assuming that

$$y_p(t) = C \sin 5t + D \cos 5t.$$

Substitute this into the original differential equation:

$$2[-25C \sin 5t - 25D \cos 5t] + 12[5C \cos 5t - 5D \sin 5t] + 50[C \sin 5t + D \cos 5t] = 60 \sin 5t.$$

Equating coefficients of sin-terms and then cos-terms:

$$-50C - 60D + 50C = 60. \qquad \therefore D = 1.$$

$$-50D + 60C + 50D = 0. \qquad \therefore C = 0.$$

The two solutions are superposed to give

$$y(t) = (c_1 \sin 4t + c_2 \cos 4t)\, e^{-3t} + \cos 5t.$$

1.8 Probability and Statistics

Events are independent if the probability of occurrence of one event does not influence the probability of occurrence of other events. The number of permutations of n things taken r at a time is

$$p(n,r) = \frac{n!}{(n - r)!}\,. \tag{1.8.1}$$

If the starting point is unknown, as in a ring, the *ring permutation* is

$$p(n,r) = \frac{(n - 1)!}{(n - r)!}\,. \tag{1.8.2}$$

The number of *combinations* of n things taken r at a time (it is not an order-conscious arrangement) is given by

$$C(n,r) = \frac{n!}{r!(n - r)!}\,. \tag{1.8.3}$$

For independent events of two sample groups A_i and B_i the following rules are necessary:

1. The probability of A_1 or A_2 occurring equals the sum of the probability of occurrence of A_1 and the probability of occurrence of A_2; that is,
$$p(A_1 \text{ or } A_2) = p(A_1) + p(A_2). \tag{1.8.4}$$
2. The probability of A_1 and B_1 occurring is given by the product
$$p(A_1 \text{ and } B_1) = p(A_1)\, p(B_1). \tag{1.8.5}$$
3. The probability of A_1 not occurring equals
$$p(\text{not } A_1) = 1 - p(A_1). \tag{1.8.6}$$
4. The probability of A_1 or B_1 occurring is given by
$$p(A_1 \text{ or } B_1) = p(A_1) + p(B_1) - p(A_1)p(B_1). \tag{1.8.7}$$

The data gathered during an experiment can be analyzed using statistical quantities defined by the following:

1. The *arithmetic mean* $\bar{x}$ is the average of the observations; that is,
$$\bar{x} = \frac{x_1 + x_2 + x_3 + \cdots + x_n}{n}$$
2. *The median* is the middle observation when all the data is ordered by magnitude; half the values are below the median.
3. The *mode* is the observation that occurs most frequently.
4. The *standard deviation* σ is a measure of variability. It is defined as
$$\sigma = \left[\frac{(x_1 - \bar{x})^2 + (x_2 - \bar{x})^2 + \cdots + (x_n - \bar{x})^2}{n - 1} \right]^{1/2} \tag{1.8.8}$$
$$= \left[\frac{x_1^2 + x_2^2 + \cdots + x_n^2 - n\,\bar{x}^2}{n - 1} \right]^{1/2}. \tag{1.8.9}$$
For large (over 50) observations, it is customary to simply use n, rather than $(n - 1)$, in the denominators of the above.
5. The *variance* is defined to be σ^2.

EXAMPLE 1.31

How many different ways can seven people be arranged in a lineup? In a circle?

Solution. This is the number of permutations of seven things taken seven at a time. The answer is

$$p(7,7) = \frac{n!}{(n - r)!} = \frac{7!}{(7 - 7)!} = 5040.$$

In a circle we use the ring permutation:

$$p(7,7) = \frac{(n - 1)!}{(n - r)!} = \frac{(7 - 1)!}{(7 - 7)!} = 720.$$

EXAMPLE 1.32

How many different collections of eight people can fit into a six-passenger vehicle? (Only six will fit in!)

Solution. The answer does not depend on the seating arrangement. If it did, it would be a permutation. Hence, we use the combination relationship and find

$$C(8,6) = \frac{n!}{(n-r)!r!}$$

$$= \frac{8!}{(8-6)!6!} = 28.$$

EXAMPLE 1.33

A carnival booth offers $10 if you pick a red ball and then a white ball (the first ball is re-inserted) from a bin containing 60 red balls, 15 white balls, and 25 blue balls. If $1 is charged for an attempt, will the operator make money?

Solution. The probability of drawing a red ball on the first try is 0.6. If it is then re-inserted, the probability of drawing a white ball is 0.15. The probability of accomplishing both is then given by

$$p(\text{red and white}) = p(\text{red})\ p(\text{white})$$

$$= 0.6 \times 0.15 = 0.09,$$

or 9 chances out of 100 attempts. Hence, the entrepeneur will pay out $90 for every $100 taken in and will thus make money.

EXAMPLE 1.34

If the operator of the bin of balls in Example 1.33 offers a $1.00 prize for $0.75 to contestants who pick a red ball or a white ball from the bin on the first attempt, will the operator make money?

Solution. The probability of selecting either a red ball or a white ball on the first attempt is

$$p(\text{red or white}) = p(\text{red}) + p(\text{white})$$

$$= 0.6 + 0.15 = 0.75.$$

Consequently, 75 out of 100 gamblers will win and the operator must pay out $75 for every $75 taken in. The offer would not make any money.

EXAMPLE 1.35

If the operator of Example 1.34 had two identical bins and offered a $1.00 prize for $0.75 to successfully withdraw a red ball from the first bin or a white ball from the second bin, will the operator make money?

Solution. The probability of selecting a red ball from the first bin (sample group A_i) or a white ball from the second bin (sample group B_i) is

$$p(\text{red or white}) = p(\text{red}) + p(\text{white}) - p(\text{red})\ p(\text{white})$$

$$= 0.6 + 0.15 - 0.6 \times 0.15 = 0.66.$$

For this situation the owner must pay $66 to every 100 gamblers who pay $75 to participate. The profit is $9.

EXAMPLE 1.36

The temperature at 2 p.m. on August 10 for 25 consecutive years at a given location in the south was measured, in degrees Celsius, to be 33, 38, 34, 26, 32, 31, 28, 39, 29, 36, 32, 29, 31, 24, 35, 34, 32, 30, 31, 32, 26, 40, 27, 33, 39. Calculate the arithmetic mean, the median, the mode and the standard deviation.

Solution. Using the appropriate equations, we calculate the arithmetic mean to be

$$\bar{T} = \frac{T_1 + T_2 + \cdots + T_{25}}{25}$$

$$= \frac{33 + 38 + 34 + \cdots + 39}{25} = \frac{801}{25} = 32.04°C$$

The median is found by first arranging the values in order. We have 24, 26, 26, 27, 28, 29, 30, 31, 31, 31, 32, 32, 32, 32, 33, 33, 34, 35, 36, 38, 39, 39, 40. Counting 12 values in from either end, the median is found to be 32°C.

The mode is the observation that occurs most often; it is 32°C.

The standard deviation is found to be

$$\sigma = \left[\frac{T_1^2 + T_2^2 + \cdots + T_{25}^2 - n\bar{T}^2}{n - 1}\right]^{1/2}$$

$$= \left[\frac{33^2 + 38^2 + \cdots + 39^2 - 25 \times 32.04^2}{25-1}\right]^{1/2} = \sqrt{\frac{26{,}099 - 25{,}664}{24}} = 4.26.$$

Practice Problems

ALGEBRA

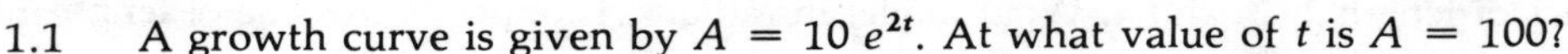

1.1 A growth curve is given by $A = 10\,e^{2t}$. At what value of t is $A = 100$?

a) 5.261 b) 3.070 c) 1.151 d) 0.726 e) 0.531

1.2 If $\ln x = 3.2$, what is x?

a) 18.65 b) 24.53 c) 31.83 d) 64.58 e) 126.7

1.3 If $\ln_5 x = -1.8$, find x.

a) 0.00483 b) 0.0169 c) 0.0552 d) 0.0783 e) 0.1786

1.4 One root of the equation $3x^2 - 2x - 2 = 0$ is

a) 1.215 b) 1.064 c) 0.937 d) 0.826 e) 0.549

1.5 $\sqrt{4 + x}$ can be written as the series

a) $2 - x/4 + x^2/64 + \cdots$
b) $2 + x/8 - x^2/128 + \cdots$
c) $2 - x^2/4 - x^4/64 + \cdots$
d) $2 + x^2/8 + x^4/128 + \cdots$
e) $2 + x/4 - x^2/64 + \cdots$

1.6 Resolve $\dfrac{2}{x(x^2 - 3x + 2)}$ into partial fractions.

a) $1/x + 1/(x - 2) - 2/(x - 1)$
b) $1/x - 2/(x - 2) + 1/(x - 1)$
c) $2/x - 1/(x - 2) - 2/(x - 1)$
d) $-1/x + 2/(x - 2) + 1/(x - 1)$
e) $-1/x - 2/(x - 2) + 1/(x - 1)$

1.7 Express $\dfrac{4}{x^2(x^2 - 4x + 4)}$ as the sum of fractions.

a) $1/x + 1/(x - 2)^2 - 1/(x - 2)$
b) $1/x + 1/x^2 - 1/(x - 2) + 1/(x - 2)^2$
c) $1/x^2 + 1/(x - 2)^2$
d) $1/x + 1/x^2 + 1/(x - 2) + 1/(x - 2)^2$
e) $1/x^2 - 1/(x - 2) + 1/(x - 2)^2$

1.8 A germ population has a growth curve of $Ae^{0.4t}$. At what value of t does it double?

a) 9.682 b) 7.733 c) 4.672 d) 1.733 e) 0.5641

TRIGONOMETRY

1.9 If $\sin\theta = 0.7$, what is $\tan\theta$?

a) 0.98 b) 0.94 c) 0.88 d) 0.85 e) 0.81

1.10 If the short leg of a right triangle is 5 units long and the long leg is 7 units long, find the angle opposite the short leg, in degrees.

a) 26.3 b) 28.9 c) 31.2 d) 33.8 e) 35.5

1.11 The expression $\tan\theta \sec\theta (1 - \sin^2\theta)/\cos\theta$ simplifies to

a) $\sin\theta$ b) $\cos\theta$ c) $\tan\theta$ d) $\sec\theta$ e) $\csc\theta$

1.12 A triangle has sides of length 2, 3, and 4. What angle, in radians, is opposite the side of length 3?

a) 0.55 b) 0.61 c) 0.76 d) 0.81 e) 0.95

1.13 The length of a lake is desired. A distance of 850 m is measured from one end to a point x on the shore. A distance of 732 m is measured from x to the other end. If an angle of 154° is measured between the two lines connecting x, what is the length of the lake?

a) 1542 b) 1421 c) 1368 d) 1261 e) 1050

1.14 Express $2\sin^2\theta$ as a function of $\cos 2\theta$.

a) $\cos 2\theta - 1$ b) $\cos 2\theta + 1$ c) $\cos 2\theta + 2$ d) $2 - \cos 2\theta$ e) $1 - \cos 2\theta$

GEOMETRY

1.15 The included angle between two successive sides of a regular eight-sided polygon is

a) 150° b) 135° c) 120° d) 75° e) 45°

1.16 A large 15-m-dia cylindrical tank, that sits on the ground, is to be painted. If one liter of paint covers 10 m², how many liters are required if it is 10 m high? (Include the top.)

a) 65 b) 53 c) 47 d) 38 e) 29

1.17 The equation of a line that has a slope of -2 and intercepts the x-axis at $x = 2$ is

a) $y + 2x = 4$
b) $y - 2x = 4$
c) $y + 2x = -4$
d) $2y + x = 2$
e) $2y - x = -2$

1.18 The equation of a line that intercepts the x-axis at $x = 4$ and the y-axis at $y = -6$ is

a) $2x - 37 = 12$
b) $3x - 2y = 12$
c) $2x + 3y = 12$
d) $3x + 2y = 12$
e) $3y - 2x = 12$

1.19 The shortest distance from the line $3x - 4y = 3$ to the point (6,8) is

a) 4.8 b) 4.2 c) 3.8 d) 3.4 e) 2.6

1.20 The equation $x^2 + 4xy + 4y^2 + 2x = 10$ represents which conic section?

a) circle b) ellipse c) parabola d) hyperbola e) plane

1.21 The x- and y-axis are the asymtotes of a hyperbola that passes through the point (2,2). Its equation is

a) $x^2 - y^2 = 0$
b) $xy = 4$
c) $y^2 - x^2 = 0$
d) $x^2 + y^2 = 4$
e) $x^2y = 8$

1.22 A 100-m-long track is to be built 50 m wide. If it is to be elliptical, what equation could describe it?

a) $50x^2 + 100y^2 = 1000$
b) $2x^2 + y^2 = 250$
c) $4x^2 + y^2 = 2500$
d) $x^2 + 2y^2 = 250$
e) $x^2 + 4y^2 = 10000$

1.23 The cylindrical coordinates (5,30°,12) are expressed in spherical coordinates as

a) (13, 30°, 67.4°)
b) (13, 30°, 22.6°)
c) (15, 52.6°, 22.6°)
d) (15, 52.6°, −22.6°)
e) (13, 30°, 67.40)

1.24 The equation of a 4-m-radius sphere using cylindrical coordinates is

a) $x^2 + y^2 + z^2 = 16$
b) $r^2 = 16$
c) $r^2 + z^2 = 16$
d) $x^2 + y^2 = 16$
e) $r^2 + y^2 = 16$

COMPLEX NUMBERS

1.25 Divide $3 - i$ by $1 + i$.

a) $1 - 2i$ b) $1 + 2i$ c) $2 - i$ d) $2 + i$ e) $2 + 2i$

1.26 Find $(1 + i)^6$.

a) $1 + i$ b) $1 - i$ c) $8i$ d) $-8i$ e) $-1 - i$

1.27 Find the first root of $(1 + i)^{1/5}$.

a) $0.17 + 1.07i$
b) $1.07 + 0.17i$
c) $1.07 - 0.17i$
d) $0.17 - 1.07i$
e) $-1.07 - 0.17i$

1.28 Express $(3 + 2i)\, e^{2it} + (3 - 2i)\, e^{-2it}$ in terms of trigonometric functions.

a) $3 \cos 2t - 4 \sin 2t$
b) $3 \cos 2t - 2 \sin 2\text{t}$
c) $6 \cos 2t - 4 \sin 2t$
d) $3 \sin 2t + 2 \sin 2t$
e) $6 \cos 2t + 4 \sin 2t$

1.29 Subtract $5e^{0.2i}$ from $6e^{2.3i}$.

a) $-8.90 + 5.48i$
b) $-0.90 + 3.48i$
c) $-0.90 - 3.48i$
d) $8.90 - 5.48i$
e) $-8.90 + 3.48i$

LINEAR ALGEBRA

1.30 Find the value of the determinant $\begin{vmatrix} 3 & 2 & 1 \\ 0 & -1 & -1 \\ 2 & 0 & 2 \end{vmatrix}$.

a) 8 b) 4 c) 0 d) −4 e) −8

1.31 Evaluate the determinant $\begin{vmatrix} 1 & 0 & 1 & 1 \\ 2 & -1 & 0 & 1 \\ 0 & 0 & 2 & 0 \\ 3 & 2 & 1 & 1 \end{vmatrix}$.

a) 8 b) 4 c) 0 d) −4 e) −8

1.32 The cofactor A_{21} of the determinant of Prob. 1.30 is

a) −5 b) −4 c) 3 d) 4 e) 5

1.33 The cofactor A_{34} of the determinant of Prob. 1.31 is

a) 4 b) 6 c) 0 d) −4 e) −6

1.34 Find the adjoint matrix of $\begin{bmatrix} 0 & 1 \\ -4 & 2 \end{bmatrix}$.

a) $\begin{bmatrix} 4 & 2 \\ 0 & 1 \end{bmatrix}$ b) $\begin{bmatrix} 1 & 0 \\ 4 & 2 \end{bmatrix}$ c) $\begin{bmatrix} 2 & 4 \\ 1 & 0 \end{bmatrix}$ d) $\begin{bmatrix} 1 & 0 \\ -4 & 2 \end{bmatrix}$ e) $\begin{bmatrix} 1 & 4 \\ 0 & 2 \end{bmatrix}$

1.35 The inverse matrix of $\begin{bmatrix} 2 & 3 \\ 1 & 1 \end{bmatrix}$ is

a) $\begin{bmatrix} -1 & 1 \\ 3 & -2 \end{bmatrix}$ b) $\begin{bmatrix} 1 & -1 \\ -3 & 2 \end{bmatrix}$ c) $\begin{bmatrix} -1 & 1 \\ -3 & 2 \end{bmatrix}$ d) $\begin{bmatrix} -2 & 3 \\ 1 & -1 \end{bmatrix}$ e) $\begin{bmatrix} 2 & 3 \\ -1 & -1 \end{bmatrix}$

1.36 Calculate $\begin{bmatrix} 2 & -1 \\ 3 & 2 \end{bmatrix} \begin{bmatrix} 2 \\ 1 \end{bmatrix}$.

a) $\begin{bmatrix} 8 \\ 3 \end{bmatrix}$ b) $\begin{bmatrix} 3 \\ 8 \end{bmatrix}$ c) $\begin{bmatrix} -3 \\ -8 \end{bmatrix}$ d) [3,8] e) [8,3]

1.37 Determine $\begin{bmatrix} 1 & 2 \\ 2 & 1 \end{bmatrix} \begin{bmatrix} -1 & 0 \\ 1 & 2 \end{bmatrix}$.

a) $\begin{bmatrix} 1 & 4 \\ -1 & 2 \end{bmatrix}$ b) $\begin{bmatrix} 1 & -1 \\ 4 & 2 \end{bmatrix}$ c) $\begin{bmatrix} 1 \\ -1 \end{bmatrix}$ d) $\begin{bmatrix} 4 \\ 2 \end{bmatrix}$ e) [1,4]

1.38 Solve for $[x_i]$.

$$\begin{aligned} 3x_1 + 2x_2 \qquad &= -2 \\ x_1 - x_2 + x_3 &= 0 \\ 4x_1 \qquad + 2x_3 &= 4 \end{aligned}$$

a) $\begin{bmatrix} 2 \\ -4 \\ -6 \end{bmatrix}$ b) $\begin{bmatrix} -2 \\ 4 \\ 12 \end{bmatrix}$ c) $\begin{bmatrix} 2 \\ 8 \\ 4 \end{bmatrix}$ d) $\begin{bmatrix} -6 \\ 8 \\ 14 \end{bmatrix}$ e) $\begin{bmatrix} 6 \\ 4 \\ 3 \end{bmatrix}$

CALCULUS

1.39 The slope of the curve $y = 2x^3 - 3x$ at $x = 1$ is

a) 3 b) 5 c) 6 d) 8 e) 9

1.40 If $y = \ell n\, x + e^x \sin x$, find dy/dx at $x = 1$.

a) 1.23 b) 3.68 c) 4.76 d) 6.12 e) 8.35

1.41 At what value of x does a maximum of $y = x^3 - 3x$ occur?

a) 2 b) 1 c) 0 d) -1 e) -2

1.42 Where does an inflection point occur for $y = x^3 - 3x$?

a) 2 b) 1 c) 0 d) -1 e) -2

1.43 Evaluate $\lim_{x \to \infty} \dfrac{2x^2 - x}{x^2 + x}$.

a) 2 b) 1 c) 0 d) -1 e) -2

1.44 If a quantity η and its derivatives η' and η'' are known at a point, its approximate value at a small distance h is

a) $\eta + h^2\eta''/2$
b) $\eta + h\eta/2 + h^2\eta''$
c) $\eta + h\eta' + h^2\eta''/2$
d) $\eta + h\eta' + h^2\eta''$
e) $\eta + h\eta'$

1.45 Find an approximation to $e^x \sin x$ for small x.

a) $x - x^2 + x^3$
b) $x + x^2 + x^3/3$
c) $x - x^2/2 + x^3/6$
d) $x + x^2 - x^3/6$
e) $x + x^2 + x^3/2$

1.46 Find the area between the y-axis and $y = x^2$ from $y = 4$ to $y = 9$.

a) 29/3 b) 32/3 c) 34/3 d) 38/3 e) 43/3

1.47 The area contained between $4x = y^2$ and $4y = x^2$ is

a) 10/3 b) 11/3 c) 13/3 d) 14/3 e) 16/3

1.48 Rotate the shaded area of Ex. 1.24 about the x-axis. What volume is formed?

a) 4π b) 6π c) 8π d) 10π e) 12π

1.49 Evaluate $\int_0^2 (e^x + \sin x)\, dx$.

a) 7.81 b) 6.21 c) 5.92 d) 5.61 e) 4.21

1.50 Evaluate $2 \int_0^1 e^x \sin x\, dx$.

a) 1.82 b) 1.94 c) 2.05 d) 2.16 e) 2.22

1.51 Find an expression $\int x \cos x\, dx$.

a) $x \cos x - \sin x + C$
b) $-x \cos x + \sin x + C$
c) $x \sin x - \cos x + C$
d) $x \cos x + \sin x + C$
e) $x \sin x + \cos x + C$

DIFFERENTIAL EQUATIONS

1.52 The differential equation $y'' + x^2y' + y + 2 = 0$ is

a) linear and homogeneous.
b) linear and nonhomogeneous.
c) nonlinear and homogeneous.
d) nonlinear and nonhomogeneous.
e) not a differential equation.

1.53 Given: $y' + 2xy = 0$, $y(0) = 2$. Find: $y(2)$.

a) 0.0366 b) 0.127 c) 0.936 d) 2.36 e) 27.3

1.54 Given: $y' + 2x = 0$, $y(0) = 1$. Find: $y(10)$.

a) -100 b) -99 c) -91 d) -86 e) -54

1.55 A spring-mass system is represented by $2\ddot{y} + \dot{y} + 50y = 0$. What frequency, in hertz, is contained in the solution?

a) 0.79 b) 1.56 c) 2.18 d) 3.76 e) 4.99

1.56 Find the solution to $\ddot{y} + 16y = 0$.

a) $c_1 \cos 4t + c_2 \sin 4t$
b) $c_1e^{4t} + c_2e^{-4t}$
c) $c_1e^{4t} + c_2te^{4t}$
d) $c_1 \cos 4t + c_2t \cos 4t$
e) $c_1 \sin 4t + c_2t \sin 4t$

1.57 Find the solution to $\ddot{y} + 8\dot{y} + 16y = 0$.

a) $c_1 \cos 4t + c_2 \sin 4t$
b) $c_1t \cos 4t + c_2 \cos 4t$
c) $c_1 \sin 4t + c_2t \sin 4t$
d) $c_1e^{4t} + c_2te^{4t}$
e) $c_1e^{-4t} + c_2te^{-4t}$

1.58 Solve the equation $\ddot{y} - 5\dot{y} + 6y = 4e^t$.

a) $c_1e^{2t} + c_2e^{3t}$
b) $c_1e^{-2t} + c_2e^{-3t} + 2e^t$
c) $c_1e^{2t} + c_2te^{3t} + 2e^t$
d) $c_1e^{2t} + c_2e^{3t} + 2e^t$
e) $c_1e^{2t} + c_2e^{3t} + c_3e^t$

1.59 Solve $\ddot{y} + 16y = 8 \sin 4t$.

a) $c_1 \sin 4t + c_2 \cos 4t - t \cos 4t$
b) $c_1 \sin 4t + c_2 \cos 4t + t \cos 4t$
c) $c_1 \sin 4t + c_2 \cos 4t - \cos 4t$
d) $c_1 \sin 4t + c_2 \cos 4t + \cos 4t$
e) $c_1 \sin 4t + c_2 \cos 4t - \sin 4t$

PROBABILITY AND STATISTICS

1.60 You reach in a jelly bean bag and grab one bean. If the bag contains 30 red, 25 orange, 15 pink, 10 green, and 5 black beans, the probability that you will get a black bean or a red bean is nearest to

a) 0.6 b) 0.5 c) 0.4 d) 0.3 e) 0.2

1.61 Two jelly bean bags are identical to that of Prob. 1.60. The probability of selecting a black bean from the first bag and a red bean from the other bag is nearest to

a) 1/100 b) 2/100 c) 3/100 d) 4/100 e) 5/100

1.62 The probability of selecting 5 beans, three of which are red and two of which are orange, from the bag of Prob. 1.60 is nearest to

a) 3/1000 b) 4/1000 c) 5/1000 d) 6/1000 e) 7/1000

1.63 Two bags each contain 2 black balls, 1 white ball, and 1 red ball. What is the probability of selecting the white ball from the first bag or the red ball from the other bag?

a) 1/2 b) 7/16 c) 1/4 d) 3/8 e) 9/16

1.64 Two professors give the following scores to their students. What is the mode?

frequency	1	3	6	11	13	10	2
score	35	45	55	65	75	85	95

a) 65 b) 75 c) 85 d) 11 e) 13

1.65 For the data of Prob. 1.64, what is the arithmetic mean?

a) 68.5 b) 68.9 c) 69.3 d) 70.0 e) 73.1

1.66 Calculate the standard deviation for the data of Prob. 1.64.

a) 9.27 b) 10.11 c) 11.56 d) 13.42 e) 13.77

Mathematics

$\ln e^x = x$

$\ln x^a = a \ln x$

$\ln x = 2.303 \log x$

$$x = \frac{-b \pm \sqrt{b^2-4ac}}{2a}$$

$$(x+a)^n = a^n + na^{n-1}x + \frac{n(n-1)}{2!}a^{n-2}x^2 + \cdots$$

$\sin\theta = \frac{y}{r}$ $\quad\cos\theta = \frac{x}{r}$ $\quad\tan\theta = \frac{y}{x}$ $\quad r = \sqrt{x^2+y^2}$

$\sin^2\theta + \cos^2\theta = 1$

$\sin 2\theta = 2\sin\theta\cos\theta$

$\cos 2\theta = \cos^2\theta - \sin^2\theta$

$= 2\cos^2\theta - 1$

$= 1 - 2\sin^2\theta$

$\sin(\alpha+\beta) = \sin\alpha\cos\beta + \sin\beta\cos\alpha$

$\cos(\alpha+\beta) = \cos\alpha\cos\beta - \sin\alpha\sin\beta$

$$\frac{\sin\alpha}{a} = \frac{\sin\beta}{b} = \frac{\sin\gamma}{c}$$

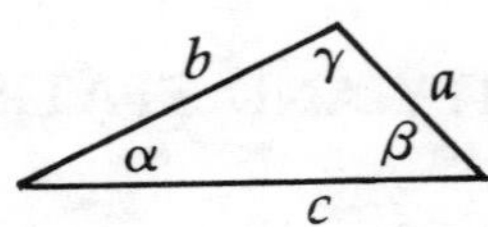

$$a^2 = b^2 + c^2 - 2bc\cos\alpha$$

$\sinh x = \frac{e^x - e^{-x}}{2}$

$\cosh x = \frac{e^x + e^{-x}}{2}$

$\cosh^2 x - \sinh^2 x = 1$

$\sinh(x+y) = \sinh x\cosh y + \sinh y\cosh x$

$\cosh(x+y) = \cosh x\cosh y + \sinh x\sinh y$

St. line: $y = mx + b$ or $\frac{x}{a} + \frac{y}{b} = 1$ **Ellipse:** $\frac{x^2}{a^2} + \frac{y^2}{b^2} = 1$ **foci:** $(\pm c, o)$ *where* $c^2 = a^2 - b^2$

Circle: $(x-a)^2 + (y-b)^2 = r^2$ **Parabola:** $y^2 = 4px$ **focus:** (p, o) **directrix:** $x = -p$

Hyperbola: $\frac{x^2}{a^2} - \frac{y^2}{b^2} = 1$ **asymtotic lines:** $y = \pm\frac{b}{a}x$

Polar coordinates: $x = r\cos\theta$ $\quad y = r\sin\theta$

Cylindrical coordinates: $x = r\cos\theta$ $\quad y = r\sin\theta$ $\quad z = z$

Spherical coordinates: $x = r\sin\phi\cos\theta$ $\quad y = r\sin\phi\sin\theta$ $\quad z = r\cos\phi$

Complex numbers: $x + iy = re^{i\theta}$ $\quad e^{i\theta} = \cos\theta + i\sin\theta$

$$\sin\theta = \frac{e^{i\theta} - e^{-i\theta}}{2i} \qquad \cos\theta = \frac{e^{i\theta} + e^{-i\theta}}{2}$$

$\frac{d}{dx}(\ln x) = \frac{1}{x}$

$\frac{d}{dx}(e^{kx}) = ke^{kx}$

$\frac{d}{dx}(\sin x) = \cos x$

$\frac{d}{dx}(\cos x) = -\sin x$

$\sin x = x - \frac{x^3}{3!} + \frac{x^5}{5!} + \cdots$

$\cos x = 1 - \frac{x^2}{2!} + \frac{x^4}{4!} + \cdots$

$e^x = 1 + x + \frac{x^2}{2!} + \cdots$

$\ln(1+x) = x - \frac{x^2}{2} + \frac{x^3}{3} - \cdots$

$\frac{1}{1+x} = 1 + x + \frac{x^2}{2} + \cdots$

$f(x) = f(a) + (x-a)f'(a) + \frac{(x-a)^2}{2!}f''(a) + \cdots$

$\int x^n dx = \frac{1}{n+1}x^{n+1} + C \qquad n \neq 1$

$\int \frac{dx}{x} = \ln x + C$

$\int e^{ax}\,dx = \frac{1}{a}e^{ax} + C$

$\int \cos x\,dx = \sin x + C$

$\int \cos^2 x\,dx = \frac{x}{2} + \frac{1}{2}\sin 2x + C$

2. Atomic and Nuclear Physics

Our understanding of solids, liquids and gases has been greatly advanced by the use of a microscopic particle model for matter. This particle model allows us to better understand the macroscopic world. For example, the macroscopic properties of pressure and temperature can be defined and better understood using the kinetic theory of gases, a theory based on the motion of microscopic particles. In the same way, the electron is actually a microscopic particle useful in our study of electricity. The process of diffusion in liquids is derivable from the microscopic particle model of the density gradient. The crystal lattice picture of a solid results from the microscopic particle view of matter in bulk. The wide-spread success of the particle model at the macroscopic level leads one to expect similar success in the description of nature by exploring the particles themselves. The atom and its nucleus will be defined and described along with the mathematical model in this microscopic domain; it is also referred to as the quantum domain.

2.1 Atomic Theory

2.1.1 Atomic Radiations

Measuring the radiations from an incandescent solid (Blackbody) established a relationship between the temperature of the solid and the total energy radiated. The relationship is called the *Stefan-Boltzmann* law, given by

$$W = \sigma T^4 \tag{2.1.1}$$

where $\sigma = 5.67 \times 10^{-8}$ watt/m$^2\cdot$K^4, T is the absolute temperature measured in kelvins, and W has units of watt/m^2. When the power per unit area per unit wavelength interval is measured for an incandescent solid Fig. 2.1 results. The area under the curve is given by the Stefan-Boltzmann law. The curve illustrates a perfect radiator at a temperature T: the so-called Blackbody radiator.

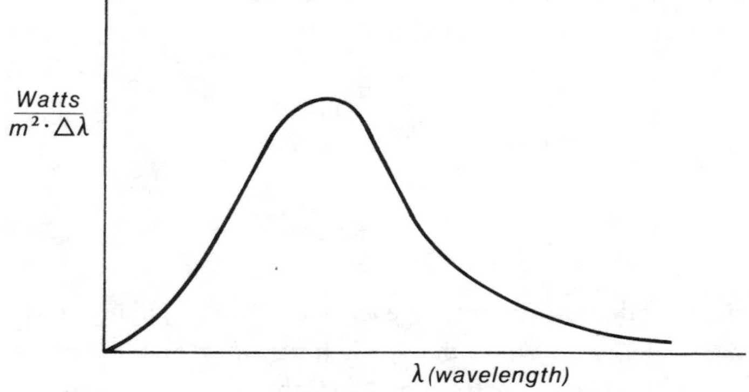

Figure 2.1. Blackbody radiation curve.

In 1900, Max Planck derived the equation of this curve. Planck's theory of Blackbody radiation requires that the radiation be treated as discrete bundles of energy. The energy exchange in emission and absorption is related to quantum oscillators within the material. Each bundle, or quantum, or photon, has energy — in joules — given by

$$E = h\nu = \frac{hc}{\lambda} \tag{2.1.2}$$

where

$h = 6.63 \times 10^{-34}$ J·s (Planck's constant)

$c = 3 \times 10^{8}$ m/s (velocity of light $= \nu\lambda$)

$\nu =$ frequency (s^{-1})

$\lambda =$ wavelength (m)

The electron volt (eV) is often used in atomic and nuclear physics as a unit of energy; it is the energy gained by an electron when accelerated through the potential difference of one volt. The electron volt is a small energy unit and is related to the Joule by

$$1 \text{ eV} = 1.6 \times 10^{-19} \text{ J} \tag{2.1.3}$$

Planck's quantum of energy can now be expressed in electron volts as

$$E = \frac{hc}{\lambda} = \frac{6.63 \times 10^{-34}\,(3 \times 10^{8})}{\lambda\ (1.6 \times 10^{-19})}$$

$$= \frac{1.24 \times 10^{-6}}{\lambda} \text{ eV} \tag{2.1.4}$$

if λ is measured in meters, or

$$E = \frac{1.24 \times 10^{4}}{\lambda} \text{ eV} \tag{2.1.5}$$

if λ is measured in angstroms Å (= 10^{-10} meters).

Albert Einstein made the first direct use of the Planck radiation quantum in order to derive an equation for the photoelectric effect. He postulated that

$$h\nu = \frac{hc}{\lambda} \tag{2.1.6}$$

$$= E_k + \phi$$

which means that the incident radiation photon of energy $h\nu$ on the photoelectric surface will eject an electron of maximum kinetic energy E_k when the work function ϕ of the surface is overcome. It is a conservation of energy equation and takes the form of a straight line as shown in Fig. 2.2. Below the cutoff frequency ν_o no photoelectrons are ejected.

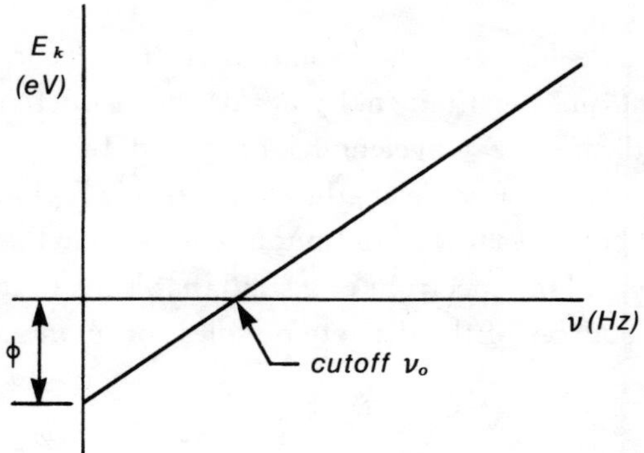

Figure 2.2. Photoelectron energy curve.

EXAMPLE 2.1

A photo surface is irradiated by photons of wavelength 4000 Å. If the surface has a work function of 2.0 eV, what is the maximum kinetic energy of the ejected electron? What is the cutoff frequency for this surface?

Solution. The maximum kinetic energy is found using Eq. 2.1.6 as follows:

$$\frac{hc}{\lambda} = E_k + \phi$$

$$\frac{1.24 \times 10^4}{4000} = E_k + 2.0$$

$$3.1 = E_k + 2.0$$

$$\therefore E_k = 1.1 \text{ eV}$$

Fig. 2.2 shows that the maximum kinetic energy is zero at the cutoff frequency; therefore,

$$h\nu = \frac{hc}{\lambda} = 0 + \phi$$

$$\frac{1.24 \times 10^4}{\lambda} = 2.0$$

$$\therefore \lambda = \frac{1.24 \times 10^4}{2.0} = 6200 \text{ Å}.$$

The speed of light is related to the wavelength and frequency by

$$c = \lambda \nu_o.$$

Thus,

$$\nu_o = \frac{3 \times 10^8}{6.2 \times 10^{-7}} = 4.84 \times 10^{14} \text{ Hz}$$

where λ is expressed in meters because of the units on c.

The energy of radiation in each region of the electromagnetic spectrum is given in Table 2.1

A comparison can be made between the radiation from a hydrogen gas discharge tube and the radiation from an incandescent tungsten filament by noting the distinctive colors emitted. The colors in the spectrum of each source indicate the wavelengths produced by exciting each source material. The hydrogen gas will give a discrete set of wavelengths in the visible region of the electromagnetic spectrum whereas the hot tungsten filament will show a continuous spectrum in the visible region. It is this difference between the spectrum of a gas and the spectrum of a solid that leads to the properties of individual atoms. The individual atoms are the sources of the discrete bundles, or quanta, of radiant energy introduced by Planck.

Wavelengths emitted by a gaseous source are measured with a spectrometer which usually records

TABLE 2.1. The Electromagnetic Spectrum

	E(eV)	λ(Å)	
	10,000	1.24	
			X-ray
	100	1.24×10^2	
	10	1.24×10^3	
			ultraviolet
blue	3.1	4×10^3	visible
red	1.8	7×10^3	
	1.0	12.4×10^3	
			infrared
	0.1	12.4×10^4	
			microwave
	0.01	12.4×10^6	

information on a photographic film. The series of lines (images of the spectrometer slit) displayed in this way (see Fig. 2.3) show the distinctive nature of emission spectra. The emission spectra of each element can be thought of as its signature.

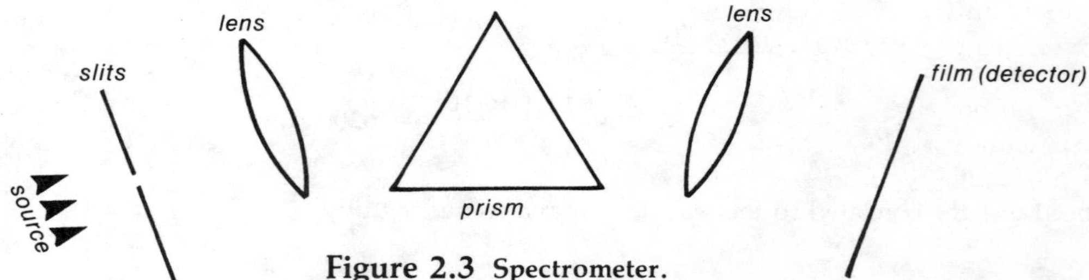

Figure 2.3 Spectrometer.

Rydberg wrote a mathematical expression to show the regularity of the wavelengths measured for hydrogen; it is

$$\frac{1}{\lambda} = R_H \left(\frac{1}{m^2} - \frac{1}{n^2}\right) \tag{2.1.7}$$

where $R_H = 1.09737 \times 10^7\ m^{-1}$ (Rydberg's constant). The measured wavelengths emitted by the excited hydrogen gas can be calculated by allowing $m = 1,2,3,\cdots$, while n takes on values of $m + 1$, $m + 2$, $m + 3,\cdots$, as indicated in Table 2.2.

TABLE 2.2. Hydrogen Emission Spectrum

Series Name	Values of m	Values of n	Type
Lyman	1	2,3,4,···	ultraviolet
Balmer	2	3,4,5,···	visible
Paschen	3	4,5,6,···	infrared
Brackett	4	5,6,7,···	infrared
Pfund	5	6,7,8,···	infrared

EXAMPLE 2.2

Determine the shortest and the longest wavelengths in the Balmer series of hydrogen.

Solution. The Balmer series has $m = 2$. Using $n = 3$ results in the longest wavelength. It is found from

$$\frac{1}{\lambda} = R_H\left(\frac{1}{2^2} - \frac{1}{3^2}\right)$$

$$\therefore \lambda = 6561\ \text{Å}.$$

The shortest is calculated for $n = \infty$:

$$\frac{1}{\lambda} = R_H\left(\frac{1}{2^2} - \frac{1}{\infty}\right)$$

$$\therefore \lambda = 3645\ \text{Å}.$$

The longest wavelength of the Balmer series is at the red end of the visible spectrum and the shortest wavelength is in the ultraviolet region of the spectrum.

2.1.2 Atomic Structure

In 1911 Rutherford concluded that all the positive charge in a heavy element, such as gold, is within a region of radius 10^{-14} meters. Since the elements are electrically neutral in their natural state, the negative charge must be outside this region identified by Rutherford. Hydrogen, which is the lightest gas, has the simplest structure to model electrically. The regularity of the measured emission spectrum wavelengths from hydrogen gas discharges indicates that it has one of the least complicated spectra.

Bohr proposed a model for the hydrogen atom in 1913. The basic principle of this model is the quantization of the angular momentum L of the electron as it moves in a circle around a positive proton charge. The angular momentum is quantized by

$$L = \frac{nh}{2\pi} = mVR \qquad (2.1.8)$$

where

$n = 1,2,3,\cdots$ (total quantum number)
m = mass of the electron $= 9.109 \times 10^{-31}$ kg
V = velocity of the electron
R = radius of the orbit
h = Planck's constant $= 6.63 \times 10^{-34}\ J \cdot s$

The quantization of the angular momentum L also quantizes the radius R and the total energy E of the one electron atom. The negative electron moving in a circle about the positive proton will experience a mechanical and electrical force to maintain this circular motion, that is,

Centripetal Force = Coulomb Force

$$\frac{mV^2}{R} = \frac{ke^2}{R^2} \tag{2.1.9}$$

where

$$e = 1.6 \times 10^{-19}\ \mathrm{C}$$
$$k = (4\pi\varepsilon_o)^{-1} = 9 \times 10^9\ \mathrm{N \cdot m^2/C^2}$$
$$\varepsilon_o = 8.85 \times 10^{-12}\ \mathrm{C^2/(N \cdot m^2)}$$

The kinetic energy K and the Coulomb potential energy P of the electron are given by

$$K = \frac{1}{2}mV^2 = \frac{ke^2}{2R} \tag{2.1.10}$$

$$P = -\frac{ke^2}{R} \tag{2.1.11}$$

The total energy E of the electron is given as the sum

$$E = K + P$$

$$= \frac{ke^2}{2R} - \frac{ke^2}{R} = -\frac{ke^2}{2R} \tag{2.1.12}$$

The negative total energy represents the attraction of negative and positive charges and, therefore, is referred to as a binding energy. The angular momentum equation and the force equation are combined to yield a useful form of the total energy:

$$E_n = -\frac{me^4}{8\varepsilon_o^2h^2}\left(\frac{1}{n^2}\right) \tag{2.1.13}$$

$$= -\frac{13.6}{n^2}\ \mathrm{eV}$$

using the values for m, e, ε_o and h listed earlier. This total energy equation establishes the quantized energy state or levels for the hydrogen atom, as illustrated in Fig. 2.4.

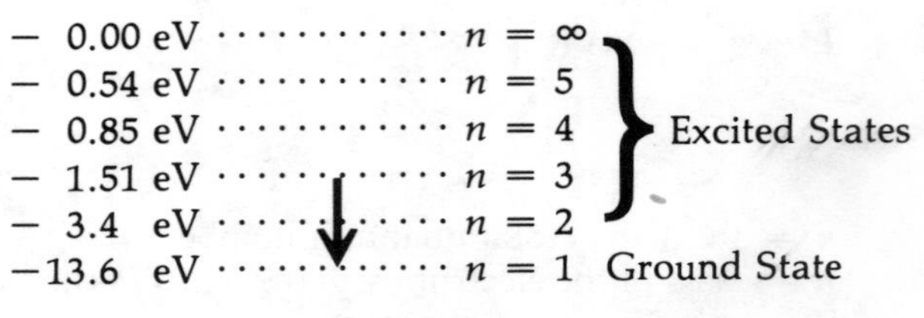

− 0.00 eV	$n = \infty$	Excited States
− 0.54 eV	$n = 5$	Excited States
− 0.85 eV	$n = 4$	Excited States
− 1.51 eV	$n = 3$	Excited States
− 3.4 eV	$n = 2$	Excited States
−13.6 eV	$n = 1$	Ground State

Figure 2.4. **Hydrogen energy level diagram.**

The emission of a Planck quantum of radiation ($h\nu$) can now be represented as a transition from the excited state $n = 3$ to the ground state $n = 1$ as shown by the arrow in Fig. 2.4.

EXAMPLE 2.3

Hydrogen atoms in a gas discharge tube are excited to the $n = 3$ state, or level. What are the wavelengths of the photons emitted by these atoms as measured with a spectrometer?

Solution. Using the energy level diagram and the total energy equation we have

$$E_3 = -\frac{13.6}{n^2} = -\frac{13.6}{9} = -1.51 \text{ eV}$$

$$E_2 = -\frac{13.6}{4} = -3.4 \text{ eV}$$

$$E_1 = -13.6 \text{ eV}.$$

The transitions, or quantum jumps, possible are $n = 3$ to $n = 1$, $n = 3$ to $n = 2$, and $n = 2$ to $n = 1$, which mean the emission of three (3) wavelengths:

$$\begin{aligned}
&1: \quad E_3 - E_1 = [-1.51 - (-13.6)] = 12.09 \text{ eV} \\
&2: \quad E_3 - E_2 = [-1.51 - (-\ 3.4)] = \ 1.89 \text{ eV} \\
&3: \quad E_2 - E_1 = [-3.4 \ \ - (-13.6)] = 10.02 \text{ eV}
\end{aligned}$$

Each energy difference is the energy of the emitted photon; thus, using Eq. 2.1.5, there results

$$\begin{aligned}
&1: \quad 12.09 = 1.24 \times 10^4/\lambda_1; \ \therefore \lambda_1 = 1025 \text{ Å} \\
&2: \quad 1.89 = 1.24 \times 10^4/\lambda_2; \ \therefore \lambda_2 = 6560 \text{ Å} \\
&3: \quad 10.2 \ = 1.24 \times 10^4/\lambda_3; \ \therefore \lambda_3 = 1215 \text{ Å}
\end{aligned}$$

The absorption of radiation by the atom is basically the reverse of the emission process. Hydrogen atoms in the ground state when radiated by a photon of wavelength 1215 Å will be excited to the $n = 2$ state or level, as shown in Ex. 2.3 above. This absorption will not take place unless the incident photon can excite the ground state electron to a specific energy level above the ground state. The emission and absorption of radiation by the atom are both quantized processes.

The hydrogen atom does exhibit all the essential features of atomic structure even though the Bohr model is an incomplete description of the atom. The deficiencies of the Bohr model are primarily the spectral line intensities, the observed fine structure of several lines and the effect of an external magnetic field on the emission spectrum.

The Bohr model does give the basic principles of atomic magnetism and the initial form of the vector model of the atom. The concept of the magnetic moment of an atom depends on the definition of the magnetic moment μ from classical electricity and magnetism; it is

$$\mu = NiA \tag{2.1.14}$$

where

N = number of turns of wire on loop
i = current in wire
A = area of current loop

Apply this to the hydrogen atom for $N = 1$ and the current will be the frequency f in the orbit multiplied by the electronic charge e. The defined current direction would be counterclockwise when the electron, rotating with orbital velocity V, moves clockwise. Using Eq. 2.1.14, we have

$$\mu = fe\pi R^2 = \frac{V}{2\pi R}\, e\pi R^2 \qquad (2.1.15)$$

$$= eVR/2$$

Using Eq. 2.1.8 we can eliminate VR and express μ as

$$\mu = \frac{ehn}{4\pi m} \qquad (2.1.16)$$

The unit magnetic moment defined for $n = 1$ is a Bohr magneton. Substituting the known values for m, h, and e there results

$$\mu = 9.27 \times 10^{-24}\ \frac{\text{Joule}}{\text{Tesla}}\ . \qquad (2.1.17)$$

The angular momentum vector $\vec{L}$ and the magnetic moment vector $\vec{\mu}$ point in opposite directions as shown in Fig. 2.5. The angular momentum vector is defined according to the vector cross product $\vec{L} = \vec{R} \times m\vec{V}$

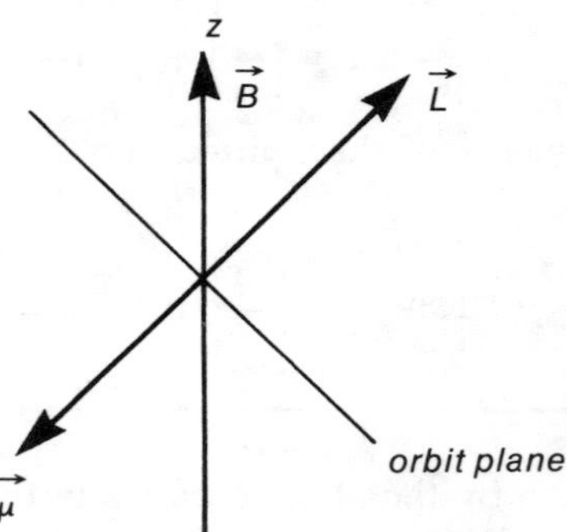

Figure 2.5. Atomic vector model.

and the magnetic moment vector $\vec{\mu}$ is defined with respect to the positive current direction. The external magnetic field $\vec{B}$ is given as a reference direction along with the positive z-axis. The vectors are related by

$$\vec{\mu} = -\frac{e}{2m}\vec{L} \qquad (2.1.18)$$

where L has the magnitude in the Bohr model of $nh/2\pi$.

The neutrality of the natural elements requires that there be as much positive charge as there is negative charge for the atom. The positive protons at the nucleus are matched by an equal number of negative electron charges outside the nucleus. The number of nuclear charges is called the atomic number Z. A hydrogen atom has Z = 1, a helium atom Z = 2, and so on throughout the periodic table. Since the Bohr model was not completely successful, a new model for many electron atoms was developed by Schroedinger called "quantum mechanics" (see Section 2.3).

The theoretical and experimental analysis of electronic structure resulted in a set of four quantum numbers, presented in Table 2.3 on page 2-9.

TABLE 2.3. The Quantum Numbers

Description	Values	Related to
Total quantum number n	$1,2,3,\cdots$	Total energy
Orbital quantum number ℓ	$0,1,2\cdots(n-1)$	Magnitude of the angular momentum vector
Magnetic quantum number m_ℓ	$0,\pm1\cdots\pm\ell$	Direction of the angular momentum vector
Spin quantum number m_s	$\pm 1/2$	Spin angular momentum vector

The shell model of the electron atom is indicated by $n = 1$, K-shell; $n = 2$, L-shell; $n = 3$, M-shell; and so on, to accommodate the number of electrons available per atom. The bookkeeping required for many electron atom systems is called the Pauli Exclusion Principle. The principle states that no two electrons in an atom can have the same set of four quantum numbers. Therefore, when $n = 1$ (K-shell) there can be only 2 electrons. Table 2.3 allows us to determine the following for the two electrons:

Electron 1	Electron 2	
$n = 1$	$n = 1$	
$\ell = 0$	$\ell = 0$	filled K-shell
$m_\ell = 0$	$m_\ell = 0$	
$m_s = +1/2$	$m_s = -1/2$	

EXAMPLE 2.4

According to the Pauli Exclusion Principle, what is the maximum number of electrons that the L-shell can contain? What element in the periodic table is complete with its L-shell filled?

Solution. The L-shell has $n = 2$ so all other determinations depend on this. From Table 2.3 we determine the following:

Possible Electrons

n	2	2	2	2	2	2	2	2
ℓ	0	0	1	1	1	1	1	1
m_ℓ	0	0	0	0	+1	−1	−1	+1
m_s	+1/2	−1/2	+1/2	−1/2	+1/2	−1/2	+1/2	−1/2

The total number of electrons in the L-shell is 8. Since the K-shell is filled first, the element must have $Z = 10$, which is Neon.

2.1.3 Atomic Reactions

The common fluorescent lamp is an application of the principles of atomic structure. In general, the gas inside the fluorescent tube is mercury vapor. The electrical discharge excites the mercury atoms to many levels above the ground state. These atoms return to the ground state in several steps giving up their energy of excitation primarily as wavelengths in the ultraviolet. The inside surface of the tube is coated with a fluorescent material which is excited by this short wavelength radiation from mercury. The excited atoms of the material return to the ground state by re-radiating this energy of excitation in the visible portion of the electromagnetic spectrum. The fluorescent material can be viewed as a wave-shifter; it receives short wavelength radiation and transforms that energy into longer wavelength radiation.

The laser represents a fine tuned atomic system. The word "laser" is an acronym for Light Amplification by Stimulated Emission of Radiation. A simplified description of the laser process follows. Excited atoms usually return to ground state in about 10^{-8} seconds with the emission of a photon. If a photon of the proper wavelength encounters this atom during this period, it is probable that the incident photon can stimulate the excited atom to give up its photon with the two photons going off together completely in phase. If these two photons stimulate additional excited atoms to give up their photons, the number of photons in phase increases, or is amplified, and lasing may result.

This process requires the manipulation of atomic energy levels, since the incident photon would normally be absorbed if the atoms were in their ground state. This manipulation of atomic states is called "optical pumping." The pumping produces more atoms in the excited state than there are atoms in the ground state, resulting in an atomic level population inversion. Therefore, the probability of the incident photon causing stimulated emission is greatly increased. Many ingenious methods have been devised to produce lasers from both solids and gases.

2.2 Nuclear Theory

2.2.1. Nuclear Radiations

The discovery of nuclear radiations identified the existence of naturally occurring radioactive elements. The components of these radiations are alpha particles, beta particles and gamma rays. The alpha particle (α) is the nucleus of the helium atom, the beta particles (β) are electrons, and the gamma rays (γ) are short wavelength electromagnetic radiation.

The regularity of these emissions from the naturally occurring heavy elements is represented by a radioactive decay law

$$N = N_o e^{-\lambda t} \tag{2.2.1}$$

where

N = number of remaining atoms after time t
N_o = initial number of atoms
λ = disintegration constant (sec^{-1})

The activity (A) of a radioactive material is given by

$$A = N\lambda \text{ curies} \tag{2.2.2}$$

where

$$1 \text{ curie} = 3.7 \times 10^{10} \text{ atom disintegrations/second}$$

The decay law can then be written as

$$A = A_o e^{-\lambda t} \tag{2.2.3}$$

These equations represent the decay of element M into element Q which is stable to further decay. The decay curves are sketched in Fig. 2.6. The half-life $T_{1/2}$ is defined as the time necessary to reduce the activity to one-half of the original activity, thus,

$$T_{1/2} = \frac{0.693}{\lambda} \tag{2.2.4}$$

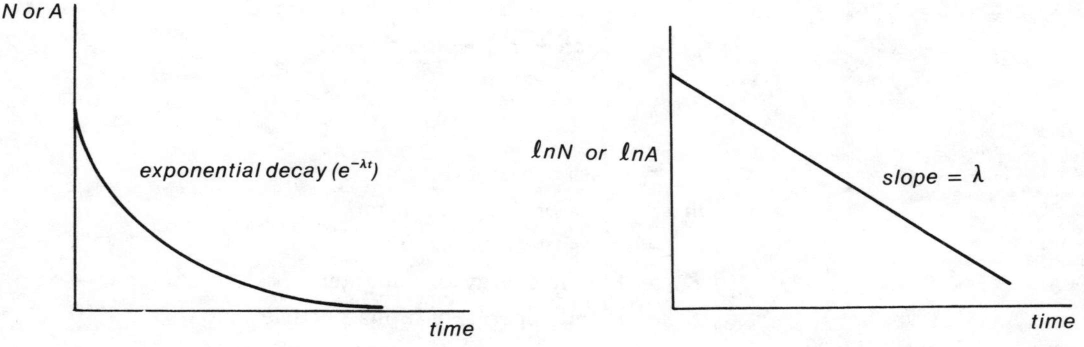

Figure 2.6. Radioactive decay curves.

Nuclear processes are designated by the symbols

$$^{A}_{Z}E_{N} \tag{2.2.5}$$

where

E = the element
A = atomic mass number (protons + neutrons)
Z = atomic charge (protons)
N = number of neutrons

The alpha, beta and gamma decay can then be represented in this shorthand form:

$$^{226}_{88}Ra \rightarrow {}^{222}_{86}Rn + {}^{4}_{2}He + Q \quad \text{Alpha decay}$$

$$^{64}_{29}Cu \rightarrow {}^{64}_{30}Zn + {}^{0}_{-1}e + Q \quad \text{Beta decay}$$

$$^{57}_{26}Fe^{*} \rightarrow {}^{57}_{26}Fe + {}^{0}_{0}\gamma \quad \text{Gamma decay}$$

These typical nuclear equations are balanced when the charge Z and the atomic mass number A each adds to the same value on both sides of the symbolic arrow. Since the gamma ray ${}^{0}_{0}\gamma$ does not change charge or mass, the gamma decay expression indicates only that the *Fe*-57 nucleus was in an excited state and subsequently returned to the ground state by the emission of this quantum of energy.

The Q-value of the disintegration represents the energy available in the nuclear transformation of mass. The basic equation of this transformation is an expression of the mass-energy conservation from special relativity:

$$\underset{\text{total energy}}{E} = \underset{\text{kinetic energy}}{E_k} + \underset{\text{rest energy}}{E_o} \qquad (2.2.6)$$

$$E_o = m_o c^2 \qquad (2.2.7)$$

where m_o represents the rest mass.

The Q-value for an alpha decay can be derived in the following manner to show the general principle:

$$m_p c^2 = m_d c^2 + K_d + m_\alpha c^2 + K_\alpha \qquad (2.2.8)$$

where

m_p = mass of parent atom
m_d = mass of daughter atom
m_α = mass of alpha particle
K_d = kinetic energy of daughter
K_α = kinetic energy of alpha particle

The rearranged equation yields

$$Q = K_\alpha + K_d = (m_p - m_d - m_\alpha)c^2 \qquad (2.2.9)$$

The Q-value of the nuclear process can be calculated from the kinetic energy or from the known masses of the atoms.

EXAMPLE 2.5

A one milligram sample of radioactive *Po*-210, which has a half-life of 138.4 days, decays into *Pb*-206 by alpha emission. (a) Determine the disintegration constant. (b) Determine the number of atoms of *Po*-210 which remain after 300 days of decay. (c) What is the activity of the sample in curies after 300 days? (d) What is the Q-value for this disintegration?

Solution.

a) The disintegration constant is

$$\lambda = \frac{0.693}{T_{1/2}} = \frac{0.693}{138.4 \times 24 \times 3600} = 5.8 \times 10^{-8} \ \text{sec}^{-1}$$

b) The number of atoms of *Po*-210 is found using Eq. 2.2.1 as follows:

$$N_o = 1 \times 10^{-3} \text{ grams} \times 6.023 \times 10^{23} \frac{\text{atoms}}{\text{gram}} \ / \ 210$$

$$N = N_o \exp(-\frac{0.693}{138.4} \times 300)$$

$$= 6.39 \times 10^{17} \text{ atoms}$$

c) At $t = 0$ the activity is given by Eq. 2.2.2 to be

$$A_o = N_o\lambda = 1.663 \times 10^{11} \text{ atom dis./s}$$

$$\therefore A = A_o e^{-\lambda t} = 1.663 \times 10^{11}\,(0.22) = 3.7 \times 10^{10} \quad \text{or 1.0 curie.}$$

d) The Q-value is found as follows:

$$Q = (m_p - m_d - m_\alpha)c^2$$

or

$$\frac{Q}{c^2} = m_p - m_d - m_\alpha = \Delta m$$

The nuclear equation is

$$^{210}_{84}Po \rightarrow {}^{206}_{82}Pb + {}^{4}_{2}He + Q.$$

The tabulated masses in isotopic mass units u are:

Po-210	Pb-206	He-4
209.9829 u	205.9745 u	4.00216 u

This gives a mass difference of

$$\Delta m = 0.0058\ u = \frac{Q}{c^2}$$

The mass difference can be expressed as an energy (MeV) by using the energy equivalent of one isotopic mass unit u; that is,

$$1\ u = 1.6604 \times 10^{-27}\text{ kg} = \frac{931.5}{c^2}\text{ MeV}$$

Finally,

$$Q = \Delta m\, c^2 = (0.0058)\, c^2 \times 931.5/c^2$$

$$= 5.402 \text{ MeV}$$

2.2.2 Nuclear Reactions

The first nuclear reactions were performed by using naturally radioactive elements as sources of alpha particles. These alpha particles ranged in energy from about 4.0 MeV to about 9.0 MeV. The early reactions were used to probe the nucleus and the particles it contained. Typical of these reactions are

$$^{14}_{7}N + {}^{4}_{2}He \rightarrow {}^{1}_{1}H + {}^{17}_{8}O + Q$$

$$^{11}_{5}B + {}^{4}_{2}He \rightarrow {}^{14}_{7}N + {}^{1}_{0}n + Q$$

The nitrogen ${}^{14}_{7}N$ reaction gave early evidence of a proton ${}^{1}_{1}H$ resulting from nuclear reactions. The boron ${}^{11}_{5}B$ reaction is one that played a role in the discovery of the neutron ${}^{1}_{0}n$. In the early 1930's the generation of neutrons and protons from such nuclear reactions indicated that these were basic particles of all nuclei.

The development of particle accelerators greatly increased the possible nuclear projectiles available and the reactions that might be expected to occur. Particles such as alpha particles α, ${}^{4}_{2}He$; protons p, ${}^{1}_{1}H$; deuterons d, ${}^{2}_{1}H$; tritons t, ${}^{3}_{1}H$; as well as electrons and heavier nuclei, were used to explore nuclear structure.

The features of the accelerated particle reaction are indicated through the conservation of charge, conservation of nucleons (neutrons and/or protons) and the conservation of energy and momentum. The conservation of charge (Z) and the conservation of nucleons (A) will be the same as the nuclear decay processes discussed earlier. The conservation of energy expression follows the general nuclear reaction nomenclature

$$X\ (a,b)\ Y$$
$$X + a \rightarrow b + Y \tag{2.2.10}$$

where X is the heavier target nucleus at rest; a is the light incident projectile; Y is the heavier resulting nucleus, and b the resulting light projectile. The energy conservation equation becomes

$$m_x c^2 + m_a c^2 + K_a = m_b c^2 + K_b + m_y c^2 + K_y \tag{2.2.11}$$

The rearranged equation defines the reaction Q-value:

$$Q = (m_x c^2 + m_a c^2) - (m_b c^2 + m_y c^2) \tag{2.2.12}$$
$$= K_b + K_y - K_a$$

The Q-value can be calculated either from the known masses or the measured kinetic energy.

The nuclear power reactions for fission and fusion are determined on this principle. The fission reaction

$${}^{235}_{92}U + {}^{1}_{0}n \rightarrow {}^{140}_{54}Xe + {}^{94}_{38}Sr + 2\,{}^{1}_{0}n + Q$$

is typical of many possible slow neutron reactions, with ${}^{235}_{92}U$ giving a Q-value of about 200 MeV per fission reaction.

The fusion reaction

$${}^{2}_{1}H + {}^{3}_{1}H \rightarrow {}^{4}_{2}He + {}^{1}_{0}n + Q$$

has a Q-value of 17.6 MeV and is the reaction most likely to be used for fusion power.

EXAMPLE 2.6

Protons of 0.7 MeV are used to irradiate a ${}^{7}_{3}Li$ target. (a) Write the complete nuclear reaction. (b) Find the Q-value for this reaction. (c) Determine the kinetic energy of the product particles if they emerge at 90° to the incident beam.

Solution.

a) The nuclear reaction is written as:

$$X + a \rightarrow b + Y$$

$$^{7}_{3}Li + {}^{1}_{1}H \rightarrow {}^{4}_{2}He + {}^{4}_{2}He + Q$$

b) The Q-value is found as follows:

$$Q = (m_x c^2 + m_a c^2) - (m_b c^2 + m_y c^2)$$

The respective masses are listed:

Li:	7.016004 u	He:	4.002603 u
H:	+ 1.007825 u	He:	+ 4.002603 u
	8.023829		8.005206

$$Q = (8.023829 - 8.005206)\, c^2 \left(\frac{931.5}{c^2}\right)$$

$$= +\ 17.3 \text{ MeV}$$

c) The kinetic energy is related to the Q-value by

$$Q = K_b + K_y - K_a.$$

Using $K_a = 0.7$ MeV we see that

$$K_b + K_y = 17.3 + 0.7 = 18.0.$$

If the alpha particles $^{4}_{2}He$ are measured at 90°, it is expected that $K_b = K_y$; therefore,

$$2K_b = 18$$

or

$$K_b = 9 \text{ MeV}.$$

2.2.3 Nuclear Structure

Through nuclear decay and nuclear reactions the basic constituents of the nucleus can be identified. The strong nuclear force which binds the constituents together has specific characteristics: it

1. is short range attractive within 10^{-14} meters.
2. is charge independent.
3. is non-central, that is, not an inverse square law force.
4. shows saturation, that is, a group of 2 protons and 2 neutrons is very stable.

Neutrons and protons have the following properties:

Property	Proton	Neutron
Charge	$+1.6 \times 10^{-19}\ C$	0
Mass	1.007276 u	1.008665 u
Spin	1/2	1/2
Decay	Stable	Half-life ~ 11 min.
Magnetic Moment	Positive	Negative

These particles are bound together as nuclei of the elements. The binding energy per particle indicates the relative strength of the nuclear forces. The binding energy of the deuteron, which has one proton and one neutron, can be calculated from known masses as follows:

$$\text{Measured deuteron } {}^{2}_{1}H_1 \text{ mass} = 2.013553\ u \text{ (nuclear mass)}$$

$$\text{Proton } p \text{ mass} = 1.007276\ u$$

$$\text{Neutron } n \text{ mass} = 1.008665\ u$$

$$\therefore \text{Binding energy} = E_b = c^2\,(m_p + m_n - m_d) = (0.002388\ u)c^2$$

$$= (0.002388)c^2 \times 931.5/c^2$$

$$= 2.22 \text{ MeV}$$

Since there are two particles in this nucleus the binding energy per nucleon is

$$\frac{E_b}{A} = \frac{2.22}{2} = 1.11 \text{ MeV/particle}$$

When this same calculation is done for all the elements, the binding energy per nucleon results as shown in Fig. 2.7. The average binding energy per nucleon is 8 MeV.

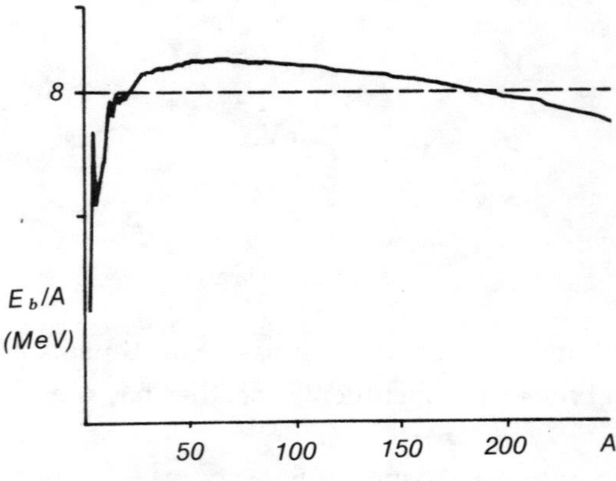

Figure 2.7. Binding energy per nucleon.

EXAMPLE 2.7

Determine the binding energy per nucleon for U-235.

Solution. The mass balance follows:

$$\text{Measured mass, } {}^{235}_{92}U = 235.043920\ u \text{ (atomic mass)*}$$

$$\text{92 protons: } 92(1.007825\ u) = 92.71990\ u$$

$$\text{143 neutrons: } 143(1.008665\ u) = 144.23910\ u$$

$$\text{Total: } 236.95900\ u$$

The binding energy is then found to be:

$$E_b = (236.95900 - 235.04392)c^2 \times \frac{931.5}{c^2} = 1784 \text{ MeV}$$

and

$$\frac{E_b}{A} = \frac{1784}{235} = 7.59 \text{ MeV/nucleon.}$$

***Note:** The atomic mass includes 92 electrons whose masses are accounted for when the mass of the "protons" are given as ${}^1_1H = 1.007825\ u$.

The basic nuclear theories advanced to explain decay, nuclear reactions, nuclear forces, binding energies, and magnetic moments are the liquid drop model, the nuclear shell model and the collective or unified model.

The liquid drop model treats nuclei in terms of parameters characteristic of a liquid drop. The three basic aspects of the drop are the volume effect, the surface effect, and the coulomb effect. The total binding energy of the nucleus can be viewed as due principally to:

1. *Volume effect*: The binding energy is directly proportional to the nuclear volume and the mass number A.
2. *Surface effect*: A negative energy which is proportional to the nuclear surface.
3. *Coulomb effect*: A negative energy which is proportional to the number of pairs of protons in the nucleus.

The liquid drop model is acceptable when determining the binding energies and for describing some nuclear reactions such as the fission process.

The shell model treats the neutrons and protons as essentially independent in their contributions to the nuclear volume. The neutrons and protons are particles of spin 1/2 and combine in their separate energy levels. The deuteron, which is composed of one proton and one neutron, has a nuclear spin of 1. The next hydrogen isotope, tritium, has 2 neutrons and one proton and has a nuclear spin of 1/2. The helium-3 nucleus has 2 protons and one neutron but also has a spin of 1/2. The odd neutron or proton defines the nuclear spin. In the case of helium-4, 2 protons and 2 neutrons, the nuclear spin is zero. The neutron and proton nuclear shells are closed at what have been called "magic numbers": 2, 8, 20, 28, 50, 82 and 126. This property is similar to completely filling the atomic shells of the noble (inert) gases. The shell model is acceptable when describing many nuclear properties such as nuclear spin, magnetic moments and nuclear scattering.

The collective or unified model of the nucleus combines the most successful features of the liquid drop and the shell models. The nucleus is described as possessing properties of vibration and rotation as well as the effects of individual particle contributions. The collective or unified model has the broadest range of success in bringing together experimental and theoretical nuclear physics.

2.3 Quantum Theory

The success of atomic and nuclear physics depends on the introduction of the Schroedinger Wave Equation (*SWE*) to describe quantized energy states. The general principles of the *SWE* are outlined here as they apply to first the hypothetical problem of a particle in an infinite square well, then the hydrogen atom, and—finally—nuclear decay.

2.3.1 Infinite Square Well

The one dimensional, time independent *SWE* is

$$\frac{d^2\psi}{dx^2} + \frac{8\pi^2 m}{h^2}(E - U)\psi = 0 \qquad (2.3.1)$$

where ψ is the wave function of the system to be determined, h is Planck's constant, m is the mass of the particle, E the total energy, U the potential energy, and x the independent variable. The solution of this *SWE* is determined using the boundary conditions of the particular problem. The "infinite square well" means the potential energy U goes to infinity at $x = 0$ and $x = L$. The particle placed inside the well shown in Fig. 2.8 has the probability of 1 of being inside and the probability of zero of being outside. These conditions must be identified in the solution of the *SWE*. For $U = 0$ inside the well, the *SWE* takes the form

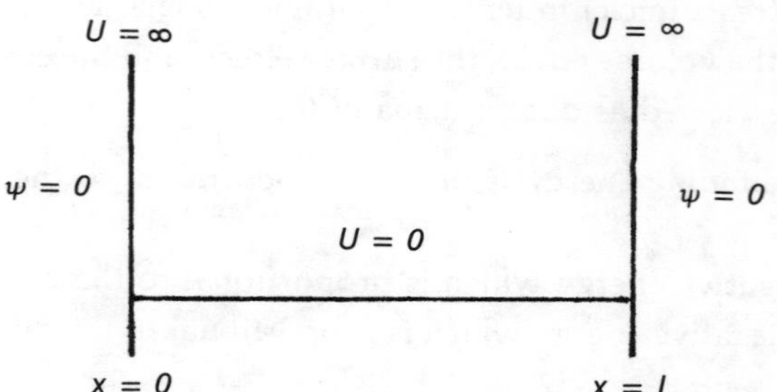

Figure 2.8. **The infinite square well.**

$$\frac{d^2\psi}{dx^2} + \frac{8\pi^2 m}{h^2}E\psi = 0 \qquad (2.3.2)$$

The standard solution of this differential equation is

$$\psi(x) = A \sin \sqrt{\frac{8\pi^2 mE}{h^2}}\, x. \qquad (2.3.3)$$

To satisfy the conditions that $\psi(0) = 0$ and $\psi(L) = 0$, we demand that

$$\sqrt{\frac{8\pi^2 mE}{h^2}} = \frac{n\pi}{L} \qquad (2.3.4)$$

providing the solution

$$\psi(x) = A \sin \frac{n\pi x}{L} \qquad n = 1,2,3,\cdots \qquad (2.3.5)$$

The quantized total energy for the particle in the well is given from Eq. 2.3.4 as

$$E = \frac{n^2h^2}{8mL^2} \tag{2.3.6}$$

The constant A in the wave function ψ is determined by normalizing the wave function between $x = 0$ and $x = L$, that is,

$$\int_0^L \psi^*\psi dx = 1 \tag{2.3.7}$$

Since the wave function is often a complex function, it is multiplied by its complex conjugate ψ^* which in the present case results in the square of the wave function. The wave function ψ is called the probability amplitude function and the $\psi^*\psi$ product the probability density function. Therefore,

$$\int_0^L A^2 \sin^2 \frac{n\pi x}{L} dx = 1 \tag{2.3.8}$$

results in

$$A = \sqrt{\frac{2}{L}} \tag{2.3.9}$$

so that

$$\psi(x) = \sqrt{\frac{2}{L}} \sin \frac{n\pi x}{L} \tag{2.3.10}$$

which is the normalized wave function.

The probability amplitude function and the probability density function are represented for $n = 2$ in Fig. 2.9. The energy quantization, the wave function, and the probability distribution are determined for a particle in an infinite square well by the *SWE*.

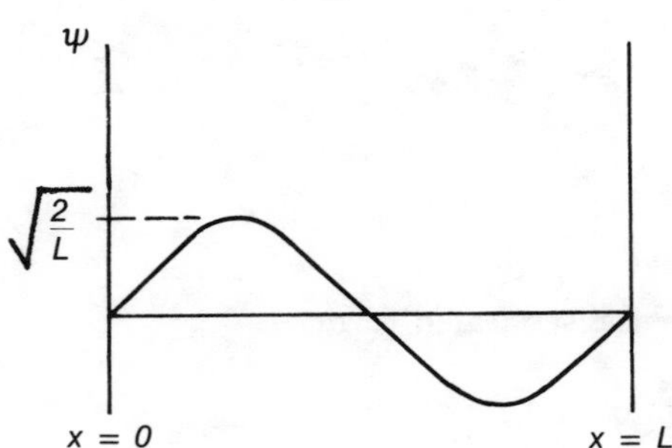

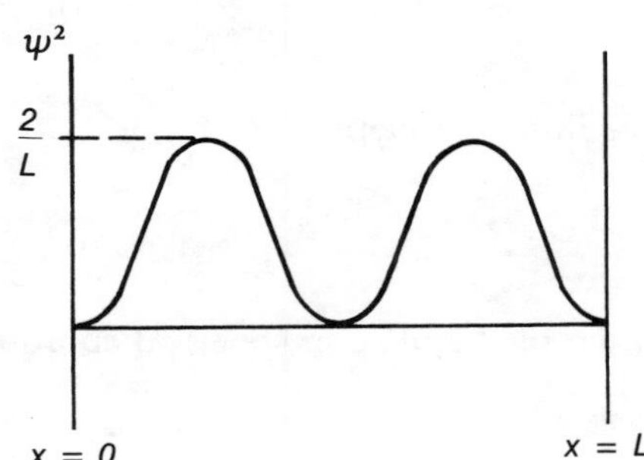

Figure 2.9. Wave function plot.

EXAMPLE 2.8

The particle in an infinite square well is an electron and the width of the well is 1 Å. (a) What is the permitted total energy of this electron for $n = 3$? (b) What is the wave function for this electron when $n = 3$? (c) Sketch the wave function and the square of the wave function for $n = 3$. (d) What is the probability of finding the particle between $x = 0$ and $x = L/3$ for $n = 3$?

Solution.

a) The total energy is found from Eq. 2.3.6 to be

$$E = \frac{n^2h^2}{8mL^2} = \frac{3^2(6.63 \times 10^{-34})^2}{8(9.11 \times 10^{-31})(1 \times 10^{-10})^2 \times 1.6 \times 10^{-19}}$$

$$= 5.44 \times 10^{-17}\ \text{J} \quad \text{or} \quad 340\ \text{eV}.$$

b) The wave function, with $n = 3$, is given by Eq. 2.3.10 as

$$\psi(x) = \sqrt{\frac{2}{L}} \sin \frac{3\pi x}{L}$$

c) The wave function and the probability density function are sketched below.

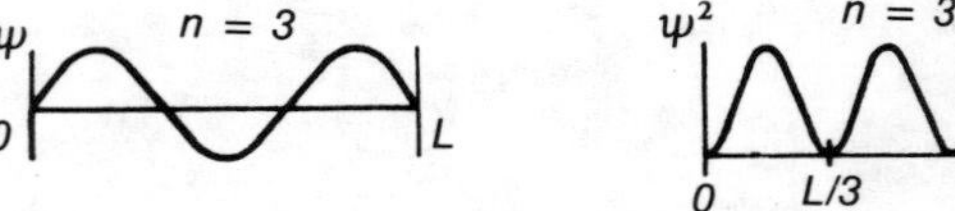

d) The area under the curve between $x = 0$ and $x = L/3$ is the probability of finding the particle in this one region of the well; that is,

$$\text{Probability} = \int_0^{L/3} \psi^* \psi dx = \frac{2}{L} \int_0^{L/3} \sin^2 \frac{3\pi x}{L}\, dx = \frac{1}{3}.$$

2.3.2. Hydrogen Atom

The application of the *SWE* to the hydrogen atom requires a three-dimensional expression and a coulomb potential U. Introducing the Laplacian ∇^2, the time independent *SWE* becomes

$$\nabla^2\psi + \frac{8\pi^2 m}{h^2}(E - U)\psi = 0 \tag{2.3.11}$$

where (see Bohr's model)

$$U = -\frac{ke^2}{r} \tag{2.3.12}$$

The solution of the *SWE* in spherical coordinates takes the separated form

$$\psi(r,\theta,\phi) = R(r)\ \Theta(\theta)\ \Phi(\phi). \tag{2.3.13}$$

Each separated function has a particular constant in its solution which is given the physical interpretation of a quantum number:

$R(r)$: Total Quantum Number $n = 1,2,3,\cdots$,
$\Theta(\theta)$: Orbital Angular Momentum Quantum Number $\ell = 0,1,\cdots(n-1)$
$\Phi(\phi)$: Magnetic Quantum Number $m_\ell = 0, \pm 1 \cdots \pm \ell$

The complete time dependent relativistic solution of the *SWE* determines the fourth spin quantum number $m_s = \pm 1/2$. The radial wave function $R(r)$ is shown as a probability distribution function in Fig. 2.10 on page 2.21 where r_o is the first Bohr orbit for the hydrogen atom which corresponds to the most probable position of the electron for $n = 1$.

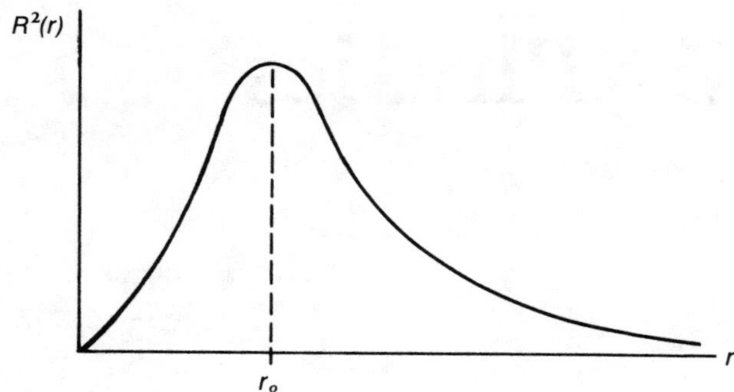

Figure 2.10. **The probability density function.**

2.3.3 Nuclear Decay

One of the earliest applications of quantum mechanics was to the alpha particle emission from the nucleus in radioactive decay. The nucleus can be viewed as a potential well similar to the infinite square well but with a potential height of about 30 MeV. Fig. 2.11 shows the potential well and the alpha particle's approximate position. Classically, the alpha particle cannot rise over the barrier and, therefore, could not

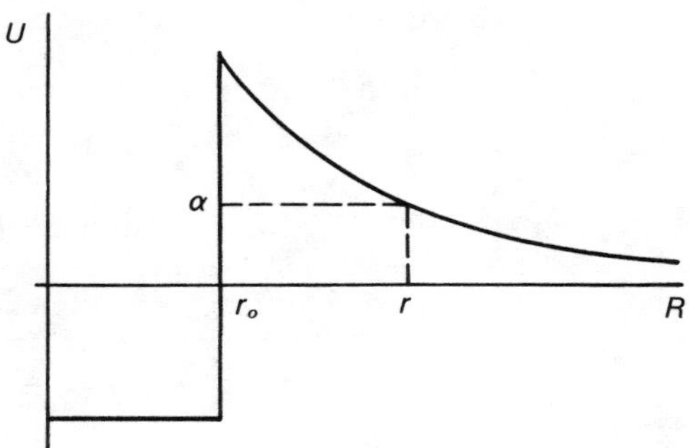

Figure 2.11. **Nuclear potential well.**

be emitted from the nucleus. In terms of quantum mechanics though, the probability of emission is not zero. If the alpha particle's wave function matches at the boundaries r_o and r, the probability of emission is greater than zero. The solution of the *SWE* will give a wave function for the alpha particle inside r_o, between r_o and r, and outside r. The ratio of the square of the amplitude of the inside wave function to the square of the amplitude of the outside wave function is the transmission probability. The alpha particle emission from the nucleus (transmission probability > 0), as explained by quantum mechanics, is referred to as the tunneling effect.

The success of the particle model of matter in atomic and nuclear physics has been very extensive. However, the search for order in natural systems still remains an unfinished task.

Physics

$c = 3.0\times 10^{8}$ m/s

$hc = 12400$ eV·Å

$e = 1.6\times 10^{-19}$ C

$1\text{Å} = 10^{-10}$ m

$h = 6.63\times 10^{-34}$ J·s

$\sigma = 5.76\times 10^{-8}$ W/m²·K⁴

$k = \frac{1}{4\pi\epsilon_o}\ 9\times 10^{9}$ N·m²/C²

1 eV $= 1.6\times 10^{-19}$ J

1 curie $= 3.70\times 10^{10}$ dis/s

$A_v = 6.023\times 10^{23}$ (gm·mol)$^{-1}$

$m_e = 9.109\times 10^{-31}$ kg $= 0.000548\ u = 0.511$ MeV/c^2

$m_p = 1.672\times 10^{-27}$ kg $= 1.007276\ u = 938.25$ MeV/c^2

$m_n = 1.675\times 10^{-27}$ kg $= 1.008665\ u = 939.55$ MeV/c^2

$1\ u = 1.6604\times 10^{-27}$ kg $= 931.478$ MeV/c^2

$E_k = \frac{1}{2}\text{mV}^2$

$E_n = -13.6/n^2$ eV

$W = \sigma T^4$

half life $= 0.693/\lambda$

$A = A_o e^{-\lambda t}$

$\text{h}\nu = E_k + \phi$

$E = h\nu = hc/\lambda$

$E = mc^2$

$A = N\lambda$

Practice Problems

2.1 Which one of the following statements is a consequence of Planck's Quantum Hypothesis?

a) Radiation flows continuously and without interruption.
b) Radiation from a Blackbody is independent of temperature.
c) Radiation is emitted in discrete bundles of energy called quanta.
d) Radiation emission and absorption have no common physical explanation.
e) Radiation from a Blackbody is proportional to the fifth power of the temperature.

2.2 What is the energy, in eV, of a photon of radiation which has a wavelength of 5000 Å?

a) 2.5 b) 3.7 c) 4.2 d) 5.6 e) 12.0

2.3 A solid sphere (Blackbody) is maintained at a temperature of 4000 K. If the sphere has a radius of 0.5 cm, what input power (watts) would be required to maintain the sphere at this temperature?

a) 1629 b) 1679 c) 2210 d) 4560 e) 5670

2.4 A star is determined to have a temperature of 4000 K and radiates 600 times as much energy per unit time as the sun whose temperature is 6000 K. The star's radius is how many times greater than the sun's?

a) 20 b) 25 c) 35 d) 50 e) 55

2.5 Which one of the following statements is correct when applied to the photoelectric effect?

a) The energy of photoelectrons emitted will increase proportionately with light intensity.
b) The stopping potential is dependent on the light intensity.
c) For a fixed frequency and a fixed stopping potential, the photo-current is directly proportional to the intensity I.
d) The photo-current increases for increasing stopping potential.
e) A photo-electron is emitted from the surface for each incident light wave after at least 1 millisecond has elapsed.

2.6 Light of a wavelength 2000 Å falls on an aluminum surface. The work function of aluminum is 4.2 eV. The kinetic energy of the slowest emitted photoelectrons is, in eV,

a) 0.0 b) 2 c) 4.2 d) 10.4 e) 20.4

2.7 Under favorable conditions, a human eye can detect about 1.5×10^{-18} J of electromagnetic energy. The number of photons of wavelength 6625 Å which can supply this amount of energy is

a) 5 b) 50 c) 500 d) 5000 e) 50 000

2.8 A wavelength of 6000 Å is incident upon a surface whose threshold wavelength is 8000 Å. What is the work function, in eV, of this surface?

a) 0.24 b) 1.55 c) 2.06 d) 3.22 e) 4.32

2.9 Light of a wavelength 4000 Å is incident upon a photoelectric material having a work function of 2.1 electron volts. The maximum speed, in m/s, of the ejected electrons is

a) 3.5×10^3 b) 4.8×10^3 c) 5.9×10^3 d) 4.8×10^4 e) 5.9×10^5

2.10 A uniform beam of red light (6000 Å wavelength) has an intensity of 3.31×10^{-8} watts/m². The number of photons crossing a lcm² surface normal to the beam in one second is

a) 10^{15} b) 10^{13} c) 10^{11} d) 10^9 e) 10^7

2.11 A faint star is just visible if 2000 photons/sec enter the eye. What energy per second (W) does this represent in terms of $\lambda = 5890$ Å?

a) 8.2×10^{-8} b) 7.9×10^{-9} c) 4.6×10^{-10} d) 5.3×10^{-12} e) 6.8×10^{-16}

2.12 The stopping potential for the fastest photoelectrons emitted by a tantalum surface is 1.6 volts. What was the frequency of the absorbed light? (Work function = 4.12 eV)

a) 2.7×10^{12} b) 9.3×10^{13} c) 7.3×10^{14} d) 1.37×10^{15} e) 6.5×10^{15}

2.13 The threshold wavelength for photoelectric emission from a certain material is 6525 Å. What is the work function, in eV, for this material?

a) 1.9 b) 2.7 c) 3.9 d) 6.0 e) 7.2

2.14 The photoelectric work function of potassium is 2.0 eV. If light having a wavelength of 3600 Å falls on potassium, what will be the required stopping potential, in volts?

a) 0.95 b) 1.25 c) 1.44 d) 1.58 e) 2.71

2.15 A typical demonstration laser beam is red light of 6500 Å wavelength and a power of 0.1 W. How many photons are emitted into this beam per sec?

a) 5.1×10^3 b) 3.2×10^5 c) 3.3×10^{17} d) 4.1×10^{20} e) 6.6×10^{30}

2.16 Which one of the following statements concerning the Bohr theory of atomic structure is true? The theory correctly

a) accounts for the intensities of the spectral lines.
b) yields the effect of the relativistic mass.
c) yields the value of the Rydberg constant for hydrogen.
d) assumes the existence of transient states.
e) assumes that the energy states of unbound particles are quantized.

2.17 All the lines in the Balmer series for hydrogen are transitions to the state $n = 2$. What is the longest wavelength, in Å, in this series?

a) 1216 b) 3650 c) 5304 d) 6563 e) 8220

2.18 Which one of the following cannot be considered a "success" of the Bohr theory?

a) It accounted for the Balmer and other hydrogen series.
b) It explained the Zeeman effect for many electron atoms.
c) It derived the Rydberg constant from known physical quantities.
d) It gave a satisfactory value for the radius of the hydrogen atom.
e) It accounted quantitatively for ionization and excitation potentials.

2.19 Calculate the energy of the $n = 4$ state of the hydrogen atom (in eV).

a) −3.4 b) −0.85 c) +0.85 d) +3.4 e) +54.3

2.20 Calculate the wavelength, in Å, of the photon emitted when a hydrogen atom undergoes a transition from the $n = 4$ state to the first excited state.

a) 1028 b) 2055 c) 3640 d) 4870 e) 9120

2.21 The electron of the hydrogen atom makes a transition from the $n = 3$ state to the $n = 5$ state. What is the energy change (in eV) in the atom during the change in energy state?
(Answer: + energy released; − energy absorbed)

a) + 0.96 b) − 0.96 c) + 1.6 d) − 1.6 e) none of these

2.22 The one electron hydrogen atom is in its second excited state. If the electron makes a transition from this state to its ground state, what will be the frequency of the emitted photon? (Answer: Hz $\times 10^{15}$)

a) 1.72 b) 2.92 c) 3.37 d) 5.21 e) 7.43

2.23 A hydrogen atom in a state having a binding energy of − 0.85 eV makes a transition to a state with an excitation energy of 10.2 eV. Find the wavelength, in angstroms, of the emitted photon.

a) 1216 b) 3650 c) 4102 d) 4863 e) 7235

2.24 Singly ionized helium ($Z = 2$) is a "hydrogen-like" atom. Calculate the energy, in eV, of the ground state of singly ionized helium.

a) + 0.85 b) − 3.4 c) − 13.58 d) − 27.2 e) − 54.3

2.25 Which of the following are possible states for the hydrogen atom?

1. $n = 3,\ \ell = 2,\ m_\ell = -1$
2. $n = 0,\ \ell = 0,\ m_\ell = 0$
3. $n = 1,\ \ell = 0,\ m_\ell = 1$

a) 1 b) 2 c) 3 d) all e) none

2.26 Choose the one correct set of possible quantum numbers for an atom.

a) $n = 0,\ \ell = 1,\ m_\ell = -1,\ m_s = -1/2$
b) $n = 1,\ \ell = 1,\ m_\ell = -1,\ m_s = -1/2$
c) $n = 2,\ \ell = 1,\ m_\ell = -1,\ m_s = -1/2$
d) $n = 3,\ \ell = -1,\ m_\ell = 1,\ m_s = 1/2$
e) $n = 4,\ \ell = -1,\ m_\ell = 0,\ m_s = 0$

2.27 Which one of the following statements is correct concerning Schrodinger's Quantum Mechanics?

a) The state of a physical system is defined by an observable called a wave function.
b) The postulates of Schrodinger's Quantum Mechanics are derivable from a more basic theory of matter.
c) The wave function in Schrodinger's Quantum Mechanics has the dimensions of an energy density.
d) The wave function and its first derivatives are everywhere continuous, single valued, and finite.
e) Schrodinger's Quantum Mechanics when applied to a "free" particle ($v = 0$) shows that this particle can exist in only certain discrete energy states.

2.28 The normalized wave functions for a particle in a one-dimensional "box" of width L is given by

$$\psi_n = \sqrt{\frac{2}{L}} \sin \frac{n\pi x}{L}$$

Which of the following curves represent ψ for $n = 5$?

a) 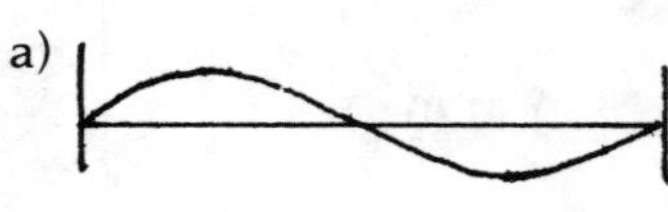
b) 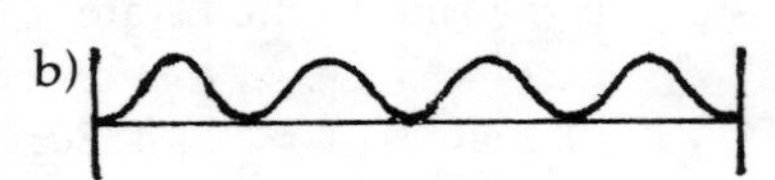
c) 

d) 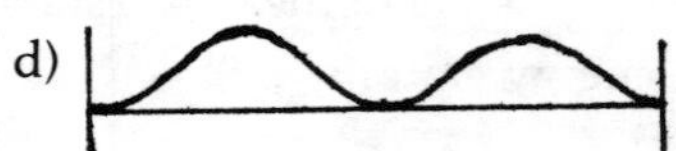
e) 

2.29 Using the wave function in the previous problem, calculate the probability of finding the particle between $L/5$ and $2L/5$ for $n = 5$.

a) 1/10 b) 1/5 c) 1/2 d) 2/3 e) 3/4

2.30 The wave function of a particle trapped between infinitely high potential walls is $\psi_n = \sqrt{\frac{2}{L}} \sin \frac{n\pi x}{L}$. If this particle is in the state $n = 4$, the probability that the particle is between $x = 0$ and $x = L/4$ is:

a) 0.13 b) 0.25 c) 0.33 d) 0.50 e) 0.75

2.31 The normalized wave function for an electron in a one-dimensional box is given by

$$\psi_n = \sqrt{\frac{2}{L}} \sin \frac{n\pi x}{L}$$

Which curve below would allow the direct calculation of the probability of finding the electron between $2L/4$ and $3L/4$ for $n = 3$?

a) 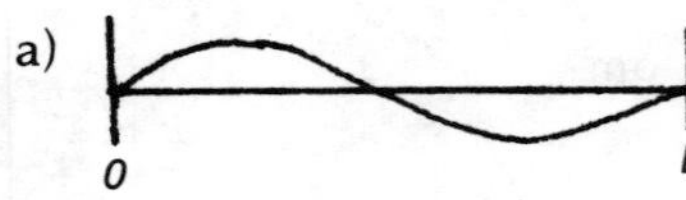

b) 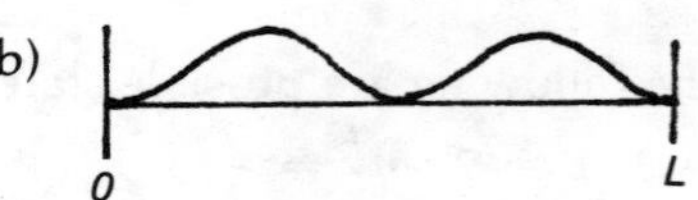

c) 

d) 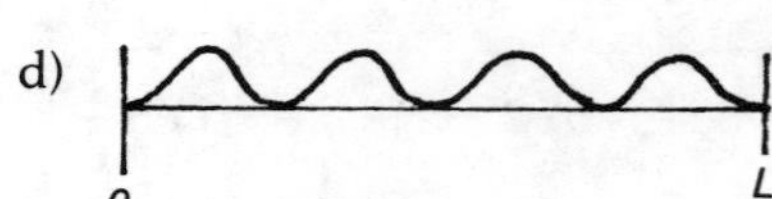

e) 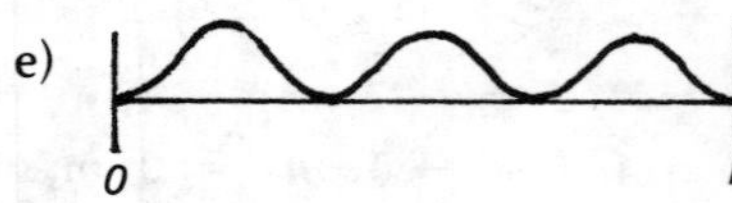

2.32 The normalized wave function for a particle in a one-dimensional box is given by

$$\psi_n = \sqrt{\frac{2}{L}} \sin \frac{n\pi x}{L} \qquad n = 5$$

Calculate the probability of finding the particle between $2L/5$ and $4L/5$.

a) 0.1 b) 0.2 c) 0.4 d) 0.6 e) 0.9

2.33 For an infinite square well potential, the allowed energies depend on the quantum number in the following manner:

a) $1/n^2$ b) $1/n$ c) n^o d) n e) n^2

2.34 Consider an electron trapped between infinitely high potential walls 1.0 Å apart. In its lowest energy state this electron has an energy, in eV, of:

a) 14 b) 17 c) 21 d) 37 e) 42

2.35 An electron is confined to a "potential well" and moves back and forth across the width of the box of 4 Å. The energy of the first state is 2.3 eV. What is the energy difference between the second and third energy states? (Answer in eV)

a) 4.8 b) 5.0 c) 9.0 d) 11.5 e) 16.4

2.36 Nuclei which have equal number of neutrons are called:

a) Isomers b) Isotopes c) Isotones d) Isobars e) Isonaws

2.37 If the natural uranium series (U-238) is referred to as the $4n + 2$ series, what is the initial value of n?

a) 38 b) 48 c) 57 d) 59 e) 61

2.38 A beta emitter has a half-life of 6.98 days. Assume that a 20 millicurie sample of this element is present at $t = 0$. At what time, in days, will the activity be 5 millicuries?

a) 8.0 b) 12.34 c) 13.96 d) 20.79 e) 69.3

2.39 If a radioactive element disintegrates for a period of time equal to its average life, what fraction of the original amount remains?

a) 0.368 b) 0.500 c) 0.632 d) 0.693 e) 0.866

2.40 The activity of a certain radioactive sample decreases by a factor of 10 in a time interval of 30 minutes. The half-life of this sample is nearest to (in minutes):

a) 31 b) 27 c) 23 d) 17 e) 9

2.41 Due to C-14 activity the charcoal from an old fire pit averages 11.3 counts per minute per gram of carbon (cpm/g) whereas in living trees it's 15.3 cpm/g. What is the age of the charcoal in years? (Half-life of C-14 = 5730 years)

a) 2506 b) 5376 c) 7758 d) 11195 e) 11462

2.42 A low power reactor containing U^{235} is operating at a power level of 2 watts. Approximately 200 MeV of useful energy is released in each fission of U^{235}. The rate of fission in this reactor is, in s^{-1},

a) 6.25×10^{10} b) 1.6×10^{11} c) 6.25×10^{13} d) 6.25×10^{16} e) 2.5×10^{27}

2.43 Only one of the following nuclear reactions is wrong. Which one is it?

a) ${}^{32}_{16}S_{(n,2n)}\ {}^{31}_{16}S$
b) ${}^{32}_{16}S_{(n,2n)}\ {}^{32}_{15}P$
c) ${}^{32}_{16}S_{(n,d)}\ {}^{31}_{15}P$
d) ${}^{32}_{16}S_{(n,\alpha)}\ {}^{29}_{16}S$
e) ${}^{32}_{16}S_{(n,\gamma)}\ {}^{29}_{16}S$

2.44 ${}^{235}_{92}U$ will fission with a slow neutron producing ${}^{143}_{57}La$ and ${}^{87}Br$ and some neutrons. Determine the atomic number (Z) for Br and the number of neutrons (n) released. (Answer: (Z, n))

a) (35,6) b) (33,7) c) (47,4) d) (36,5) e) (29,8)

2.45 Assume a neutron causes fission in ${}^{235}_{92}U$ producing ${}^{97}_{40}Zr$, ${}^{134}Te$ and some neutrons. Determine the atomic number (Z) for Te, and the number of neutrons (n) released. (Answer: (Z, n))

a) (53,4) b) (33,7) c) (57,2) d) (47,3) e) (52,5)

2.46 A reaction equation for the discovery of the neutron was ${}^{11}_{5}B(\alpha, n)$. If the masses involved are known to be respectively; ${}^{4}_{2}He = 4.0012\ u$, ${}^{11}_{5}B = 11.0083\ u$, and neutron $1.0087\ u$ and if all the nuclide kinetic energies are neglected (including the neutron) calculate the value of the atomic number of the product nuclide multiplied by its masses.

a) 14.1218 b) 28.0016 c) 91.0923 d) 98.0056 e) 196.0112

2.47 Which one of the following nuclear reactions is written correctly?

a) ${}^{10}_{5}B + {}^{1}_{0}n \rightarrow {}^{11}_{5}B + {}^{1}_{1}H$
b) ${}^{14}_{7}N + {}^{4}_{2}He \rightarrow {}^{18}_{9}F + {}^{0}_{-1}e$
c) ${}^{7}_{3}Li + {}^{1}_{1}H \rightarrow {}^{4}_{2}He + {}^{2}_{1}H$
d) ${}^{14}_{7}N + {}^{1}_{0}n \rightarrow {}^{14}_{6}C + {}^{1}_{1}H$
e) ${}^{59}_{27}Co + {}^{1}_{0}n \rightarrow {}^{58}_{27}Co + {}^{2}_{1}H$

3. Materials Science

Materials science is a very broad field of study covering the structures and properties of metallic materials, ceramics, polymers and composites. Hence, it is necessary to discuss the electronic, magnetic, optical, mechanical and chemical properties of various materials in the materials science discipline.

A review of past Fundamentals of Engineering exams indicates that there are two major areas emphasized in materials science. These are related to the fundamentals of 1.) strength, deformation and plasticity of crystalline solids, and 2.) phase equilibria in metallic systems. Typically the latter is troublesome to engineers who do not have a metallurgy-materials science background. In this review, therefore, these two subject areas are thoroughly treated. (For a more complete understanding of the topic, we recommend that you simply refer to appropriate texts.)

3.1 Mechanical Properties of Metals and Alloys

There are several standard experimental techniques used to determine how a material responds to an applied state of stress. The specific responses and the corresponding tests may be summarized as follows:

- Capacity to withstand static load → Tension or compression test.
- Resistance to permanent deformation → Hardness test.
- Toughness of a matrial under shock loading → Impact test.
- The useful life of a material under cyclic loading → Fatigue test.
- Elevated temperature behavior → Creep and stress rupture tests.

In tension testing, there are two distinct stages of deformation:

1) Elastic deformation (reversible)
2) Plastic deformation (irreversible)

The elastic range is characterized by *Hook's law,* which is the linear equation

$$\sigma = E\varepsilon \tag{3.1.1}$$

where σ and ε are the stress and strain, respectively, and E (the slope of the line) is the Young's modulus of the material, also called the modulus of elasticity.

Beyond the elastic range, the material undergoes plastic deformation. In this range, the stress-strain relation is nonlinear. It is important to note that, unlike elastic deformation, the volume of the material remains constant during plastic deformation (we will see later why this is so). The stress at which this nonlinearity begins is called the *yield stress* σ_y.

Very frequently, it is difficult to obtain a reliable value of σ_y from experimental data because the linear to nonlinear transition of the stress-strain diagram is rather gradual. To avoid this difficulty an *off-set yield stress* $\sigma_{0.2}$ at a specified percentage of plastic strain is defined (usually 0.2% plastic strain).

The engineering stress-strain curve (Fig. 3.1) shows a maximum value of stress called the *ultimate tensile strength* σ_{uts}. Beyond this value, the engineering stress decreases, and the sample fails in tension at a *fracture stress* σ_f, which is (usually) less than σ_{uts}. The diagram in Fig. 3.1 shows various features of the engineering stress-strain diagram.

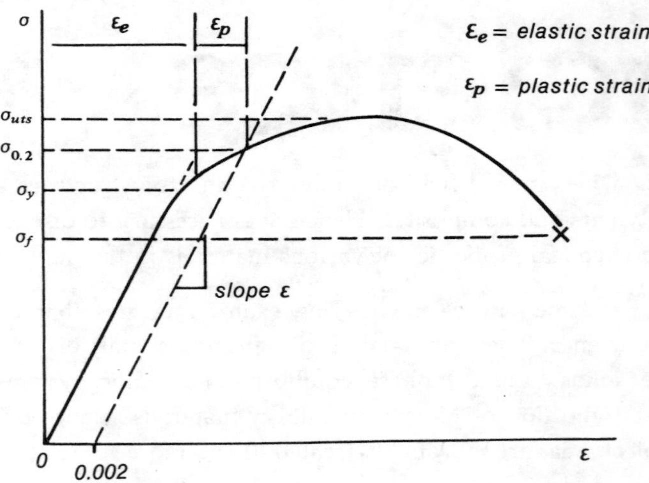

Figure 3.1. Engineering stress-strain diagram.

The reason $\sigma_f < \sigma_{uts}$ is that the engineering stress is obtained by dividing the load by the "original" cross-sectional area of the sample. At stresses $\geqslant \sigma_{uts}$ the sample starts to neck locally, tri-axial stresses develop and eventually the sample breaks. Since the cross-section is less at the necking, the load bearing capacity goes down. This load divided by the original area is less than the "true" stress beyond this point.

The extent of necking depends on the *ductility* of the material. If the material is very *brittle*, that is, has little or no ductility, then there will be very little necking strain and in that case $\sigma_f \equiv \sigma_{uts}$. Often it is possible to make a qualitative assessment of ductility from the nature of the tensile fracture. Fig. 3.2 shows three types of necking which might be observed.

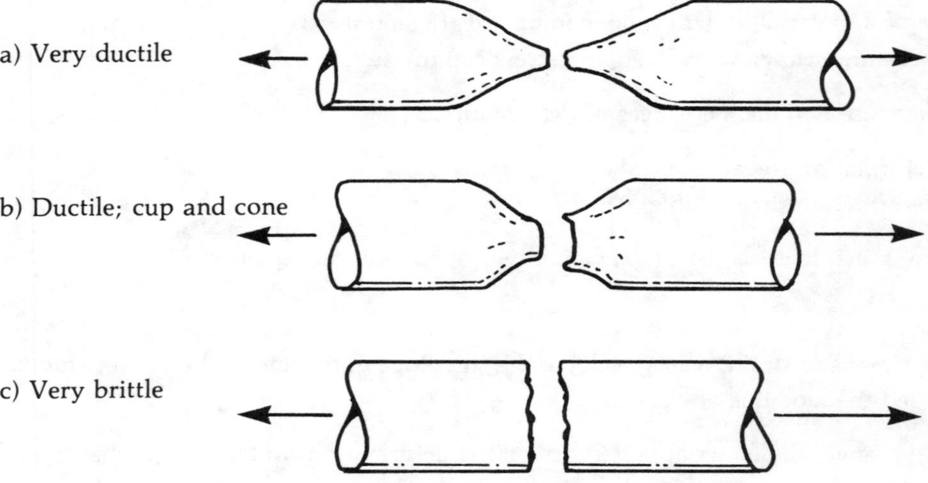

Figure 3.2 Examples of tensile failure.

Plastic deformation, at ordinary temperature, introduces some additional strength to many metals and alloys. This is called *work-hardening* or *strain-hardening*. The nature of this hardening can be understood by studying Fig. 3.3. If the material is loaded in the plastic range to strain ε_1 and then unloaded,

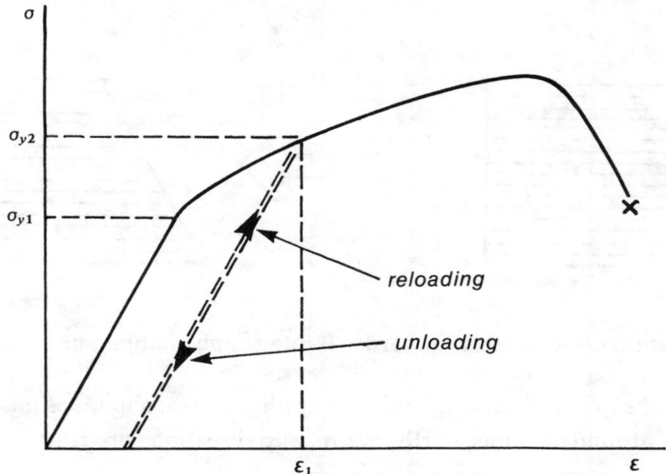

Figure 3.3 **Tensile loading-unloading curve.**

the "apparent" yield stress σ_{y2} on reloading is greater than σ_{y1}. Such a hardness is imparted on cold forged tools, for example. It must be noted that although the yield strength has apparently increased, the material loses some amount of its plastic flow property. This "extra strength" is removed if the material is annealed at an elevated temperature.

In steel and other ferrous alloys, the stress strain curve might look (depending on carbon and other alloy content) like the one shown in Fig. 3.4

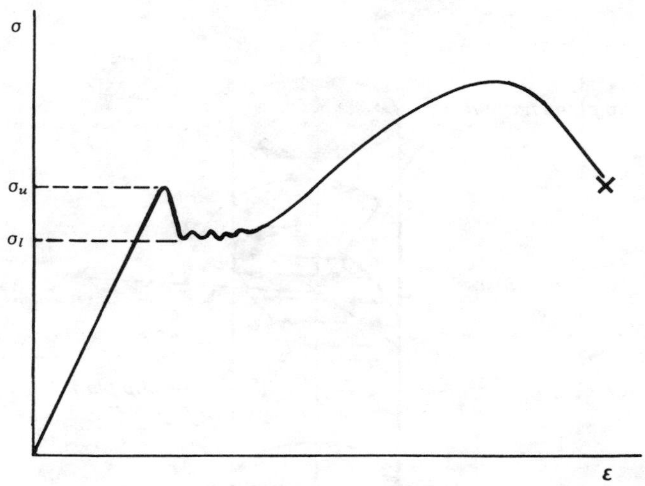

Figure 3.4 **Stress-strain diagram of mild steel.**

The quantities σ_u and σ_l in Fig. 3.4 are called *upper and lower yield points*. This peculiarity of yielding occurs due to the interaction of carbon atoms with atomic scale defects called *dislocations*.

3.1.1 Nature of Plastic Flow

In crystalline materials (all metals and alloys are crystalline) plastic flow or plastic deformation involves the *sliding* of atomic planes analogous to shearing a deck of cards, as shown in Fig. 3.5. The sliding

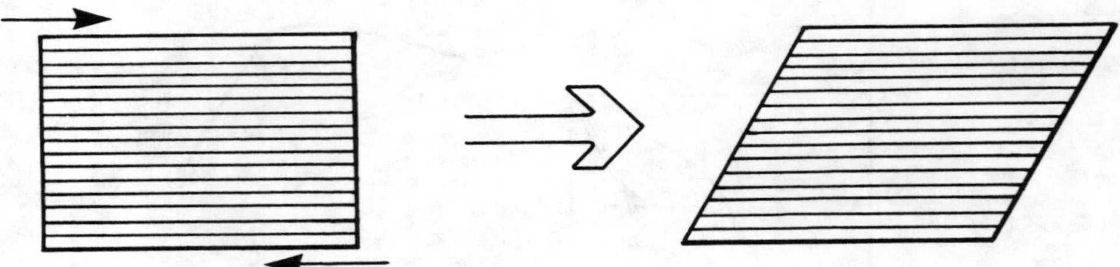

Figure 3.5 **Shearing of a deck of cards. (Note shape change but no volume change.)**

of atomic planes is called *slip deformation*. Under an applied stress, slip takes place on those crystal planes which have the densest atomic packing. Furthermore, slip directions are restricted to the crystallographic directions along which the atoms are most closely packed.

The combination of a close-packed plane and a close-packed direction is called a *slip system*. Depending on the crystal structure, some metals and alloys will have more slip-systems than others. The higher the number of planes and directions along which slip can take place, the easier it is to produce plastic deformation without brittle fracture.

Slip occurs when the resolved component of shear stress τ_R, given by the expression

$$\tau_R = \frac{P}{A} \cos \phi \cos \lambda, \tag{3.1.2}$$

on the slip plane and along the slip direction exceeds a critical value, called *critical resolved shear stress*, $(\tau_R)_{crit}$; see Fig. 3.6. This critical value is a property of the material.

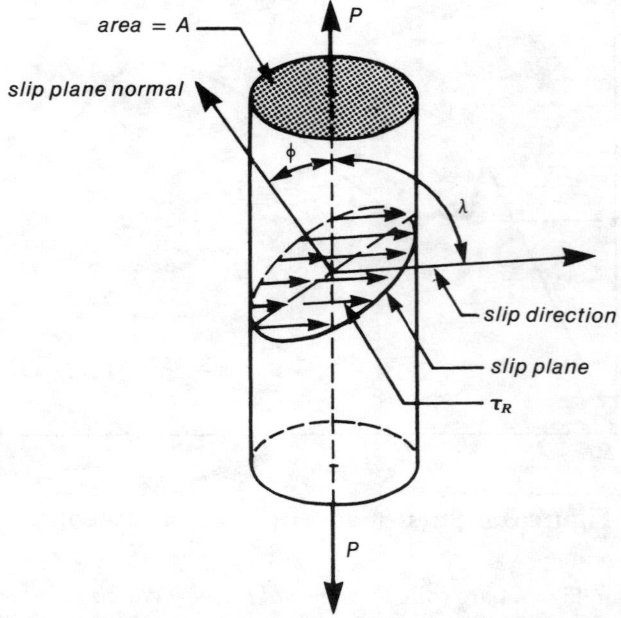

Figure 3.6. **Resolved shear stress on the slip plane.**

It is found that the experimental value of $(\tau_R)_{crit}$ is at least five times smaller than the theoretically calculated value based on the force necessary to slide close packed planes of atoms. The apparent anomaly here was resolved when dislocations were discovered. Slip becomes easier if dislocations are present. Let us consider an analogy. A long narrow rug on a hallway is to be moved x meters to the right, at shown in Fig. 3.7(a.) This could be done by pulling, as shown. It must be noted that considerable friction must be overcome to do this. Alternatively, a bulge could be made as shown in Fig. 3.7(b) and then this bulge could be "walked" to the right. When the bulge exits on the right, the rug has moved x meters to the right. Considerably less effort is required in the second procedure.

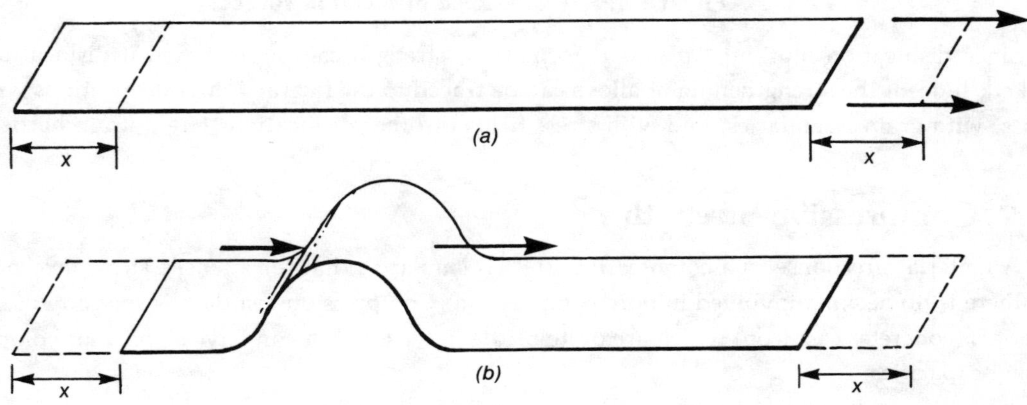

Figure 3.7. Analogy of a dislocation motion.

Analogous to the bulge, there is an extra half-plane of atoms in an edge dislocation as shown in Fig. 3.8. The row of atoms at the end of the half-plane is situated in between two equilibrium sites. Thus, it takes less force to move this plane. When the extra half-plane emerges on the surface, we have one elementary slip step.

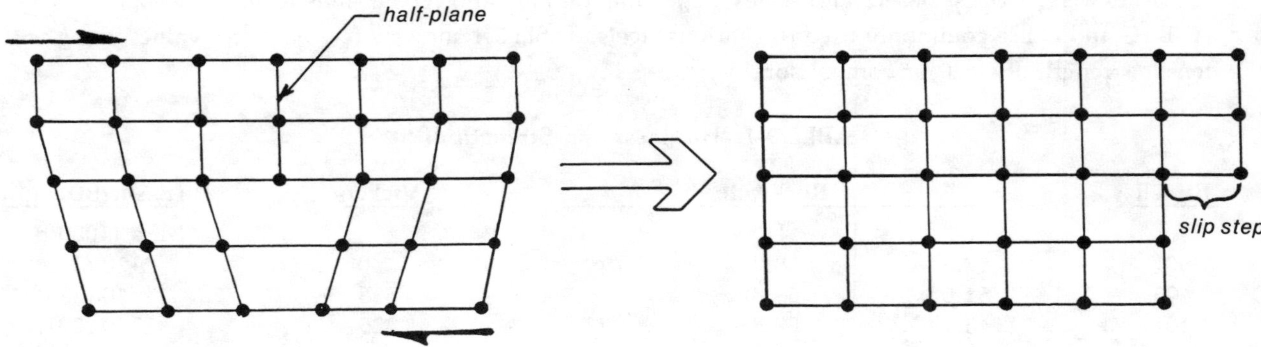

Figure 3.8. An edge dislocation.

The row of atoms at the end of the half-plane can be viewed as a line called a *dislocation line*. When a dislocation line moves, plastic deformation occurs.

It has been established that existing dislocation lines in a material can multiply under an applied stress, as shown in Fig. 3.9. This mechanism is called a *Frank-Reed source* and is experimentally verifiable.

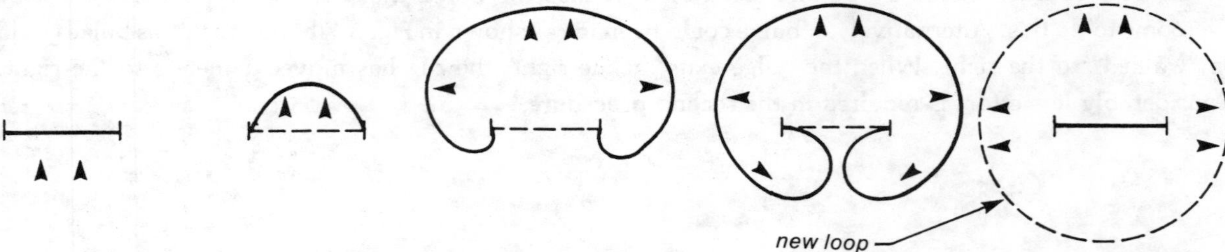

Figure 3.9. **Frank-Reed dislocation source.**

Since dislocation motion ≡ plastic deformation, strength can be increased if dislocation motion is blocked. Indeed, the strengthening of alloys can be traced to the interaction of dislocations with dispersed phases, with grain boundaries, and with stress fields of other dislocations (e.g., strain hardening).

3.1.2 Compressive Strength

A material's response to a compressive stress is similar to that for a tensile stress, except for the fact that there is no necking involved in pure compression. Compression test data is specially useful for those materials (concrete, cast iron) which are quite brittle in tension but can have significant compressive load bearing capacities.

3.1.3 Hardness of Materials

A hardness test is an empirical method which determines the resistance of a material to the penetration of an indenter. Hardness measurement can be useful for obtaining a qualitative estimate of service wear, strength and toughness. Further, for steel, an empirical correlation exists between hardness and tensile strength. The most commonly used hardness tests are:

1) Brinell.
2) Rockwell.
3) Vickers.
4) Microhardness (Vickers or Koop indentors).

Rockwell hardness has several scales (depending on load and type of indenter): A through V. Scales A, B, C and D are commonly used for various steels. Table 3.1 shows a few hardness values along with tensile strength of steel for comparison.

TABLE 3-1. Hardness and Strength of Steel.*

Brinell	Rockwell			Vickers	Tensile Strength
	C	D	A		MPa (1000 psi)
601	57.3	68.7	79.8	640	2120 (308)
495	51.0	63.8	76.3	528	1740 (253)
401	43.1	57.8	72.0	425	1380 (201)
302	32.1	49.3	66.3	319	1030 (150)
229	20.5	40.5	60.8	241	760 (111)

*A more complete table can be found in "The Testing and Inspection of Engineering Materials": H. F. Danis, G. E. Trowell and C. T. Wiskocil, McGraw Hill Book Co., NY.

3.1.4 Fatigue Test

Life in cyclic loading is important in many applications. Fatigue life is determined from experimental data relating number of cycles (N) to failure with cyclic stress amplitude (S). A schematic S vs N curve is shown in Fig. 3.10. Note in Fig. 3.10 that for steel, there is a critical value of stress S_{crit} below which fatigue life is virtually infinity. This limit is called the *endurance limit*. Notice that no such endurance limit exists for *Al* (or many other nonferrous metals and alloys). Fatigue fractures are progressive, beginning as minute cracks that grow under the action of the fluctuating stress. Fatigue strength is defined as the maximum cyclic stress amplitude for a specified number of cycles until failure.

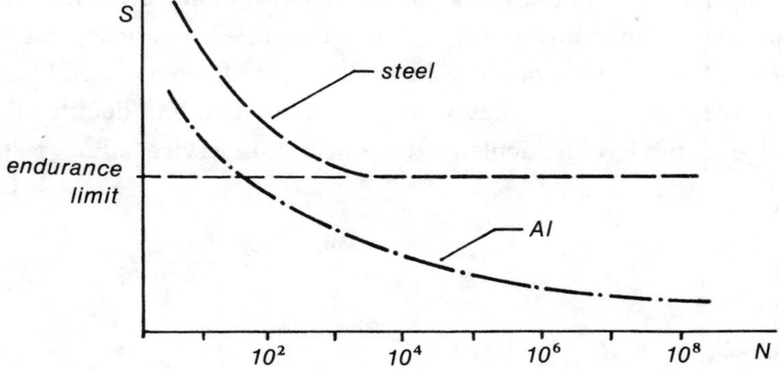

Figure 3.10. **S-N diagram (schematic) of aluminum and steel.**

Fatigue is a surface active failure. Fatigue cracks start at the surface; surface defects such as notches can initiate a crack. A rough surface may reduce fatigue strength by as much as 25%. Cold rolling or shot peening (which introduce surface compressive stress) can increase fatigue strength by as much as 25%. *Corrosion-fatigue* is an important cause of service failure if a corrosive-environment and cyclic stresses co-exist. For example, it has been shown that the endurance limit of a steel (tested in air) is altogether eliminated when the sample is tested in pure water. Fatigue life or fatigue strength can be improved by:

a) A highly polished surface.
b) Surface hardening (carburizing, nitriding, etc.)
c) Surface compressive stresses (shot peening, cold rolling, etc.)

3.1.5 Toughness and Impact Testing

The *impact value* is a simple evaluation of the notch toughness of the material. And *toughness* is a measure of energy absorbed by the material before fracture. Two types of machines are commonly used to test these qualities, the Charpy and the Izod, both of which use swinging-pendulum loading with notched-bar samples. In tension, the area under the stress-strain curve is the energy/unit volume to fracture (see Fig. 3.11). The area under the stress-strain diagram of material A is less than that of material B.

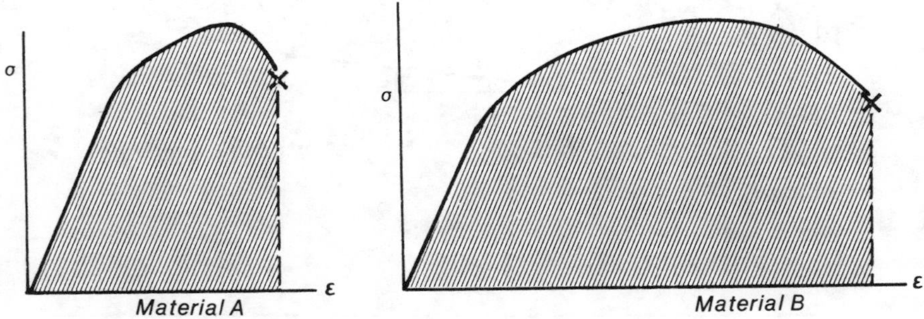

Figure 3.11. **Comparison of toughness of two materials.**

The area has the dimension of stress (force/unit area), which if multiplied and divided by length, results in energy/volume. Hence, material B can be analyzed as tougher than material A.

The presence of a notch introduces tri-axial stresses. Many materials become more brittle under such a state of stress than under uni-axial tension or compression.

Temperature has a pronounced effect on the energy absorption and fracture behavior of steel in a notch condition. The sharpness of the transition from tough to brittle fracture depends on the material and also on the notch geometry. Frequently, Charpy values will change from 35 or 40 to as low as 7 N·m in a temperature interval of 4 to 10°C. Fig. 3.12 shows the qualitative nature of ductile-brittle transition (also the behavior of *Ni*, which does not have a ductile-brittle transition). Service failures testify to the increased

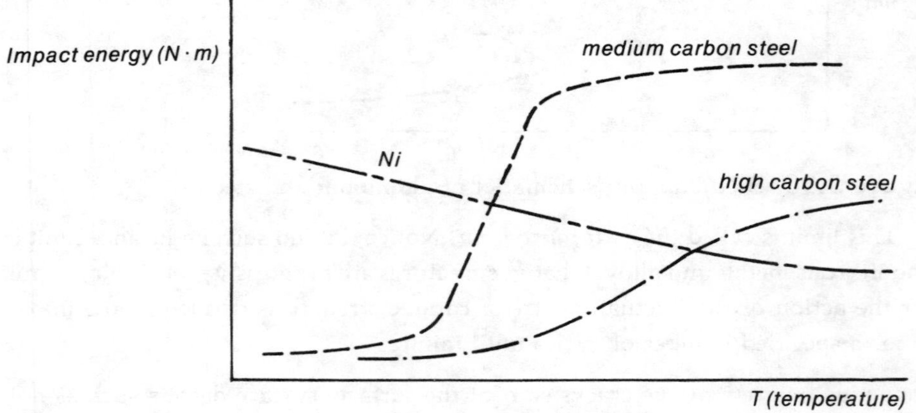

Figure 3.12. Impact energy vs temperature.

hazards of subnormal temperature; and so determination of transition temperature has come to be an important criterion for materials selection. It must be noted that in a large rigid structure, the transition temperature would be considerably higher than that for a standard Charpy sample.

3.1.6 High-Temperature Property: Creep and Stress Rupture

The progressive deformation of a material at constant stress is called *creep*. Below about 40% of the absolute melting temperature, creep strain is negligible for most structural metals and alloys. An idealized shape of a creep curve is shown in Fig. 3.13. Andrade's empirical formula for the creep curve is given by

$$\varepsilon = \varepsilon_o(1 + \beta t^{1/3})\, e^{kt} \qquad (3.1.3)$$

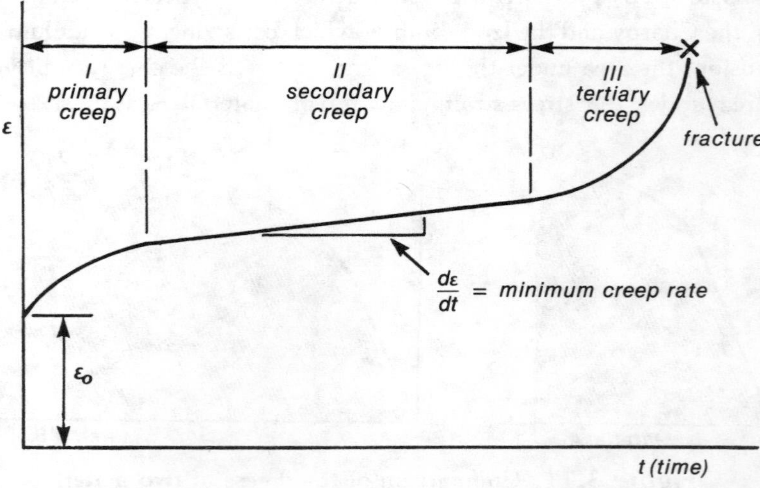

Figure 3.13. A typical creep curve showing three stages of creep.

where ε is the strain in time t, ε_o is the initial elastic strain, and β and k are material constants.

The stress-rupture test is basically similar to a creep test except that the test is always carried out to failure. Elongation, time to failure, applied load, and temperature are all reported for the purpose of design data.

One particular mode of failure of polycrystalline metals and alloys at elevated temperatures is *grain boundary sliding*. The influence of grain size on creep resistance is not clear-cut. There is some evidence, however, to indicate that creep rate is lower in large grain materials. Because grain boundaries are the nucleation sites for high-temperature fracture, the control (or elimination) of grain boundaries will suppress fracture and increase rupture life. It must be noted that the environment plays an important role in high temperature mechanical properties. The nature of oxides, for example, can influence creep and stress-rupture.

3.1.7 Metallurgical Variables in Materials Response to Stresses

Microstructural conditions, heat treatment, processing variables, and service conditions, can all influence mechanical properties of metals and alloys.

A. *Microstructural Conditions*

1) Grain size effect: At ordinary temperature, fine grain is better for strength. At high temperatures, perhaps a larger grain size is desirable.

2) Single-phase vs. multiphase alloys: Many times a second phase might have a profound effect on the mechanical properties; e.g., the retained austenite may be a problem in fatigue. Deformation behavior of phases may be different, and a simple averaging of properties might not be appropriate.

3) Porosity and inclusions: Poor mechanical properties result from high porosity and inclusions.

4) Directionality of microstructure: Rolling direction vs. transverse direction will have different mechanical properties. These also introduce anisotropy of properties.

B. *Effects of Heat Treatment*

1) Annealing: Softening, ductile behavior (depending on alloy).

2) Quenching of steel: Martensite formation, strong but brittle. In high carbon steel quench-cracks may form.

3) Tempering of martensite: Hardness decreases but toughness increases. Strength is sacrificed to avoid brittle failure.

4) Age hardening: Depending on alloy composition, fine scale ($\sim 10^{-7}$ M) precipitation may be formed which interacts with dislocations. Increased strength is thus obtained.

5) Case hardening: A hard surface and soft core combination is obtained by carburizing and nitriding. Fatigue strength can be increased by this method. A better wear resistance surface can be produced.

C. *Effects of Some Processing Variables*

1) Welding: The heat-affected zone with large grain size will have poorer mechanical properties. Local chemical composition changes can occur, including a loss of carbon in steel. Large parts can have quench cracking due to rapid quenching effects.

2) Flame cutting: Drastic changes of microstructure occur near the flame-cut surface and these changes affect the mechanical properties.

3) Machining, grinding: Cold work results in strain hardening. Excessive cold work may produce surface cracks.

D. *Effects of Service Conditions*

1) Extreme low temperature: Ductile-brittle transition occurs in steel.

2) Extreme high temperature: Causes corrosion and oxidation of surface. Surface cracks may form. Results in problems with corrosion fatigue, creep and rupture.

3) Impact loading: Notch sensitivity, surface scratches or corrosion pits can initiate brittle fracture.

4) Corrosive environment: Stress-corrosion, pitting corrosion, and corrosion fatigue result.

These are but a few examples of the service and material variables which can influence the mechanical response of a material. Often, deterioration of material properties with time leads to a service failure. But, the most important function of an appropriate materials input in engineering design is to prevent unexpected catastrophic failure.

3.2 Equilibrium Phase Diagrams

For most practical purposes, alloy compositions are listed in weight percentage (wt.%). For example, 70-30 brass means 70 wt.% *Cu* and 30 wt.% *Zn*. In this discussion, we will consider a binary alloy (two chemical elements) of elements *A* and *B* (e.g., $A = Cu$, $B = Zn$). Sometimes, however, it is convenient to express an alloy composition in atomic percentage (at.%). Weight percent composition can be converted to atomic percent by using the following formulas:

$$\text{at.\% } A = \frac{W_A}{W_A + (M_A/M_B)\, W_B} \times 100, \tag{3.2.1}$$

$$\text{at.\% } B = \frac{W_B}{W_B + (M_B/M_A)\, W_A} \times 100, \tag{3.2.2}$$

where W_A and W_B are weight percents of elements *A* and *B*, respectively, in the alloy, and M_A and M_B are respective atomic weights. Similarly, at.% can be converted to wt.% by using the following formulas:

$$\text{wt.\% } A = \frac{P_A\, M_A}{P_A\, M_A + P_B\, M_b} \times 100, \tag{3.2.3}$$

$$\text{wt.\% } B = \frac{P_A\, M_A}{P_A\, M_A + P_B\, M_B} \times 100, \tag{3.2.4}$$

where P_A, P_B are at.% percents of *A* and *B*, respectively.

There are various methods for determining equilibrium phase diagrams: X-ray diffraction, optical microscopy, calorimetric and thermal analyses. We will consider thermal analysis, i.e., the cooling curve method, here since it is very instructive.

The so-called equilibrium phase diagrams show the existence or coexistence of phases at any given temperature and alloy composition. The term *equilibrium* implies that the alloy is cooled at such a slow rate that thermodynamic equilibrium is attained at each temperature. A *phase* is a volume of material bounded by a distinct boundary within which the chemical composition is uniform. A phase has a fixed crystal structure and thermo-physical properties at a given temperature. The equilibrium between phases is determined by Gibb's phase rule:

$$P + F = C + 2 \tag{3.2.5}$$

where P = number of phases, C = number of chemical elements in the alloy and F = the degree of freedom, i.e., the number of independent variables. Eq. 3.25 is the generalized phase rule where both pressure and temperature are independent external variables. If the pressure is kept constant, as is the usual case (usually 1 atmosphere), then we have the *condensed phase rule* given by

$$P + F = C + 1 \tag{3.2.6}$$

We will use Eq. 3.2.6 in our discussion. As an example, consider a binary alloy, that is, $C = 2$. In this case, if the number of phases P in equilibrium is 3, then Eq. 3.2.6 predicts that $F = 0$, i.e., no degree of freedom. Thus, in such a case, the composition of the phases and the temperature at which the three phases coexist are fixed: no degree of freedom.

Consider the cooling of pure molten metal in a furnace which has a cooling curve given by Fig. 3.14(a). If we now place a thermocouple in the molten metal and plot the temperature of the metal as it cools, we will obtain a curve as shown in Fig. 3.14(b).

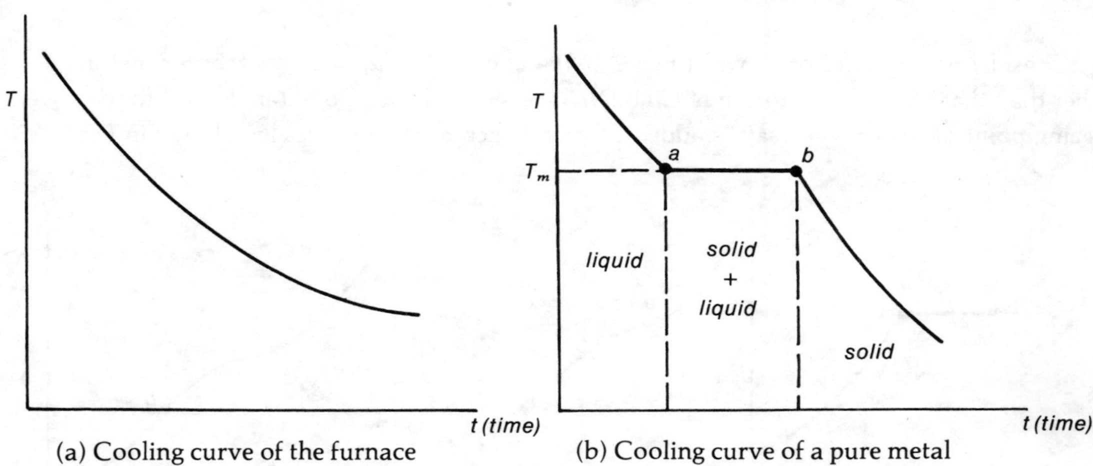

(a) Cooling curve of the furnace

(b) Cooling curve of a pure metal

Figure 3.14. The cooling of a pure metal in a furnace.

The horizontal shelf $\overline{ab}$ is the *thermal arrest*, at a temperature T_m. In this case T_m is the freezing (or melting) point of the metal where liquid and solid coexist. Until all of the molten metal is solidified, the temperature does not change. An analogy is the equilibrium between ice and water: until the ice cubes melt, the temperature of a glass of water remains at 0°C.

If we now add a small amount of an alloying element in a pure metal and repeat the cooling curve experiment, we will obtain the curve as shown in Fig. 3.15. Comparing Fig. 3.15 with Fig. 3.14(b), we note

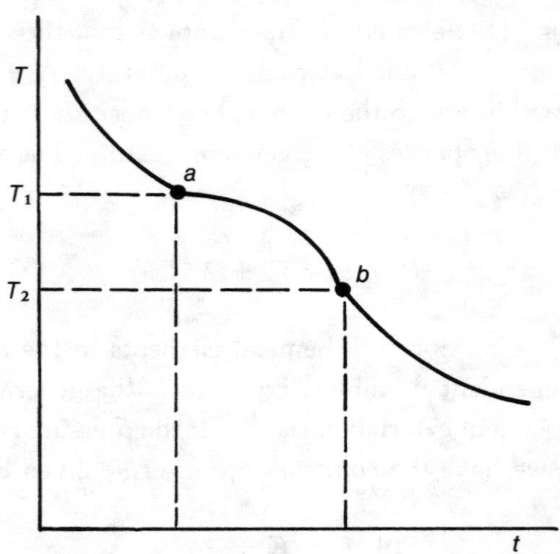

Figure 3.15. Cooling curve of liquid dilute alloy.

that, unlike a single arrest temperature, we have a change in slope at point *a*. Actually freezing begins at point *a* and is completed at point *b*. Thus, within the range of temperatures T_1 and T_2, we have a mixture of solid and liquid. If the alloying element has a higher melting point (or freezing point) than the host pure metal, then at temperature T_1 this element starts to freeze and the remaining liquid has less of the alloying element in solution. Thus, the new liquid has a lower freezing temperature. This process continues until at temperature T_2 an almost pure host element freezes.

Consider now a series of alloys of nickel (*Ni*) and copper (*Cu*); we note that the melting point of *Ni* is higher than that of *Cu* and addition of *Cu* to *Ni* lowers the freezing point (analogous to the lowering of the freezing point of water when salt is added). Several such cooling curves are shown in Fig. 3.16.

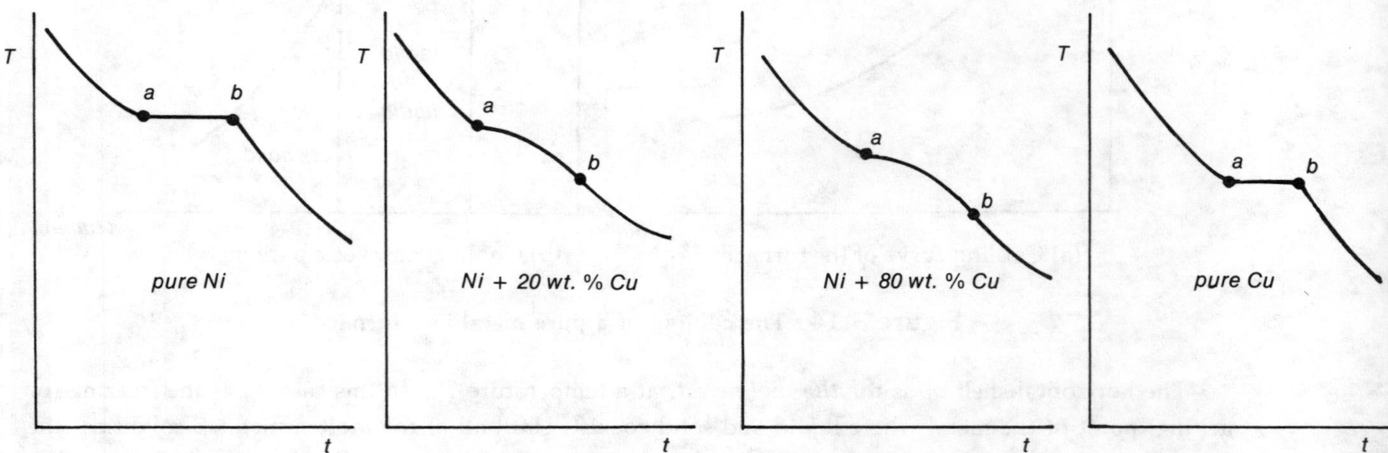

Figure 3.16. Schematic cooling for *Ni*, *Ni* + 20 wt.% *Cu*, *Ni* + 80 wt.% *Cu*, and pure *Cu*.

In all these curves, point *a* corresponds to the beginning of freezing and point *b* the end of freezing. We now plot these temperatures of beginning and end of freezing in a diagram with temperature and alloy composition as coordinates. Such a diagram is shown in Fig. 3.17.

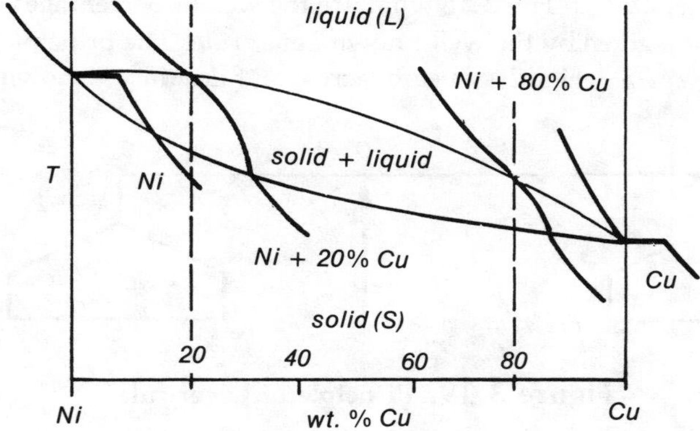

Figure 3.17. Arrest temperatures vs. composition.

Fig. 3.17 is indeed an *equilibrium phase diagram* of the *Ni-Cu* alloy system. The upper curve defines the temperature above which the alloy is liquid and this curve is called the *liquidus*; the lower curve defines the temperature below which the alloy is solid and it is called the *solidus*. The area bounded by the liquidus and the solidus is the region of two phases: solid + liquid.

Let us now analyze the cooling of an alloy of composition C as shown in Fig. 3.18. At temperature T_1 — which is above the liquidus — the alloy of composition C is entirely liquid. At temperature T_2 an extremely minute quantity of solid forms; and at T_3, we are in the two-phase $S + L$ region. At T_3 we have a mixture of solid and liquid, and the composition of the solid is given by the intersection of the temperature-horizontal with the solidus curve. The composition of the liquid phase is given by the intersection of the temperature line with the liquidus line. The compositions can be obtained by drawing vertical lines

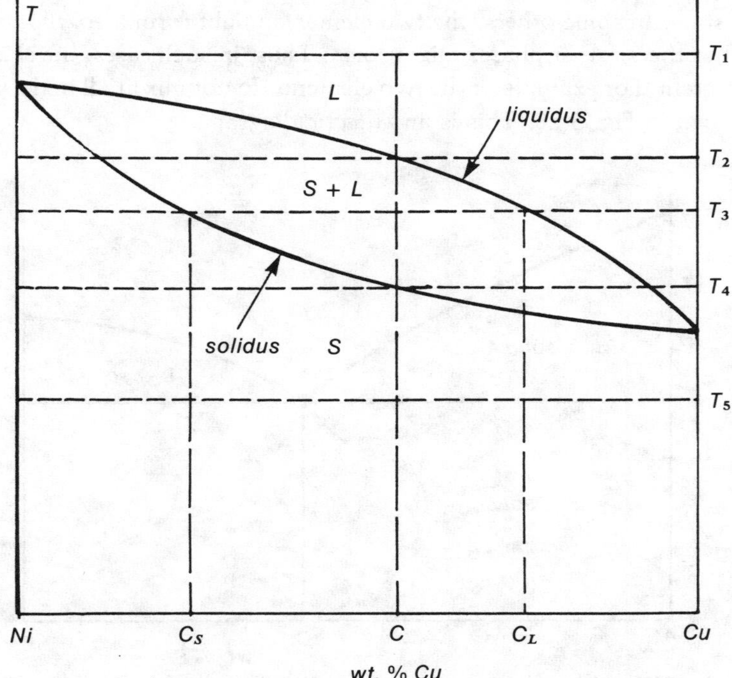

Figure 3.18. Analysis of cooling an alloy of composition C, S = solid; L = liquid.

through these intersection points — that is, C_L and C_S in the figure are the compositions of the liquid and solid phases respectively.

The next situation we should work through is this: If two phases are present, what are the proportions of these phases? For example, at T_3 in Fig. 3.18 what are the weight percentages of the solid and the liquid phases? These percentages are given by the well-known *Lever rule*. The principle of the Lever rule is based on balancing two weights on a weightless beam across a fulcrum as shown in Fig. 3.19. The alloy

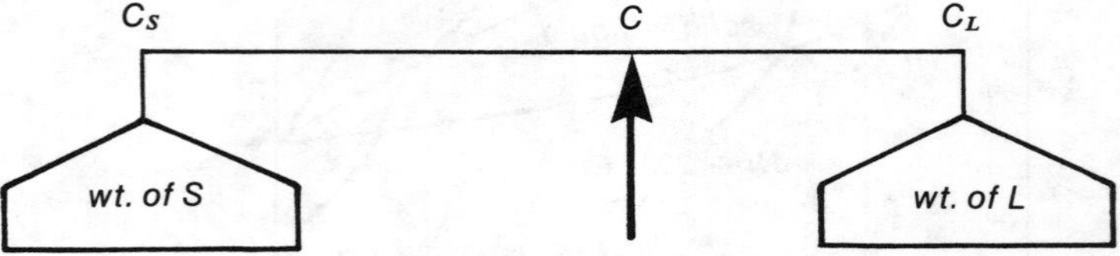

Figure 3.19. Principle of Lever rule.

composition is the fulcrum, and weights of phases (S and L) are suspended to balance the beam. The beam will balance if the moments of the weights are equal, i.e.,

$$W_S \times \overline{CC_S} = W_L \times \overline{CC_L} \tag{3.2.7}$$

It is easy to see from the above that

$$\text{wt.\% } S = \frac{C_L - C}{C_L - C_S} \times 100 \tag{3.2.8}$$

$$\text{wt.\% } L = \frac{C - C_L}{C_L - C_S} \times 100 \tag{3.2.9}$$

The *Ni-Cu* equilibrium diagram is a very simple case where *Ni* and *Cu* remain in solution in the solid phase. Such an alloy is called a *solid-solution* alloy. Some elements, however, do not like to remain in solution in the solid state. In some others, the two elements might form a *compound* at a fixed (or nearly fixed) composition. In others, other phases might form. Thus, in such cases, the shapes of the equilibrium diagram are quite different. For example, if the two elements do not mix at all in the solid state, we obtain a phase diagram as shown in Fig. 3.20. This is an *eutectic* diagram.

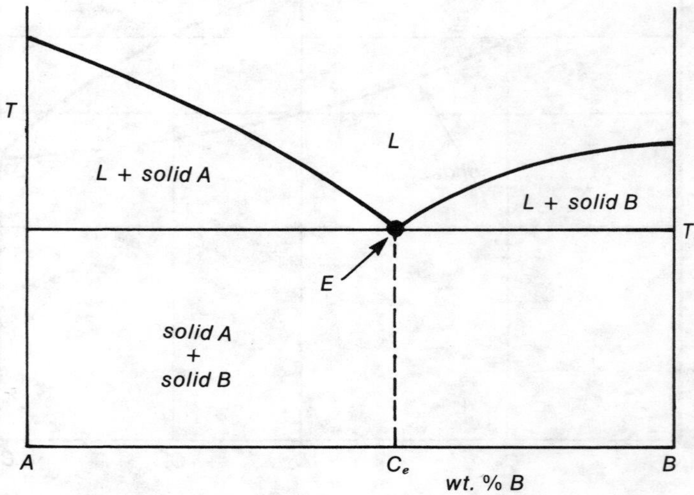

Figure 3.20. An eutectic diagram.

In this diagram (Fig. 3.20), elements A and B do not remain in solution below temperature T_e, the *eutectic temperature*. If we choose an alloy of composition C_e (as shown in the figure) at T_e, there is a three-phase equilibrium for this composition at point E. Point E is known as the *eutectic point*. At a temperature infinitesimally below T_e, solidification starts. Since A and B do not mix in the solid state, the solidification process must separate out pure A and B. In the previous diagram we assumed that the melting point of A is higher than that for pure B, and thus, first a speck of pure A solidifies. This makes the remaining liquid richer in B and a speck of pure B solidifies. In such a way alternate layers of plates of pure A and B solidify to give rise to an *eutectic microstructure*.

In some cases, there might be a limited solid solubility of the two elements. For such a case, we obtain a diagram as shown in Fig. 3.21. Below T_e, we have two solid phases, which we have designated as α and β.

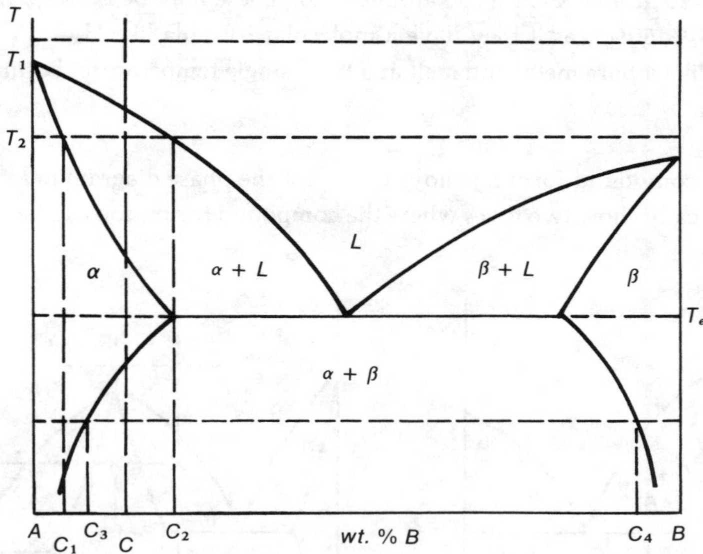

Figure 3.21. **Eutectic diagram with limited solid solubility.**

Although there are some standard notations — α, β, γ , depending on the crystal structure of phases — it is important to note that α in this case is a solid solution which primarily consists of element A with some small amount of B dissolved in it. Similarly β is primarily B with some small amount of A in it.

Consider the freezing of an alloy of composition C, as shown in Fig. 3.21. At temperature T_1, the alloy is in the single-phase liquid region. At temperature T_2 the alloy is in the two-phase liquid + α region. Note: the intersection point between the composition-vertical and temperature-horizontal lies in the two-phase region. This region is defined by the liquidus line which delineates the boundary of the liquid and the solidus line which defines the boundary of the solid α-phase. Thus, the region consists of solid α and the liquid alloy. At T_2 the temperature horizontal intersects the solidus and the liquidus. If we drop perpendiculars from these points to the composition axis, we obtain C_1 as the composition of solid α in equilibrium with a liquid of composition C_2 at the temperature.

Applying the Lever rule at temperature T_2:

$$\text{wt.\%} = \frac{C_2 - C}{C_2 - C_1} \times 100 \tag{3.2.10}$$

$$\text{wt.\% liquid} = \frac{C - C_1}{C_2 - C_1} \times 100 \tag{3.2.11}$$

Similarly, at temperature T_3 the temperature horizontal intersects lines which separate α and β. Thus, at T_3 the two equilibrium phases, for the alloy composition C, are solid α of composition C_3 and solid β of composition C_4, respectively. The proportions of phases are

$$\text{wt.\%}\ \alpha = \frac{C_4 - C}{C_4 - C_3} \times 100 \tag{3.2.12}$$

$$\text{wt.\%}\ \beta = \frac{C - C_3}{C_4 - C_3} \times 100 \tag{3.2.13}$$

Some elements might form a compound-like mixture at a fixed composition, such as 75 atomic % A + 25 atomic % B, or 50 atomic % A + 50 atomic % B. These may be correspondingly designated as A_3B (75%:25%), AB (50%:50%), as if they have a molecular formula like H_2O. When such compounds form they behave more like a pure metal and melt at a fixed single temperature. Formation of a compound splits the phase diagram.

It is easier to depict compound-forming alloys if we plot the phase diagram in at.% composition. Fig. 3.22 (a) and (b) schematically show two cases where the compound formation can be viewed as a separation

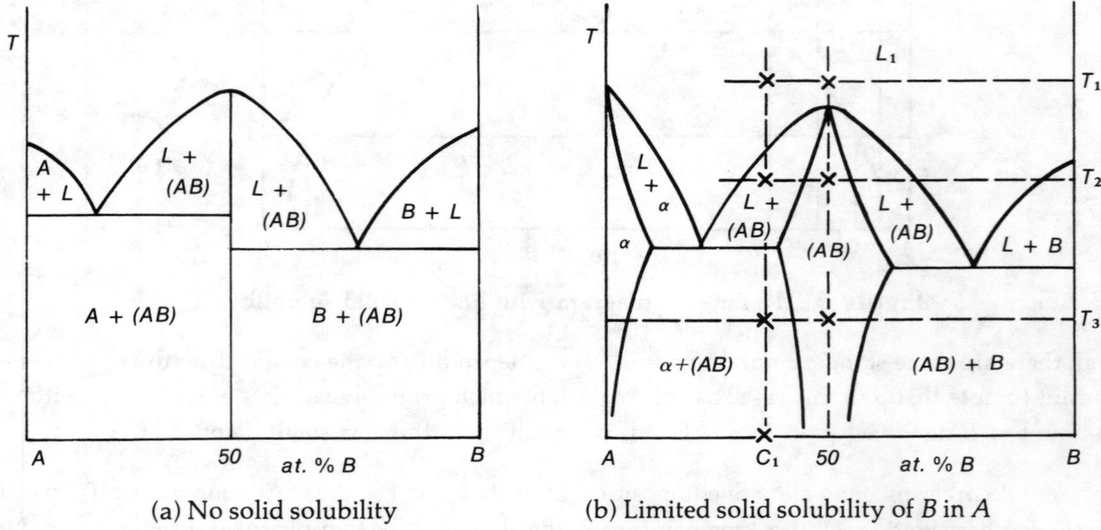

(a) No solid solubility

(b) Limited solid solubility of B in A

Figure 3.22. Schematic phase diagrams showing an AB compound.

of the phase diagram into two regions of eutectic type diagrams. Consider alloys of composition C_1 and A + 50% B in Fig. 3.22(b). At temperature T_1 alloy C_1 is liquid, at T_2 it is in two-phase L + (AB) regions (note (AB) is a compound), and at T_3 it is in two-phase α + (AB) regions. Note the intersection points of the temperature-horizontals with the various phase boundaries. For the alloy with A + 50 at.% B, T_1 is the region of liquid phase but both at T_2 and T_3 the alloy is in a single-phase AB compound.

The eutectic diagram, as discussed earlier, can be viewed as if a chemical reaction occurs at the eutectic point; i.e., in melting an eutectic solid, two solids react to form a liquid of a fixed composition. Thus, an eutectic phase separation is sometimes referred to as an *eutectic reaction*.

Other reactions in various binary alloys are similar to the eutectic reaction. For example, a *peritectic* reaction is one in which a liquid L_I reacts with a solid S_I to form a second solid phase S_{II}. Fig. 3.23 shows a schematic diagram with a peritectic reaction. The point P is the peritectic point, and T_p and C_p are the

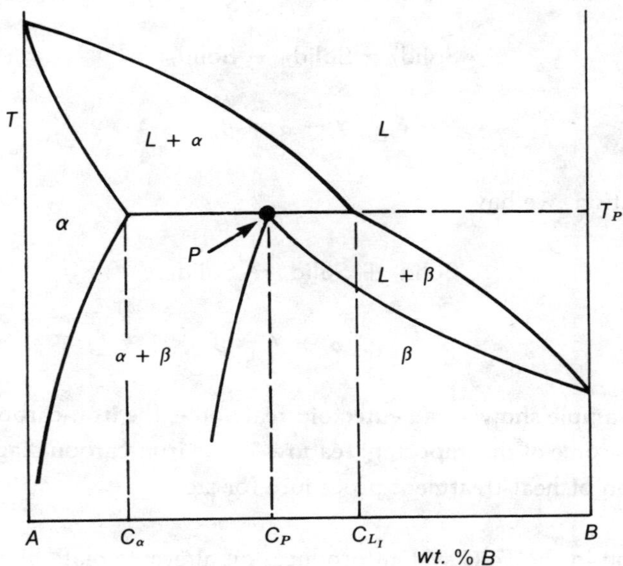

Figure 3.23. Schematic diagram showing a peritectic reaction.

peritectic temperature and composition, respectively. At T_p, liquid of composition C_{L_I} reacts with solid α of composition C_α to form a new solid β; for example,

$$L_I + \alpha \rightarrow \beta \qquad (3.2.14)$$

Another type of reaction, known as the *monotectic* reaction is shown in Fig. 3.24. In this diagram M is the monotectic point. At the monotectic temperature T_M, liquid of composition C_{L_I} reacts with pure A to form another liquid of composition $C_{L_{II}}$. Note that the diagram shows a region where two immissible liquids L_I and L_{II} coexist. Thus, the monotectic reaction in the above diagram could be written as:

$$\text{Solid } A + \text{liquid } L_I \rightarrow L_{II}$$

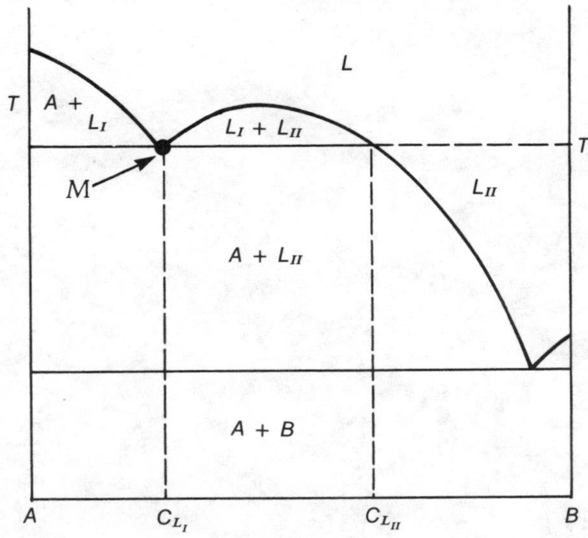

Figure 3.24. Schematic presentation of a monotectic reaction.

Analogous to the eutectic and peritectic reactions during the freezing of an alloy, we find two other reactions in the solid state, diffusion controlled phase separation. These are called *eutectoid* and *peritectoid* reactions. For an eutectoid reaction, we can write

$$\text{Solid}_I \rightarrow \text{Solid}_{II} + \text{Solid}_{III}$$

$$\text{e.g., } \gamma \rightarrow \alpha + \beta,$$

and for a peritectoid reaction, we have

$$\text{Solid}_I + \text{Solid}_{II} \rightarrow \text{Solid}_{III}$$

$$\text{e.g., } \alpha + \gamma \rightarrow \beta$$

An important practical example showing an eutectoid reaction is the iron-carbon diagram. On page 3-19, Fig. 3.25 (a) and (b) shows some of the important features of the iron-carbon diagram. This diagram is very useful in the determination of heat-treatment procedure for steels.

The eutectoid reaction in the *Fe-C* system produces an alternate plate-like microstructure known as *pearlite*. The alternate plates consist of *ferrite* and *cementite*. Ferrite, denoted by α, is almost pure iron with a small amount of carbon ($\leqslant 0.02\%$). Cementite, also known as *carbide*, has a composition: $Fe + 6.7$ wt.% *C* and in atomic % it has a formula Fe_3C. The phase, designated as γ in this diagram, is known as *austenite*. Since the eutectoid reaction at $\approx 723°C$ is controlled by solid state diffusion, this reaction can be suppressed by a rapid quenching of γ. In that case, austenite (γ) can transform to a metastable phase known as *martensite*. Martensite is the phase responsible for the dramatic increase of hardness of steels upon quenching. A steel with $< 0.8\%$ C is called a *hypoeutectoid* steel and one with $> 0.8\%$ is called a *hypereutectoid* steel.

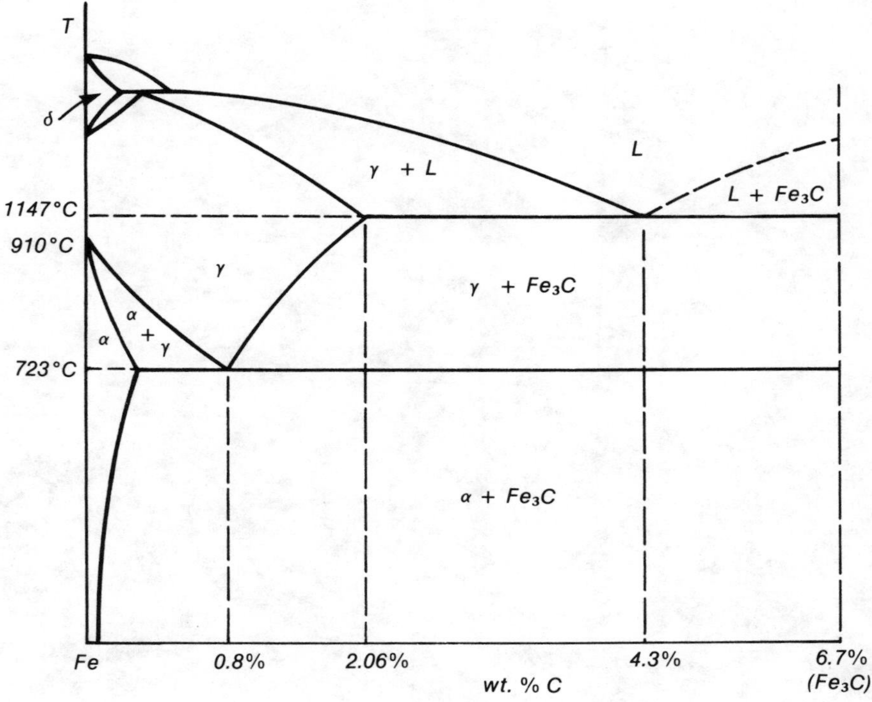

(a) *Fe-C* phase diagram.

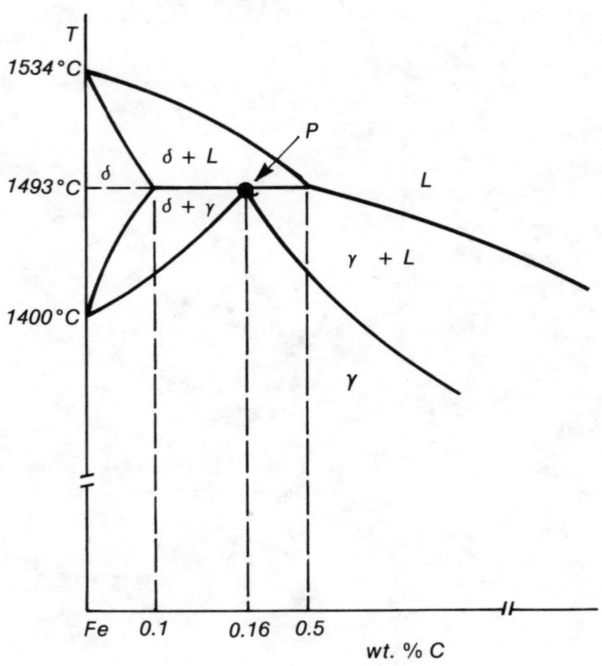

(b) A magnified view of the δ-*Fe* region. Point *P* is a peritectic point.

Figure 3.25 The iron-carbon diagram.

(This page is intentionally blank.)

Practice Problems

3.1 Deformation is irreversible if

a) Hooke's law holds.
b) the deformation rate is slow.
c) applied stress is greater than the yield stress.
d) there is no work hardening.
e) modulus of elasticity is low.

3.2 Off-set yield stress is defined as

a) stress at which yielding starts.
b) stress required for inhomogeneous deformation.
c) stress corresponding to a specified plastic strain.
d) stress at the proportional limit.
e) a specified percent of fracture stress.

3.3 Ultimate tensile strength is

a) the theoretical strength of a metal.
b) the stress to initiate plastic flow.
c) the fracture stress.
d) the maximum engineering stress.
e) the stress at 0.2% plastic strain.

3.4 All of the following statements about fracture are correct, except:

a) extent of necking depends on ductility.
b) ductile fracture produces a cup and cone fracture surface.
c) in a ductile fracture the reduction in area is zero.
d) a low carbon steel can be more brittle at low temperatures.
e) high strain rate is more conducive to brittle fracture.

3.5 All of the following statements about yield-point phenomenon are true, except:

a) dislocation motion is responsible for yield point.
b) presence of carbon in iron is responsible for yield point in steels.
c) upper and lower yield points are found in many aluminum alloys.
d) yield point indicates inhomogeneous deformation.
e) high strain rate accentuates yield point.

3.6 Plastic deformation is caused by

a) stored elastic energy.
b) dislocation motion.
c) low value of modulus.
d) complex elastic stresses.
e) motion of vacancies.

3.7 All of the following statements regarding slip deformation are correct, except:

a) a volume change is associated with slip.
b) slip occurs along close-packed directions.
c) some crystals have more slip systems than others.
d) dislocation motion is necessary for slip.
e) slip is irreversible.

3.8 All of the following statements regarding metal fatigue are correct, except:

a) surface roughness decreases fatigue life.
b) thermal cycling of a metal can produce fatigue failure.
c) carburizing treatment can improve fatigue life.
d) a fatigue endurance limit is observed in pure aluminum.
e) fatigue failure can occur at stresses below the yield stress.

3.9 A solid solution alloy is

a) an alloy produced by melting two metals together which do not mix in the solid state.
b) an alloy of two solid phases.
c) an alloy which is homogeneous and single phase at ordinary temperatures.
d) an alloy which is quickly attacked by an acid.
e) an alloy which does not have any grain boundaries.

3.10 All of the following statements regarding an eutectic alloy are correct, except:

a) eutectic temperature is invariant.
b) an alloy of eutectic composition solidifies within a range of temperatures.
c) two elements must be partially or totally insoluble in the solid state to form an eutectic.
d) eutectic microstructure is easily detectible under a microscope.
e) in an eutectic reaction a liquid decomposes into two solids.

3.11 A peritectic reaction is defined as:

a) two solids reacting to form a liquid.
b) two liquids reacting to form a solid.
c) a liquid and a solid reacting to form another solid.
d) two solids reacting to form a third solid.
e) a liquid separating into two solids.

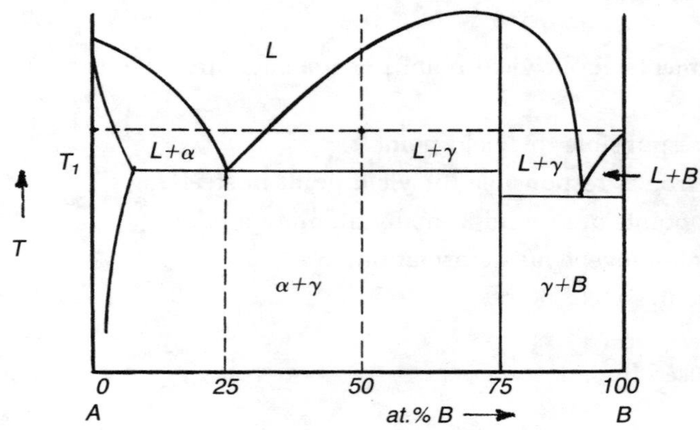

3.12 In the above diagram which of the following compound or compounds are present?

a) AB b) AB_2 c) A_3B_2 d) AB_3 e) A_2B_3

3.13 For an alloy of A + 50 at.% B as shown in the diagram for Problem 3.12, which phases are present at the temperature T_1?

a) single phase liquid (L)
b) liquid + γ
c) liquid + α
d) $\alpha + \gamma$
e) $\gamma + B$

3.14 For the iron-carbon diagram shown in Fig. 3.25 all of the following are true, except:

a) an eutectoid reaction at 1147°C.
b) a peritectic reaction at 1493°C.
c) maximum carbon content of austenite is 2.06%.
d) γ to δ transformation temperature is 1400°C.
e) eutectoid carbon composition is 0.8%.

3.15 A binary alloy of composition: 50 wt.% A and 50 wt.% B has two equilibrium phases, α and β at a temperature T_1. The composition of α-phase is 75 wt.% A and 25 wt.% B. The composition of β-phase is 25 wt.% A and 75 wt.% B. The proportions of α- and β-phases are

a) 25 wt.% α, 75 wt.% β.
b) 75 wt.% α, 25 wt.% β.
c) 50 wt.% α, 50 wt.% β.
d) 80 wt.% α, 20 wt.% β.
e) 20 wt.% α, 80 wt.% β.

(This page is intentionally blank.)

4. Chemistry

Chemistry attempts to describe and predict the macroscopic behavior of matter in terms of the properties of the submicroscopic *atoms* and *molecules* of which it consists. Although the physical properties of matter are of some interest to chemists because of what they reveal about the nature of atoms and molecules, the chemical properties or *chemical reactions* of matter are far more important. Thus, an understanding of the definition, structure, and properties of atoms and molecules, the definition of a chemical reaction, and the effect of variables on a chemical reaction are fundamental to understanding chemistry.

4.1. Atoms

An *atom* is the smallest subdivision or particle of an element that enters into the chemical reactions of that element. An *element* is a pure substance that cannot be decomposed into other pure substances. Each element is composed of only one kind of atom, and is represented by a *chemical symbol*. Thus, gold is an element represented by the symbol *Au*, and is composed entirely of atoms of gold. No chemical or physical process can separate a sample of gold into any other pure substance.

4.1.1 The Periodic Table of the Elements

The chemical and physical properties of the elements follow a pattern that is revealed in the *periodic table* of the elements (Table 4.1). In the table, the elements are ordered according to their atomic weights. **Note:** ordering according to atomic weights — an experimental observation — results in ordering according to atomic numbers — an inferred quantity. The significance of this fact will be discussed when atomic structure is discussed. The *atomic weight* of an element is the weight of an atom of that element in atomic mass units. One *atomic mass unit (amu)* is one-twelfth the mass of one normal *C* atom, very close to that of one *H* atom. Thus, the approximate atomic weight of an element is the weight of one atom (or 10 atoms, or 6×10^{23} atoms) of that element relative to the weight of one (or 10, or 6×10^{23}) atom of hydrogen. When ordered in the periodic table according to their atomic weights, the elements fall into families of related physical and chemical properties. A thorough understanding of the periodic table enables the chemist to infer the properties of an unknown element with some reliability from its determined position in the table. Perhaps more important, it is not necessary to memorize the properties of all 106 known elements to have a reasonable grasp of them. It is sufficient to know the properties of a few representative elements and understand the trends within the table. Thus, if you know that the element sodium (*Na*), is a soft, malleable solid with a metallic luster, that can be cut with a knife, conducts heat and electricity, and reacts vigorously with water with bubbling, you can expect that cesium (*Cs*) — in the same column or group — will show the same properties, more or less. The principles discussed here and in Section 4.3 on Chemical Reactions, when combined with practice and experience, will allow the skillful use of the periodic table.

TABLE 4.1. The Periodic Chart of the Elements

IA	IIA	IIIB	IVB	VB	VIB	VIIB	VIIIB			IB	IIB	IIIA	IVA	VA	VIA	VIIA	O
1 **H** 1.0080																	2 **He** 4.003
3 **Li** 6.940	4 **Be** 9.013											5 **B** 10.82	6 **C** 12.011	7 **N** 14.008	8 **O** 16.000	9 **F** 19.00	10 **Ne** 20.183
11 **Na** 22.991	12 **Mg** 24.32											13 **Al** 26.98	14 **Si** 28.09	15 **P** 30.975	16 **S** 32.066	17 **Cl** 35.457	18 **Ar** 39.944
19 **K** 39.100	20 **Ca** 40.08	21 **Sc** 44.96	22 **Ti** 47.90	23 **V** 50.95	24 **Cr** 52.01	25 **Mn** 54.94	26 **Fe** 55.85	27 **Co** 58.94	28 **Ni** 58.71	29 **Cu** 63.54	30 **Zn** 65.38	31 **Ga** 69.72	32 **Ge** 72.60	33 **As** 74.91	34 **Se** 78.96	35 **Br** 79.916	36 **Kr** 83.80
37 **Rb** 85.48	38 **Sr** 87.63	39 **Y** 88.92	40 **Zr** 91.22	41 **Nb** 92.91	42 **Mo** 95.95	43 **Tc** (99)	44 **Ru** 101.1	45 **Rh** 102.91	46 **Pd** 106.4	47 **Ag** 107.880	48 **Cd** 112.41	49 **In** 114.82	50 **Sn** 118.70	51 **Sb** 121.87	52 **Te** 127.61	53 **I** 126.91	54 **Xe** 131.30
55 **Cs** 132.91	56 **Ba** 137.36	57 ***La** 138.92	72 **Hf** 178.50	73 **Ta** 180.95	74 **W** 183.86	75 **Re** 186.22	76 **Os** 190.2	77 **Ir** 192.2	78 **Pt** 195.09	79 **Au** 197.0	80 **Hg** 200.61	81 **Tl** 204.39	82 **Pb** 207.21	83 **Bi** 209.00	84 **Po** (210)	85 **At** (210)	86 **Rn** (222)
87 **Fr** (223)	88 **Ra** (226)	89 **†Ac** (227)	104 **Ku** (261)	105 **Ha** (260)													

* Lanthanides	58 **Ce** 140.13	59 **Pr** 140.92	60 **Nd** 144.27	61 **Pm** (147)	62 **Sm** 150.35	63 **Eu** 152.0	64 **Gd** 157.26	65 **Tb** 158.93	66 **Dy** 162.51	67 **Ho** 164.94	68 **Er** 167.27	69 **Tm** 168.94	70 **Yb** 173.04	71 **Lu** 174.99
† Actinides	90 **Th** (232)	91 **Pa** (231)	92 **U** 238.07	93 **Np** (237)	94 **Pu** (242)	95 **Am** (243)	96 **Cm** (247)	97 **Bk** (249)	98 **Cf** (251)	99 **Es** (254)	100 **Fm** (253)	101 **Md** (256)	102 **No** (253)	103 **Lw** (257)

4.1.2 General Trends in the Periodic Table

There is a greater change in the properties of the elements across the table (the rows or *periods*) than there is down the table (the columns, *families*, or *groups*). **Note:** Hydrogen, the simplest element, is an exception. It exhibits some of the properties of both groups IA and VIIA as well as some unique ones. Thus lithium (*Li*, group IA) is much more like francium (*Fr*, group IA) than it is like fluorine (*F*, group VIIA). As a result, the elements can be divided into two general classes, the metals and non-metals, by a more or less vertical line. The *metals* are conducting and have low *electron affinities* (tendency to accept an electron). They therefore tend to give up electrons easily to form positive ions, and are reducing agents. The *non-metals* are non-conducting and have high electron affinities. They therefore tend to accept electrons to form negative ions, and are oxidizing agents. Metallic properties tend to increase for the elements of higher atomic number. As a result, polonium (*Po*, element number 84) is more metallic than tellurium (*Te*, element number 52), and, even though these elements are in the same family (group VIA), polonium is classified as a metal and tellurium as a non-metal. The increase in metallic properties with increasing atomic number thus accounts for the diagonal division separating the metals from the non-metals. The *transition metals* (elements number 21-29, 39-47, 57-79, 89-106) are classified together since their properties are very similar. The changes across the table are dramatically attenuated for the transition metals and they represent a gradual transition from the group IIA metals to the group IIB metals.

The common names of some of the groups, or families, of elements in the periodic table follow:

Group IA: *alkali metals*
Group IIA: *alkaline earth metals*
Group VIIA: *halogens*
Group 0: *noble gases*
Elements 57-71: *lanthanides* or *rare earths*
Elements 89-106: *actinides*
Groups IB-VIIB, VIII: *transition metals*

4.1.3 Atomic Structure and the Properties of the Elements

The periodicity in the properties of the elements can be understood in terms of the structures of their atoms. An atom consists of an extremely dense *nucleus* of *protons* (particles with a unit positive charge and a mass very close to 1 *amu*) and *neutrons* (neutral particles of essentially the same mass as the proton), and a very diffuse surrounding electron cloud containing enough *electrons* (entities with a unit negative charge and a comparitively negligible mass — about 5.5×10^{-4} *amu*) to exactly balance the nuclear charge. Thus, for any atom, the total number of protons equals the total number of electrons. The chemistry of an element is essentially determined by the electron cloud in the atoms of that element. The periodicity of the properties of the elements is thus a reflection of the periodicity of the structure of the electron clouds in atoms. It should now be apparent that the atomic number, which corresponds to the number of protons in the atomic nucleus (hence the number of electrons in the electron cloud), is actually the parameter which determines the position of an element in the periodic table, rather than the atomic weight (which is the sum of the masses of the protons, neutrons, and electrons in the atom).

4.1.3.1 Electron Orbitals

The theory of wave mechanics applied to atomic structure tells us that electrons (which appear to have properties of both waves and particles) occupy regions in space of specific shape and size called *orbitals*.

Note: The older term, *orbit*, derives from a planetary theory of the atom which did not take account of the wave nature of electrons. An orbital can be defined by three *quantum numbers*: $n = 1,2 \cdots n$, which defines the size of the orbital *shell*; $\ell = 0, 1, 2 \cdots n - 1$, which defines the shape of the orbital *subshell*; and $m = 0, \pm 1, \pm 2 \cdots \pm \ell$, which defines the orientation of the orbital subshell in space. The larger the

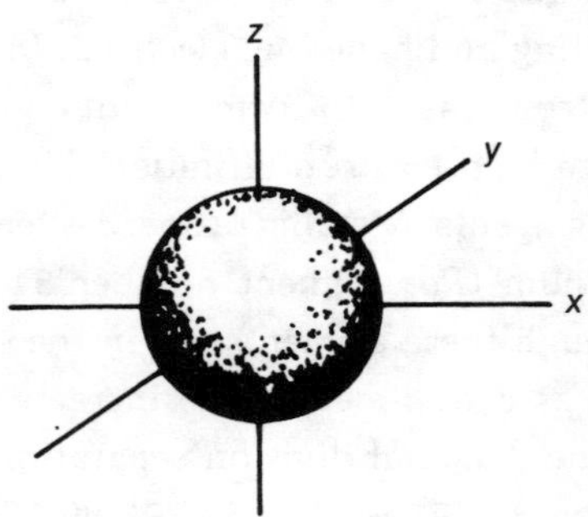

Figure 4.1. Boundary surface representation of an electron in the $n = 1$ state of the hydrogen atom. (Volume encloses 90% of the electron density. Nucleus is at the origin.)

principle quantum number, n, the larger the orbital, and the further the average negative charge of the electron from the positive nucleus, hence, the higher the energy. For a principal quantum number $n = 4$, there are different orbital shapes corresponding to the *subsidiary quantum numbers* $\ell = 0, 1, 2, 3$. These shapes are often identified by letter symbols instead of number symbols. An *s* orbital has $\ell = 0$ and is spherical in shape (Fig. 4.1). A *p* orbital has $\ell = 1$ and is dumbbell shaped Fig. 4.2). A *d* orbital has $\ell = 2$ and is still more complex in shape (Fig. 4.3). An *f* orbital has $\ell = 3$, and so on. There is only one orientation in space for a spherical shape, and therefore only one value (*0*) for *m*, the *magnetic orbital*

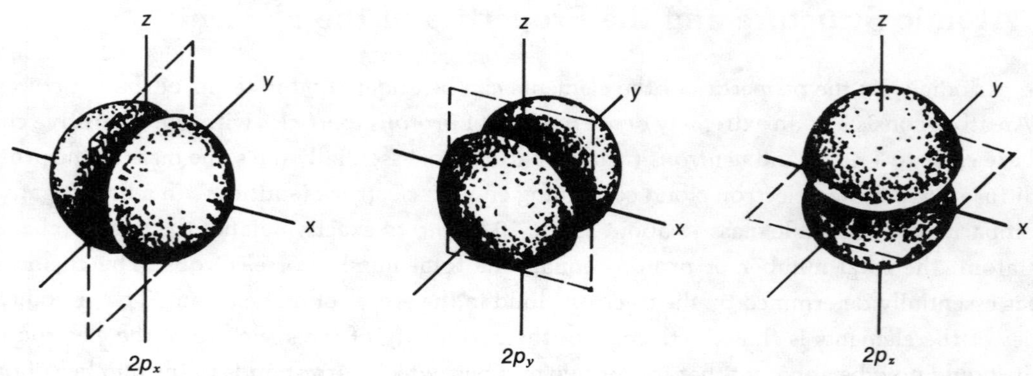

Figure. 4.2. Boundary surface diagrams for the *2p* orbitals.

quantum number, for an *s* orbital. However, the dumbbell-shaped *p* orbital has three perpendicular orientations in space (see Fig. 4.2), and has three values for $m = +1, 0, -1$. That is, for a given principal quantum number *n* there is only one *s* orbital, but there are three *p* orbitals. For the *d* orbital, there are five values for $m = +2, +1, 0, -1, -2$, hence five different *d* orbitals. The quantum numbers for the thirty

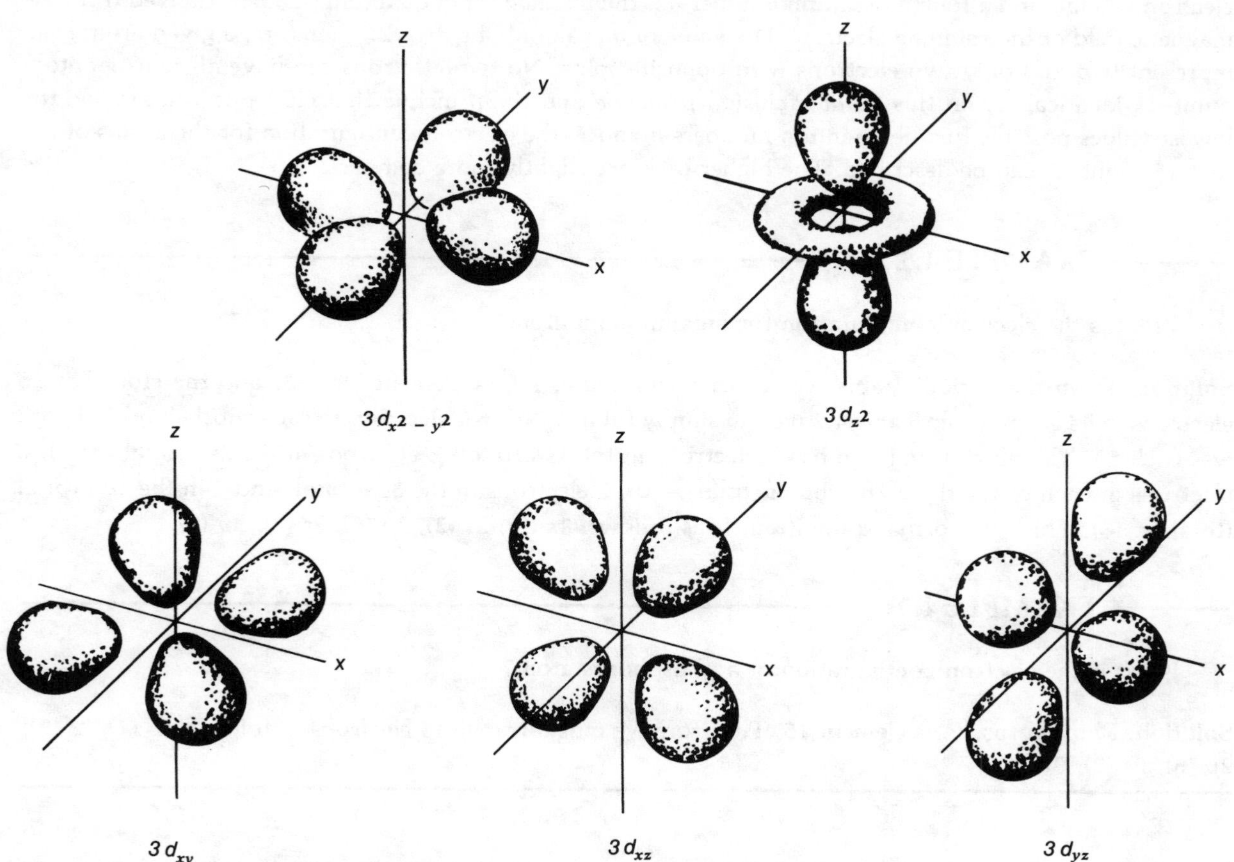

Figure 4.3. Boundary surface diagrams for the *3d* orbitals.

atomic orbitals of the first four shells are presented in Table 4.2.

TABLE 4.2. The Orbitals of the First Four Shells

Shell n	Subshell ℓ	Orbital m	Subshell Notation	Number of Orbitals per Subshell
1	0	0	1*s*	1
2	0	0	2*s*	1
	1	+ 1, 0, − 1	2*p*	3
3	0	0	3*s*	1
	1	+ 1, 0, − 1	3*p*	3
	2	+ 2, + 1, 0, − 1, − 2	3*d*	5
4	0	0	4*s*	1
	1	+ 1, 0, − 1	4*p*	3
	2	+ 2, + 1, 0, − 1, − 2	4*d*	5
	3	+ 3, + 2, + 1, 0, − 1, − 2, − 3	4*f*	7

4.1.3.2 Electron Configurations of the Elements

The state of an electron in an atom is completely described by the three quantum numbers defining the electron orbital, and a fourth quantum number m_s, the magnetic spin quantum number, derived from the magnetic field of the spinning electron. The value of m_s can only be $\pm$ 1/2. Therefore, a given orbital can represent two and only two electrons with opposite spins. No two electrons can have all four quantum numbers identical. From this *Pauli exclusion principle* and the principle that electrons will assume the lowest values possible for the quantum numbers n and ℓ, the electron configuration for the atoms of the first 18 elements can be described. The higher ones are slightly more complex.

EXAMPLE 4.1

What is the electron configuration for an aluminum atom?

Solution. From the periodic table we see that aluminum, *Al*, is element No. 13, and therefore has 13 electrons in its atom. From Table 4.2 we can simply fill in up to two electrons in each subshell orbital until we reach 13. Therefore aluminum has 2 electrons in the 1*s* orbital, 2 electrons in the 2*s* orbital (= 4), 2 electrons in each of the three 2*p* orbitals (total =10), 2 electrons in the 3*s* orbital, and 1 in the 3*p* orbital (total = 13). This, in shorthand notation, is as follows: 1*s* (2), 2*s* (2), 2*p* (6), 3*s* (2), 3*p* (1).

EXAMPLE 4.2

What is the electron configuration of a phosphorus atom?

Solution. Phosphorus, *P*, is element 15. Therefore we must arrange 15 electrons as follows: 1*s* (2), 2*s* (2), 2*p* (6), 6*s* (2), 3*p* (3).

In Example 4.2 there is an ambiguity about the state of the 3*p* electrons. We don't know whether to put two of them in one 3*p* orbital or put each of the three in a different 3*p* orbital. This ambiguity is removed by *Hund's rule of maximum multiplicity* which states that, whenever possible, the electrons in a subshell will have the same value for the magnetic spin quantum number, m_s. For phosphorus, this is possible for the 3*p* subshell electrons if each electron is in a different orbital with m_s = + 1/2 (or − 1/2). Since the spins of electrons with opposite signs for m_s are said to be paired, an alternate way of stating Hund's rule is that the electrons are distributed among the orbitals of a subshell in a way that gives the maximum number of unpaired electrons with parallel spins. We might, therefore, write the electron configuration of a phosphorus atom in more detail as follows: 1*s* (2), 2*s* (2), 2*p* (6), 3*s* (2), $3p_x$ (1), $3p_y$ (1), 3*pz* (1). The maximum numbers of electrons that can occupy each subshell for the first four shells are given in Table 4.3.

TABLE 4.3. Maximum Number of Electrons for the Subshells of the First Four Shells

Subshell Notation	Orbitals per Subshell	Electrons per Subshell	Electrons per Shell ($2n^2$)
1*s*	1	2	2
2*s*	1	2	8
2*p*	3	6	
3*s*	1	2	18
3*p*	3	6	
3*d*	5	10	
4*s*	1	2	32
4*p*	3	6	
4*d*	5	10	
4*f*	7	14	

The periodicity of the properties of the elements revealed in the periodic table can now be seen to have its origins in the periodicity of the electronic configurations of the atoms. The chemical properties of an atom are determined by the electrons with the largest value of n, that is the outer shell or *valence electrons*. Thus, the properties of lithium are determined by its lone 2*s* electron, those of sodium by its lone 3*s* electron, those of potassium by its lone 4*s* electron, and those of francium by its lone 7*s* electron. All of the atoms of the elements of Group 0, the noble gases, have *closed-shell configurations*, that is, all of the orbitals of the highest principle quantum number are full of electrons. The closed-shell configuration represents a particularly stable state. Any further electrons would have to occupy higher principle quantum number (and therefore higher energy, less stable) orbital states. All of the atoms of the halogens (Group VIIA) are one electron short of closed-shell configurations, hence can accept an electron very readily.

EXAMPLE 4.3

Determine the electron configuration of the valence electrons of an atom of *N*, and compare it with that for *As*.

Solution. *N* has 7 electrons. The electron configuration is 1*s* (2), 2*s* (2), $2p_x$ (1), $2p_y$ (1), $2p_z$ (1). The valence electrons are those with $n = 2$, i.e.: 2*s* (2), $2p_x$ (1), $2p_y$ (1), $2p_z$ (1).

As has 33 electrons. The first three shells ($n = 1, 2, 3$) are completely filled to account for $2 + 8 + 18 = 28$ electrons (see Table 4.3). That leaves 5 electrons for the fourth shell, $n = 4$. Therefore the configuration of the valence electrons of *As* is 4*s* (2), $4p_x$ (1), $4p_y$ (1), $4p_z$ (1). Note the similarity to *N* in the filled *s* orbital and three singly occupied *p* orbitals.

4.1.4 The Properties of the Families of Elements

The similarities in the properties of the elements in each group or family in the periodic table are simply a manifestation of the similarities in the electronic configurations of the valence electrons of the corresponding atoms. Some of those properties will now be examined for the more common families of elements.

4.1.4.1 Group IA — The Alkali Metals

The alkali metal atoms of Group IA all have one *s* electron beyond a closed-shell configuration. They therefore tend to lose an electron very readily to the environment to give a positively charged species — a cation with a closed-shell configuration — as shown in the following simple chemical equation for sodium:

$$Na \rightarrow Na^{+} + 1e^{-}$$

One sodium atom yields one sodium cation plus one electron.

They are thus said to have a low electron affinity, or desire to accept electrons, and a low *ionization energy*, that is, it doesn't take much energy to make them lose an electron to form a cation. **Note:** This does not mean that the converse reaction, $Na + 1e^{-} \rightarrow Na^{-}$, is impossible, only that it is extraordinarily difficult. This atomic property of the alkali metals is reflected in their metallic lustre and their conductivity (electrons are readily available to form a *conducting band*). It is also reflected in their chemical reactivity as reducing agents. *Reducing agents* give up electrons easily to other substances and are themselves oxidized in the process. A simple example is the chemical reaction of a sodium atom and a chlorine atom:

$$Na + Cl \rightarrow Na^{+} + Cl^{-}$$

One sodium atom plus one chlorine atom yields one sodium cation plus one chloride anion.

In the above example, an electron is transferred from sodium to chlorine, and the sodium atom is oxidized and the chlorine atom is reduced. The sodium atom acts as a reducing agent and the chlorine atom acts as an oxidizing agent.

EXAMPLE 4.4

Complete the following chemical reaction:

$$Rb + F \rightarrow ?$$

Solution. Since rubidium *Rb* is in the same family as sodium, it is a reducing agent. Since fluorine *F* is in the same family as chlorine, it is an oxidizing agent. Hence an electron will transfer from rubidium to fluorine:

$$Rb + F \rightarrow Rb^{+} + F^{-}$$

Since the higher atomic number elements in a family have larger electron clouds which screen the nuclear positive charge, electron affinity decreases as atomic number increases within a family. Thus, potassium (atomic No. 19) is a more powerful reducing agent than lithium (atomic No. 3).

4.1.4.2 Group IIA — The Alkaline Earth Metals

The lightest of the Group IIA elements, beryllium (*Be*, atomic No. 4), has the electronic configuration 1*s* (2), 2*s* (2). The *s* electrons in the second shell are easily lost to give Be^{+2} with a closed-shell configuration in the first shell. The alkaline earth metals are thus typically reducing agents that give up two electrons. They are said to be *divalent*.

EXAMPLE 4.5

Complete the following chemical reaction:

$$Ca + 2\ Br \rightarrow ?$$

Solution. Calcium (*Ca*, atomic No. 20) has the following electron configuration: 1*s* (2), 2*s* (2), 2*p* (6), 3*s* (2), 3*p* (6), 4*s* (2). It is an alkaline earth with two valence electrons in the fourth shell. It is, therefore, a reducing agent with a tendency to give up two electrons. Bromine, like chlorine, can accept one electron to give a closed-shell anion. Therefore, two bromine atoms can accept one electron each from one calcium atom.

$$Ca + 2\ Br \rightarrow Ca^{+2} + 2\ Br^{-}$$

The alkaline earths are less powerful reducing agents than the alkali metals since the higher nuclear charge tends to hold the valence electrons more tightly. In fact, electron affinity increases across the periodic table and decreases down the periodic table; i.e., reducing power decreases across the periodic table and increases down the table.

The alkaline earth metals have a metallic luster and are harder than the alkali metals. They are good conductors of electricity.

EXAMPLE 4.6

Assign the ionization energies in the column on the right to the elements in the column on the left:

Element	Ionization Energy (kJ/mol)
Cs	738
Mg	503
Ca	590
Ba	376

Solution. Magnesium (*Mg*, No. 12), calcium (*Ca*, No. 20), and barium (*Ba*, No. 56) are all alkaline earths. The first ionization energy, the energy required to remove the first valence electron, should decrease with increasing atomic number. Cesium (*Cs*, No. 55) is an alkali metal and should have a lower ionization energy than the next alkaline earth, barium. It therefore has the lowest ionization energy of the four and the correct answer is:

Cs	376
Mg	738
Ca	590
Ba	503

4.1.4.3 Group VIIA — The Halogens

The halogen atoms of Group VIIA are all one electron short of a closed-shell configuration. Because they have high nuclear charges which hold their valence electrons very tightly, it is no surprise that they have high electron affinities. They therefore tend to accept electrons readily to give closed-shell configuration anions, and are oxidizing agents. Chlorine (*Cl*, No. 17) is typical with electron configuration 1*s* (2), 2*s* (2), 2*p* (6), 3*s* (2), 3*p* (5):

$$Cl + 1e \rightarrow Cl^-$$

Reactions of halogen atoms as oxidizing agents with metals acting as reducing agents have already been illustrated above. An example is

$$I + K \rightarrow I^- + K^+$$

As you might expect, the oxidizing power of the halogens is greatest for fluorine (*F*, No. 9) and decreases down the table, the inverse of the reducing power of the metals.

The halogens tend to be colored. Thus, chlorine is a green gas, bromine is a red, volatile liquid, and iodine is a deep purple, volatile solid.

EXAMPLE 4.7

Which of the following chemical reactions has the greatest tendency to take place?

$$Cl + K \rightarrow Cl^- + K^+$$

$$Br + K \rightarrow Br^- + K^+$$

$$Cl + Li \rightarrow Br^- + Li^+$$

$$Br + Li \rightarrow Br^- + Li^+$$

Solution. Since oxidizing power decreases down the table, chlorine *Cl* is a more powerful oxidizing agent than bromine *Br*. Since reducing power increases down the table, potassium *K* is a more powerful reducing agent than lithium *Li*. The reaction with the greatest tendency to take place is the one between the best oxidizing and best reducing agents, i.e.,

$$Cl + K \rightarrow Cl^- + K^+$$

4.1.4.4 Group 0 — The Noble Gases

The noble gases of Group 0 all have closed-shell configurations. They therefore have both low electron affinities and high ionization potentials. An added electron would have to go into the next orbital shell of much higher energy, and removal of an electron would have to take place from the closed shell. Because of these properties, the noble gases enter into chemical reactions with extreme reluctance. Some of the properties of the noble gases are given in Table 4.4.

TABLE 4.4. Some Properties of the Noble Gases

Gas	Melting Point (°C)	Boiling Point (°C)	Ionization Energy (kJ/mol)	Abundance in Atmosphere (Volume %)
He	—	− 268.9	2.37×10^3	5×10^{-4}
Ne	− 248.6	− 245.9	2.08×10^3	2×10^{-3}
Ar	− 189.3	− 185.8	1.52×10^3	0.93
Kr	− 157	− 152.9	1.35×10^3	1×10^{-4}
Xe	− 112	− 107.1	1.17×10^3	8×10^{-6}
Rn	− 71	− 61.8	1.04×10^3	trace

The noble gases are the only non-metallic elements that exist as single atoms in the elemental state. All of the others exist as clusters of atoms bonded together as molecules. Thus, the low melting points and boiling points of the noble gases reflect the stability of the atoms. However, in order to understand the chemical and physical properties of the rest of the elements, it is necessary to understand the molecule and the nature of chemical bonds between atoms.

4.2 Molecules

A *molecule* is the smallest subdivision or particle of a chemical compound that enters into the chemical reactions of that compound. A *chemical compound* is a substance containing more than one element combined in definite proportions. The molecules of a chemical compound are all identical clusters of atoms held together by chemical bonds. Thus, the chemical compound water consists of molecules containing two hydrogen atoms and one oxygen atom joined together by chemical bonds, symbolized by the structure shown in Fig. 4.4.

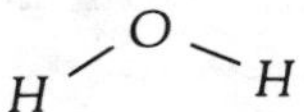

Figure 4.4 **The structure of water.**

4.2.1 Molecular Structure

The term *structure* can be used in chemistry on a number of levels, expressed by symbols of different complexity. At the simplest level, the structure of water is defined by its molecular formula, H_2O. A *molecular formula* defines the number of atoms of each element in the molecule of a compound. Thus, water has two hydrogen atoms and one oxygen atom in each molecule. At the next level, the structure of water is further defined by *HOH*, which defines the order in which the atoms are attached to one another. Thus, the two hydrogen atoms are attached to the one oxygen atom. At the next level, the shape of the molecule might be defined as either linear ($H-O-H$), or bent ($\overset{O}{\underset{H\quad H}{/\ \ \backslash}}$). The water molecule is bent. The shape might then be more precisely defined in terms of the actual bond angles and bond lengths involved. The structure of water (see Fig. 4.5): bond angle 104.5°, $O-H$ bond length = 0.96 Å = 0.096 nm = 96 pm.

$H - O$
104.5° \ 0.96Å
H

Figure 4.5 **The detailed structure of water.**

Note: Molecular distances are often given in Angstrom units (Å).
One Å = 10^{-8} centimeters (*cm*) = 0.1 nanometers (*nm*) = 100 picometers (*pm*).

For more complex molecules, a structure might include a description of the three-dimensional shape of the molecule. For example, the molecule of methane (the main component of marsh gas and natural gas) has the molecular formula CH_4. It is pyramidal in shape with each hydrogen atom attached to the single central carbon atom. The $H-C-H$ bond angles are all exactly 109°28′, and the $C-H$ bond lengths are all 1.09 Å. A qualitative representation of the methane structure might look like that of Fig. 4.6.

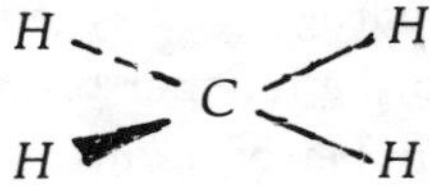

Figure 4.6 **The structure of methane.**

Thus the carbon atom and the two hydrogens attached to it with solid lines are implied to be in the plane of the paper. The wedged bond (*C◄H*) is intended to project above the paper, while the dashed bond (*C--H*) is intended to project below the paper.

Perhaps the most complete, though not always the most useful, definition of the structure of a molecule would be a mathematical description of the geometry and motions of the atoms and the states of all the electrons in it. Molecular structures thus require different approaches for each level of complexity.

4.2.1.1 Determination of the Molecular Formula

If all molecules of a pure compound contain the same elements in the same ratios, then the compound will always contain the same elements combined in the same proportions by mass. This is the *law of definite proportions.* Therefore, if we knew the mass ratio of oxygen to hydrogen in water, and knew the atomic masses of oxygen and hydrogen, we could determine the ratio of oxygen atoms to hydrogen atoms in the water molecule, that is, the *empirical formula.* If we knew the mass of a molecule of water, the *molecular weight,* then we could determine the actual number of oxygen and hydrogen atoms in a water molecule, the *molecular formula.*

EXAMPLE 4.8

Water can be broken down into elemental hydrogen and oxygen by electrolysis. Every 100 g of water gives 88.8 g of oxygen and 11.2 g of hydrogen. What is the empirical formula of water?

Solution. The atomic weight of oxygen is 16.0, while that of hydrogen is 1.008. The ratio of hydrogen atoms to oxygen atoms is therefore,

$$\frac{11.2/1.008}{88.8/16.0} = 2.00$$

The empirical formula of water is therefore H_2O.

The empirical formula can also be determined in some instances by application of *Avogadro's law,* which states: equal volumes of all gases at the same temperature and pressure contain equal numbers of molecules.

EXAMPLE 4.9

Electrolysis of 18 g of water gives 22.4 ℓ of hydrogen gas and 11.2 ℓ of oxygen gas at 22°C and 1 atmosphere pressure. What is the empirical formula of water?

Solution. Since the ratio of the volumes of hydrogen to oxygen is 2:1, Avogadro's law tells us that the ratio of the number of molecules of hydrogen to the number of molecules of oxygen is 2:1. If hydrogen and oxygen molecules each contain the same number of atoms, then the empirical formula of water is H_2O.

There are several ways of determining the molecular weight of a pure compound. At the present time, an instrument called a mass spectrometer is most often used. However, if the compound is a gas at known temperature and pressure, then the volume of a given mass of the gas can be used to determine the molecular weight. For this purpose, the gram-atom, gram-mole, and mole are useful and important concepts. A *gram-atom* of an element is an amount in grams numerically equal to the atomic weight. Thus, 1.000 gram-atom of sodium contains 22.9 grams of sodium. A *gram-mole* is that amount of a pure compound numerically equivalent in grams to the molecular weight of the compound. Thus 1.000 gram-mole of water contains $2 \times 1.088 + 16.00 = 18.02$ grams of water. Since the weight ratio of a gram-atom of sodium to a gram-atom of hydrogen is the same as the ratio of the atomic weights of the elements (23.0 : 1.0) a gram-atom of sodium contains the same number of atoms of sodium as a gram-atom of hydrogen contains of hydrogen. Similarly, a gram-mole of water contains the same number of water molecules as a gram-mole of methane contains methane molecules. In fact, the number of atoms in a gram-atom of any element and the number of molecules in a gram-mole of any compound is a constant. The constant is called *Avogadro's number* and has been experimentally determined to be 6.02205×10^{23}. The amount of any substance that contains Avogadro's number of elementary units is called a *mole* (abbreviated *mol*).

With the exception of the noble gases, which are monatomic, all elements which are gases at ordinary temperatures and pressures are diatomic, that is, contain two atoms per molecule (e.g., N_2, O_2, Cl_2). Note that one mole of any gas will occupy 22.4 ℓ at 22°C and 1 atm — the same as one mole of helium (Avogadro's law). This temperature and pressure is called Standard Temperature and Pressure, or STP. Hence, 1 mol of any gas at STP contains 22.4 ℓ.

EXAMPLE 4.10

A sample of helium gas weighing 4.00 g occupies 22.4 ℓ. The same volume of hydrogen gas weighs 2.02 g. What is the molecular formula of hydrogen gas?

Solution. Since the atomic weight of helium is 4.00, the sample of helium is 1.00 gram-atom. Since the atomic weight of hydrogen is 1.008, the sample of hydrogen is 2.00 gram-atoms and must contain twice as many atoms as the helium sample. Since both occupy the same volume, both must contain the same number of molecules (Avogadro's law). Therefore, a molecule of hydrogen contains twice as many atoms as a molecule of helium. If we assume that a molecule of helium contains only one atom (reasonable, since helium is an unreactive noble gas), then a molecule of hydrogen contains two atoms and has the molecular formula H_2.

EXAMPLE 4.11

A sample of steam weighing 27.0 g occupies the same volume as a sample of hydrogen weighing 3.02 g at 110°C and 1 atm. What is the molecular formula for water?

Solution. Since the two samples occupy the same volume under the same conditions, they have the same number of molecules. Therefore, the ratio of their weights, 27.0:3.02, is equal to the ratio of their molecular weights. If the molecular weight of hydrogen (H_2) is 2.02, the molecular weight of water is 18.0. Thus, the molecular formula for water is H_2O (see Ex. 4.9).

EXAMPLE 4.12

A sample of ethane is burned in air to give carbon dioxide (molecular formula CO_2) and water. The ratio of the volume of water vapor to that of carbon dioxide at 110° and 1 atm is 3:2. If 1 ℓ of ethane gas at STP gives 2 ℓ of CO_2 gas at STP, how many carbon and hydrogen atoms are there in a molecule of ethane?

Solution. The burning of ethane can be represented symbolically by the following expression:

$$\text{ethane } (g) + O_2\ (g) \qquad CO_2\ (g) + H_2O\ (\ell)$$

Note: The letters in parenthesis simply indicate whether the substance is a gas (g), liquid (ℓ), or solid (s) at STP.

Since equal volumes of CO_2 gas and H_2O gas at the same temperature and pressure will contain equal numbers of molecules, there are twice as many hydrogen atoms per unit volume in the water as there are carbon atoms per unit volume in the carbon dioxide. Since burning ethane gives 3 volumes of water vapor to 2 volumes of carbon dioxide, the ratio of hydrogen atoms to carbon atoms in ethane is $(3 \times 2):(2 \times 1) = 3:1$.

Since 1 ℓ of ethane gives 2 ℓ of carbon dioxide, each molecule of ethane must give two molecules of carbon dioxide. There must therefore be two carbon atoms in each molecule of ethane. A molecule of ethane therefore contains 2 carbon atoms and 6 hydrogen atoms. Note that we do not yet know the molecular formula of ethane completely, since it may contain other atoms as well (e.g., oxygen).

EXAMPLE 4.13

The density of ethane gas at STP is 1.34 g/ℓ. What is the molecular weight and molecular formula for ethane?

Solution. The molecular weight is $(22.4\ \ell \times 1.34\ \text{g}/\ell) = 30$ g. The molecular formula is therefore C_2H_6. Any additional atoms in the molecule would increase the molecular weight.

We can now convert the symbolic expression for the burning or combustion of ethane in Ex. 4.12 into a balanced chemical equation. A *balanced chemical equation* is a symbolic representation of a chemical change in which the number of each kind of atom on the left side of the equation, the reactants, is equal to the number of each kind of atom on the right side of the equation, the products:

$$C_2H_6 + \frac{7}{2} O_2 \qquad 2\ CO_2 + 3\ H_2O$$

4.2.1.2 Determination of Chemical Structure

The determination of the structure of a complex substance of known molecular formula is often a very difficult research task requiring years of effort including the analysis of chemical reactions and the use of sophisticated instruments. Perhaps the most powerful technique is *x-ray crystallography* in which an x-ray picture is taken of the arrangement of the atoms in the regularly spaced molecules of a crystalline substance; pictures taken from several angles allow an accurate three-dimensional structure for the molecule to be constructed. For simpler molecules, a knowledge of *chemical bonding* principles can at least limit the number of possible structures for a given molecular formula.

EXAMPLE 4.14

Given that a carbon atom can bond with no more than four other atoms, and hydrogen with only one, what is the order of connection of atoms in ethane, C_2H_6?

Solution. There is only one logically possible arrangement:

```
      H  H
      |  |
   H—C—C—H
      |  |
      H  H
```

4.2.1.3 Chemical Bonds

There are two limiting types of chemical bonds which hold atoms together — the ionic bond and the covalent bond. Metal bonding, a third and very different type of bonding, is found in the metals where the atoms are bonded together as cations in a "sea" of conducting valence electrons.

The ionic bond, represented by sodium chloride, $Na^+ Cl^-$, occurs whenever there is complete transfer of one or more electrons from one atom to another to give oppositely charged ions. The force holding the two particles together is then the electrostatic attraction between the oppositely charged ions. A crystal of sodium chloride, or table salt, might be regarded as a single enormous molecule in which all of the positive sodium ions and neighboring negative chloride ions are bonded together by ionic bonds in the crystal lattice. Ionic bonds are formed between atoms of very different electron affinities and the electron transfer usually brings both atoms to their closed-shell configurations.

The covalent bond, present in the hydrogen molecule (H_2), water (H_2O), methane (CH_4), and ethane (C_2H_6) occurs whenever one or more electrons are shared between two atoms. The force holding the atoms together is then the electrostatic attraction between the shared electrons and the positively charged nuclei. The hydrogen molecule might be regarded as two positively charged protons suspended at equilibrium distance in a diffuse elliptical cloud of two electrons. The covalent bond is formed between atoms of similar electron affinities, the shared electrons generally being sufficiently numerous to bring both atoms to their noble gas configurations. Thus, the two shared electrons in the hydrogen molecule can be considered to fill the 1*s* shell of each hydrogen atom. Although most bonds can be easily classified as ionic or covalent, in practice there is a spectrum of bond types from pure covalent through highly polar covalent to ionic bonds.

EXAMPLE 4.15

Arrange the following chemical bonds in order of increasing ionic character: *CCl, LiCl, LiI, CC, KCl, CAl, CMg*.

Solution. The most covalent bond is, of course, the *C-C* bond, between identical atoms. The most ionic is the K^+Cl^- bond between the atom of lowest electron affinity, *K*, and the one of highest electron affinity, *Cl*. Li^+Cl^- will be less ionic (*Li* has a higher electron affinity than *K*), Li^+I^- still less (*I* has a lower electron affinity than *Cl*), then *C-Mg*, *C-Cl*, and *C-Al* (reflecting the relative differences in electron affinities). Note that the partial electron transfer is from *C* to *Cl* leaving carbon partially positive for *C-Cl*, while it is from metal to carbon for *C-Mg* and *C-Al*, leaving carbon partially negative. The correct order of *increasing* ionic character is then: *C-C* < *C-Al* < *C-Cl* < *C-Mg* < *LiI* < *LiCl* < *KCl*.

The number of electrons "transferred" in an ionic bond, and the number of covalent bonds formed to a given atom is limited by the number of electrons in the valence shell. A convenient, though somewhat arbitrary, way of keeping track of the number of electrons transferred is the oxidation number, or oxidation state, or valence. (Valence is an older term whose meaning has become confused and is therefore rarely used now.) The oxidation number may be defined as the total number of electrons transferred or partially transferred to or from an atom in the bonds it forms with other atoms in the compound. If the electrons are transferred from the atom, its oxidation number is positive. If they are transferred to the atom, its oxidation number is negative. For example, in potassium chloride, one electron is transferred from potassium to chlorine. The oxidation number of potassium is therefore 1+, while that of chlorine is 1−. It is obvious that the oxidation numbers of monatomic ions in ionic compounds are equivalent to the charges on the ions. The oxidation number of the atoms in a covalently bonded molecule or ion can be determined by assuming that the electrons are transferred to the atom with the higher electron affinity. For example, in the covalent molecule *H-Cl*, we assume that one electron is transferred from hydrogen to chlorine. The oxidation number of hydrogen is therefore 1+ and that of chlorine 1−. For a neutral molecule, the sum of the oxidation numbers of all the atoms must, of course, be zero. For a charged molecule ion or complex ion, the sum of the oxidation numbers of all the atoms must equal the charge on the ion. A covalent bond between two atoms of the same element results in no electron transfer and no change in oxidation number.

The concept of oxidation number is a powerful tool for the analysis of a type of chemical reaction called an oxidation-reduction reaction or *Redox reaction.*

EXAMPLE 4.16

Assign oxidation numbers to all of the atoms in each of the following compounds: $NaBr$, $BaCl_2$, CaO, NH_3, CCl_4, OF_2, O_2, C_2H_6, CO, CO_2, Na_2SO_4, $K_2Cr_2O_7$.

Solution.

Compound	Atom	Oxidation Number	Explanation
$NaBr$	*Na*	1+	One electron is transferred from *Na* to *Br*. Oxidation number = charge.
	Br	1−	
$BaCl_2$	*Ba*	2+	Two electrons transferred from *Ba*. One to each *Cl*.
	Cl	1−	
CaO	*Ca*	2+	Two electrons transferred from the alkaline earth, *Ca*, to oxygen.
	O	2−	
NH_3	*H*	1+	*H* is *1+*. Therefore, *N* must be *3−* for a net *0*.
	N	3−	
CCl_4	*Cl*	1−	One electron considered transferred to each *Cl* (higher *e.a.*).
	C	4+	
OF_2	*F*	1−	*F* has the highest *e.a.* of all. It is *1−* except in F_2.
	O	2+	

Ex. 4.16 Solution (continued)

Compound	Atom	Oxidation Number	Explanation
O_2	O	0	Same element. No *e* transfer.
C_2H_6	H	1+	*H* always 1+ with non-metals. *C* must be 3− for net = 0.
	C	3−	
CO	O	2−	*O* generally 2−. *C* must be 2+.
	C	2+	
CO_2	O	2−	*C* must be 4+ for net = 0.
	C	4+	
Na_2SO_4	Na	1+	*Na* must be 1+. The anion must be SO_4^{-2}. If O is 2−, then *S* must be 6+.
	O	2−	
	S	6+	
$K_2Cr_2O_7$	K	1+	The complex dichromate ion must be $Cr_2O_7^{-2}$. If *O* is 2−, the two *Cr* atoms must total 12+. Assuming they are in the same oxidation state, they are each 6+.
	O	2−	
	Cr	6+	

4.3 Chemical Reactions

A *chemical reaction* occurs whenever a chemical compound is formed from the elements or from another chemical compound. Thus, a chemical reaction involves the transformation of the arrangement of atoms in the molecules of the starting materials, or *reactants*, into the arrangement of atoms in the molecules of the *products*. The two main types of variables that affect the course of a reaction are the quantities of reactants, and the conditions (e.g., temperature and either pressure for reactions of gases, or concentration for reactions in solution). The two main characteristics of the reaction which affect its course are the *rate* and the *equilibrium*.

4.3.1 Balancing Chemical Equations

In order to understand the effect of the quantities of reactants on the course of a chemical reaction, it is necessary to have a *balanced chemical equation* for the reaction. In a chemical equation, the structures or condensed structures of the reactants and products are written on the left and right, respectively, with an arrow between the two. In a balanced chemical reaction, the numbers of molecules of reactants and products are adjusted such that both the total number of atoms and the total charge of each element on the left is equal to that of the same element on the right. Thus, reaction of magnesium with bromine to give magnesium bromide is symbolized: $Mg + Br_2 \rightarrow MgBr_2$. Reaction of potassium with chlorine to give potassium chloride is symbolized: $2K + Cl_2 \rightarrow 2KCl$. Note that the smallest integers possible are used. We would not write $2Mg + 2Br_2 \rightarrow 2MgBr_2$. However, one might sometimes write $K + \frac{1}{2}Cl_2 \rightarrow KCl$. For the reaction of sodium with water to give sodium hydroxide and hydrogen gas, we might write either $2Na + 2H_2O \rightarrow 2NaOH + H_2$, or $Na + H_2O \rightarrow 1/2H_2$.

Slightly more complex reactions can often be balanced by trial and error. For example, rust formation is illustrated by the (unbalanced) equation:

$$Fe + H_2O \rightarrow Fe_2O_3 + H_2$$

In order to balance this equation, we might first locate the molecule that has the largest number of atoms of the elements — the most complex molecule. In this case, it is Fe_2O_3. We must therefore have at least two atoms of iron in the reactants and three molecules of water:

$$2Fe + 3H_2O \rightarrow Fe_2O_3 + 3H_2.$$

EXAMPLE 4.17

Balance the chemical equation for the complete combustion of ethane (C_2H_6) to carbon dioxide and water.

Solution. Since ethane has two carbon atoms and six hydrogen atoms, we know that each molecule of ethane must give two molecules of CO_2 and three of H_2O

$$C_2H_6 + \frac{7}{2}O_2 \rightarrow 2\ CO_2 + 3\ H_2O$$

or

$$2\ C_2H_6 + 7\ O_2 \rightarrow 4\ CO_2 + 6\ H_2O$$

The balanced chemical equation of Ex. 4.17 tells us a number of useful facts. It tells us that two molecules of ethane require seven molecules of oxygen for complete combustion to give four molecules of carbon dioxide and six molecules of water. It tells us that 2*ℓ* of ethane requires 7*ℓ* of oxygen for complete combustion; less would give incomplete combustion. It tells us that 60 g of ethane (2 mol) will give 108 g of water (6 mol) upon complete combustion.

More complex chemical equations can often be balanced by the use of oxidation numbers. For example, consider the reaction:

$$H\overset{5+}{N}O_3 + H_2\overset{2-}{S} \rightarrow \overset{2+}{N}O + \overset{0}{S} + H_2O$$

The oxidation numbers of N and S in reactants and products have been identified. N undergoes a change of -3 (5+ to 2+) from reactants to products. Sulfur undergoes a change of -2 (2− to 0). The total change in oxidation number for the reaction must be zero. Therefore, all we need do is find the common denominator for 2 and 3, namely 6. Hence we can write

$$2\ HNO_3 + 3\ H_2S \rightarrow 2\ NO + 3\ S + H_2O$$

The remainder of the equation can be balanced by inspection. Thus, there are eight hydrogen atoms on the left, requiring four water molecules on the right

$$2\ HNO_3 + 3\ H_2S \rightarrow 2\ NO + 3\ S + 4\ H_2O$$

Note that the oxygen atoms now balance also.

EXAMPLE 4.18

Balance the following equation:

$$I_2 + H_2O + ClO_3^- \rightarrow IO_3^- + Cl^- + H^+$$

Solution. First identify the oxidation numbers:

$$H_2O + \overset{0}{I_2} + \overset{5+}{Cl}O_3^- \rightarrow \overset{5+}{I}O_3^- + \overset{1-}{Cl^-} + H^+$$

The change for I_2 is +10 (there are 2 I atoms in I_2). The change for Cl is −6. The common denominator is 30.

$$H_2O + 3\ I_2 + 5\ ClO_3^- \rightarrow 6\ IO_3^- + 5\ Cl^- + H^+$$

The remainder is balanced by inspection. There are 18 oxygen atoms on the right, requiring 3 from the H_2O on the left, and 6 H^+ on the right:

$$3\ H_2O + 3\ I_2 + 5\ ClO_3^- \rightarrow 6\ IO_3^- + 5\ Cl^- + 6\ H^+$$

Note that the charge also balances.

4.3.2 Chemical Equilibrium

If a sample of the colorless gas, N_2O_4 is obtained and placed in a container at 25°C and 1 atm pressure, it will slowly turn orange because of the formation of the orange gas, NO_2. The balanced chemical reaction can be written as

$$N_2O_4 \rightleftharpoons 2\ NO_2$$

Note the arrows in both directions. This implies that the reaction can proceed in either direction. In fact, if a pure sample of the orange gas, NO_2, is placed in a container at 25°C and 1 atm, the color will become lighter as N_2O_4 is formed. Whether we start from N_2O_4 or NO_2, the reaction mixture will eventually stabilize at the same color, the same final concentration of N_2O_4, and the same final concentration of NO_2. The reaction is then said to be at *chemical equilibrium*. At chemical equilibrium the reaction mixture is not static on a microscopic scale, but there is exactly the same amount of N_2O_4 being formed per second from NO_2 as there is dissociating to give NO_2.

4.3.2.1 The Equilibrium Constant

The position of equilibrium for a given reaction can be defined by the equilibrium constant K, a constant for the reaction which is dependent only upon the temperature. The form of the equilibrium constant depends on the *stoichiometry* of the reaction as shown for each of the following reaction types:

Reaction Type	Form of K
$A \rightleftharpoons B$	$[B]/[A]$
$A + B \rightleftharpoons C$	$[C]/[A][B]$
$A + B \rightleftharpoons C + D$	$[C][D]/[A][B]$
$A \rightleftharpoons 2\ B$	$[B]^2/[A]$
$A + 2\ B \rightleftharpoons C + D$	$[C][D]/[A][B]^2$

The numerator in the expression for the equilibrium constant can be obtained by writing the product of the concentrations of each of the substances on the right-hand side of the equation (the products) raised to the power of the number of molecules of that substance in the equation. The denominator is obtained by the same operation for the substances on the left-hand side of the equation (the reactants).

EXAMPLE 4.19

Write the expression for the equilibrium constant for each of the following chemical reactions (each equation is balanced):

1. $N_2O_4\ (g) \rightleftharpoons 2\ NO_2\ (g)$
2. $H_2\ (g) + I_2\ (g) \rightleftharpoons 2\ HI\ (g)$
3. $2\ CO\ (g) + O_2\ (g) \rightleftharpoons 2\ CO_2\ (g)$
4. $N_2\ (g) + 3\ H_2\ (g) \rightleftharpoons 2\ NH_3\ (g)$

Solution.

1. $K = [NO_2]^2/[N_2O_4]$
2. $K = [HI]^2/[H_2][I_2]$
3. $K = [CO_2]^2/[CO]^2[O_2]$
4. $K = [NH_3]^2/[N_2][H_2]^3$

EXAMPLE 4.20

At 1 atm and 25°C, the concentrations of the components of an equilibrium mixture of NO_2 and N_2O_4 are $[N_2O_4] = 4.27 \times 10^{-2}$ mol/ℓ, $[NO_2] = 1.41 \times 10^{-2}$ mol/ℓ. What is the equilibrium constant for the reaction at 25°C (expressed as $N_2O_4 \rightleftharpoons 2\ NO_2$)?

Solution. The equilibrium constant is calculated as follows:

$$K_{25°} = [NO_2]^2/[N_2O_4]$$

$$= (1.41 \times 10^{-2})^2\ (\text{mol}/\ell)^2/4.27 \times 10^{-2}\ (\text{mol}/\ell)$$

$$= 4.66 \times 10^{-3}\ \text{mol}/\ell$$

The chemical reactions in Ex. 4.19 have all reactants and products in the gas phase. For reactions in which one or more of the components are pure solids or liquids, the derivation of the equilibrium constant expression is modified. Since the concentration of pure liquids and solids remains constant, it is not necessary to include them in the expression for the equilibrium constant; their values are absorbed in the constant *K*. For example, consider the following reaction:

$$CaCO_3 \rightleftharpoons CaO + CO_2$$

Since $CaCO_3$ and CaO are pure solids, their concentrations do not appear in the equilibrium constant expression, which is

$$K = [CO_2]$$

That is to say, the concentration of CO_2 over a mixture of $CaCO_3$ and CaO at a given temperature is always the same. It does not depend on the quantities of $CaCO_3$ and CaO present. Similarly, for the reaction of HNO_3 in dilute aqueous solution, the H_2O concentration does not appear in the expression for the equilibrium constant:

$$HNO_3 + H_2O \rightleftharpoons H_3O^+ + NO_3^-$$

$$K = [H_3O^+][NO_3^-]/[HNO_3]$$

Since water is the solvent, it is present in large excess and its essentially constant concentration is included in the value of *K*.

EXAMPLE 4.21

Write the expression for the equilibrium constant for the following reaction:

$$C(s) + CO_2\,(g) \rightleftharpoons CO(g)$$

Solution. First we must balance the equation:

$$C(s) + CO_2(g) \rightleftharpoons 2\,CO\,(g)$$

Then the equilibrium constant may be written as

$$K = [CO]^2/[CO_2]$$

Note that solid carbon does not appear in the expression.

Since the partial pressure of a gas is a measure of its concentration, equilibrium constants for reactions involving gases may be written in terms of partial pressures. An equilibrium constant expressed this way is designated K_p. For example, the equilibrium constant expressed in partial pressures for the reaction of Ex. 4.19-1 is

$$K_p = (p_{NO_2})^2/(p_{N_2O_4})$$

K_p and K are related by the following expression:

$$K_p = K(RT)^{\Delta n}$$

where Δn is the change in the number of moles of gases between reactants and products. For the reaction under consideration,

$$N_2O_4(g) \rightleftharpoons 2\,NO_2\,(g)$$

One mole of reactants gives two moles of products, $\Delta n = +1$. The equilibrium constant is then given by

$$K_p = K\,(RT)^{+1} = KRT$$

For a reaction with no change in the number of moles of gases, i.e., $\Delta n = 0$, $K_p = K$.

EXAMPLE 4.22

K_p is 167.5 atm at 1000°C for the following reaction:

$$C(s) + CO_2C(g) \rightleftharpoons 2\,CO\,(g)$$

What is the partial pressure of $CO(g)$ in equilibrium when the partial pressure of $CO_2\,(g)$ is 1.0 atm? 0.10 atm?

Solution. The equilibrium constant is

$$K_p = (p_{co})^2/(p_{co_2}) = 167.5 \text{ atm}$$

There then follows

$$(p_{co})^2/1 \text{ atm} = 167.5 \text{ atm}$$

When $p_{co_2} = 1.0$ atm, the partial pressure of carbon monoxide is then

$$p_{co} = 12.9 \text{ atm}$$

Likewise, for 0.1 atm

$$(p_{co})^2/0.1 \text{ atm} = 167.5 \text{ atm}$$

$$\therefore p_{co} = 4.10 \text{ atm}$$

4.3.2.2 Factors Affecting Chemical Equilibrium

The value of the equilibrium constant is characteristic of the particular reaction at a particular temperature. It depends on the relative stability of reactants and products. If the products are very much more stable than the reactants, then K will be much greater than 1 and the reaction will tend to proceed to complete formation of products, with the generation of a large amount of heat. Such a reaction is said to be exothermic as indicated by the following:

An exothermic reaction, $K >> 1$

$$A + B \rightleftharpoons C + D + \text{Heat}$$

If the reactants are very much more stable than the products, then K will be much less than 1 and the reaction will tend to favor reactants and will generally proceed with the absorption of heat. Such a reaction, said to be endothermic, is indicated by the following:

An endothermic reaction, $K << 1$

$$A + B \rightleftharpoons C + D - \text{Heat}$$

Although external conditions do not generally affect the equilibrium constant significantly, they can affect the relative concentrations of reactants and products dramatically. In Ex. 4.22, for instance, the ratio p_{co}/p_{co_2} is 12.9 when p_{co_2} is 1.0 atm, but changes to 4.1 when p_{co_2} is 0.10 atm. Note that reducing the pressure on the system increases the relative amount of CO. Concentration changes can have the same effect for certain reactions.

EXAMPLE 4.23

Calculate the ratio of the concentration of N_2O_4 to NO_2 at equilibrium at 25°C if the concentration of NO_2 is 1 mol/ℓ (see Ex. 4.20). Compare it to the ratio in Ex. 4.20.

Solution. The equilibrium constant is calculated to be

$$K = [NO_2]^2/[N_2O_4] = 4.66 \times 10^{-3} \text{ mol}/\ell$$

Hence, there results

$$\frac{1 \text{ mol}^2/\ell^2}{[N_2O_4]} = 4.66 \times 10^{-3} \text{ mol}/\ell$$

This leads to

$$[N_2O_4] = 2.15 \times 10^2 \text{ mol}/\ell$$

or

$$[N_2O_4]/[NO_2] = 2.15 \times 10^2$$

From Ex. 4.20

$$[N_2O_4]/[NO_2] = 4.27 \times 10^{-2}/1.41 \times 10^{-2}$$

$$= 3.03$$

Note that at the higher concentration of NO_2, the ratio $[N_2O_4]/[NO_2]$ is higher.

Observations of the effect of concentration, pressure, and temperature on the equilibrium composition of chemical reactions led to the formulation of *LeChatelier's Principle:* a system at equilibrium responds to stress to establish a new equilibrium composition that reduces the stress. For instance, in the CO-CO_2 equilibrium, an increase in pressure results in the formation of less CO which reduces the pressure (since there are 2 equivalents of CO formed from each equivalent of CO_2). In the $N_2O_4 - NO_2$ system, an increase in the concentration of NO_2 results in the formation of more N_2O_4 thus decreasing the concentration of NO_2. The extension of LeChatelier's Principle to cover temperature effects suggests that increasing the temperature of an exothermic reaction should cause the equilibrium to shift to the left (more reactants) so that less heat is evolved. Increasing the temperature of an endothermic reaction should cause the equilibrium to shift to the right (more products) so that more heat is absorbed. These predictions are generally true. However, note the important difference that the effect of temperature actually changes the value of the equilibrium constant, while the effect of pressure and concentration changes the composition without changing the value of K.

EXAMPLE 4.24

For each of the following reactions give the effect of the indicated stress on the stated quantity.

1. $2\ SO_2(g) + O_2(g) \rightleftharpoons 2\ SO_3\ (g)$ + Heat

   What is the effect on p_{SO_3}/p_{SO_2} of

   a. Increased total pressure?

   b. Decreased temperature?

2. $CO_2(g) + H_2(g) \rightleftharpoons CO(g) + H_2O(g)$ − Heat

   What is the effect on p_{CO}/p_{CO_2} of

   a. Increased total pressure?

   b. Decreased temperature?

3. $2\ Pb_3O_4\ (s) \rightleftharpoons 6\ PbO(s) + O_2\ (g)$ − Heat

   What is the effect on p_{O_2} of

   a. Increased T?

   b. Increased total pressure?

   c. Added Pb_3O_4?

   What is the effect on the ratio of the total quantity of O_2 to that of Pb_3O_4 of

   d. Increased T?

   e. Increased total pressure?

   f. Added Pb_3O_4?

Solutions.

1a. p_{SO_3}/p_{SO_2} will increase. Since there are three equivalents of gases on the left and two on the right, the system will respond to an increase in total pressure by shifting to the right, thereby reducing the number of molecules and the pressure.

1b. p_{SO_3}/p_{SO_2} will increase. The reaction is exothermic. Therefore, decreasing the temperature will cause a shift to the right to liberate more heat.

2a. p_{CO}/p_{CO_2} will not change. Since there is no difference in the number of gaseous molecules on the two sides of the equation, pressure changes will not change the composition.

2b. p_{CO}/p_{CO_2} will decrease. The reaction is endothermic. A decrease in temperature will result in a shift to the left to liberate more heat (i.e., the reverse reaction is exothermic).

3a. p_{O_2} will increase. The reaction is endothermic. An increase in T will cause a shift to the right with the absorption of more heat.

3b. p_{O_2} will decrease. Since O_2 is the only gas present, an increase in total pressure will reduce the concentration of O_2 in the gas phase in order to reduce the pressure.

3c. p_{O_2} will not change. The concentration of solids remains constant.

3d. O_2/Pb_3O_4 will increase. Although the *concentration* of Pb_3O_4 is constant, more of it will be converted to PbO and O_2 at higher temperatures for this endothermic reaction.

3e. O_2/Pb_3O_4 will decrease. At higher total pressure O_2 and PbO will combine to form more Pb_3O_4, thus reducing the pressure.

3f. O_2/Pb_3O_4 will decrease. The added Pb_3O_4 will not affect the total quantity of O_2 present.

4.3.3 Reaction Rate

The fact that a chemical reaction is exothermic and has a large, favorable equilibrium constant does not ensure that products will be formed. For instance, the reaction of the hydrocarbons (compounds containing only carbon and hydrogen, e.g., ethane, C_2H_6) in gasoline with the oxygen in air to give carbon dioxide and water is an exothermic reaction with a large, favorable equilibrium constant. Yet, gasoline can stand in air for years without transforming into carbon dioxide and water. However, let someone strike a match or spark, and the favorable equilibrium constant and exothermicity is immediately and dramatically revealed. The problem is, of course, in the reaction rate. Unless there is an available chemical path (a mechanism) to get from reactants to products, the rate of conversion of reactants to products will be too slow to be observed. The effect of variables on reaction rate can be understood in terms of reaction rate theories.

4.3.3.1 Collision Theory of Reaction Rate

Consider a reaction

$$A + B \rightarrow \text{products}$$

Since A and B must encounter each other in order to react, the rate will depend upon the number of collisions per second between A and B. This collision frequency or rate will depend upon the concentrations $[A]$ and $[B]$, that is,

$$\text{Rate} \propto [A][B]$$

The collision frequency will also depend upon the temperature, since increasing temperature increases the velocity of the molecules and therefore the collision frequency:

$$\text{Rate} \propto [A][B] \cdot f(T)$$

Not all collisions will be effective, however, since not all will be energetic enough to result in reaction. We must therefore introduce a factor, *A (the Ahrennius factor)*, for the fraction of effective collisions. This factor will, of course, also be temperature dependent, since an increase in temperature will increase the fraction of sufficiently energetic collisions. Our rate expression then becomes

$$\text{Rate} = A \cdot f(T) \cdot [A][B]$$

The first part of the expression, $A\, f(T)$, represents a temperature dependent rate constant characteristic of the reaction; the second part, $[A][B]$, represents the effect of the concentration variable. Thus, increasing the temperature of a reaction increases the rate by affecting the rate constant, while increasing the pressure of a gas phase reaction, or the concentration of reactants in a solution reaction, increases the rate by increasing the total collision frequency. For the combustion of gasoline, the match provides the high temperature necessary to obtain an initial high rate of reaction; the heat necessary to maintain a high rate of reaction is provided by the exothermicity of the reaction itself.

4.3.3.2 Transition State Theory of Reaction Rate

An alternative way of coming to the same qualitative conclusions about reaction rates is to regard the reactants as being in equilibrium with a transition state (*TS*). The transition state is the most energetic configuration through which the reactants must pass in order to be converted to products. It is represented by

$$A + B \rightleftharpoons TS \rightleftharpoons \text{products}$$

The rate of the reaction will depend on the concentration of the transition state, which can be expressed in terms of an equilibrium constant

$$K = [TS]/[A][B]$$

by the equation

$$\text{Rate} \propto [TS]$$

$$= K[A][B]$$

Since the transition state is much higher in heat content than the reactants, its formation is highly endothermic, and increasing the temperature will increase its concentration, expressed as

$$A + B \rightleftharpoons TS - \text{Heat}$$

The rate at which the transition state goes on to products is taken to be a constant. (It is also temperature dependent.) Therefore, the rate expression becomes

$$\text{Rate} = K \cdot f(T) \cdot K \cdot f(T) \cdot [A][B]$$

The form is very similar to the expression derived from collision theory and again implies an increase in rate with increasing temperature, pressure, and concentration.

Practice Problems

ATOMS

4.1 Which statement is incorrect?

a) Solutions may be homogeneous or heterogeneous.
b) Matter may be homogeneous or heterogeneous.
c) Both elements and compounds are composed of atoms.
d) All substances contain atoms.
e) Substances are always homogeneous.

4.2 Which of the following statements is not correct?

a) An element may be separated into atoms.
b) An element may be a gas, a liquid, or a solid.
c) A compound can be separated into its elements by chemical means.
d) An element is always heterogeneous.
e) A compound may be a gas, a liquid or a solid.

4.3 In relation to the proton, the electron is

a) about the same mass and of opposite charge.
b) about the same mass and of the same charge.
c) about the same mass and with no charge.
d) much lighter and of opposite charge.
e) much heavier and with no charge.

4.4 A negative ion of a certain element can be formed by

a) subtraction of a proton from an atom of that element.
b) subtraction of an electron from an atom of that element.
c) subtraction of a neutron from an atom of that element.
d) addition of an electron to an atom of that element.
e) addition of a neutron to an atom of that element.

4.5 Metallic conduction involves

a) migration of cations toward a positively charged electrode.
b) migration of cations toward a negatively charged electrode.
c) migration of anions toward a positively charged electrode.
d) passage of electrons from one atom of a metal to another.
e) migration of anions toward a negatively charged electrode.

4.6 Which of the following statements is true?

a) Within a group of elements in the periodic table, the largest atom has the highest ionization potential.
b) Within a period of elements in the periodic table, the noble gas has the highest ionization potential.
c) When all valence p orbitals of an atom are half filled, the ionization potential of that atom is lower than the ionization potential of an atom with only two electrons in the valence p orbitals.
d) It is easier to form a 2+ *ion* than a 1+ ion.
e) Ionization potential is the same as electronegativity.

4.7 Which one of the following elements has the largest atomic radius?

a) lithium b) sodium c) beryllium d) magnesium e) phosphorus

4.8 In the series of elements *B*, *Al*, *Ga*, *In*,

a) metallic character increases from *B* to *In*.
b) electronegativity increases from *B* to *In*.
c) ionization energy increases from *B* to *In*.
d) nonmetallic character increases from *B* to *In*.
e) none of the above trends is correct.

4.9 Which of the following lists contains only nonmetals?

a) beryllium (*Be*), hydrogen (*H*), osmium (*Os*)
b) germanium (*Ge*), palladium (*Pd*), silicon (*Si*)
c) carbon (*C*), sulfur (*S*), fluorine (*F*)
d) calcium (*Ca*), chlorine (*Cl*), boron (*B*)
e) zinc (*Zn*), gallium (*Ga*), germanium (*Ge*)

4.10 In an element

a) the atomic number is equal to the number of neutrons in the atom.
b) the number of protons always equals the number of neutrons in the atom.
c) the mass number is equal to the number of electrons in the atom.
d) the atomic number is equal to the number of protons in the atom.
e) the number of electrons can never equal the number of neutrons in the atom.

4.11 What is the ground state electron configuration of aluminum (*Al*, $Z = 13$)?

a) $1s^2\ 2s^2\ 2p^5\ 3s^2\ 3p^1$
b) $1s^2\ 2s^2\ 2p^6\ 3s^2\ 3p^1$
c) $1s^2\ 2s^2\ 2p^6\ 3s^2\ 4s^1$
d) $1s^2\ 2s^2\ 2p^6\ 3s^2\ 3p^2$
e) $1s^2\ 2s^2\ 2p^6\ 3s^2\ 3d^1$

4.12 Which of the following electron configurations is *inconsistent* with Hund's rule (the principle of maximum multiplicity)?

a) $[Kr]\ 5s^2\ 4d^{10}\ 5p_x^2\ 5p_y^1\ 5p_z^0$
b) $[Kr]\ 5s^2\ 4d^{10}\ 5p_x^1\ 5p_y^0\ 5p_z^0$
c) $[Kr]\ 5s^2\ 4d^{10}\ 5p_x^1\ 5p_y^1\ 5p_z^1$
d) $[Kr]\ 5s^2\ 4d^{10}\ 5p_x^2\ 5p_y^1\ 5p_z^1$
e) $[Kr]\ 5s^2\ 4d^{10}\ 5p_x^1\ 5p_y^0\ 5p_z^1$

4.13 Which of these electron configurations is found in periodic Group VI?

a) $\cdots ns^2\ np^6$ b) $\cdots np^6$ c) $\cdots ns^6$ d) $\cdots ns^2\ np^4$ e) $\cdots ns^5\ np^{-1}$

4.14 From a consideration of electron configurations, which of the following elements would you expect to be most similar in chemical properties to strontium (*Sr*; $Z = 38$)?

a) *Rb* b) *Y* c) *Sc* d) *Ba* e) *Ti*

4.15 The principal quantum number designates the

a) shape of an orbital.
b) main energy level in which an electron is found.
c) sublevel of energy in which an electron is found.
d) number of electrons allowed in a main energy level.
e) orientation of the orbital in space.

4.16 In any atom what is the total number of electrons which can have a principal quantum number of 5 and a secondary quantum number (ℓ) of zero?

a) 2 b) 4 c) 5 d) 6 e) 18

4.17 For a neutral atom of an element in its ground state, 35 electrons occupy the energy levels up to and including the $n = 4$ energy level. If all electrons in the valence (outermost) p-orbitals are removed by ionization, how many electrons remain in the resulting ion?

a) 18 b) 28 c) 20 d) 30 e) 35

4.18 How many electrons does a phosphorus atom have in its set of valence shell *p* orbitals?

a) 0 b) 1 c) 2 d) 3 e) 10

4.19 An atom of an unknown element Q has a mass number of 31 and the nucleus contains 15 protons. The element is

a) gallium *Ga*
b) sulfur *S*
c) phosphorus *P*
d) palladium *Pd*
e) Scandium *Sc*

4.20 An ion of an unknown element has an atomic number of 15 and contains 18 electrons. The ion is

a) P^{3-} b) *Ar* c) O^{2-} d) Si^{3-} e) S^{+}

MOLECULES

4.21 Consider the following statements about ionic and covalent bonds. Which statement is true?

a) In a covalent molecule, each atom is bonded to only two other atoms.
b) An ionic bond is an electrostatic interaction localized between two definite ions of identical electrical charge.
c) A covalent bond occurs when electrons are completely transferred from one atom to another.
d) When a covalent bond forms between two atoms with different electronegativities, the bond is always polar.
e) A compound never contains both ionic and covalent bonds.

4.22 Which one of the following compounds is classified as an alkane?

a) ethylene b) benzene c) propane d) acetylene e) ethanol

4.23 Which one of the following bonds is most covalent?

a) *MgCl* b) *AlP* c) *NaCl* d) *MgS* e) *NaP*

4.24 The sum of the oxidation states of all the atoms in a neutral molecule

a) must be a small positive number.
b) must be a small negative number.
c) must be zero.
d) can be either positive or negative, but not zero.
e) can have any value, including zero.

4.25 The oxidation state of an element bonded only to itself

a) must be a small positive number.
b) must be a small negative number.
c) can be either positive or negative, but zero.
d) can have any value, including zero.
e) must be zero.

4.26 The oxidation state of sulfur (*S*) in the ion SO_3^{2-} is

a) 1+ b) 2+ c) 3+ d) 4+ e) 6−

4.27 A mole

a) is a unit of measurement applicable only to molecules.
b) equals the number of atoms in one gram of carbon-12.
c) equals the number of molecules in 20 liters of air.
d) is Avogadro's number of anything.
e) equals the number of atoms in 22.4 liters of a diatomic gas.

4.28 Which statement is incorrect?

a) Avogadro's number equals the number of molecules in one mole of nitrogen molecules.
b) Avogadro's number equals the number of atoms in one mole of nitrogen atoms.
c) Avogadro's number equals the number of atoms in one mole of nitrogen molecules.
d) Avogadro's number equals 6.02×10^{23}.
e) Avogadro's number equals the number of one faraday of electricity (one faraday equals 96 500 coulombs — the charge carried by one mole of electrons).

4.29 An empty aluminum Coke can weighs 50 grams. How many moles of aluminum does one Coke can contain? (Atomic weight of $Al = 27$)

a) 1350 b) 1.85 c) 1.0×10^{25} d) 3.0×10^{25} e) 27

4.30 A 27 gram sample of oxygen difluoride, OF_2, contains how many molecules? (Atomic weights: $O = 16$, $F = 19$; Avogadro's number: 6.0×10^{23})

a) 3.0×10^{23}
b) 2 times 6.0×10^{23}
c) 6.0×10^{23} divided by 4
d) 3.0×10^{23} times 54
e) 12.0×10^{23}

4.31 How many grams are there in 0.01 mole of Na_2SO_4?

a) 7.1 g b) 14.2 g c) 9.6 g d) 1.42 g e) 0.71 g

4.32 What is the volume at standard temperature and pressure of 16 grams of gaseous sulfur dioxide, SO_2?

a) 22.4 liters b) 11.2 liters c) 5.6 liters d) 16.8 liters e) 64 liters

4.33 What is the percentage by weight of aluminum, Al, in alumina, Al_2O_3? (Atomic weights: $Al = 27$, $O = 16$)

a) 63 b) 37 c) 23 d) 53 e) 64

4.34 A certain compound consists only of sulfur (S) and chlorine (Cl). It contains 47.5 percent by weight of sulfur and has a molecular weight of 135. What is its molecular formula? (Atomic weights: $S = 32$, $Cl = 35.5$)

a) SCl_2 b) SCl c) S_2Cl_2 d) S_2Cl e) S_3Cl

4.35 An unknown organic compound was analyzed and found to contain 34.6 percent carbon, 3.8 percent hydrogen, and 61.5 percent oxygen. Which one of the following compounds could the unknown be?

a) methanol CH_3OH
b) oxalic acid CO_2H-CO_2H
c) acetic acid CH_3-CO_2H
d) malonic acid $CO_2H-CH_2-CO_2H$
e) propionic acid $CH_3-CH_2-CO_2H$

REACTIONS

4.36 What is the expression for the equilibrium constant for the following system?

$$2NOCl(g) \rightleftharpoons 2NO(g) + Cl_2(g)$$

a) $K = [NO]^2[Cl_2]^2/[NOCl]^2$
b) $K = 2[NO][Cl_2]/2[NOCl]$
c) $K = [NO]^2[Cl_2]/[NOCl]^2$
d) $K = [NO]^2[Cl_2]^2/[NOCl]^2$
e) $K = [NOCl]^2/[NO]^2[Cl_2]$

4.37 For the reaction of solid BaO with carbon dioxide according to the equation $BaO(s) + CO_2(g) \rightleftharpoons BaCO_3(s)$, the equilibrium expression may be represented as

a) $[BaCO_3]/[BaO]$
b) $1/[CO_2]$
c) $[BaO][CO_2]/[BaCO_3]$
d) $[CO_2]$
e) $[BaCO_3]/[BaO][CO_2]$

4.38 Assume excess oxygen reacts with methane to form 14 grams of carbon monoxide according to the equation $2CH_4 + 3O_2 \rightleftharpoons 2CO + 4H_2O$. How many moles of methane will be consumed?

a) 2.0 moles methane
b) one-third mole methane
c) 0.25 mole methane
d) 0.5 mole methane
e) 4.0 moles of methane

4.39 What coefficient is required for NO_2 in order to balance the equation?

$$2Pb(NO_3)_2 \rightarrow 2PbO + NO_2 + O_2$$

a) 0.5 b) 1 c) 1.5 d) 4 e) 10

4.40 What coefficient is required for H_2O in order to balance the equation?

$$Be_3N_2 + H_2O \rightarrow 3Be(OH)_2 + 2NH_3$$

a) 1 b) 2 c) 3 d) 6 e) 18

4.41 In the following reaction determine the change, if any, that occurs in the oxidation number of the underlined element, and whether the element is oxidized, reduced, or unchanged:

$$3Mg + \underline{N}_2 \rightarrow Mg_3\underline{N}_2$$

a) from 0 to +3; oxidized
b) from 0 to −3; oxidized
c) from +3 to +5; oxidized
d) from 0 to −3; reduced
e) from +5 to +3; reduced

4.42 According to the equation $2Al + 6HCl \rightarrow 2AlCl_3 + 3H_2$,

a) production of 1 mole of H_2 requires 3 moles of HCl.
b) production of 1 mole of $AlCl_3$ requires 3 moles of HCl.
c) production of 2 moles of H_2 requires 2 moles of HCl.
d) production of 2 moles of H_2 requires 5 moles of HCl.
e) production of 4 moles of H_2 requires 2 moles of HCl.

4.43 When crystals of sodium sulfate are dissolved in water, the resulting solution feels warmer. The solubility of Na_2SO_4 could be increased by

a) increasing the temperature.
b) increasing the pressure.
c) decreasing the temperature.
d) adding more solute to the solution.
e) stirring the solution.

4.44 Which of the following statements is false?

a) An exothermic reaction always goes faster than an endothermic reaction.
b) A catalyst provides a different route by which the reaction can occur.
c) Some reactions may never reach completion (100% products).
d) The rate of a reaction depends upon the height of the energy barrier (energy of activation).
e) The activation energy is independent of the energy of reaction.

4.45 In which one of the following reactions would an increase in the volume of the container cause an increase in the amount of products at equilibrium? (All substances are gases unless marked otherwise.)

a) $2NO + 5H_2 \rightleftharpoons 2NH_3 + 2H_2O$
b) $CH_3CHO + \text{heat} \rightleftharpoons CH_4 + CO$
c) $SO_2 \rightleftharpoons S(s) + O_2$
d) $SO_3 + HF \rightleftharpoons HSO_3(\ell)$
e) $C + H_2O \rightleftharpoons CO + H_2$

(This page is intentionally blank.)

5. Engineering Economy

Engineering designs are intended to produce good results. In general, the good results are accompanied by undesirable effects including the costs of manufacturing or construction. Selecting the best design from a set of technologically feasible alternatives, or deciding whether or not to implement a proposed design, requires the engineer to anticipate and compare the good and bad outcomes. If outcomes are evaluated in dollars and if "good" is defined as positive monetary value, then design decisions may be guided by the techniques known as engineering economy. Decisions based solely on engineering economy may be guaranteed to result in maximum goodness only if all outcomes are anticipated and can be monetized (measured in dollars).

5.1 Value and Interest

"Value" is not synonymous with "amount." The value of an amount of money depends on when the amount is received or spent. For example, the promise that you will be given a dollar one year from now is of less value to you than a dollar received today. The difference between the anticipated amount and its current value is called "interest" and is frequently expressed as a time rate. If an interest rate of 10% per year is used, the expectation of receiving $1.00 one year hence has a value now of about $0.91. In engineering economy, interest usually is stated in percent per year. If no time unit is given, "per year" is assumed.

EXAMPLE 5.1

What amount must be paid in two years to settle a current debt of $1,000 if the interest rate is 6%?

Solution.

$$\begin{aligned}\text{Value after one year} &= 1000 + 1000 \times 0.06 \\ &= 1000\,(1 + 0.06) \\ &= \$1060\end{aligned}$$

$$\begin{aligned}\text{Value after two years} &= 1060 + 1060 \times 0.06 \\ &= 1000\,(1 + 0.06)^2 \\ &= \$1124\end{aligned}$$

Hence, $1124 must be paid in two years to settle the debt.

5.2 Cash Flow Diagrams

As an aid to analysis and communication, an engineering economy problem may be represented graphically by a horizontal time axis and vertical vectors representing dollar amounts. The cash flow diagram for Ex. 5.1 is sketched in Fig. 5.1 on the next page. Income is up and expenditures are down. It is important to pick a point of view and stick with it. For example, the vectors in Fig. 5.1 would have been reversed if the point of view of the lender had been adopted.

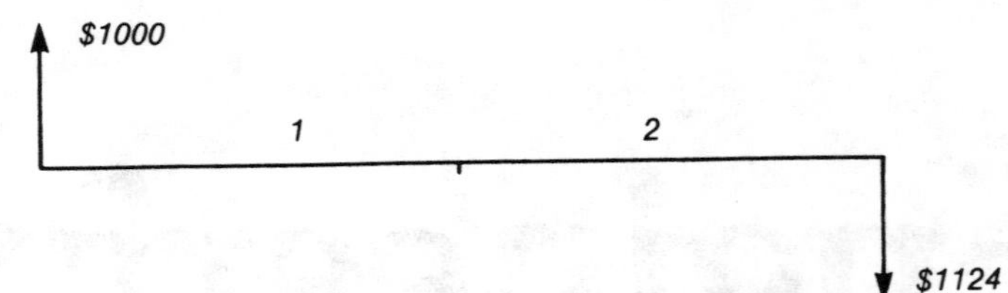

Figure 5.1. Cash flow diagram for Example 5.1.

It is a good idea to draw a cash flow diagram for every engineering economy problem that involves amounts occurring at different times.

In engineering economy, amounts are almost always assumed to occur at the ends of years. Consider, for example, the value today of the future operating expenses of a truck. The costs probably will be paid in varied amounts scattered throughout each year of operation, but for computational ease the expenses in each year are represented by their sum (computed without consideration of interest) occurring at the end of the year. The error introduced by neglecting interest for partial years usually is insignificant compared to uncertainties in the estimates of future amounts.

5.3 Cash Flow Patterns

Engineering economy problems involve the following four patterns of cash flow both separately and in combination.

P-pattern: A single amount P occurring at the beginning of n years. P frequently represents "present" amounts.

F-pattern: A single amount F occurring at the end of n years. F frequently represents "future" amounts.

A-pattern: Equal amounts A occurring at the ends of n years. The A-pattern frequently is used to represent "annual" amounts.

G-pattern: End-of-year amounts increasing by an equal annual gradient G. Note that the first amount occurs at the end of the second year. G is the abbreviation of "gradient."

The four cash flow patterns are illustrated in Fig. 5.2.

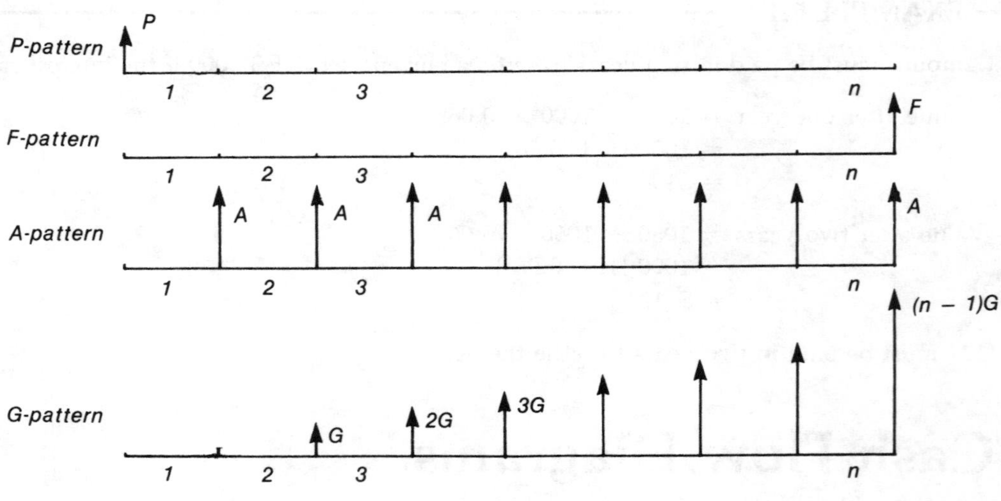

Figure 5.2. Four cash flow patterns.

5.4 Equivalence of Cash Flow Patterns

Two cash flow patterns are said to be equivalent if they have the same value. Most of the computational effort in engineering economy problems is directed at finding a cash flow pattern that is equivalent to a combination of other patterns. Ex. 5.1 can be thought of as finding the amount in an *F*-pattern that is equivalent to $1,000 in a *P*-pattern. The two amounts are proportional, and the factor of proportionality is a function of interest rate *i* and number of periods *n*. There is a different factor of proportionality for each possible pair of the cash flow patterns defined in Section 5.3. To minimize the possibility of selecting the wrong factor, mnemonic symbols are assigned to the factors. For Ex. 5.1, the proportionality factor is written $(F/P)_n^i$ and solution is achieved by evaluating

$$F = (F/P)_n^i \, P$$

To analysts familiar with the cancelling operation of algebra, it is apparent that the correct factor has been chosen. However, the letters in the parentheses together with the sub- and super-scripts constitute a single symbol; therefore, the cancelling operation is not actually performed. Table 5.1 lists symbols and formulas for commonly used factors. Table 5.2, located at the end of this chapter, presents a convenient way to find numerical values of interest factors. Those values are tabulated for selected interest rates *i* and number of interest periods *n*; linear interpolation for intermediate values of *i* and *n* is acceptable for most situations.

TABLE 5.1 Formulas for Interest Factors

Symbol	To Find	Given	Formula
$(F/P)_n^i$	*F*	*P*	$(1+i)^n$
$(P/F)_n^i$	*P*	*F*	$\dfrac{1}{(1+i)^n}$
$(A/P)_n^i$	*A*	*P*	$\dfrac{i(1+i)^n}{(1+i)^n-1}$
$(P/A)_n^i$	*P*	*A*	$\dfrac{(1+i)^n-1}{i\,(1+i)^n}$
$(A/F)_n^i$	*A*	*F*	$\dfrac{i}{(1+i)^n-1}$
$(F/A)_n^i$	*F*	*A*	$\dfrac{(1+i)^n-1}{i}$
$(A/G)_n^i$	*A*	*G*	$\dfrac{1}{i}-\dfrac{n}{(1+i)^n-1}$
$(F/G)_n^i$	*F*	*G*	$\dfrac{1}{i}\left[\dfrac{(1+i)^n-1}{i}-n\right]$
$(P/G)_n^i$	*P*	*G*	$\dfrac{1}{i}\left[\dfrac{(1+i)^n-1}{i(1+i)^n}-\dfrac{n}{(1+i)^n}\right]$

EXAMPLE 5.2

Derive the formula for $(F/P)^i_n$.

Solution. For $n = 1$,

$$F = (1 + i)\,P$$

that is,

$$(F/P)^i_1 = (1 + i)^1$$

For any n,

$$F = (1 + i)\,(F/P)^i_{n-1}\,P$$

that is,

$$(F/P)^i_n = (1 + i)\,(F/P)^i_{n-1}$$

By induction,

$$(F/P)^i_n = (1 + i)^n$$

EXAMPLE 5.3

A new widget twister, with a life of six years, would save $2,000 in production costs each year. Using a 12% interest rate, determine the highest price that could be justified for the machine. Although the savings occur continuously throughout each year, follow the usual practice of lumping all amounts at the ends of years.

Solution. First, sketch the cash flow diagram.

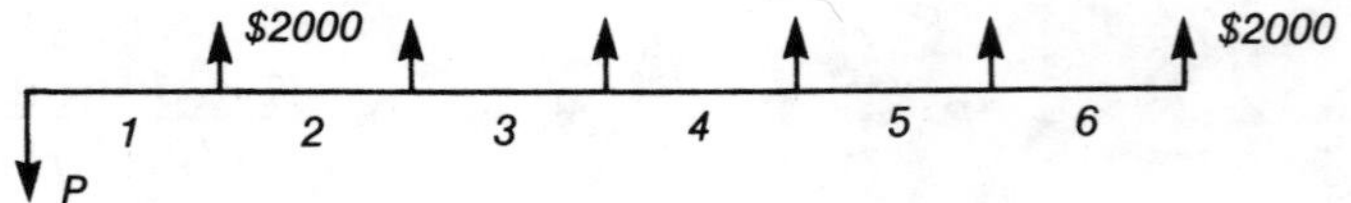

The cash flow diagram indicates that an amount in a P-pattern must be found that is equivalent to $2,000 in an A-pattern. The corresponding equation is

$$P = (P/A)^i_n\,A$$

$$= (P/A)^{12\%}_6\,2000$$

Table 5.2 is used to evaluate the interest factor for $i = 12\%$ and $n = 6$:

$$P = 4.1114 \times 2000$$

$$= \$8223$$

EXAMPLE 5.4

How soon does money double if it is invested at 8% interest?

Solution. Obviously, this is stated as

$$F = 2\,P$$

Therefore,

$$(F/P)_n^{8\%} = 2$$

In the 8% interest table, the tabulated value for (F/P) that is closest to 2 corresponds to $n = 9$ years.

EXAMPLE 5.5

Find the value in 1987 of a bond described as "Acme 8% of 2000" if the rate of return set by the market for similar bonds is 10%.

Solution. The bond description means that the Acme Company has an outstanding debt that it will repay in the year 2000. Until then, the company will pay out interest on that debt at the 8% rate. Unless otherwise stated, the principal amount of a single bond is \$1000. If it is assumed that the debt is due December 31, 2000, interest is paid every December 31, and the bond is purchased January 1, 1987, then the cash flow diagram, with unknown purchase price P, is:

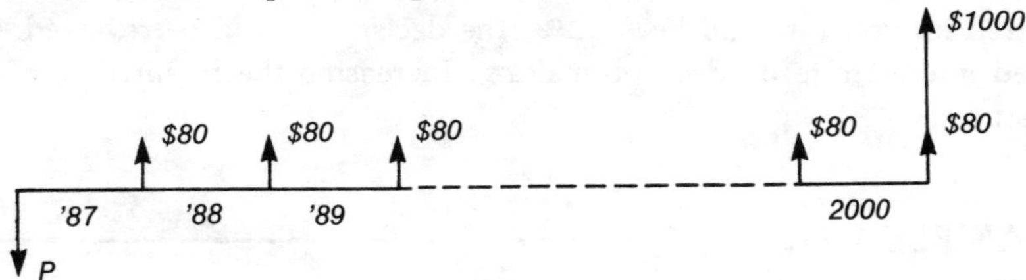

The corresponding equation is

$$P = (P/A)_{14}^{10\%}\ 80 + (P/F)_{14}^{10\%}\ 1000$$

$$= 7.3667 \times 80 + 0.2633 \times 1000$$

$$= \$853$$

That is, to earn 10% the investor must buy the 8% bond for \$853, a "discount" of \$147. Conversely, if the market interest rate is less than the nominal rate of the bond, the buyer will pay a "premium" over \$1000.

The solution is approximate because bonds usually pay interest semiannually, and \$80 at the end of the year is not equivalent to \$40 at the end of each half year. But the error is small and is neglected.

EXAMPLE 5.6

You are buying a new television. From past experience you estimate future repair costs as:

First Year	\$ 5
Second Year	15
Third Year	25
Fourth Year	35

The dealer offers to sell you a four-year repair contract for \$60. You require at least a 6% interest rate on your investments. Should you invest in the repair contract?

Solution. Sketch the cash flow diagram.

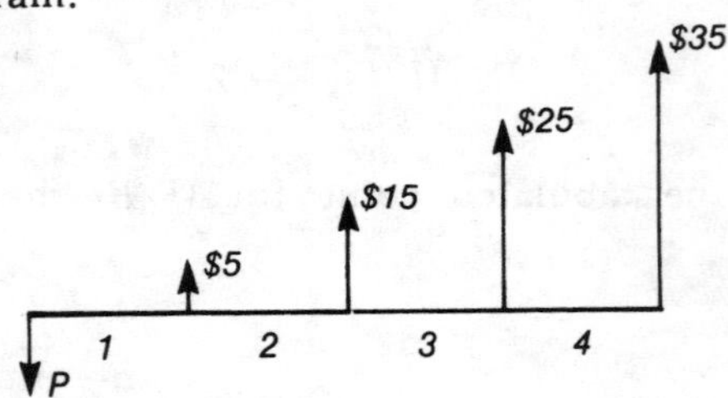

The known cash flows can be represented by superposition of a \$5 *A*-pattern and a \$10 *G*-pattern. Verify that statement by drawing the two patterns. Now it is clear why the standard *G*-pattern is defined to have the first cash flow at the end of the second year. Next, the equivalent amount *P* is computed:

$$P = (P/A)_4^{6\%}\ A + (P/G)_4^{6\%}\ G$$
$$= 3.4651 \times 5 + 4.9455 \times 10$$
$$= \$67$$

Since the contract can be purchased for less than \$67, the investment will earn a rate of return greater than the required 6%. Therefore, you should purchase the contract.

If the required interest rate had been 12%, the decision would be reversed. This demonstrates the effect of required interest rate on decision-making. Increasing the required rate reduces the number of acceptable investments.

EXAMPLE 5.7

Compute the annual equivalent repair costs over a 5-year life if a typewriter is warranted for two years and has estimated repair costs of \$100 annually. Use $i = 10\%$.

Solution. The cash flow diagram appears as:

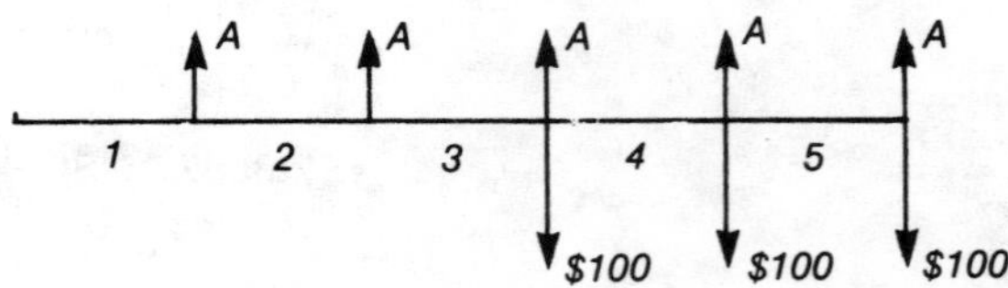

There are several ways to find the 5-year *A*-pattern equivalent to the given cash flow. One of the more efficient methods is to convert the given 3-year *A*-pattern to an *F*-pattern, and then find the 5-year *A*-pattern that is equivalent to that *F*-pattern. That is,

$$A = (A/F)_5^{10\%}\ (F/A)_3^{10\%}\ 100$$
$$= \$54$$

5.5 Unusual Cash Flows and Interest Periods

Occasionally an engineering economy problem will deviate from the year-end cash flow and annual compounding norm. The examples in this section demonstrate how to handle these situations.

EXAMPLE 5.8

PAYMENTS AT BEGINNINGS OF YEARS

Using 10% interest rate, find the future equivalent of:

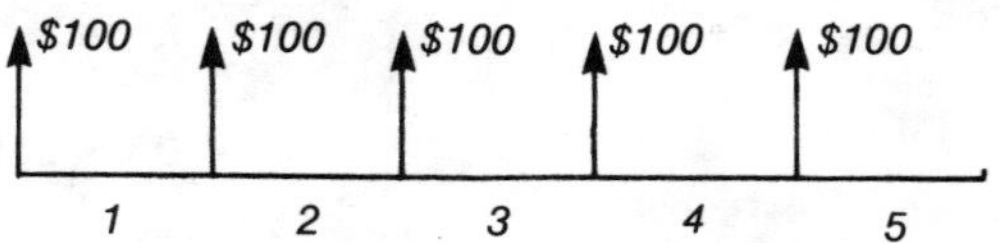

Solution. Shift each payment forward one year. That is,

$$A = (F/P)_{1}^{10\%}\ 100 = \$110$$

This converts the series to the equivalent A-pattern:

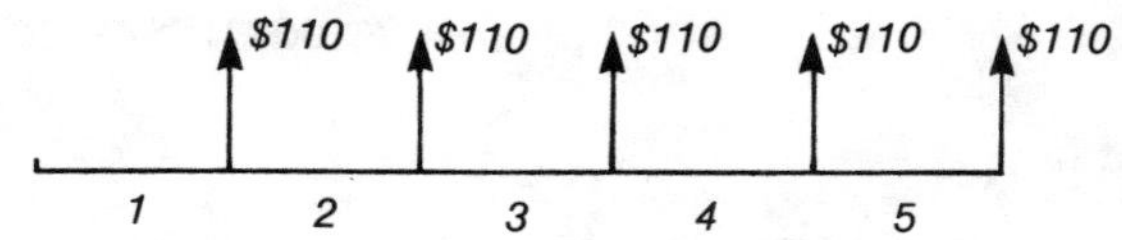

and the future equivalent is found to be

$$F = (F/A)_{5}^{10\%}\ 110 = \$672$$

Alternative Solution. Convert to a six-year series:

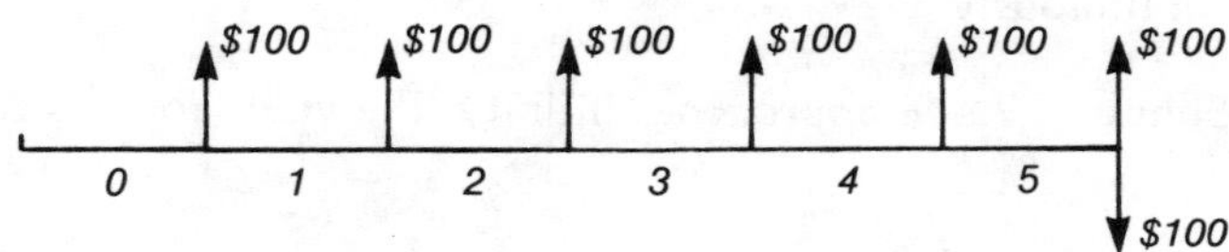

The future equivalent is

$$F = (F/A)_{6}^{10\%}\ 100 - 100 = \$672$$

EXAMPLE 5.9

SEVERAL INTEREST AND PAYMENT PERIODS PER YEAR

Compute the present value of eighteen monthly payments of $100 each, where interest is 1% per month.

Solution. The present value is computed as

$$P = (P/A)_{18}^{1\%}\ 100 = \$1640$$

EXAMPLE 5.10

ANNUAL PAYMENTS BUT INTEREST COMPOUNDED *m* TIMES PER YEAR

Compute the effective annual interest rate equivalent to 5% nominal annual interest compounded daily. There are 250 banking days in a year.

Solution. The legal definition of nominal annual interest is

$$i_n = m\,i$$

where i is the interest rate per compounding period. For the example,

$$i = i_n/m$$

$$= 0.05/250 = 0.0002 \text{ or } 0.02\% \text{ per day}$$

Because of compounding, the effective annual rate is greater than the nominal rate. By equating (F/P)-factors for one year and m periods, the effective annual rate i_e may be computed as follows:

$$(1 + i_e)^1 = (1 + i)^m$$

$$i_e = (1 + i)^m - 1$$

$$= (1.0002)^{250} - 1 = 0.051266 \text{ or } 5.1266\%$$

EXAMPLE 5.11

CONTINUOUS COMPOUNDING

Compute the effective annual interest rate i_e equivalent to 5% nominal annual interest compounded continuously.

Solution. As m approaches infinity, the value for i_e is found as follows:

$$i_e = e^{mi} - 1$$

$$= e^{0.05} - 1$$

$$= 0.051271 \text{ or } 5.1271\%$$

EXAMPLE 5.12

ANNUAL COMPOUNDING BUT *m* PAYMENTS PER YEAR

Compute the year-end amount equivalent to twelve end-of-month payments of $10 each. Annual interest rate is 6%.

Solution. The usual simplification in engineering economy is to assume that all payments occur at the end of the year, giving an answer of $120. This approximation may not be acceptable for a precise analysis of a financial agreement. In such cases, the agreement's policy on interest for partial periods must be investigated.

EXAMPLE 5.13

ANNUAL COMPOUNDING BUT PAYMENT EVERY m YEARS

With interest at 10% compute the present equivalent of

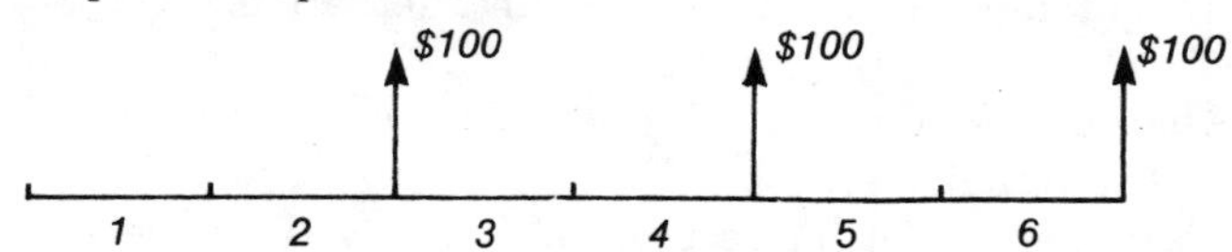

Solution. First convert each payment to an A-pattern for the m preceding years. That is,

$$A = (A/F)_{2}^{10\%} 100$$

$$= \$47.62$$

Then, convert the A-pattern to a P-pattern:

$$P = (P/A)_{6}^{10\%} 47.62$$

$$= \$207$$

5.6 Evaluating Alternatives

The techniques of engineering economy assume the objective of maximizing net value. For a business, "value" means after-tax cash flow. For a not-for-profit organization, such as a government agency, value may include non-cash benefits, such as, clean air, improved public health, recreation, to which dollar amounts have been assigned.

Sections 5.7 through 5.17 concern strategies for selecting alternatives such that net value is maximized. The logic of these methods will be clear if the following distinctions are made between two different types of interest rates, and between two different types of relationships among alternatives.

TYPES OF INTEREST RATES

Rate of Return (ROR): The estimated interest rate produced by an investment. It may be computed by finding the interest rate such that the estimated income and non-cash benefits (positive value), and the estimated expenditures and non-cash costs (negative value) sum to a net equivalent value of zero.

Minimum Attractive Rate of Return (MARR): The lowest rate of return that the organization will accept. In engineering economy problems, it is usually a given quantity and may be called, somewhat imprecisely, "interest," "interest rate," "cost of money," or "interest on capital."

TYPES OF ALTERNATIVE SETS

Mutually Exclusive Alternatives: Exactly one alternative must be selected.
Examples: "Shall Main Street be paved with concrete or asphalt?" "In which room will we put the piano?" If a set of alternatives is mutually exclusive, it is important to determine whether the set includes the null (do nothing) alternative. Serious consequences can arise from failure to recognize the null alternative.

Independent Alternatives: It is possible (but not necessarily economical) to select any number of the available alternatives.
Examples: "Which streets should be paved this year?" "Which rooms shall we carpet?"

5.7 Annual Equivalent Cost Comparisons

The estimated income and benefits (positive) and expenditures and costs (negative) associated with an alternative are converted to the equivalent *A*-pattern using an interest rate equal to *MARR*. The *A*-value is the annual net equivalent value (*ANEV*) of that alternative. If the alternatives are mutually exclusive, the one with the largest *ANEV* is selected. If the alternatives are independent, all that have positive *ANEV* are selected.

EXAMPLE 5.14

A new cap press is needed. Select the better of the two available models described below. *MARR* is 10%.

Model	Price	Annual Maintenance	Salvage Value	Life
Reliable	11,000	1,000	1,000	10 yrs.
Quicky	4,000	1,500	0	5 yrs.

Solution. The *ANEV* is calculated for each model:

$$\begin{aligned} \textit{Reliable:}\ ANEV &= -\,(A/P)_{10}^{10\%}\,11000 - 1000 + (A/F)_{10}^{10\%}\,1000 \\ &= -\,\$2730. \end{aligned}$$

$$\begin{aligned} \textit{Quicky:}\ ANEV &= -\,(A/P)_{5}^{10\%}\,4000 - 1500 \\ &= -\,\$2560. \end{aligned}$$

Negative *ANEV* indicates a rate of return less than *MARR*. However, these alternatives are mutually exclusive and the null is not available. The problem is one of finding the less costly way to perform a necessary function. Therefore, *Quicky* is selected. If *MARR* had been much lower, *Reliable* would have been selected. By setting the *MARR* relatively high, the organization is indicating that funds are not available to invest now in order to achieve savings in the future.

5.8 Present Equivalent Cost Comparisons

The estimated income and benefits (positive), and expenditures and costs (negative) associated with an alternative are converted to the equivalent *P*-pattern using an interest rate equal to *MARR*. The *P*-value is the present net equivalent value (*PNEV*) of that alternative. If the alternatives are mutually exclusive, the one with the largest *PNEV* is selected. If the alternatives are independent, all that have positive *PNEV* are selected.

The present equivalent cost method requires that alternatives be evaluated over the same span of time. If their lives are not equal, the lowest common multiple of the lives is used for the time span, with each alternative repeated to fill the span. A variation, called the "capitalized cost method," computes the *PNEV* for repeated replacement of the alternatives for an infinite time span.

PNEV is also called "life cycle cost," "present worth," "capital cost," and "venture worth."

EXAMPLE 5.15

Repeat Ex. 5.14 using the present equivalent cost method.

Solution. The *PNEV* is calculated for each model

$$\textit{Reliable}: \; PNEV = -11000 - (P/A)_{10}^{10\%}\,1000 + (P/F)_{10}^{10\%}\,1000$$

$$= -\$16{,}800$$

$$\textit{Quicky}: \; PNEV = -4000 - (P/F)_{5}^{10\%}\,4000 - (P/A)_{10}^{10\%}\,1500$$

$$= -\$15{,}700$$

Note that *Quicky* was replaced in order to fill the ten-year time span. As in Ex. 5.14, *Quicky* is selected. The two methods always will give the same decision if used correctly. Observe that for both alternatives $PNEV = (P/A)_{10}^{10\%}\, ANEV$.

5.9 Incremental Approach

For a set of mutually exclusive alternatives, only the differences in amounts need to be considered. Compute either the *ANEV* or the *PNEV* and base the decision on the sign of that value.

EXAMPLE 5.16

Repeat Ex. 5.14 using an incremental present net equivalent value approach.

Solution. *Reliable* costs $7000 more than *Quicky* but saves $500 each year in maintenance expenses and eliminates the need for a $4000 replacement after five years. In addition, *Reliable* has a $1000 salvage value whereas *Quicky* has none.

Reliable − *Quicky*:

$$PNEV = -7000 + (P/A)_{10}^{10\%}\,500 + (P/F)_{5}^{10\%}\,4000 + (P/F)_{10}^{10\%}\,1000$$

$$= -\$1060$$

The negative result dictates selection of *Quicky*. That is, the additional initial cost required to purchase *Reliable* is not justified.

5.10 Rate of Return Comparisons

The expression for *ANEV* or *PNEV* is formulated and then solved for the interest rate that will give a zero *ANEV* or *PNEV*. This interest rate is the rate of return (*ROR*) of the alternative. To apply the rate of return method to mutually exclusive alternatives requires incremental comparison of each possible pair of alternatives; increments of investment are accepted if their rates of return exceed *MARR*. For independent alternatives, all those with *ROR* exceeding *MARR* are accepted. The rate of return method permits conclusions to be stated as functions of *MARR*, which is useful if *MARR* has not been determined precisely.

EXAMPLE 5.17

A magazine subscription costs $5.00 for one year or $8.00 for two years. If you want to receive the magazine for at least two years, which alternative is better?

Solution. The two-year subscription requires an additional initial investment of $3.00 and eliminates the payment of $5.00 one year later. The rate of return formulation is

$$PNEV = 0$$

$$-3 + 5\,(P/F)_1^i = 0$$

The solution for i is as follows:

$$-3 + 5\frac{1}{(1+i)} = 0$$

$$i = 0.67 \text{ or } 67\%$$

Therefore, if your *MARR* is less than 67%, subscribe for two years.

EXAMPLE 5.18

Repeat Ex. 5.14 using the rate of return method.

Solution. Use the incremental expression derived in Ex. 5.16, but set *PNEV* equal to zero and use interest rate as the unknown.

$$-\,7000 + (P/A)_{10}^i\,500 + (P/F)_5^i\,4000 + (P/F)_{10}^i\,1000 = 0$$

By trial and error, the interest rate is found to be 6.6%. Therefore, *Reliable* is preferred if, and only if, *MARR* is less than 6.6%.

5.11 Benefit/Cost Comparisons

The benefit/cost ratio is determined from the formula:

$$\frac{B}{C} = \frac{\text{Uniform net annual benefits}}{\text{Annual equivalent of initial cost}}$$

where *MARR* is used in computing the *A*-value in the denominator. As with the rate of return method, mutually exclusive alternatives must be compared incrementally, the incremental investment being accepted if the benefit/cost ratio exceeds unity. For independent alternatives, all those with benefit/cost ratios exceeding unity are accepted.

Note that the only pertinent fact about a benefit/cost ratio is whether it exceeds unity. This is illustrated by the observation that a project with a ratio of 1.1 may provide greater net benefit than a project with a ratio of 10 if the investment in the former project is much larger than the investment in the latter. It is incorrect to rank mutually exclusive alternatives by their benefit/cost ratios as determined by comparing each alternative to the null (do nothing) alternative.

The benefit/cost ratio method will give the same decisions as the rate of return method, present equivalent cost method and annual equivalent cost method if the following conditions are met:

1. Each alternative is comprised of an initial cost and uniform annual benefit.

2. The form of the benefit/cost ratio given above is used without deviation.

EXAMPLE 5.19

A road resurfacing project costs $200,000, lasts five years and saves $100,000 annually in patching costs. *MARR* is 10%. Should the road be resurfaced?

Solution. The benefit/cost ratio is

$$\frac{B}{C} = \frac{100{,}000}{(A/P)^{10\%}_{5}\ 200{,}000} = 1.9$$

Since the ratio exceeds unity, the resurfacing is justified.

5.12 A Note on *MARR*

In engineering economy examination problems, *MARR* is a given quantity. However, the following discussion of the determination of *MARR* will help clarify the logic underlying the various comparison methods.

In general, an organization will be able to identify numerous opportunities to spend money now that will result in future returns. For each of these independent investment opportunities, an expected rate of return can be estimated. Similarly, the organization will be able to find numerous sources of funds for investment. Associated with each source of funds is an interest rate. If the source is a loan, the associated interest rate is simply that charged by the lender. Funds generated by operations of the organization, or provided by its owners (if the organization is a business), or extracted from taxpayers (if the organization is a government agency) can be thought of as being borrowed from the owners or taxpayers. Therefore, such funds can be assigned a fictitious interest rate, which should not be less than the maximum rate of return provided by other opportunities in which the owners or taxpayers might invest.

Value will be maximized if the rates of return of all the selected investments exceed the highest interest rate charged for the money borrowed, and if every opportunity has been taken to invest at a rate of return exceeding that for which money can be borrowed. The marginal dollar is invested at a rate of return equal to the interest rate at which it was borrowed. That rate is the Minimum Attractive Rate of Return. No investments should be made that pay rates of return less than *MARR*, and no loans should be taken that charge interest rates exceeding *MARR*. Furthermore, the organization should exploit all opportunities to borrow money at interest rates less than *MARR* and invest it at rates of return exceeding *MARR*.

To estimate *MARR* precisely would require the ability to foresee the future, or at least to predict all future investment and borrowing opportunities and their associated rates. A symptom of *MARR* being set too low is insufficient funds for all the investments that appear to be acceptable. Conversely, if *MARR* has been set too high, some investments will be rejected that would have been profitable.

5.13 Replacement Problems

How frequently should a particular machine be replaced? This type of problem can be approached by varying the life n. For each value of n, the annual costs and salvage value are estimated, and then the *ANEV* is computed. The value of n resulting in the smallest annual equivalent cost is the optimum, or economic, life of the machine. This approach is complicated by technological improvements in replacement machinery, which may make it advantageous to replace a machine before the end of its economic life. In practice, technological advances are difficult to anticipate.

Another form of the replacement problem asks if an existing asset should be replaced by a new (and possibly different) one. Again, the annual equivalent cost method is recommended. The *ANEV* of the replacement is computed, using its economic life for n. However, the annual cost of the existing asset is simply the estimated expense of one more year of operation. This strategy is based on the assumption that annual costs of the existing asset increase monotonically as it ages.

5.14 Always Ignore the Past

Engineering economy, and decision-making in general, deals with alternatives. But there is only one past and it affects all future alternatives equally. Therefore, past costs and income associated with an existing asset should not be included in computations that address the question of replacing the asset. Only the estimated cash flows of the future are relevant.

The mistake of counting past costs is common in everyday affairs. For example, a student may say, "I paid $40 for this textbook so I will not sell it for $20." A more rational approach would be to compare the highest offered price to the value of retaining the text.

EXAMPLE 5.20

Yesterday a machine was bought for $10,000. Estimated life is ten years, with no salvage value at that time. Current book value is $10,000. Today a vastly improved model was announced. It costs $15,000, has a ten-year life and no salvage value at that time, but reduces operating costs by $4,000 annually. The current resale value of the older machine has dropped to $1,000 due to this stunning technological advance. Should the old model be replaced with the new model at this time?

Solution. The purchase price of the old machine, its book value, and the loss on the sale of the old machine are irrelevant to the analysis. The incremental cost of the new machine is $14,000 and the incremental income is $4,000 annually. A rate of return comparison is formulated as follows:

$$-14{,}000 + (P/A)^{i}_{10}4000 = 0$$

Solving for rate of return gives $i = 26\%$, indicating that the older machine should be replaced immediately if *MARR* is less than 26%.

5.15 Break-Even Analysis

A break-even point is the value of an independent variable such that two alternatives are equally attractive. For values of the independent variable above the break-even point, one of the alternatives is preferred; for values of the independent variable below the break-even point, the other alternative is preferred. Break-even analysis is particularly useful for dealing with an independent variable that is subject to change or uncertainty since the conclusion of the analysis can be stated as a function of that variable. The rate of return method, as applied to mutually exclusive alternatives, is an example of break-even analysis. The independent variable is *MARR*.

EXAMPLE 5.21

An item can be manufactured by hand for $5. Alternatively, the item can be produced by a machine at a fixed annual equivalent cost of $4,000 plus a variable cost of $1 per item. Assume that the cost of laying off and hiring workers is zero. For each of the two manufacturing processes, answer the following questions:

a) For what production rate is one method more economical than the other?

b) If the item is sold for $6, how many must be sold to make a profit?

c) How low must the price fall, in the short term, before production is discontinued?

Solution.

a) Let P be production rate in units per year. Production costs for the two processes are equated:

$$\text{Cost by machine} = \text{Cost manually}$$

$$4000 + 1\,P = 5\,P$$

$$\therefore P = 1000$$

If annual production is expected to be less than 1000 units, the manual process is more economical. For production rates exceeding 1000 units per year, the machine process is preferred.

b) Setting profit equal to zero is expressed as

$$\text{gross income} - \text{cost} = 0$$

$$\text{Manual production: } 6\,P - 5\,P = 0$$

$$\therefore P = 0$$

$$\text{Machine production: } 6\,P - (4000 + 1\,P) = 0$$

$$\therefore P = 800$$

With price maintained at \$6, the mechanized operation will be unprofitable if production rate is less than 800 units per year, but the manual operation is profitable at all production rates.

c) Manual production becomes unprofitable if the price drops below \$5, and production will cease at that level. For the machine, the \$4,000 cost continues whether or not the machine is running. Incremental income is generated so long as the price stays above the variable (per item) cost. Therefore, production will continue at any price over \$1, even though a net loss may be sustained. Of course, if it appears that the price and production rate will not soon increase sufficiently to provide a profit, then the operation will be terminated.

5.16 Income Tax and Depreciation

Businesses pay to the federal government a tax that is a proportion of taxable income. Taxable income is gross revenue less operating costs (wages, cost of materials, etc.), interest payments on debts, and depreciation. Depreciation is different from the other deductions in that it is not a cash flow.

Depreciation is an accounting technique for charging the initial cost of an asset against two or more years of production. For example, if you buy a \$15,000 truck for use in your construction business, deducting its total cost from income during the year of purchase gives an unrealistically low picture of income for that year, and an unrealistically high estimate of income for the succeeding years during which you use the truck. A more level income history would result if you deducted \$5,000 per year for three years. In fact, the Internal Revenue Service (IRS) requires that most capital assets used in business be depreciated over a number of years rather than being deducted as expenses during the year of purchase.

An asset is depreciable if it is used to produce income, has a determinable life greater than one year, and decays, wears out, becomes obsolete, or gets used up. Examples are: tools, production machinery, computers, office equipment, buildings, patents, contracts, franchises, and livestock raised for wool, eggs, milk or breeding. Non-depreciable assets include personal residence, land, natural resources, annual crops, livestock raised for sale or slaughter, and items intended primarily for resale such as stored grain and the merchandise in a department store.

Since depreciation is not a cash flow, it will not directly enter an engineering economy analysis. However, depreciation must be considered when estimating future income taxes, which are cash flows.

For assets placed in service after 1980, the IRS requires that depreciation be computed by the accelerated cost recovery system (*ACRS*). But older methods may still show up in engineering economy problems. Therefore, the three methods permitted before 1981 are helpful to know. The following notation will be used in defining methods for computing depreciation:

B — The installed first cost, or basis.

n — Recovery period in years.

D_x — Depreciation in year x.

V_x — Undepreciated balance at the end of year x, also called book value. $V_0 = B$.

V_n — Estimated salvage at age n.

In computing depreciation there is no attempt to equate book value with resale value, productive worth, or any other real figure. A business is not obliged to keep an asset for exactly n years, nor to sell it for exactly its book value or estimated salvage value. These, then, are the four depreciation methods:

1. *Accelerated Cost Recovery System (ACRS):* An asset is classed as having a recovery period n of 3, 5, 10 or 15 years using IRS guidelines. For each class, a set of annual rates R_x is specified by the IRS. With 3-year property, for example, $R_1 = 0.25$, $R_2 = 0.38$, $R_3 = 0.37$. Depreciation is calculated by

$$D_x = R_x B.$$

By definition, the salvage value in the *ACRS* is zero.

2. *Straight Line Depreciation:* Depreciation is the same for every year and is calculated as

$$D_x = (B - V_n)/n$$

In this and the next two methods, n and V_n are estimated by the taxpayer at time of purchase and are required to be realistic.

3. *Sum of Years' Digits:* This method, the one that follows, and most instances of the *ACRS* method are said to be "accelerated" because they reduce book value more rapidly than does the straight line method. In general, accelerated depreciation is desirable because it produces larger tax deductions (and, therefore, larger after-tax cash flows) in the early years.

The sum of the years' digits method uses the relationship

$$D_x = (B - V_n)\frac{\text{(years of life remaining at start of year } x)}{\text{(sum of digits of all years of life)}}$$

$$= (B - V_n)(n - x + 1)/(0.5\,n^2 + 0.5\,n)$$

4. *Declining Balance:* Depreciation is taken as a proportion of book value:

$$D_x = V_{x-1}\,C/n$$

For values of C equaling 1.25, 1.5 and 2 the method is called, respectively: 125% declining balance, 150% declining balance, and double declining balance. The formula may result in book values less than estimated salvage value, but this is not permitted by the IRS.

EXAMPLE 5.22

The purchase price of a light truck is $15,000, its recovery period is three years, and it can be sold for an estimated $1,500 at that time. Compute the depreciation schedules using each of the methods described.

Solution.

Accelerated Cost Recovery System

Year	Depreciation	Book Value
		$15,000
1	0.25 × 15,000 = $3750	$11,250
2	0.38 × 15,000 = $5700	$ 5,550
3	0.37 × 15,000 = $5550	0

Straight Line

Year	Depreciation	Book Value
		$15,000
1	(15,000 − 1500)/3 = $4500	$10,500
2	$4500	$ 6,000
3	$4500	$ 1,500

Sum of Years' Digit

Year	Depreciation	Book Value
		$15,000
1	(15,000 − 1500) 3/6 = $6750	$ 8,250
2	13,500 × 2/6 = $4500	$ 3,750
3	13,500 × 1/6 = $2250	$ 1,500

Double Declining Balance

Year	Depreciation	Book Value
		$15,000
1	15,000 × 2/3 = $10,000	$ 5,000
2	5,000 × 2/3 = $ 3,333	$ 1,667
3	$167	$ 1,500

In the third year, the formula would have resulted in a book value less than the estimated salvage value, so the formula was abandoned.

5.17 Inflation

The "buying power" of money changes with time. A decline in "buying power" is experienced as a general increase in prices and is called "inflation."

Inflation, if it is anticipated, can be exploited by fixing costs and allowing income to increase. A manufacturing business can fix its costs by entering long-term contracts for materials and wages, by purchasing materials long before they are needed, and by stockpiling its product for sale later. Income is allowed to respond to inflation by avoiding long-term contracts for the product. Borrowing becomes more attractive if inflation is expected since the debt will be paid with the less valuable cash of the future.

EXAMPLE 5.23

A machine having a five-year life can replace a worker now earning $10,000 per year who is subject to 5% annual "cost of living" increases. Operating and maintenance costs for the machine are negligible. *MARR* is 10%. Find the maximum price that can be justified for the machine if:

a) general price inflation is 5%, and

b) general price inflation is zero.

Solution.

a) Although the worker gets a larger amount of money each year, his raises are exactly matched by increased prices, including those of his employer's product. "Buying power" of his annual wage remains equal to the current value of $10,000. Hence, the maximum justifiable price for the machine is

$$P = (P/A)_5^{10\%}\ 10{,}000 = \$37{,}908$$

b) The maximum justifiable price of the machine is equal to the present equivalent value of the annual amounts of the wage:

$$(P/F)_1^{10\%}\ (1.05)\ \ 10{,}000 = \$\ 9{,}545$$

$$(P/F)_2^{10\%}\ (1.05)^2\ 10{,}000 = \$\ 9{,}112$$

$$(P/F)_3^{10\%}\ (1.05)^3\ 10{,}000 = \$\ 8{,}697$$

$$(P/F)_4^{10\%}\ (1.05)^4\ 10{,}000 = \$\ 8{,}302$$

$$(P/F)_5^{10\%}\ (1.05)^5\ 10{,}000 = \underline{\$\ 7{,}925}$$

$$\therefore P = \$43{,}581$$

EXAMPLE 5.24

Recompute the value, in terms of 1987 "buying power," of the "Acme 8% of 2000" bond discussed in Ex. 5.5, but assume 6% annual inflation.

Solution. The cash flow for each year must be divided by an inflation factor as well as multiplied by an interest factor, and then the factored cash flows are added:

$$(P/F)^{10\%}_{1} \quad 80/(1.06) = \$\ 69$$

$$(P/F)^{10\%}_{2} \quad 80/(1.06)^2 = \$\ 59$$

$$(P/F)^{10\%}_{3} \quad 80/(1.06)^3 = \$\ 50$$

$$\vdots$$

$$(P/F)^{10\%}_{13} \quad 80/(1.06)^{13} = \$\ 11$$

$$(P/F)^{10\%}_{14} \quad 80/(1.06)^{14} = \$\ \ 9$$

$$(P/F)^{10\%}_{14} \quad 1000/(1.06)^{14} = \underline{\$116}$$

$$P = \$541$$

Note that investors can account for anticipated inflation simply by using increased values of *MARR*. A *MARR* of 16.6% gives the same conclusions as a *MARR* of 10% with 6% inflation.

TABLE 5.2. Compound Interest Factors

$i = 1.00\%$

n	(P/F)	(P/A)	(P/G)	(F/P)	(F/A)	(A/P)	(A/F)	(A/G)	n
1	.9901	0.9901	-0.0000	1.0100	1.0000	1.0100	1.0000	-0.0000	1
2	.9803	1.9704	0.9803	1.0201	2.0100	0.5075	0.4975	0.4975	2
3	.9706	2.9410	2.9215	1.0303	3.0301	0.3400	0.3300	0.9934	3
4	.9610	3.9020	5.8044	1.0406	4.0604	0.2563	0.2463	1.4876	4
5	.9515	4.8534	9.6103	1.0510	5.1010	0.2060	0.1960	1.9801	5
6	.9420	5.7955	14.3205	1.0615	6.1520	0.1725	0.1625	2.4710	6
7	.9327	6.7282	19.9168	1.0721	7.2135	0.1486	0.1386	2.9602	7
8	.9235	7.6517	26.3812	1.0829	8.2857	0.1307	0.1207	3.4478	8
9	.9143	8.5660	33.6959	1.0937	9.3685	0.1167	0.1067	3.9337	9
10	.9053	9.4713	41.8435	1.1046	10.4622	0.1056	0.0956	4.4179	10
11	.8963	10.3676	50.8067	1.1157	11.5668	0.0965	0.0865	4.9005	11
12	.8874	11.2551	60.5687	1.1268	12.6825	0.0888	0.0788	5.3815	12
13	.8787	12.1337	71.1126	1.1381	13.8093	0.0824	0.0724	5.8607	13
14	.8700	13.0037	82.4221	1.1495	14.9474	0.0769	0.0669	6.3384	14
15	.8613	13.8651	94.4810	1.1610	16.0969	0.0721	0.0621	6.8143	15
16	.8528	14.7179	107.2734	1.1726	17.2579	0.0679	0.0579	7.2886	16
17	.8444	15.5623	120.7834	1.1843	18.4304	0.0643	0.0543	7.7613	17
18	.8360	16.3983	134.9957	1.1961	19.6147	0.0610	0.0510	8.2323	18
19	.8277	17.2260	149.8950	1.2081	20.8109	0.0581	0.0481	8.7017	19
20	.8195	18.0456	165.4664	1.2202	22.0190	0.0554	0.0454	9.1694	20
21	.8114	18.8570	181.6950	1.2324	23.2392	0.0530	0.0430	9.6354	21
22	.8034	19.6604	198.5663	1.2447	24.4716	0.0509	0.0409	10.0998	22
23	.7954	20.4558	216.0660	1.2572	25.7163	0.0489	0.0389	10.5626	23
24	.7876	21.2434	234.1800	1.2697	26.9735	0.0471	0.0371	11.0237	24
25	.7798	22.0232	252.8945	1.2824	28.2432	0.0454	0.0354	11.4831	25
26	.7720	22.7952	272.1957	1.2953	29.5256	0.0439	0.0339	11.9409	26
28	.7568	24.3164	312.5047	1.3213	32.1291	0.0411	0.0311	12.8516	28
30	.7419	25.8077	355.0021	1.3478	34.7849	0.0387	0.0287	13.7557	30
∞	.0000	100.000	10 000.0	∞	∞	0.0100	0.0000	100.0000	∞

$i = 2.00\%$

n	(P/F)	(P/A)	(P/G)	(F/P)	(F/A)	(A/P)	(A/F)	(A/G)	n
1	.9804	0.9804	-0.0000	1.0200	1.0000	1.0200	1.0000	-0.0000	1
2	.9612	1.9416	0.9612	1.0404	2.0200	0.5150	0.4950	0.4950	2
3	.9423	2.8839	2.8458	1.0612	3.0604	0.3468	0.3268	0.9868	3
4	.9238	3.8077	5.6173	1.0824	4.1216	0.2626	0.2426	1.4752	4
5	.9057	4.7135	9.2403	1.1041	5.2040	0.2122	0.1922	1.9604	5
6	.8880	5.6014	13.6801	1.1262	6.3081	0.1785	0.1585	2.4423	6
7	.8706	6.4720	18.9035	1.1487	7.4343	0.1545	0.1345	2.9208	7
8	.8535	7.3255	24.8779	1.1717	8.5830	0.1365	0.1165	3.3961	8
9	.8368	8.1622	31.5720	1.1951	9.7546	0.1225	0.1025	3.8681	9
10	.8203	8.9826	38.9551	1.2190	10.9497	0.1113	0.0913	4.3367	10
11	.8043	9.7868	46.9977	1.2434	12.1687	0.1022	0.0822	4.8021	11
12	.7885	10.5753	55.6712	1.2682	13.4121	0.0946	0.0746	5.2642	12
13	.7730	11.3484	64.9475	1.2936	14.6803	0.0881	0.0681	5.7231	13
14	.7579	12.1062	74.7999	1.3195	15.9739	0.0826	0.0626	6.1786	14
15	.7430	12.8493	85.2021	1.3459	17.2934	0.0778	0.0578	6.6309	15
16	.7284	13.5777	96.1288	1.3728	18.6393	0.0737	0.0537	7.0799	16
17	.7142	14.2919	107.5554	1.4002	20.0121	0.0700	0.0500	7.5256	17
18	.7002	14.9920	119.4581	1.4282	21.4123	0.0667	0.0467	7.9681	18
19	.6864	15.6785	131.8139	1.4568	22.8406	0.0638	0.0438	8.4073	19
20	.6730	16.3514	144.6003	1.4859	24.2974	0.0612	0.0412	8.8433	20
21	.6598	17.0112	157.7959	1.5157	25.7833	0.0588	0.0388	9.2760	21
22	.6468	17.6580	171.3795	1.5460	27.2990	0.0566	0.0366	9.7055	22
23	.6342	18.2922	185.3309	1.5769	28.8450	0.0547	0.0347	10.1317	23
24	.6217	18.9139	199.6305	1.6084	30.4219	0.0529	0.0329	10.5547	24
25	.6095	19.5235	214.2592	1.6406	32.0303	0.0512	0.0312	10.9745	25
26	.5976	20.1210	229.1987	1.6734	33.6709	0.0497	0.0297	11.3910	26
28	.5744	21.2813	259.9392	1.7410	37.0512	0.0470	0.0270	12.2145	28
30	.5521	22.3965	291.7164	1.8114	40.5681	0.0446	0.0246	13.0251	30
∞	.0000	50.0000	2500.0000	∞	∞	0.0200	0.0000	50.0000	∞

TABLE 5.2. Compound Interest Factors (continued)

$i = 4.00\%$

n	(P/F)	(P/A)	(P/G)	(F/P)	(F/A)	(A/P)	(A/F)	(A/G)	*n*
1	.9615	0.9615	-0.0000	1.0400	1.0000	1.0400	1.0000	-0.0000	1
2	.9246	1.8861	0.9246	1.0816	2.0400	0.5302	0.4902	0.4902	2
3	.8890	2.7751	2.7025	1.1249	3.1216	0.3603	0.3203	0.9739	3
4	.8548	3.6299	5.2670	1.1699	4.2465	0.2755	0.2355	1.4510	4
5	.8219	4.4518	8.5547	1.2167	5.4163	0.2246	0.1846	1.9216	5
6	.7903	5.2421	12.5062	1.2653	6.6330	0.1908	0.1508	2.3857	6
7	.7599	6.0021	17.0657	1.3159	7.8983	0.1666	0.1266	2.8433	7
8	.7307	6.7327	22.1806	1.3686	9.2142	0.1485	0.1085	3.2944	8
9	.7026	7.4353	27.8013	1.4233	10.5828	0.1345	0.0945	3.7391	9
10	.6756	8.1109	33.8814	1.4802	12.0061	0.1233	0.0833	4.1773	10
11	.6496	8.7605	40.3772	1.5395	13.4864	0.1141	0.0741	4.6090	11
12	.6246	9.3851	47.2477	1.6010	15.0258	0.1066	0.0666	5.0343	12
13	.6006	9.9856	54.4546	1.6651	16.6268	0.1001	0.0601	5.4533	13
14	.5775	10.5631	61.9618	1.7317	18.2919	0.0947	0.0547	5.8659	14
15	.5553	11.1184	69.7355	1.8009	20.0236	0.0899	0.0499	6.2721	15
16	.5339	11.6523	77.7441	1.8730	21.8245	0.0858	0.0458	6.6720	16
17	.5134	12.1657	85.9581	1.9479	23.6975	0.0822	0.0422	7.0656	17
18	.4936	12.6593	94.3498	2.0258	25.6454	0.0790	0.0390	7.4530	18
19	.4746	13.1339	102.8933	2.1068	27.6712	0.0761	0.0361	7.8342	19
20	.4564	13.5903	111.5647	2.1911	29.7781	0.0736	0.0336	8.2091	20
21	.4388	14.0292	120.3414	2.2788	31.9692	0.0713	0.0313	8.5779	21
22	.4220	14.4511	129.2024	2.3699	34.2480	0.0692	0.0292	8.9407	22
23	.4057	14.8568	138.1284	2.4647	36.6179	0.0673	0.0273	9.2973	23
24	.3901	15.2470	147.1012	2.5633	39.0826	0.0656	0.0256	9.6479	24
25	.3751	15.6221	156.1040	2.6658	41.6459	0.0640	0.0240	9.9925	25
26	.3607	15.9828	165.1212	2.7725	44.3117	0.0626	0.0226	10.3312	26
28	.3335	16.6631	183.1424	2.9987	49.9676	0.0600	0.0200	10.9909	28
30	.3083	17.2920	201.0618	3.2434	56.0849	0.0578	0.0178	11.6274	30
∞	.0000	25.000	625.0000	∞	∞	0.0400	0.0000	25.0000	∞

$i = 6.00\%$

n	(P/F)	(P/A)	(P/G)	(F/P)	(F/A)	(A/P)	(A/F)	(A/G)	*n*
1	.9434	0.9434	-0.0000	1.0600	1.0000	1.0600	1.0000	-0.0000	1
2	.8900	1.8334	0.8900	1.1236	2.0600	0.5454	0.4854	0.4854	2
3	.8396	2.6730	2.5692	1.1910	3.1836	0.3741	0.3141	0.9612	3
4	.7921	3.4651	4.9455	1.2625	4.3746	0.2886	0.2286	1.4272	4
5	.7473	4.2124	7.9345	1.3382	5.6371	0.2374	0.1774	1.8836	5
6	.7050	4.9173	11.4594	1.4185	6.9753	0.2034	0.1434	2.3304	6
7	.6651	5.5824	15.4497	1.5036	8.3938	0.1791	0.1191	2.7676	7
8	.6274	6.2098	19.8416	1.5938	9.8975	0.1610	0.1010	3.1952	8
9	.5919	6.8017	24.5768	1.6895	11.4913	0.1470	0.0870	3.6133	9
10	.5584	7.3601	29.6023	1.7908	13.1808	0.1359	0.0759	4.0220	10
11	.5268	7.8869	34.8702	1.8983	14.9716	0.1268	0.0668	4.4213	11
12	.4970	8.3838	40.3369	2.0122	16.8699	0.1193	0.0593	4.8113	12
13	.4688	8.8527	45.9629	2.1329	18.8821	0.1130	0.0530	5.1920	13
14	.4423	9.2950	51.7128	2.2609	21.0151	0.1076	0.0476	5.5635	14
15	.4173	9.7122	57.5546	2.3966	23.2760	0.1030	0.0430	5.9260	15
16	.3936	10.1059	63.4592	2.5404	25.6725	0.0990	0.0390	6.2794	16
17	.3714	10.4773	69.4011	2.6928	28.2129	0.0954	0.0354	6.6240	17
18	.3503	10.8276	75.3569	2.8543	30.9057	0.0924	0.0324	6.9597	18
19	.3305	11.1581	81.3062	3.0256	33.7600	0.0896	0.0296	7.2867	19
20	.3118	11.4699	87.2304	3.2071	36.7856	0.0872	0.0272	7.6051	20
21	.2942	11.7641	93.1136	3.3996	39.9927	0.0850	0.0250	7.9151	21
22	.2775	12.0416	98.9412	3.6035	43.3923	0.0830	0.0230	8.2166	22
23	.2618	12.3034	104.7007	3.8197	46.9958	0.0813	0.0213	8.5099	23
24	.2470	12.5504	110.3812	4.0489	50.8156	0.0797	0.0197	8.7951	24
25	.2330	12.7834	115.9732	4.2919	54.8645	0.0782	0.0182	9.0722	25
26	.2198	13.0032	121.4684	4.5494	59.1564	0.0769	0.0169	9.3414	26
28	.1956	13.4062	132.1420	5.1117	68.5281	0.0746	0.0146	9.8568	28
30	.1741	13.7648	142.3588	5.7435	79.0582	0.0726	0.0126	10.3422	30
∞	.0000	16.6667	277.7778	∞	∞	0.0600	0.0000	16.6667	∞

TABLE 5.2. Compound Interest Factors (continued)

i = 8.00%

n	(P/F)	(P/A)	(P/G)	(F/P)	(F/A)	(A/P)	(A/F)	(A/G)	*n*
1	.9259	0.9259	-0.0000	1.0800	1.0000	1.0800	1.0000	-0.0000	1
2	.8573	1.7833	0.8573	1.1664	2.0800	0.5608	0.4808	0.4808	2
3	.7938	2.5771	2.4450	1.2597	3.2464	0.3880	0.3080	0.9487	3
4	.7350	3.3121	4.6501	1.3605	4.5061	0.3019	0.2219	1.4040	4
5	.6806	3.9927	7.3724	1.4693	5.8666	0.2505	0.1705	1.8465	5
6	.6302	4.6229	10.5233	1.5869	7.3359	0.2163	0.1363	2.2763	6
7	.5835	5.2064	14.0242	1.7138	8.9228	0.1921	0.1121	2.6937	7
8	.5403	5.7466	17.8061	1.8509	10.6366	0.1740	0.0940	3.0985	8
9	.5002	6.2469	21.8081	1.9990	12.4876	0.1601	0.0801	3.4910	9
10	.4632	6.7101	25.9768	2.1589	14.4866	0.1490	0.0690	3.8713	10
11	.4289	7.1390	30.2657	2.3316	16.6455	0.1401	0.0601	4.2395	11
12	.3971	7.5361	34.6339	2.5182	18.9771	0.1327	0.0527	4.5957	12
13	.3677	7.9038	39.0463	2.7196	21.4953	0.1265	0.0465	4.9402	13
14	.3405	8.2442	43.4723	2.9372	24.2149	0.1213	0.0413	5.2731	14
15	.3152	8.5595	47.8857	3.1722	27.1521	0.1168	0.0368	5.5945	15
16	.2919	8.8514	52.2640	3.4259	30.3243	0.1130	0.0330	5.9046	16
17	.2703	9.1216	56.5883	3.7000	33.7502	0.1096	0.0296	6.2037	17
18	.2502	9.3719	60.8426	3.9960	37.4502	0.1067	0.0267	6.4920	18
19	.2317	9.6036	65.0134	4.3157	41.4463	0.1041	0.0241	6.7697	19
20	.2145	9.8181	69.0898	4.6610	45.7620	0.1019	0.0219	7.0369	20
21	.1987	10.0168	73.0629	5.0338	50.4229	0.0998	0.0198	7.2940	21
22	.1839	10.2007	76.9257	5.4365	55.4568	0.0980	0.0180	7.5412	22
23	.1703	10.3711	80.6726	5.8715	60.8933	0.0964	0.0164	7.7786	23
24	.1577	10.5288	84.2997	6.3412	66.7648	0.0950	0.0150	8.0066	24
25	.1460	10.6748	87.8041	6.8485	73.1059	0.0937	0.0137	8.2254	25
26	.1352	10.8100	91.1842	7.3964	79.9544	0.0925	0.0125	8.4352	26
28	.1159	11.0511	97.5687	8.6271	95.3388	0.0905	0.0105	8.8289	28
30	.0994	11.2578	103.4558	10.0627	113.2832	0.0888	0.0088	9.1897	30
∞	.0000	12.500	156.2500	∞	∞	0.0800	0.0000	12.5000	∞

i = 10.00%

n	(P/F)	(P/A)	(P/G)	(F/P)	(F/A)	(A/P)	(A/F)	(A/G)	*n*
1	.9091	0.9091	-0.0000	1.1000	1.0000	1.1000	1.0000	-0.0000	1
2	.8264	1.7355	0.8264	1.2100	2.1000	0.5762	0.4762	0.4762	2
3	.7513	2.4869	2.3291	1.3310	3.3100	0.4021	0.3021	0.9366	3
4	.6830	3.1699	4.3781	1.4641	4.6410	0.3155	0.2155	1.3812	4
5	.6209	3.7908	6.8618	1.6105	6.1051	0.2638	0.1638	1.8101	5
6	.5645	4.3553	9.6842	1.7716	7.7156	0.2296	0.1296	2.2236	6
7	.5132	4.8684	12.7631	1.9487	9.4872	0.2054	0.1054	2.6216	7
8	.4665	5.3349	16.0287	2.1436	11.4359	0.1874	0.0874	3.0045	8
9	.4241	5.7590	19.4215	2.3579	13.5795	0.1736	0.0736	3.3724	9
10	.3855	6.1446	22.8913	2.5937	15.9374	0.1627	0.0627	3.7255	10
11	.3505	6.4951	26.3963	2.8531	18.5312	0.1540	0.0540	4.0641	11
12	.3186	6.8137	29.9012	3.1384	21.3843	0.1468	0.0468	4.3884	12
13	.2897	7.1034	33.3772	3.4523	24.5227	0.1408	0.0408	4.6988	13
14	.2633	7.3667	36.8005	3.7975	27.9750	0.1357	0.0357	4.9955	14
15	.2394	7.6061	40.1520	4.1772	31.7725	0.1315	0.0315	5.2789	15
16	.2176	7.8237	43.4164	4.5950	35.9497	0.1278	0.0278	5.5493	16
17	.1978	8.0216	46.5819	5.0545	40.5447	0.1247	0.0247	5.8071	17
18	.1799	8.2014	49.6395	5.5599	45.5992	0.1219	0.0219	6.0526	18
19	.1635	8.3649	52.5827	6.1159	51.1591	0.1195	0.0195	6.2861	19
20	.1486	8.5136	55.4069	6.7275	57.2750	0.1175	0.0175	6.5081	20
21	.1351	8.6487	58.1095	7.4002	64.0025	0.1156	0.0156	6.7189	21
22	.1228	8.7715	60.6893	8.1403	71.4027	0.1140	0.0140	6.9189	22
23	.1117	8.8832	63.1462	8.9543	79.5430	0.1126	0.0126	7.1085	23
24	.1015	8.9847	65.4813	9.8497	88.4973	0.1113	0.0113	7.2881	24
25	.0923	9.0770	67.6964	10.8347	98.3471	0.1102	0.0102	7.4580	25
26	.0839	9.1609	69.7940	11.9182	109.1818	0.1092	0.0092	7.6186	26
28	.0693	9.3066	73.6495	14.4210	134.2099	0.1075	0.0075	7.9137	28
30	.0573	9.4269	77.0766	17.4494	164.4940	0.1061	0.0061	8.1762	30
∞	.0000	10.0000	100.0000	∞	∞	0.1000	0.0000	10.0000	∞

TABLE 5.2. Compound Interest Factors (continued)

i = 12.00%

n	(P/F)	(P/A)	(P/G)	(F/P)	(F/A)	(A/P)	(A/F)	(A/G)	n
1	.8929	0.8929	-0.0000	1.1200	1.0000	1.1200	1.0000	-0.0000	1
2	.7972	1.6901	0.7972	1.2544	2.1200	0.5917	0.4717	0.4717	2
3	.7118	2.4018	2.2208	1.4049	3.3744	0.4163	0.2963	0.9246	3
4	.6355	3.0373	4.1273	1.5735	4.7793	0.3292	0.2092	1.3589	4
5	.5674	3.6048	6.3970	1.7623	6.3528	0.2774	0.1574	1.7746	5
6	.5066	4.1114	8.9302	1.9738	8.1152	0.2432	0.1232	2.1720	6
7	.4523	4.5638	11.6443	2.2107	10.0890	0.2191	0.0991	2.5515	7
8	.4039	4.9676	14.4714	2.4760	12.2997	0.2013	0.0813	2.9131	8
9	.3606	5.3282	17.3563	2.7731	14.7757	0.1877	0.0677	3.2574	9
10	.3220	5.6502	20.2541	3.1058	17.5487	0.1770	0.0570	3.5847	10
11	.2875	5.9377	23.1288	3.4785	20.6546	0.1684	0.0484	3.8953	11
12	.2567	6.1944	25.9523	3.8960	24.1331	0.1614	0.0414	4.1897	12
13	.2292	6.4235	28.7024	4.3635	28.0291	0.1557	0.0357	4.4683	13
14	.2046	6.6282	31.3624	4.8871	32.3926	0.1509	0.0309	4.7317	14
15	.1827	6.8109	33.9202	5.4736	37.2797	0.1468	0.0268	4.9803	15
16	.1631	6.9740	36.3670	6.1304	42.7533	0.1434	0.0234	5.2147	16
17	.1456	7.1196	38.6973	6.8660	48.8837	0.1405	0.0205	5.4353	17
18	.1300	7.2497	40.9080	7.6900	55.7497	0.1379	0.0179	5.6427	18
19	.1161	7.3658	42.9979	8.6128	63.4397	0.1358	0.0158	5.8375	19
20	.1037	7.4694	44.9676	9.6463	72.0524	0.1339	0.0139	6.0202	20
21	.0926	7.5620	46.8188	10.8038	81.6987	0.1322	0.0122	6.1913	21
22	.0826	7.6446	48.5543	12.1003	92.5026	0.1308	0.0108	6.3514	22
23	.0738	7.7184	50.1776	13.5523	104.6029	0.1296	0.0096	6.5010	23
24	.0659	7.7843	51.6929	15.1786	118.1552	0.1285	0.0085	6.6406	24
25	.0588	7.8431	53.1046	17.0001	133.3339	0.1275	0.0075	6.7708	25
26	.0525	7.8957	54.4177	19.0401	150.3339	0.1267	0.0067	6.8921	26
28	.0419	7.9844	56.7674	23.8839	190.6989	0.1252	0.0052	7.1098	28
30	.0334	8.0552	58.7821	29.9599	241.3327	0.1241	0.0041	7.2974	30
∞	.0000	8.333	69.4444	∞	∞	0.1200	0.0000	8.3333	∞

i = 15.00%

n	(P/F)	(P/A)	(P/G)	(F/P)	(F/A)	(A/P)	(A/F)	(A/G)	n
1	.8696	0.8696	-0.0000	1.1500	1.0000	1.1500	1.0000	-0.0000	1
2	.7561	1.6257	0.7561	1.3225	2.1500	0.6151	0.4651	0.4651	2
3	.6575	2.2832	2.0712	1.5209	3.4725	0.4380	0.2880	0.9071	3
4	.5718	2.8550	3.7864	1.7490	4.9934	0.3503	0.2003	1.3263	4
5	.4972	3.3522	5.7751	2.0114	6.7424	0.2983	0.1483	1.7228	5
6	.4323	3.7845	7.9368	2.3131	8.7537	0.2642	0.1142	2.0972	6
7	.3759	4.1604	10.1924	2.6600	11.0668	0.2404	0.0904	2.4498	7
8	.3269	4.4873	12.4807	3.0590	13.7268	0.2229	0.0729	2.7813	8
9	.2843	4.7716	14.7548	3.5179	16.7858	0.2096	0.0596	3.0922	9
10	.2472	5.0188	16.9795	4.0456	20.3037	0.1993	0.0493	3.3832	10
11	.2149	5.2337	19.1289	4.6524	24.3493	0.1911	0.0411	3.6549	11
12	.1869	5.4206	21.1849	5.3503	29.0017	0.1845	0.0345	3.9082	12
13	.1625	5.5831	23.1352	6.1528	34.3519	0.1791	0.0291	4.1438	13
14	.1413	5.7245	24.9725	7.0757	40.5047	0.1747	0.0247	4.3624	14
15	.1229	5.8474	26.6930	8.1371	47.5804	0.1710	0.0210	4.5650	15
16	.1069	5.9542	28.2960	9.3576	55.7175	0.1679	0.0179	4.7522	16
17	.0929	6.0472	29.7828	10.7613	65.0751	0.1654	0.0154	4.9251	17
18	.0808	6.1280	31.1565	12.3755	75.8364	0.1632	0.0132	5.0843	18
19	.0703	6.1982	32.4213	14.2318	88.2118	0.1613	0.0113	5.2307	19
20	.0611	6.2593	33.5822	16.3665	102.4436	0.1598	0.0098	5.3651	20
21	.0531	6.3125	34.6448	18.8215	118.8101	0.1584	0.0084	5.4883	21
22	.0462	6.3587	35.6150	21.6447	137.6316	0.1573	0.0073	5.6010	22
23	.0402	6.3988	36.4988	24.8915	159.2764	0.1563	0.0063	5.7040	23
24	.0349	6.4338	37.3023	28.6252	184.1678	0.1554	0.0054	5.7979	24
25	.0304	6.4641	38.0314	32.9190	212.7930	0.1547	0.0047	5.8834	25
26	.0264	6.4906	38.6918	37.8568	245.7120	0.1541	0.0041	5.9612	26
28	.0200	6.5335	39.8283	50.0656	327.1041	0.1531	0.0031	6.0960	28
30	.0151	6.5660	40.7526	66.2118	434.7451	0.1523	0.0023	6.2066	30
∞	.0000	6.6667	44.4444	∞	∞	0.1500	0.0000	6.6667	∞

TABLE 5.2. Compound Interest Factors (continued)

i = 20.00%

n	(P/F)	(P/A)	(P/G)	(F/P)	(F/A)	(A/P)	(A/F)	(A/G)	n
1	.8333	0.8333	-0.0000	1.2000	1.0000	1.2000	1.0000	-0.0000	1
2	.6944	1.5278	0.6944	1.4400	2.2000	0.6545	0.4545	0.4545	2
3	.5787	2.1065	1.8519	1.7280	3.6400	0.4747	0.2747	0.8791	3
4	.4823	2.5887	3.2986	2.0736	5.3680	0.3863	0.1863	1.2742	4
5	.4019	2.9906	4.9061	2.4883	7.4416	0.3344	0.1344	1.6405	5
6	.3349	3.3255	6.5806	2.9860	9.9299	0.3007	0.1007	1.9788	6
7	.2791	3.6046	8.2551	3.5832	12.9159	0.2774	0.0774	2.2902	7
8	.2326	3.8372	9.8831	4.2998	16.4991	0.2606	0.0606	2.5756	8
9	.1938	4.0310	11.4335	5.1598	20.7989	0.2481	0.0481	2.8364	9
10	.1615	4.1925	12.8871	6.1917	25.9587	0.2385	0.0385	3.0739	10
11	.1346	4.3271	14.2330	7.4301	32.1504	0.2311	0.0311	3.2893	11
12	.1122	4.4392	15.4667	8.9161	39.5805	0.2253	0.0253	3.4841	12
13	.0935	4.5327	16.5883	10.6993	48.4966	0.2206	0.0206	3.6597	13
14	.0779	4.6106	17.6008	12.8392	59.1959	0.2169	0.0169	3.8175	14
15	.0649	4.6755	18.5095	15.4070	72.0351	0.2139	0.0139	3.9588	15
16	.0541	4.7296	19.3208	18.4884	87.4421	0.2114	0.0114	4.0851	16
17	.0451	4.7746	20.0419	22.1861	105.9306	0.2094	0.0094	4.1976	17
18	.0376	4.8122	20.6805	26.6233	128.1167	0.2078	0.0078	4.2975	18
19	.0313	4.8435	21.2439	31.9480	154.7400	0.2065	0.0065	4.3861	19
20	.0261	4.8696	21.7395	38.3376	186.6880	0.2054	0.0054	4.4643	20
21	.0217	4.8913	22.1742	46.0051	225.0256	0.2044	0.0044	4.5334	21
22	.0181	4.9094	22.5546	55.2061	271.0307	0.2037	0.0037	4.5941	22
23	.0151	4.9245	22.8867	66.2474	326.2369	0.2031	0.0031	4.6475	23
24	.0126	4.9371	23.1760	79.4968	392.4842	0.2025	0.0025	4.6943	24
25	.0105	4.9476	23.4276	95.3962	471.9811	0.2021	0.0021	4.7352	25
26	.0087	4.9563	23.6460	114.4755	567.3773	0.2018	0.0018	4.7709	26
28	.0061	4.9697	23.9991	164.8447	819.2233	0.2012	0.0012	4.8291	28
30	.0042	4.9789	24.2628	237.3763	1181.8816	0.2008	0.0008	4.8731	30
∞	.0000	5.0000	25.0000	∞	∞	0.2000	0.0000	5.0000	∞

i = 25.00%

n	(P/F)	(P/A)	(P/G)	(F/P)	(F/A)	(A/P)	(A/F)	(A/G)	n
1	.8000	0.8000	0.0000	1.2500	1.0000	1.2500	1.0000	0.0000	1
2	.6400	1.4400	0.6400	1.5625	2.2500	0.6944	0.4444	0.4444	2
3	.5120	1.9520	1.6640	1.9531	3.8125	0.5123	0.2623	0.8525	3
4	.4096	2.3616	2.8928	2.4414	5.7656	0.4234	0.1734	1.2249	4
5	.3277	2.6893	4.2035	3.0518	8.2070	0.3718	0.1218	1.5631	5
6	.2621	2.9514	5.5142	3.8147	11.2588	0.3388	0.0888	1.8683	6
7	.2097	3.1611	6.7725	4.7684	15.0735	0.3163	0.0663	2.1424	7
8	.1678	3.3289	7.9469	5.9605	19.8419	0.3004	0.0504	2.3872	8
9	.1342	3.4631	9.0207	7.4506	25.8023	0.2888	0.0388	2.6048	9
10	.1074	3.5705	9.9870	9.3132	33.2529	0.2801	0.0301	2.7971	10
11	.0859	3.6564	10.8460	11.6415	42.5661	0.2735	0.0235	2.9663	11
12	.0687	3.7251	11.6020	14.5519	54.2077	0.2684	0.0184	3.1145	12
13	.0550	3.7801	12.2617	18.1899	68.7596	0.2645	0.0145	3.2437	13
14	.0440	3.8241	12.8334	22.7374	86.9495	0.2615	0.0115	3.3559	14
15	.0352	3.8593	13.3260	28.4217	109.6868	0.2591	0.0091	3.4530	15
16	.0281	3.8874	13.7482	35.5271	138.1085	0.2572	0.0072	3.5366	16
17	.0225	3.9099	14.1085	44.4089	173.6357	0.2558	0.0058	3.6084	17
18	.0180	3.9279	14.4147	55.5112	218.0446	0.2546	0.0046	3.6698	18
19	.0144	3.9424	14.6741	69.3889	273.5558	0.2537	0.0037	3.7222	19
20	.0115	3.9539	14.8932	86.7362	342.9447	0.2529	0.0029	3.7667	20
21	.0092	3.9631	15.0777	108.4202	429.6809	0.2523	0.0023	3.8045	21
22	.0074	3.9705	15.2326	135.5253	538.1011	0.2519	0.0019	3.8365	22
23	.0059	3.9764	15.3625	169.4066	673.6264	0.2515	0.0015	3.8634	23
24	.0047	3.9811	15.4711	211.7582	843.0329	0.2512	0.0012	3.8861	24
25	.0038	3.9849	15.5618	264.6978	1054.7912	0.2509	0.0009	3.9052	25
26	.0030	3.9879	15.6373	330.8722	1319.4890	0.2508	0.0008	3.9212	26
28	.0019	3.9923	15.7524	516.9879	2063.9515	0.2505	0.0005	3.9457	28
30	.0012	3.9950	15.8316	807.7936	3227.1743	0.2503	0.0003	3.9628	30
∞	.0000	4.0000	16.0000	∞	∞	0.2500	0.0000	4.0000	∞

(This page is intentionally blank.)

Practice Problems

(If only a few are selected, choose those with a star.)

5.1 Which of the following would be most difficult to monetize?

a) maintenance cost
b) selling price
c) fuel cost
d) prestige
e) interest on debt

VALUE AND INTEREST

*5.2 If $1,000 is deposited in a savings account that pays 6% annual interest and all the interest is left in the account, what is the account balance after three years?

a) $840 b) $1,000 c) $1,180 d) $1,191 e) $3,000

*5.3 Your perfectly reliable friend, Merle, asks for a loan and promises to pay back $150 two years from now. If the minimum interest rate you will accept is 8%, what is the maximum amount you will loan Merle?

a) $119 b) $126 c) $129 d) $139 e) $150

EQUIVALENCE OF CASH FLOW PATTERNS

*5.4 The annual amount of a series of payments to be made at the end of each of the next twelve years is $500. What is the present worth of the payments at 8% interest compounded annually?

a) $500 b) $3,768 c) $6,000 d) $6,480 e) $6,872

*5.5 Consider a prospective investment in a project having a first cost of $300,000, operating and maintenance costs of $35,000 per year, and an estimated net disposal value of $50,000 at the end of thirty years. Assume an interest rate of 8%.

What is the present equivalent cost of the investment if the planning horizon is thirty years?

a) $670,000 b) $689,000 c) $720,000 d) $791,000 e) $950,000

If the project replacement will have the same first cost, life, salvage value, and operating and maintenance costs as the original, what is the capitalized cost of perpetual service?

a) $670,000 b) $689,000 c) $720,000 d) $765,000 e) infinite

*5.6 Maintenance expenditures for a structure with a twenty-year life will come as periodic outlays of $1,000 at the end of the fifth year, $2,000 at the end of the tenth year, and $3,500 at the end of the fifteenth year. With interest at 10%, what is the equivalent uniform annual cost of maintenance for the twenty-year period?

a) $200 b) $262 c) $300 d) $325 e) $342

5.7 After a factory has been built near a stream, it is learned that the stream occasionally overflows its banks. A hydrologic study indicates that the probability of flooding is about 1 in 8 in any one year. A flood would cause about $20,000 in damage to the factory. A levee can be constructed to prevent flood damage. Its cost will be $54,000 and its useful life is thirty years. Money can be borrowed for 8% interest. If the annual equivalent cost of the levee is less than the annual expectation of flood damage, the levee should be built. The annual expectation of flood damage is $(1/8) \times 20{,}000 = \$2{,}500$. Compute the annual equivalent cost of the levee.

a) $1,261 b) $1,800 c) $4,320 d) $4,800 e) $6,750

5.8 If $10,000 is borrowed now at 6% interest, how much will remain to be paid after a $3,000 payment is made four years from now?

a) $7,000 b) $9,400 c) $9,625 d) $9,725 e) $10,700

*5.9 A piece of machinery costs $20,000 and has an estimated life of eight years and a scrap value of $2,000. What uniform annual amount must be set aside at the end of each of the eight years for replacement if the interest rate is 4%?

a) $1,953 b) $2,170 c) $2,250 d) $2,500 e) $2,898

*5.10 The maintenance costs associated with a machine are $2,000 per year for the first ten years, and $1,000 per year thereafter. The machine has an infinite life. If interest is 10%, what is the present worth of the annual disbursements?

a) $16,145 b) $19,678 c) $20,000 d) $100,000 e) infinite

*5.11 A manufacturing firm entered into a ten-year contract for raw materials which required a payment of $100,000 initially and $20,000 per year beginning at the end of the fifth year. The company made unexpected profits and asked that it be allowed to make a lump sum payment at the end of the third year to pay off the remainder of the contract. What lump sum is necessary if the interest rate is 8%?

a) $85,600 b) $100,000 c) $120,000 d) $200,000 e) $226,000

UNUSUAL CASH FLOWS AND INTEREST PAYMENTS

5.12 A bank currently charges 10% interest compounded annually on business loans. If the bank were to change to continuous compounding, what would be the effective annual interest rate?

a) 10% b) 10.517% c) 12.5% d) 12.649% e) infinite

*5.13 Terry bought an electric typewriter for $50 down and $30 per month for 24 months. The same machine could have been purchased for $675 cash. What nominal annual interest rate is Terry paying?

a) 7.6% b) 13.9% c) 14.8% d) 15.2% e) 53.3%

5.14 How large a contribution is required to endow perpetually a research laboratory which requires \$50,000 for original construction, \$20,000 per year for operating expenses, and \$10,000 every three years for new and replacement equipment? Interest is 4%.

a) \$70,000 b) \$640,000 c) \$790,000 d) \$1,000,000 e) infinite

ANNUAL EQUIVALENT COST COMPARISONS

*5.15 One of the two production units described below must be purchased. The minimum attractive rate of return is 12%. Compare the two units on the basis of equivalent annual cost.

	Unit A	Unit B
Initial Cost	\$16,000	\$30,000
Life	8 years	15 years
Salvage value	\$ 2,000	\$ 5,000
Annual operating cost	\$ 2,000	\$ 1,000

a) *A* — \$5,058; *B* — \$5,270
b) *A* — \$4,916; *B* — \$4,872
c) *A* — \$3,750; *B* — \$2,667
d) *A* — \$1,010; *B* — \$1,010
e) *A* — \$2,676; *B* — \$4,250

5.16 Tanks to hold a corrosive chemical are now being made of material *A*, and have a life of eight years and a first cost of \$30,000. When these tanks are four years old, they must be relined at a cost of \$10,000. If the tanks could be made of material *B*, their life would be twenty years and no relining would be necessary. If the minimum rate of return is 10%, what must be the first cost of a tank made of material *B* to make it economically equivalent to the present tanks?

a) \$30,000 b) \$40,000 c) \$51,879 d) \$58,760 e) \$100,000

PRESENT EQUIVALENT COST COMPARISONS

5.17 Compute the life cycle cost of a reciprocating compressor with first cost of \$120,000, annual maintenance cost of \$9,000, salvage value of \$25,000 and life of six years. The minimum attractive rate of return is 10%.

a) \$120,000 b) \$145,000 c) \$149,000 d) \$153,280 e) \$167,900

5.18 A punch press costs \$100,000 initially, requires \$10,000 per year in maintenance expenses, and has no salvage value after its useful life of ten years. With interest at 10%, the capitalized cost of the press is:

a) \$100,000 b) \$161,400 c) \$197,300 d) \$200,000 e) \$262,700

5.19 A utility is considering two alternatives for serving a new area. Both plans provide twenty years of service, but plan *A* requires one large initial investment, while plan *B* requires additional investment at the end of ten years. Neglect salvage value, assume interest at 8%, and determine the present cost of both plans.

	Plan A	**Plan B**
Initial Investment	$50,000	$30,000
Investment at end of 10 years	none	$30,000
Annual property tax and maintenance, years 1-10	$ 800	$ 500
Annual property tax and maintenance, years 11-20	$ 800	$ 900

a) *A* — $48,780; *B* — $49,250
b) *A* — $50,000; *B* — $30,000
c) *A* — $50,000; *B* — $60,000
d) *A* — $57,900; *B* — $50,000
e) *A* — $66,000; *B* — $74,000

*5.20 The heat loss of a bare steam pipe costs $206 per year. Insulation *A* will reduce heat loss by 93% and can be installed for $116; insulation *B* will reduce heat loss by 89% and can be installed for $60. The insulations require no additional expenses and will have no salvage value at the end of the pipe's estimated life of eight years. Determine the present net equivalent value of the two insulations if the interest rate is 10%.

a) *A* — $116; *B* — $90
b) *A* — $906; *B* — $918
c) *A* — $1,022; *B* — $978
d) *A* — $1,417; *B* — $1,406
e) *A* — $1,533; *B* — $1,467

INCREMENTAL APPROACH

5.21 A helicopter is needed for six years. Cost estimates for two copters are:

	The Whirl 2B	**The Rote 8**
Price	$95,000	$120,000
Annual maintenance	3,000	9,000
Salvage value	12,000	25,000
Life in years	3	6

With interest at 10%, what is the annual cost advantage of the Rote 8?

a) 0   b) $4,260   c) $5,670   d) $5,834   e) $6,000

5.22 A standard prime mover costs $20,000 and has an estimated life of six years. By the addition of certain auxiliary equipment, an annual savings of $300 in operating costs can be obtained, and the estimated life of the prime mover extended to nine years. Salvage value in either case is $5,000. Interest on capital is 8%. Compute the maximum expenditure justifiable for the auxiliary equipment.

a) $1,149   b) $1,800   c) $2,700   d) $7,140   e) $13,300

*5.23 An existing electrical transmission line needs to have its capacity increased, and this can be done in either of two ways. The first method is to add a second conductor to each phase wire, using the same poles, insulators and fittings, at a construction cost of $15,000 per mile. The second method for increasing capacity is to build a second line parallel to the existing line, using new poles, insulators and fittings, at a construction cost of $23,000 per mile. At some time in the future, the transmission line will require another increase in capacity, with the first alternative now requiring a second line at a cost of $32,500 per mile, and the second alternative requiring added conductors at a cost of $23,000 per mile. If interest rate is 6%, how many years between the initial expenditure and the future expenditure will make the two methods economically equal?

a) 1 b) 3 c) 5 d) 10 e) 25

REPLACEMENT PROBLEMS

5.24 One year ago machine *A* was purchased at a cost of $2,000, to be useful for five years. However, the machine failed to perform properly and costs $200 per month for repairs, adjustments and shut-downs. A new machine *B* designed to perform the same functions is quoted at $3,500, with the cost of repairs and adjustments estimated to be only $50 per month. The expected life of machine *B* is five years. Except for repairs and adjustments, the operating costs of the two machines are substantially equal. Salvage values are insignificant. Using 8% interest rate, compute the incremental annual net equivalent value of machine *B*.

a) − $877 b) $923 c) $1,267 d) $1,800 $2,677

BREAK-EVEN ANALYSIS

5.25 Bear Air, an airline serving the Arctic, buys 6000 spark plugs per year at a price of $18 each. It costs them $85 to process each order. If Bear Air's minimum attractive rate of return is 10%, what is the most economic quantity of spark plugs to buy at one time?

a) 1 b) 115 c) 532 d) 561 e) 6,000

*5.26 Bear Air has been contracting its overhaul work to Aleutian Aeromotive for $40,000 per plane per year. Bear estimates that by building a $500,000 maintenance facility with a life of 15 years and a salvage value of $100,000, they could handle their own overhauls at a variable cost of only $30,000 per plane per year. The maintenance facility could be financed with a secured loan at 8% interest. What is the minimum number of plans Bear must operate in order to make the maintenance facility economically feasible?

a) 5 b) 6 c) 10 d) 40 e) 50

5.27 It costs Bear Air $1,200 to run a scheduled flight, empty or full, from Coldfoot to Frostbite. Moreover, each passenger generates a cost of $40. The regular ticket costs $90. The plane holds 65 people, but it is running only about 20 per flight. The sales director has suggested selling special tickets for $50 to Eskimos who do not normally fly.

What is the minimum number of Eskimo tickets that must be sold in order for a flight to produce a profit?

a) 5 b) 10 c) 15 d) 20 e) 45

What would be the total profit on the flight from Coldfoot to Frostbite if all 65 passengers started wearing sealskins and buying Eskimo tickets?

a) − $800 b) − $550 c) 0 d) $400 e) $500

5.28 Two electric motors are being considered for an application in which there is uncertainty concerning the hours of usage. Motor *A* costs $3,000 and has an efficiency of 90%. Motor *B* costs $1,400 and has an efficiency of 80%. Each motor has a ten-year life and no salvage value. Electric service costs $1.00 per year kW of demand and $0.01 per kWh of energy. The output of the motors is to be 74.6 kW, and interest rate is 8%. At how many hours usage per year would the two motors be equally economical? If the usage is less than this amount, which motor is preferable?

a) 1800, *A* b) 1800, *B* c) 2200, *A* d) 2200, *B* e) 2500, *A*

INCOME TAX AND DEPRECIATION

5.29 A drill press is purchased for $10,000 and has an estimated life of twelve years. The salvage value at the end of twelve years is estimated to be $1,300. Using straight-line depreciation, compute the book value of the drill press at the end of eight years.

a) $1,300 b) $3,333 c) $3,475 d) $4,200 e) $4,925

*5.30 A grading contractor owns earth-moving equipment that cost $90,000. At the end of its eight-year estimated life, it will have a salvage value of $18,000. Compute the depreciation for the first two years and the book value at the end of five years by the straight-line method.

a) $9,000; $9,000; $45,000
b) $11,250; $11,250; $33,750
c) $16,000; $14,000; $30,000
d) $18,000; $18,000; $90,000
e) $22,500; $16,875; $21,358

*5.31 Rework Problem 5.30 using the sum of years' digits method.

a) $9,000; $9,000; $45,000
b) $11,250; $11,250; $33,750
c) $16,000; $14,000; $30,000
d) $18,000; $18,000; $90,000
e) $22,500; $16,875; $21,358

*5.32 Rework Problem 5.30 using the double-declining balance method.

a) $9,000; $9,000; $45,000
b) $11,250; $11,250; $33,750
c) $16,000; $14,000; $30,000
d) $18,000; $18,000; $90,000
e) $22,500; $16,875; $21,358

5.33 An asset has an initial cost of $80,000, an expected life of six years, and no salvage value. Assume a minimum attractive rate of return of ten percent and an incremental tax rate of fifty percent. Compute the present equivalent value of the tax savings achieved by using the sum of the years' digits method rather than the straight-line method for computing depreciation.

a) $2,300 b) $4,600 c) $29,000 d) $31,300 e) $58,000

6. Electrical Theory

A review of concepts in Electrical Engineering of interest to all engineers is presented in this chapter. The subjects include basic circuits, basic electromagnetics, the fundamentals of operational electronics, and the basics of electric power distribution and machinery. Each section begins with a brief statement of the concepts to be covered and a listing of pertinent formulas with units. These are followed by example problems and, in conclusion, a collection of review problems.

6.1 Circuits

Electric circuits are interconnected electrical components which can be either passive or active. Passive devices such as resistors, inductors and capacitors absorb or store electrical energy. Active devices such as signal generators and amplifiers are actually energy converters and have the capability of changing the form in which electrical energy is handled. Essentially, in making calculations pertaining to electrical circuits, one is interested in simply keeping track of the electrical energy or power involved and the form in which it exists.

The primary quantities of interest in making circuit calculations are presented in Table 6.1.

Table 6.1 Quantities Used in Electric Circuits.

Quantity	*Symbol*	*Unit*	*Defining Equation*	*Definition*
Charge	Q	coulomb	$Q = \int Idt$	
Current	I	ampere	$I = \frac{dQ}{dt}$	Time rate of flow of charge past a point in the circuit.
Voltage	V	volt	$V = \frac{dW}{dt}$	Energy per unit charge either gained or lost through a circuit element
Energy	W	joules	$W = \int VdQ = \int Pdt$	
Power	P	watts	$P = \frac{dW}{dt} = IV$	Power is the time rate of energy flow.

The circuits reviewed in this section will contain one or more sources interconnected with passive components. These passive circuit components include:

a) Resistors, R, are energy absorbing components and have a resistance value measured in ohms:

$$I = \frac{V}{R} \tag{6.1.1}$$

b) Inductors, L, are energy storage components and have an inductance value measured in henries:

$$I = \frac{1}{L}\int V dt \tag{6.1.2}$$

c) Capacitors, C, are energy storage components and have a capacitance value measured in farads:

$$I = C\frac{dV}{dt} \tag{6.1.3}$$

Sources in electric circuits can be either independent of current and/or voltage values elsewhere in the circuit, or they can be dependent upon them. In this section only independent sources will be considered. Fig. 6.1 shows both ideal and linear models for current and voltage sources.

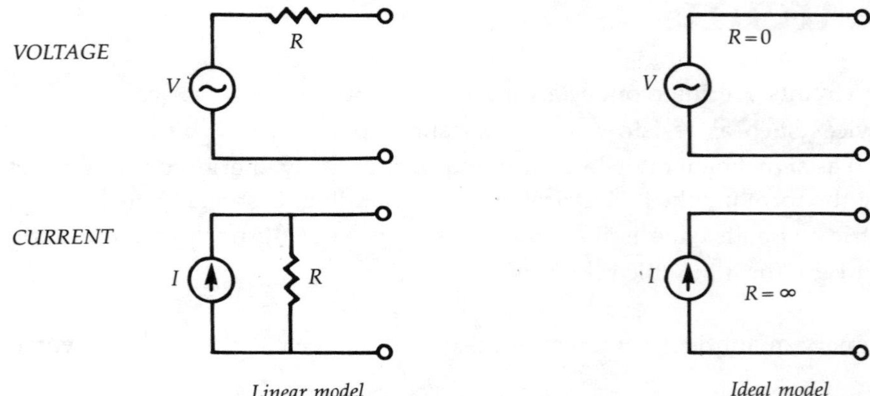

Figure 6.1 Ideal and linear models of current and voltage sources.

Two laws of conservation govern the behavior of all electrical circuits:

a) Kirchoff's Voltage Law (KVL), for the conservation of energy, states that the sum of voltage rises or drops around any closed path in an electrical circuit must be zero.

$$\underset{\text{around closed path}}{\sum V_{DROPS} = 0} \qquad \underset{\text{around closed path}}{\sum V_{RISES} = 0} \tag{6.1.4}$$

b) Kirchoff's Current Law (KCL), for the conservation of charge, states that the flow of charges either into or out of any node in a circuit must add to zero.

$$\underset{\text{at node}}{\sum I_{IN} = 0} \qquad \underset{\text{at node}}{\sum I_{OUT} = 0} \tag{6.1.5}$$

6.1.1 DC Circuits

In a DC circuit the only crucial components are resistors. Another component, the inductor, appears as a zero resistance connection and a third component, a capacitor, appears as an infinite resistance, or open circuit in a DC circuit. The three circuit components are summarized in Table 6.2.

Table 6.2 DC Circuit Components

Component	*Impedance*	*Current*	*Power*
Resistor	R	$I = V/R$	$P = I^2R = V^2/R$
Inductor	Zero	Depends on series R	None dissipated
Capacitor	Infinite	Zero	None dissipated

EXAMPLE 6.1

Compute the current in the 10Ω resistor.

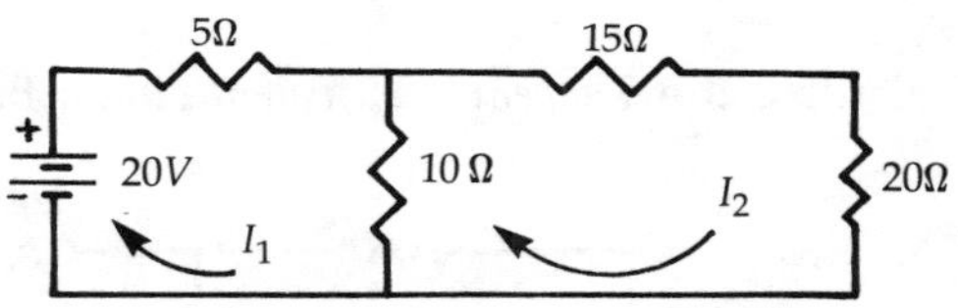

Solution. Assume loop currents I_1 and I_2. Write KVL around both meshes:

$$\sum V_{DROPS} = 0$$

Mesh 1: $-20 + 5I_1 + 10I_1 - 10I_2 = 0$

Mesh 2: $-10I_1 + 10I_2 + 15I_2 + 20I_2 = 0$

These are arranged as

$$15I_1 - 10I_2 = 20$$

$$-10I_1 + 45I_2 = 0$$

The solution is

$$I_1 = 1.57\text{A}, \qquad I_2 = 0.35\text{A}.$$

The current in the 10Ω resistor is

$$I = I_1 - I_2$$

$$= 1.57 - 0.35 = 1.22\text{A}.$$

EXAMPLE 6.2

Compute the power delivered to the 6Ω resistor.

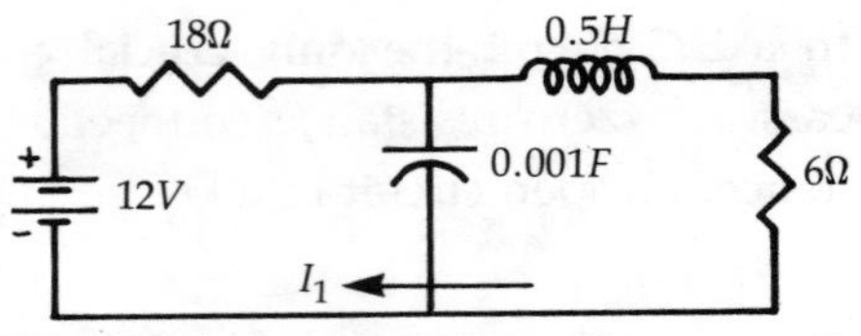

Solution. Only one current path exists since no DC current flows through a capacitor; the voltage drop across the inductor is zero. Write KVL:

$$\sum V_{DROPS} = 0$$

$$-12 + 18I_1 + 6I_1 = 0$$

$$\therefore\ I_1 = 0.5\text{A}.$$

The power is then

$$P = I^2R$$

$$= 0.5^2 \times 6 = 1.5\text{W}.$$

EXAMPLE 6.3

Calculate the *Thevenin equivalent circuit* between points a and b for the circuit shown.

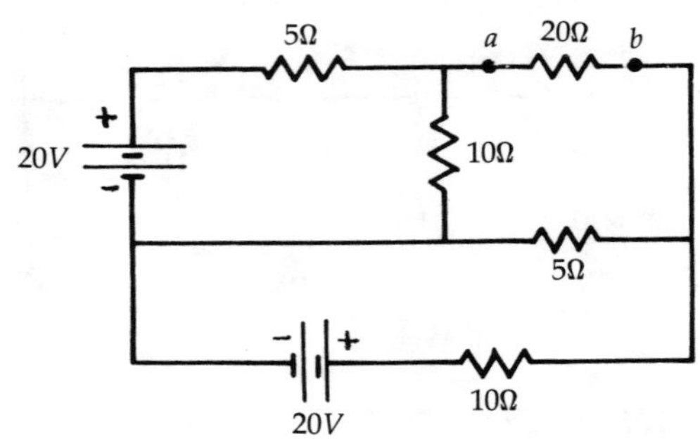

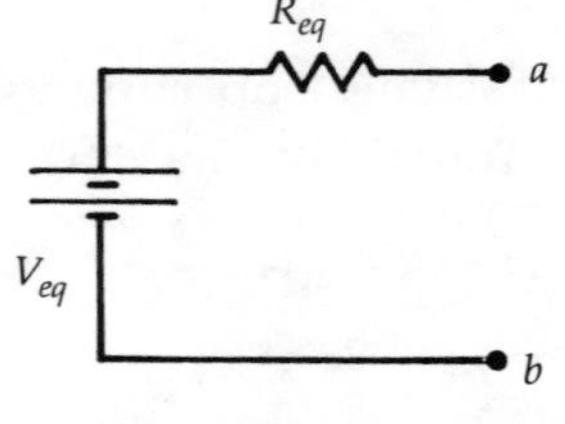

Thevenin equivalent circuit

Solution. V_{eq} is the voltage across a–b with the 20Ω resistor out of the circuit. R_{eq} is the resistance across a–b with the 20Ω resistor removed and the 20V sources replaced by a short circuit.

The voltage V_{eq} is formed as follows:

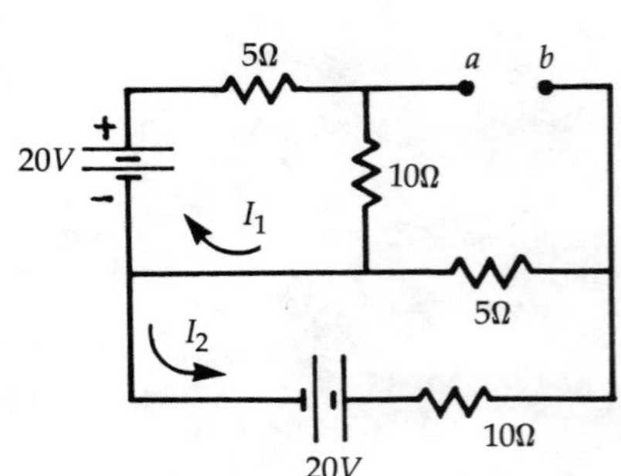

$$I_1 = \frac{20}{15} = \frac{4}{3}\text{ A.}$$

$$I_2 = \frac{20}{15} = \frac{4}{3}\text{ A.}$$

$$-V_{ab} + 10I_1 - 5I_2 = 0.$$

$$\therefore\ V_{ab} = 10 \times \frac{4}{3} - 5 \times \frac{4}{3} = 6.67\text{ V.}$$

$$\therefore\ V_{eq} = 6.67\text{ V.}$$

To find R_{eq}, the voltage sources are replaced by short circuits:

$$R_{eq} = \frac{1}{\frac{1}{5} + \frac{1}{10}} + \frac{1}{\frac{1}{10} + \frac{1}{5}}$$

$$= 6.67\Omega.$$

6.1.2 AC Circuits—Single Phase

Single phase AC circuits operate with sinusoidal signal sources and are assumed to be operating steady state. That is, all transients have long since died out. Sinusoidal currents and voltages can be represented either as trigonometric expressions or as complex numbers where the explicit time dependence has been dropped. Consider the following two notations representing a sinusoidal voltage:

Trigonometric *Complex*

$$V(t) = V_m\cos(\omega t + \phi) \qquad V = \frac{V_m}{\sqrt{2}}\, e^{j\phi} \tag{6.1.6}$$

where V_m is the maximum voltage. The conversion from trigonometric to complex notation is given by the expression:

$$V(t) = Re[V_m e^{j\phi} e^{j\omega t}] \tag{6.1.7}$$

where "*Re*" means "the real part of". Once sinusoidal currents and voltages are expressed in complex notation the solution of AC circuits becomes nearly identical to the solution of DC circuits. The difference being that the inductors and capacitors can no longer be considered simple short, or open circuit components.

The following table presents the definitions and relationships between current, voltage, and power for the passive circuit components *R*, *L*, and *C*.

Table 6.3 AC Circuit Components

Component	*Impedance* (*ohms*)	*Admittance* (*mhos*)	*Current* (*amperes*)	*Power* (*watts*)
Resistor	R	$G = \frac{I}{R}$	$I = \frac{V}{R}$	$P = I^2R = \frac{V^2}{R}$ real/dissipated
Inductor	$jX_L = j\omega L$	$-jB_L = \frac{-j}{X_L}$	$I = \frac{-jV_L}{\omega L}$	$Q = I^2X_L = \frac{V_L^2}{X_L}$ reactive/stored
Capacitor	$-jX_C = \frac{-j}{\omega C}$	$jB_C = \frac{j}{X_C}$	$I = jV_C\omega C$	$Q_C = I^2X_C = \frac{V_C^2}{X_C}$ reactive/stored
R, L, C	Z	Y	$I = V/Z$	–

For a series connection of components, the impedances add to give a total impedance for the combination; there results

$$\text{Series } R\text{–}L: \quad Z = R + jX_L = \sqrt{R^2+X_L^2}\ \underline{/\theta_L} = |Z|\underline{/\theta_L}$$
$$\text{Series } R\text{–}C: \quad Z = R - jX_C = \sqrt{R^2+X_C^2}\ \underline{/\theta_C} = |Z|\underline{/\theta_C} \tag{6.1.8}$$

For a parallel connection of components, the admittances add to give a total admittance as follows:

$$\text{Parallel } R\text{–}L: \quad Y = G - jB_L = \sqrt{G^2+B_L^2}\ \underline{/\theta_L} = |Y|\underline{/\theta_L}$$
$$\text{Parallel } R\text{–}C: \quad Y = G + jB_C = \sqrt{G^2+B_C^2}\ \underline{/\theta_C} = |Y|\underline{/\theta_C} \tag{6.1.9}$$

The relationship between Z and Y is always $Z = 1/Y$.

Each circuit parameter in an AC circuit can be represented by a complex number, where a subscript "r" denotes the real part, a subscript "x" denotes the imaginary part, and an asterisk denotes the complex conjugate:

$$\text{Current:} \quad I = I_r + jI_x = |I|\ \underline{/\theta}$$
$$\text{Voltage:} \quad V = V_r + jV_x = |V|\ \underline{/\theta} \tag{6.1.10}$$
$$\text{Power:} \quad S = VI^* = P + jQ$$

where

$$P = \text{real dissipated power} = Re[VI^*] = |V|\,|I| \cos\theta \tag{6.1.11}$$

$$Q = \text{reactive stored power} = Im[VI^*] = |V|\,|I| \sin\theta \tag{6.1.12}$$

and θ in these power equations is as displayed in Fig. 6.2. The quantity $\cos\theta$ is often referred to as the *power factor, pf*. These complex quantities can be visualized using the phasor diagrams and triangles shown in Fig. 6.2.

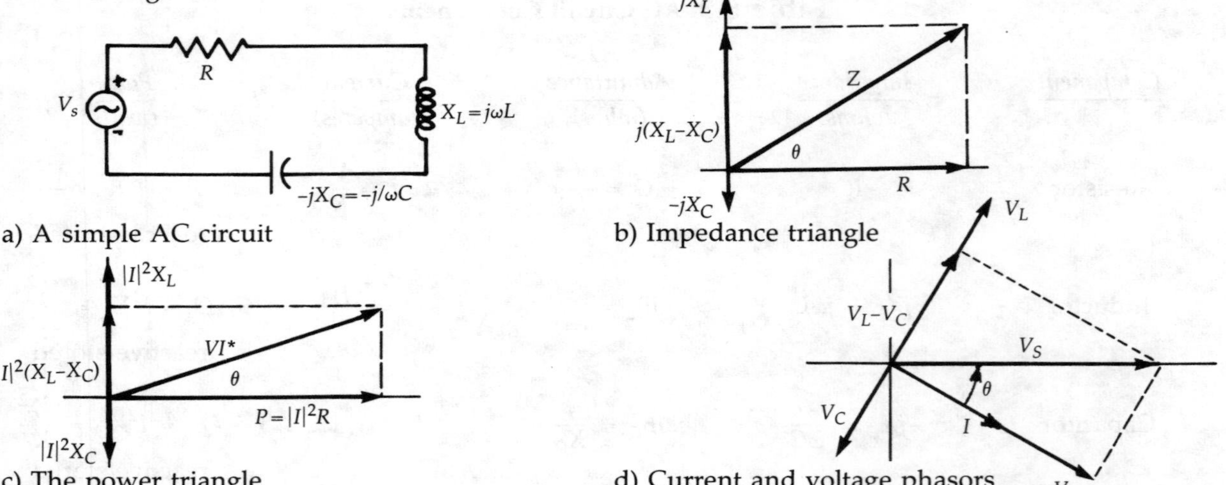

Figure 6.2 Diagrams for the simple AC circuit.

EXAMPLE 6.4

Compute the power delivered to the 25Ω resistor.

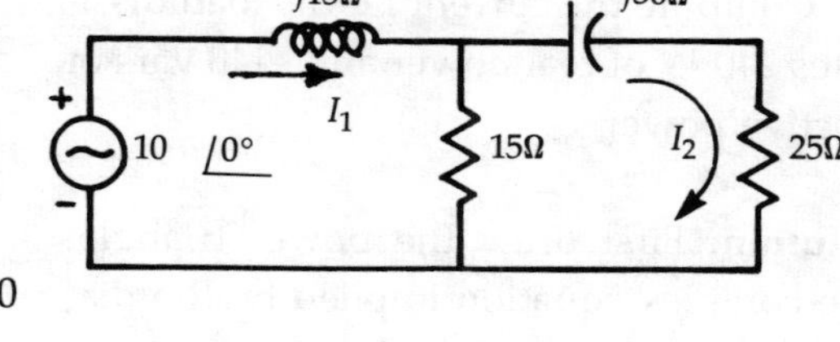

Solution. Use the KVL as follows:

$$\Sigma V_{DROPS} = 0$$

$$10 + j15I_1 + 15I_1 - 15I_2 = 0$$

$$-15I_1 + 15I_2 - j30I_2 + 25I_2 = 0.$$

These are written as

$$(15 + j15)I_1 - 15I_2 = -10$$

$$-15I_1 + (40 - j30)I_2 = 0.$$

A simultaneous solution yields

$$I_2 = -\,0.179 \angle{-10.3^\circ}\text{ A}.$$

The power delivered is then

$$P = |I_2|^2R = 0.179^2 \times 25 = 0.800\text{ W}.$$

EXAMPLE 6.5

Compute the value of X_C so that I is in phase with V_S.

Solution. The impedance of the parallel load is

$$Z_{load} = \frac{-jX_C\,(j20)}{-jX_C + j20} = \frac{20X_C}{j(20 - X_C)}$$

To make I in phase with V_S, the total impedance seen by the generator must have a zero imaginary part:

$$j60 + \frac{-j20X_C}{20 - X_C} = 0.$$

$$\therefore\ X_C = 15\Omega.$$

EXAMPLE 6.6

Compute the current I if the load dissipates 240 W of real power and –140 vars of reactive power.

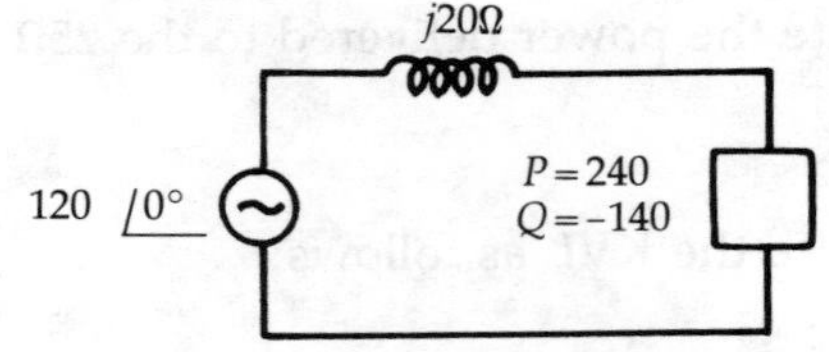

Solution. First draw the power triangle. The complex equation implied by the diagram is:

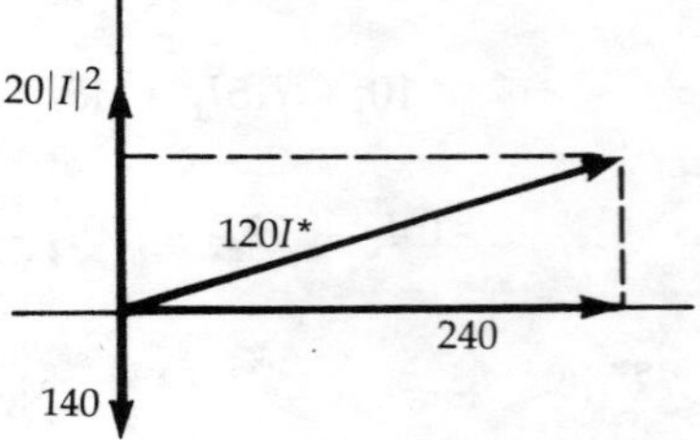

$$120\,(I_r - jI_x) = 240 + j\,[20(I_r^2 + I_x^2) - 140]$$

$$\therefore\ 120I_r = 240 \quad\text{and}\quad -120I_x = 20\,(I_r^2 + I_x^2) - 140$$

The solution, then, is

$$I = 2 + j\,0.464 \qquad\text{or}\qquad I = 2 - j\,6.464.$$

Additional information would be needed to narrow the selection to one of the above answers.

6.1.3 AC Circuit—Three Phase

Three-phase circuits are composed of three single-phase circuits where the source voltages for each phase are 120° apart. The load impedances for three-phase circuits can be connected in either a wye (Y) connection or a delta (Δ) connection as shown in Fig. 6.3. Phasor diagrams for a typical set of source voltages are given in Fig. 6.4. One may further specify the source voltage to be either line-to-line or line-to-neutral. Fig. 6.4 shows both the line-to-neutral voltages, which can be used for wye-connected load, and the line-to-line voltages, which can be used for either a delta- or wye-connected load. The advantage of three-phase circuits is that the instantaneous power flow is uniform and not pulsating as it is in a single-phase circuit.

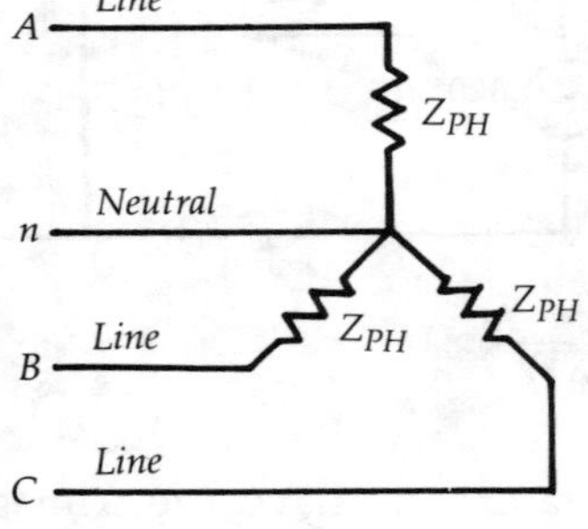

wye-connected load

Line
A
Z_{PH} Z_{PH}
Line
B
Z_{PH}
Line
C

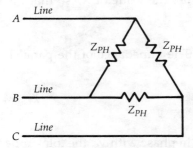

delta-connected load

Figure 6.3 **Load impedances.**

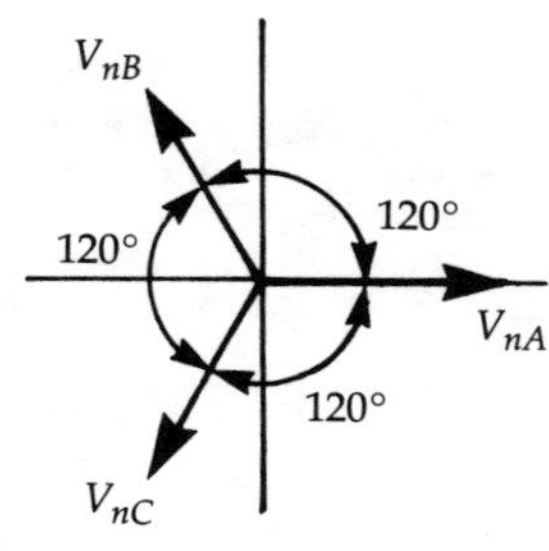

line-to-neutral voltages

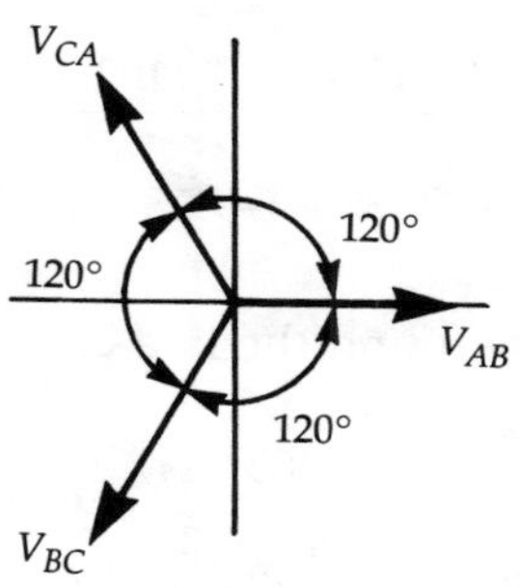

line-to-line voltages

Figure 6.4 **Phasor diagrams.**

Kirchoff's circuit laws apply to three-phase circuits just as they apply to all other circuits; however, it is typically not necessary to write complex sets of circuit equations to solve the simple balanced circuits being considered here. Just remember that where all three-phase impedances are equal the circuits are balanced. The expressions used in solving balanced circuits are tabulated in Table 6.4.

Table 6.4 **AC Circuit Expressions (Three-Phase)**

Load	*Current*	*Voltage*	*Power*
wye load	$I_L = I_{PH} = \dfrac{V_{PH}}{Z_{PH}}$	$V_L = \sqrt{3}\, V_{PH}$	$P = 3P_{PH} = 3\lvert V_{PH}\rvert\, \lvert I_{PH}\rvert \cos\theta$ $= \sqrt{3}\, \lvert V_L\rvert\, \lvert I_L\rvert \cos\theta$ $Q = 3Q_{PH} = 3\lvert V_{PH}\rvert\, \lvert I_{PH}\rvert \sin\theta$ $= \sqrt{3}\, \lvert V_L\rvert\, \lvert I_L\rvert \sin\theta$
delta load	$I_L = \sqrt{3}\, I_{PH}$	$V_L = V_{PH} = I_{PH}Z_{PH}$	same as above

EXAMPLE 6.7

A balanced delta-connected load is driven from a 208-volt line to a line system as shown. Calculate the line current I_A.

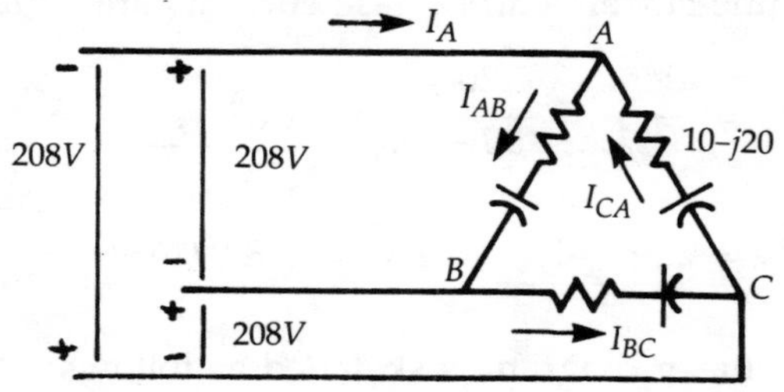

Solution. Sketch the phasor diagram to assist in the calculations. The currents are calculated as follows:

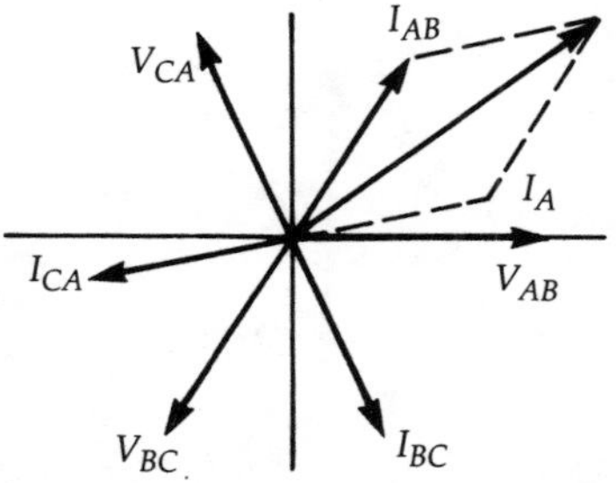

$$I_{AB} = \frac{V_{AB}}{10 - j20} = \frac{208 \angle 0^\circ}{10 - j20} = 9.30 \angle 63.4^\circ$$

$$\therefore\ I_{CA} = 9.30 \angle 183.4^\circ$$

Kirchoff's Current Law allows us to find the line current as

$$I_A = I_{AB} - I_{CA}$$

$$= 9.30 \angle 63.4^\circ - 9.30 \angle 183.4^\circ$$

$$= 4.17 + j8.32 - (-9.28 - j\,0.55) = 13.45 + j8.87 = 16.11 \angle 33.41^\circ \text{ A}$$

We could have found the line current magnitude as follows:

$$I_{PH} = \frac{V_{PH}}{Z_{PH}} = \frac{208}{\sqrt{10^2+20^2}} = 9.30 \text{ A}$$

$$\therefore I_L = \sqrt{3}\, I_{PH} = \sqrt{3} \times 9.30 = 16.11\text{A}$$

EXAMPLE 6.8

A balanced Y-connected load is attached to a 3ϕ four-wire line with a 120 V line to neutral voltages as shown. Sketch a phasor diagram showing all line and phase currents and voltages. Assume V_{nA} to be at zero phase angle.

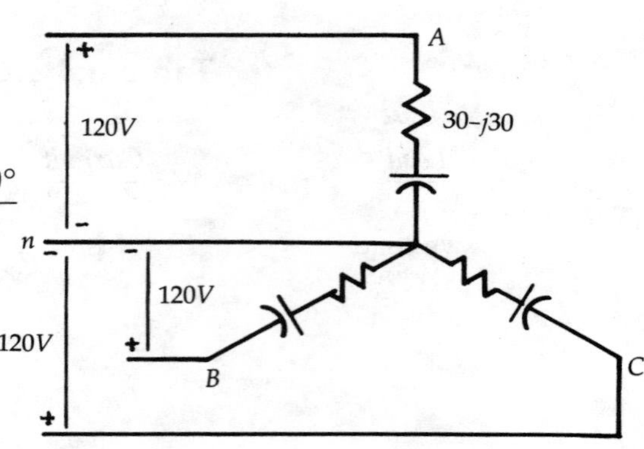

Solution. The phase voltages are expressed as

$$V_{nA} = 120 \angle 0^\circ \qquad V_{nB} = 120 \angle -120^\circ \qquad V_{nC} = 120 \angle 120^\circ$$

The line voltages are then

$$V_{AB} = V_{An} + V_{nB} = 208 \angle -150^\circ$$

$$V_{BC} = V_{Bn} + V_{nC} = 208 \angle 90^\circ$$

$$V_{CA} = V_{Cn} + V_{nA} = 208 \angle -30^\circ$$

The line currents and phase currents are equal, so that

$$I_A = \frac{V_{nA}}{Z_{PH}} = \frac{120 \angle 0^\circ}{30 - j30} = 2.83 \angle 45^\circ$$

$$I_B = 2.83 \angle -75^\circ \qquad I_C = 2.83 \angle 165^\circ$$

The phasor diagram is sketched as follows:

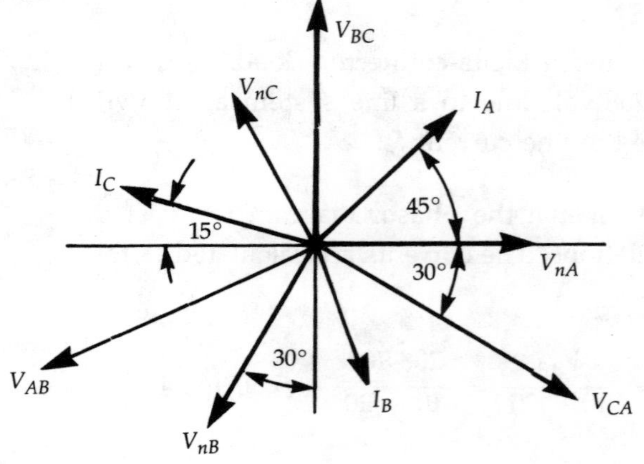

EXAMPLE 6.9

A balanced Y-connected load is connected to a three-phase source as shown. Determine the power delivered to the load.

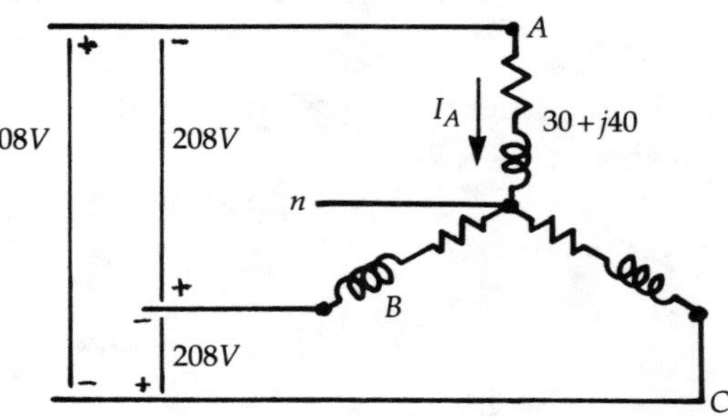

Solution. First calculate the phase voltage V_{PH}:

$$V_{PH} = \frac{V_L}{\sqrt{3}} = V_{nA} = \frac{208 \angle 150°}{\sqrt{3}} = 120 \angle 150 \text{ V}.$$

Note: If V_{AB} is assumed to be at 0°, then V_{nA} is at -150°; see Example 6.8. The phase current is

$$I_{PH} = \frac{V_{PH}}{Z_{PH}} = \frac{120 \angle 150°}{50 \angle 53.1°} = 2.4 \angle -96.9° \text{ A}$$

The power dissipated in phase A is

$$P_{PH} = |V_{PH}|\ |I_{PH}| \cos\theta$$

$$= 120 \times 2.40 \times \cos 53.1° = 172.9 \text{ W}$$

$$\therefore P_{total} = 3 \times 172.9 = 519 \text{ W}$$

This could also have been computed as

$$P_{total} = \sqrt{3}\ |V_L|\ |I_L| \cos\theta$$

$$= \sqrt{3} \times 208 \times 2.40 \cos 53.1° = 519 \text{ W}$$

6.1.4 DC Transients

Transients in a DC circuit can exist any time a sudden change is made in the source voltage(s) applied to the circuit. In the cases to be considered for this review, the transient will be due to the initial connection of the source to the circuit. The problem that engineers will face, in their test-taking and in their work, is one of computing the current in the circuit at any specified time.

Unlike the steady state solutions of DC circuits, transient solutions require the inclusion of capacitive and inductive effects as functions of time. If only a single inductor or a single capacitor are contained in a circuit, the current will either grow or decay exponentially. If both an inductor and a capacitor are included in the circuit, oscillations can occur. Only the single capacitor and single inductor cases are considered here.

For exponentially changing functions a time constant (T) can be defined. The time constant is the time it takes to complete 63.2% of the change that will ultimately be made. This is a practical number, since in a time corresponding to four or five time constants most transients have disappeared from a circuit.

Several general expressions can be written for the voltage and current in a DC transient circuit. Two of these are:

$$I(t) = I_\infty + (I_o - I_\infty)\, e^{-t/T}$$
$$V(t) = V_\infty + (V_o - V_\infty)\, e^{-t/T} \qquad (6.1.8)$$

where

I_o and V_o are the values at the instant of a sudden change.

I_∞ and V_∞ are the values that will exist in the circuit after the transient has died away (let $t = \infty$).

$T = L/R$ for series R–L circuits.

$T = RC$ for series R–C circuits.

EXAMPLE 6.10

The switch is closed at $t = 0$. Determine the current in the circuit at $t = 2T$, two time constants after the switch is closed.

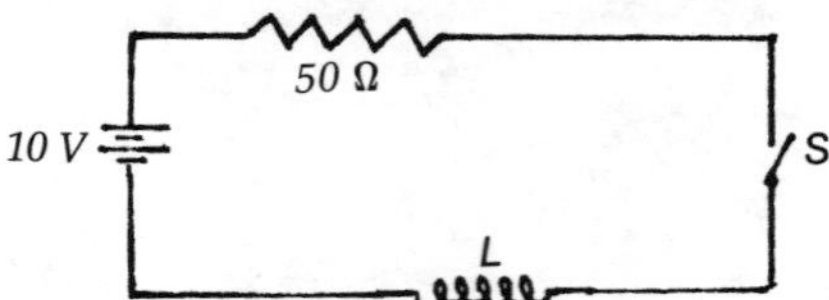

Solution. The general expression for the transient current in a series R–L or R–C circuit is

$$I(t) = I_\infty + (I_o - I_\infty)\, e^{-t/T}$$

For the case being considered

$$I_o = 0, \qquad I_\infty = \frac{10}{50} = 0.2$$

Thus

$$I(2T) = 0.2 - 0.2e^{-2} = 0.173 \text{ A}$$

6.2 Static Electric Fields

Electric fields exist any time an electric charge is present in the region. The presence of an electric field causes many things to happen. It produces forces on stationary charges, forces on charges moving in conductors and, therefore, forces on the conductors themselves. The forces on the charges in a conductor also cause the motion of those charges to produce current. Forces on the charges in an atom can cause the atom to come apart, creating a material breakdown. In addition to their usefulness in calculating forces, electric field quantities can be used to calculate stored energy and capacitance for any shape of electrodes on which the original charge may exist, as well as the potential difference between those electrodes.

Here we will examine only the most basic concepts of electric field theory and attempt to do so with minimal use of mathematics.

The two most basic field quantities are *electric flux density* $\vec{D}$ (coulombs/meter^2) and *electric field intensity* $\vec{E}$ (newtons/coulomb or volts/meter). These are defined as follows:

$$\vec{E} = \vec{F}/Q$$

$$E = dV/dL \qquad (6.2.1)$$

For the derivative definition, $\vec{E}$ is in the direction of the maximum rate of change. The electric flux density is a vector quantity whose magnitude is the maximum value of the derivative

$$D = dQ/dA \qquad (6.2.2)$$

and $\vec{D}$ is perpendicular to the surface dA.

The relationship between D and E depends on the material in which the field exists through a material parameter called the *permittivity* ϵ; the relationship is

$$E = D/\epsilon \qquad (6.2.3)$$

where $\epsilon = \epsilon_o\epsilon_r$, $\epsilon_o = 8.85 \times 10^{-12}$ $C/n{\bullet}m^2$ (the permittivity of free space). The *relative permittivity* ϵ_r depends on the material and will be specified in a particular problem. The field quantities for some simple geometries are especially useful. They follow.

Point charge q:

$$D = \frac{q}{4\pi r^2} \qquad E = \frac{q}{4\pi\epsilon r^2} \qquad (6.2.4)$$

with vectors $\vec{D}$ and $\vec{E}$ directed radially away from the point charge.

Uniform line of charge ρ_L C/m:

$$D = \frac{\rho_L}{2\pi r} \qquad E = \frac{\rho_L}{2\pi\epsilon r} \qquad (6.2.5)$$

with vectors $\vec{D}$ and $\vec{E}$ normal to the line of charge.

Uniform plane of charge ρ_S C/m^2:

$$D = \frac{\rho_S}{2} \qquad E = \frac{\rho_S}{2\epsilon} \qquad (6.2.6)$$

with vectors $\vec{D}$ and $\vec{E}$ directed normally away from the plane.

If D and E are not known they must be calculated using Eqs. 6.2.1 and 6.2.2 from the known voltages or charges. Once D and E are known they can be used to calculate the following quantities:

Voltage:

$$V = \int \vec{E}{\bullet}\vec{dL} \qquad (6.2.7)$$

Charge: $$Q = \iint \vec{D} \cdot \vec{dA} \tag{6.2.8}$$

Stored electrical energy: $$W_E = \frac{1}{2} \oint \vec{D} \cdot \vec{E} dV = \frac{1}{2} CV^2 \tag{6.2.9}$$

Capacitance: $$C = Q/V = 2W_E/V^2 \tag{6.2.10}$$

Current: $$I = \iint \sigma \vec{E} \cdot \vec{dA} \tag{6.2.11}$$

Force on a charge q: $$\vec{F} = q\vec{E} \tag{6.2.12}$$

The following examples will illustrate the use of several of the above relationships.

EXAMPLE 6.11

A point charge of $q = 5 \times 10^{-9}$ coulombs is located in rectangular coordinates at (1,3,5) and a line charge $\rho_L = 1.5 \times 10^{-9}$ coulomb/m is located parallel to the x-axis at $y = 3$ and $z = 0$. Find the point where the net electric field is zero.

Solution. The field intensity from the line charge is

$$E\rho = \frac{\rho_L}{2\pi\epsilon_o r}$$

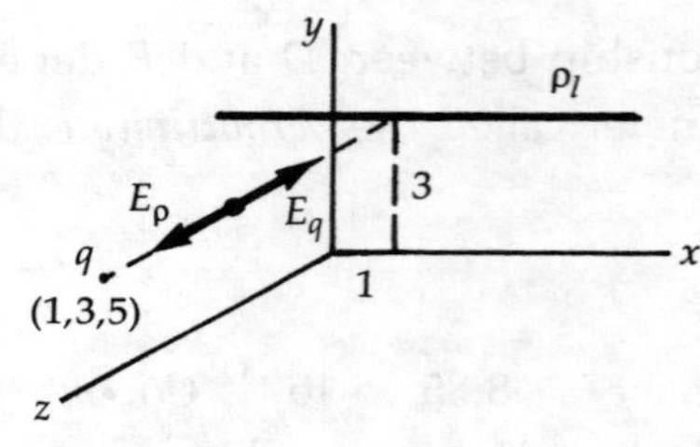

The field intensity from the point charge is

$$E_q = \frac{q}{4\pi\epsilon_o r^2}$$

Since the E fields from the sources are in opposite directions, we simply equate the above two expressions, recognizing that r is different in each equation:

$$\frac{1.5 \times 10^{-9}}{2\pi z\epsilon_o} = \frac{5 \times 10^{-9}}{4\pi\epsilon_o(5-z)^2}$$

The quadratic $3z^2 - 35z + 75 = 0$ results. The appropriate root is $z = 2.829$ m.

EXAMPLE 6.12

Concentric coaxial cylinders 50 cm long are separated by air with permittivity ϵ_o. If the radius of the inner conductor is 0.2 cm, and the inner radius of the outer conductor is 1.0 cm, compute the capacitance per meter between the cylinders.

Solution. Assume a charge ρ_L on the inner conductor, $-\rho_L$ on the outer conductor and compute the potential difference V. Combining Eqs. 6.2.5 and 6.2.7 we have

$$V = \int_{0.2}^{1.0} \vec{E} \cdot \vec{dl} = \int_{0.2}^{1.0} \frac{\rho_L}{2\pi r \epsilon_o} dr$$

$$= \frac{\rho_L}{2\pi \epsilon_o} \ ln \ \frac{1.0}{0.2}$$

To find the capacitance per meter between the cylinders we use Eq. 6.2.10 and obtain

$$C = \frac{\rho_L}{V} = \frac{2\pi \epsilon_o}{ln \ 5}$$

$$\frac{2\pi \times 8.85 \times 10^{-12}}{ln \ 5} = 34.55 \times 10^{-12} \text{ F/m}$$

EXAMPLE 6.13

The voltage distribution between two coaxial cylinders is given by

$$V = 50 \ ln \ \frac{2.5}{r} \text{ volts}$$

where r varies from 0.5 cm to 2.5 cm. At what radius r between the cylinders does the largest magnitude of the E field exist?

Solution. The E field is given by

$$E = \frac{dV}{dr}$$

$$= 50 \frac{1}{2.5/r} \left(-\frac{2.5}{r^2} \right) = -\frac{50}{r}$$

By observation the largest magnitude of the E field occurs at the minimum value of r, i.e., $r = 0.5$ cm.

6.3 Static Magnetic Fields

Magnetic fields exist any time there is a moving charge or electric current in the region. Just as electric fields interact with charges, magnetic fields interact with moving charges or currents. A current-carrying conductor in a magnetic field will have a force exerted on it due to its presence in that magnetic field. In addition to their usefulness in calculating forces, magnetic field quantities can be used to calculate energy stored in a magnetic field and inductance for any configuration of current-carrying conductors.

The two most basic magnetic field quantities are *magnetic field intensity* $\vec{H}$ (amperes/meter) and *magnetic flux density* $\vec{B}$ (webers/meter^2 or tesla).

Magnetic field intensity is defined by

$$H = dI/dL \tag{6.3.1}$$

where $\vec{H}$ is in the direction of the maximum rate of change.

Magnetic flux density is defined by

$$B = F/IL \tag{6.3.2}$$

where $\vec{B}$ is perpendicular to both $\vec{F}$ and $\vec{IL}$. $\vec{F}$ is the force on the current element $\vec{IL}$ residing in a field with flux density $\vec{B}$. The relationship between B and H depends on the material in which the field exists through a material parameter called the *permeability* μ. The relationship is

$$H = B/\mu \tag{6.3.3}$$

where $\mu = \mu_o\mu_r$, $\mu_o = 4\pi \times 10^{-7}$ H/m (the permeability of free space). The *relative permeability* μ_r depends on the material and will be specified in a particular problem.

The expression for the magnetic field produced by a straight conductor carrying a current I is especially useful. It is

$$B = \frac{\mu I}{4\pi r}(\sin\theta_1 + \sin\theta_2) \tag{6.3.4}$$

where $\vec{B}$ is in a direction perpendicular to a plane containing the point and the straight wire and where r is the perpendicular distance shown (see Fig. 6.5). A general expression, called *Amperes Rule*, for computing the field B given a known current distribution is

$$\vec{B} = \frac{\mu}{4\pi}\int \frac{I\vec{dl}\times\hat{a}_R}{R^2} \tag{6.3.5}$$

The parameters for Eq. 6.3.5 are defined in Fig. 6.6; $\hat{a}_R$ is a unit vector.

Once B is known, the energy stored in the magnetic field of an inductor can be found, using the following:

$$\begin{aligned} W &= \frac{1}{2}\int \vec{B}\cdot\vec{H}\, dV \\ &= \frac{1}{2}LI^2 \end{aligned} \tag{6.3.6}$$

where I is the current producing the field B. Knowing B, the total force on a current-carrying conductor in the field can be computed using

$$\vec{F} = \vec{IL} \times \vec{B} \tag{6.3.7}$$

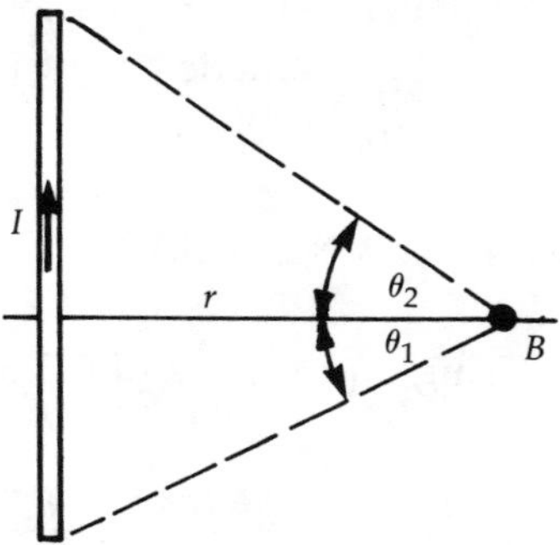

Figure 6.5 **Magnetic field due to a straight conductor.**

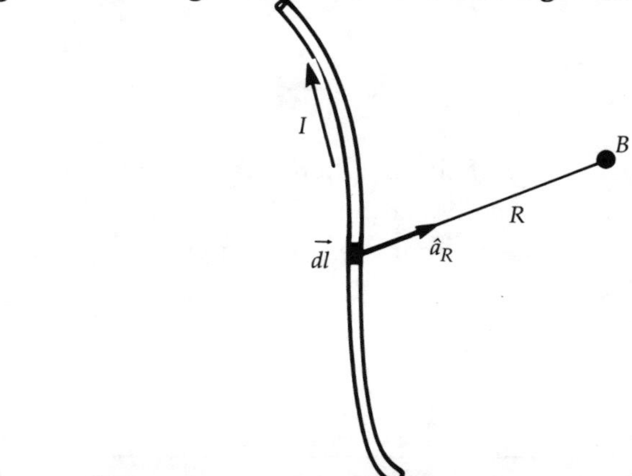

Figure 6.6 **Magnetic field due to a curved conductor.**

A magnetic circuit is formed when a magnetic field is confined inside a closed circuit of magnetic material. If one recognizes the analogies, the solution of a magnetic circuit is similar to the solution of an electric circuit. Fig. 6.7 shows a simple magnetic circuit along with an analogous electric circuit. The equation similar to Kirchoff's voltage law is

$$NI = HL \tag{6.3.8}$$

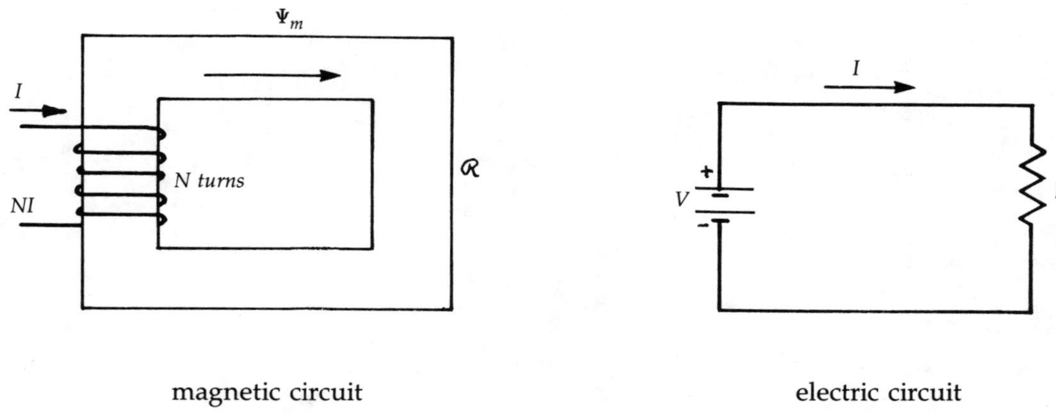

Figure 6.7 **A magnetic circuit with its analogous electric circuit.**

NI is the *mmf* due to the current (*I*) in the coil and is similar to the *emf* of the battery in the electric circuit. *H L* is the magnetic potential drop in the magnetic circuit and is similar to the *I R* voltage drop in the electric circuit. The magnetic flux Ψ_m in the magnetic circuit is similar to the current *I* in the electric circuit and is related to the flux density *B* by

$$\Psi_m = BA \tag{6.3.9}$$

where *A* is the cross-sectional area. Using Eqs. 6.3.3 and 6.3.8 this can be put in the form

$$\Psi_m = \frac{\mu NIA}{L} = \frac{NI}{\mathcal{R}} \tag{6.3.10}$$

where we have used

$$\mathcal{R} = \frac{L}{\mu A} \tag{6.3.11}$$

The quantity $\mathcal{R}$ is called the *reluctance* of the magnetic circuit and is similar to the resistance of the electric circuit. The primary difference between the electric circuit and the magnetic circuit is that *R* is linear whereas $\mathcal{R}$ is nonlinear. This means that the solution of a magnetic circuit is best done graphically.

EXAMPLE 6.14

Compute the value of the magnetic field intensity at the point (1, –1) in the figure shown.

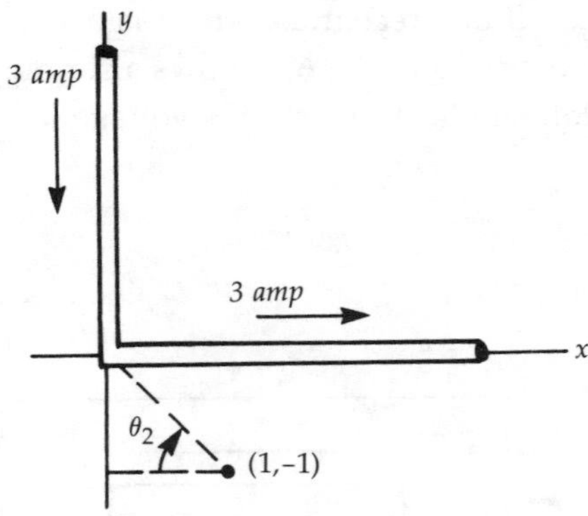

Solution. The field from a straight wire is given by

$$\vec{B} = \frac{\mu I}{4\pi r}(\sin\theta_1 + \sin\theta_2)\,\hat{a}$$

For the length along the y-axis

$$\vec{B} = \frac{\mu_o 3}{4\pi\times 1}(\sin 90° - \sin 45°)\,\hat{z}$$

$$= \frac{3\mu_o}{4\pi}(1 - 1/\sqrt{2})\,\hat{z}$$

For the length along the *x*-axis

$$\vec{B} = \frac{3\mu_o}{4\pi\times 1}(\sin 90° + \sin 45°)\,(-\hat{z})$$

$$= -\frac{3\mu_o}{4\pi}(1 + 1/\sqrt{2})\,\hat{z}$$

Adding:

$$\vec{B}_{total} = -\frac{3\mu_o}{2\pi\sqrt{2}}\ \hat{z}\ \text{webers/m}^2$$

EXAMPLE 6.15

Two parallel conductors each carry 50 amperes, but in opposite directions. If the conductors are one meter apart, compute the force in newtons per meter on either conductor.

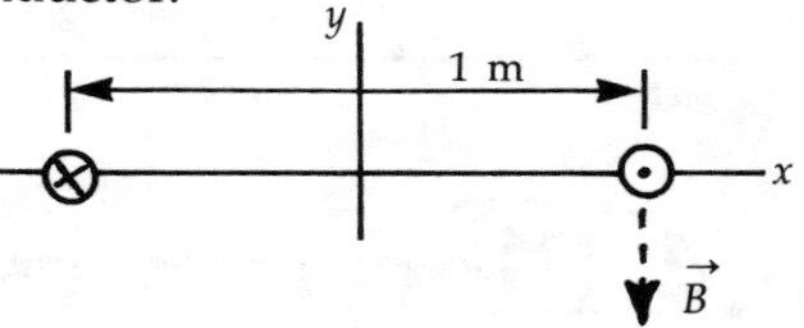

Solution. First, it is necessary to compute the field B that one conductor produces at the location of the other conductor; it is, using μ_o for air,

$$B = \frac{\mu_o I}{2\pi r}$$

$$= \frac{4\pi \times 10^{-7} \times 50}{2\pi \times 1} = 10^{-5}\ \text{Wb/m}$$

If the wires are aligned parallel to the z-axis, as shown, then $\vec{B} = -10^{-5}\,\hat{y}$ where $\hat{y}$ is a unit vector in the y-direction. This is the field at the right conductor due to the current in the left conductor. Then, using Eq. 6.3.7 there results

$$\vec{F} = I\vec{L} \times \vec{B}$$

$$50(1)\,\hat{z} \times (-10^{-5}\,\hat{y}) = 50 \times 10^{-5}\,\hat{x}\ \text{N}$$

Note that the conductors repel one another.

EXAMPLE 6.16

A magnetization curve for the cast steel of an inductor core is shown. The magnetic flux in the iron is $\Psi_m = 8 \times 10^{-5}$ Wb. Find the current I.

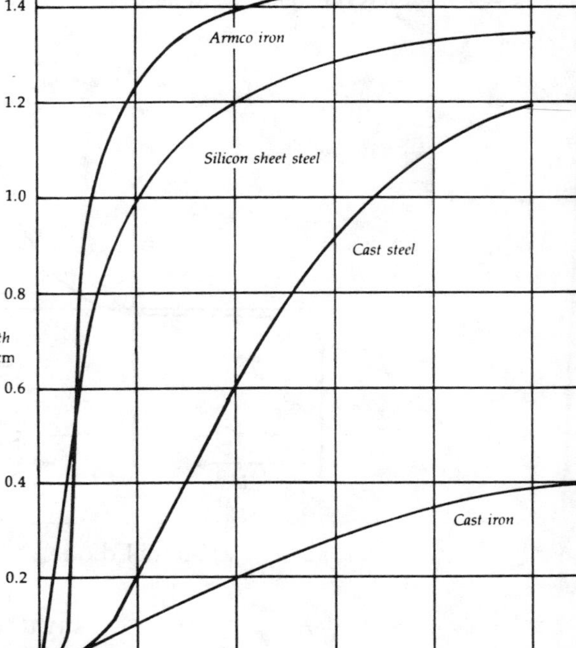

Solution. The flux density B is

$$B = \frac{\Psi_m}{A}$$

$$= \frac{8 \times 10^{-5}}{0.0004} = 0.2\ \text{Wb/m}^2$$

Using the appropriate curve we find H to be

$$H = 200 \text{ A} \cdot \text{t/m}$$

Using Eq. 6.3.8, the current is found to be

$$I = \frac{HL}{N}$$

$$= \frac{200 \times 0.2}{50} = 0.8 \text{ A}$$

6.4 Electronics

Electronics is the study of circuits which are used to process electrical signals (voltages or currents) which contain information. It involves the use of both passive circuit components—which have been previously discussed—as well as active and/or nonlinear circuit components, which have not been discussed until this point in the review.

This review of electronics will include simple diodes and operational amplifiers. Diodes are nonlinear devices which are used in power supplies, wave shaping and logic circuits. Operational amplifiers are active devices which are used to build circuits which can be used to integrate and differentiate signals, sum signals together, and filter portions of signals out, as well as amplify signals.

6.4.1 Diodes

The forward and reverse characteristics of an actual diode are shown in Fig. 6.8 along with the characteristics of an ideal diode. An ideal diode has zero current flow in the reverse direction and zero

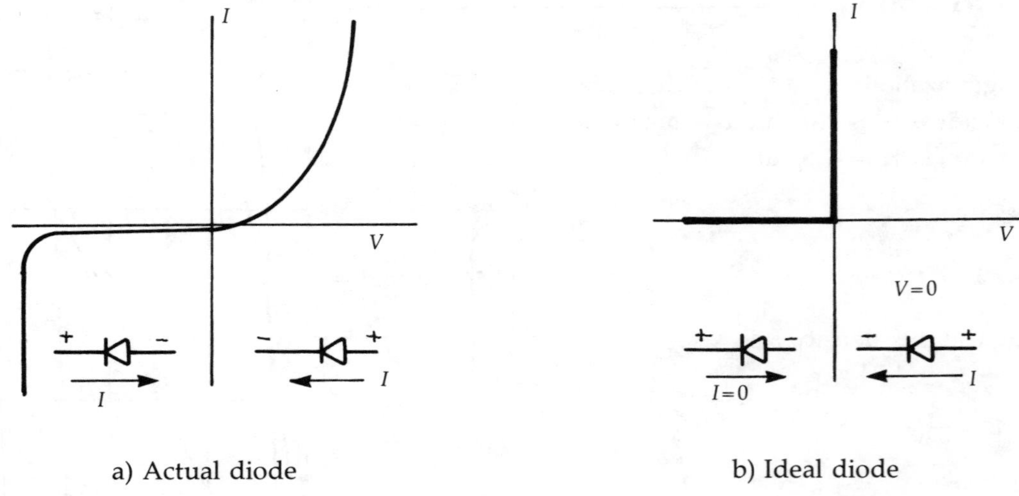

a) Actual diode

b) Ideal diode

Figure 6.8 **Characteristics of a Diode.**

voltage drop across the diode when current is flowing in the forward direction. The analysis of diode circuits is greatly simplified when ideal diodes are assumed. Table 6.5 gives the characteristics of some simple diode circuits, assuming that they are ideal.

Table 6.5 Some Simple Diode Circuits

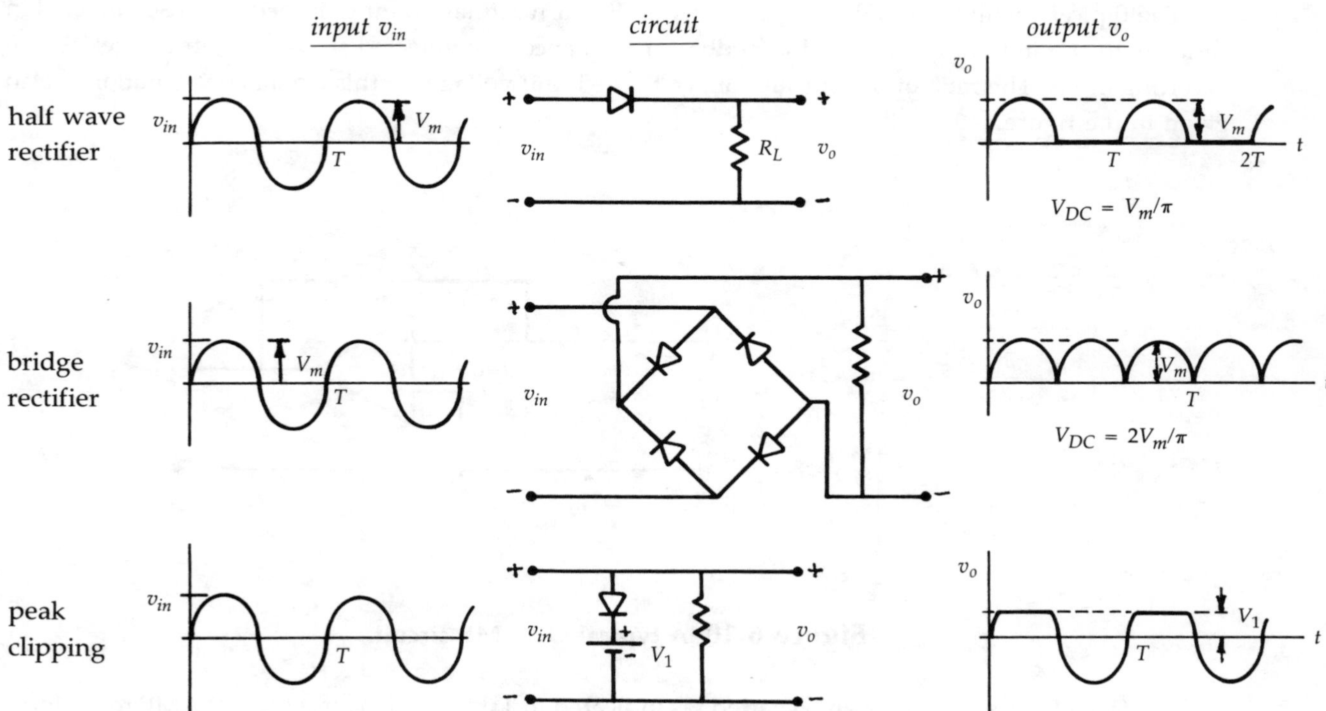

6.4.2 Operational Amplifiers

The symbolic representation of an operational amplifier is shown in Fig. 6.9. The OP-AMP, as it is often called, has two inputs. One is marked with a "+" sign and is called the non-inverting input, which means that the output is of the same polarity as the input. If the input is positive with respect to ground, then the output is also positive. The other input is marked with a "–" and is called the inverting input, which means that the output is opposite in polarity to the input with respect to ground.

The most important feature of the OP-AMP is the extremely high gain A that it possesses. *Gain* is defined as the ratio of the output voltage to the input voltage. With an extremely high gain it only

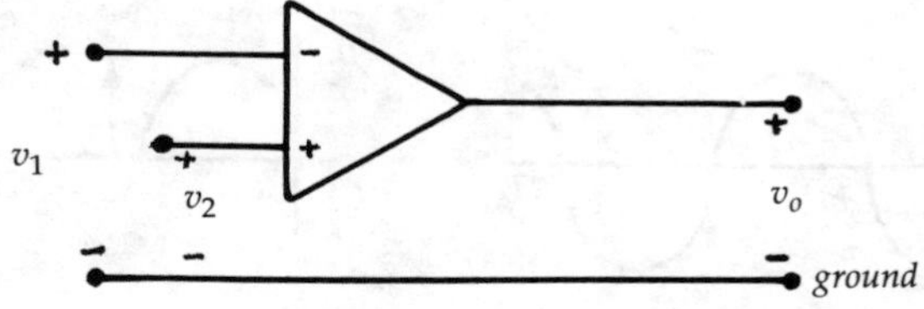

Figure 6.9 **An operational amplifier.**

requires a very, very small input signal to realize a finite output signal. The input current is, in fact, so small that it may safely be assumed to be zero. In the analysis of OP-AMP circuits the input current to either of the input ports is assumed to be zero and the two inputs are assumed to be at the same voltage. Normally, the non-inverting input is tied to ground, so the inverting input is assumed to be at ground potential also.

The typical circuit for an OP-AMP is shown in Fig. 6.10. It has an input impedance Z_i connected in series with the inverting input and a feedback impedance Z_f connected from the output back to the inverting input. The ratio of the output voltage to the input voltage for this circuit configuration is also given in the figure.

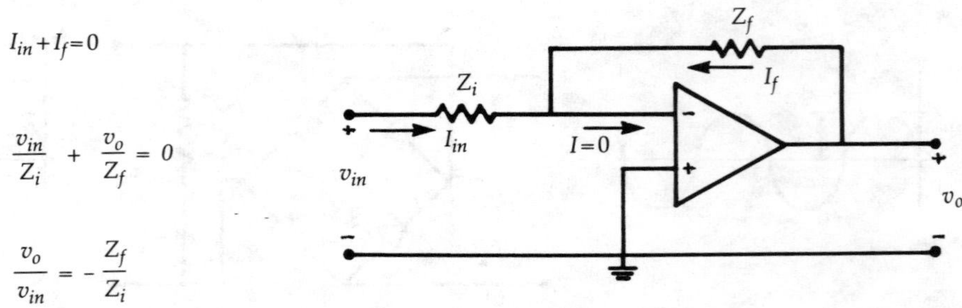

Figure 6.10 **A typical OP-AMP circuit.**

The impedances Z_i and Z_f can be varied, as indicated in Table 6.6, to cause the OP-AMP to perform many different functions. In this table, the lower case *v's* represent arbitrary functions of time and the upper case *V's* represent *rms* values of sinusoidal voltages.

EXAMPLE 6.17

A half-wave rectifier circuit uses an ideal diode. If the desired DC load voltage is 9 V, what is the *rms* value of the source?

Solution. The time-dependent voltages v_{AC} and v_{DC} appear as sketched below.

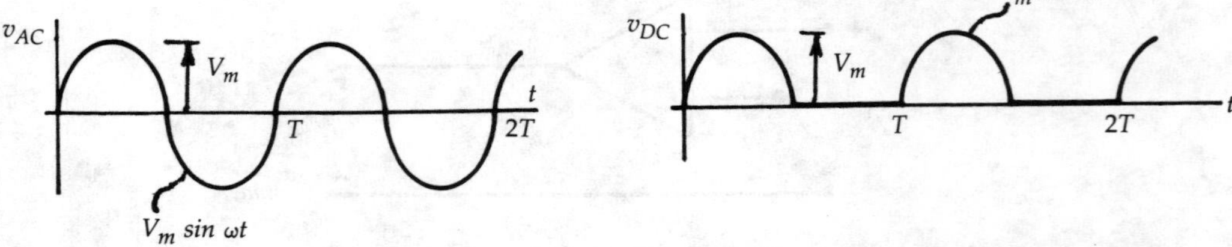

The average, or DC value, of the rectified voltage is given by the integral

$$V_{DC} = \frac{1}{T}\int_0^T v_{DC}(t)dt = \frac{V_m}{\pi}$$

$$\therefore\ V_m = \pi V_{DC} = 9\pi \text{ volts}$$

The required *rms* value of the source is therefore

$$V_{rms} = \frac{V_m}{\sqrt{2}} = \frac{9\pi}{\sqrt{2}} = 19.99 \text{ volts}$$

EXAMPLE 6.18

Determine the gain of the OP-AMP circuit shown.

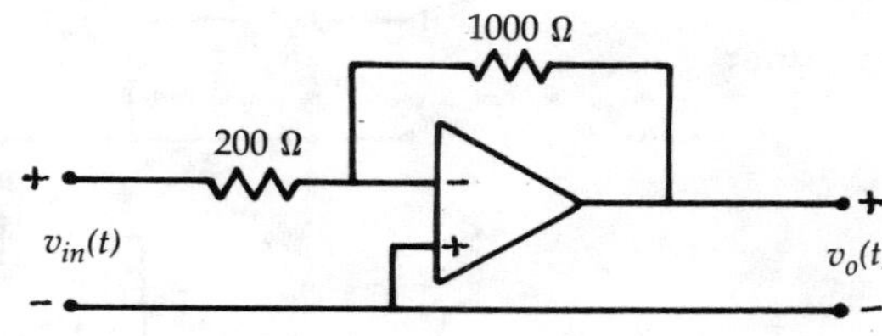

Solution. The gain expression for an OP-AMP circuit is

$$A = -Z_f/Z_i = -R_f/R_i$$

For this example

$$A = -1000/200 = -5$$

so that

$$v_o(t) = -5\ v_{in}(t)$$

EXAMPLE 6.19

A square wave with extrema ± 1 is input to the OP-AMP circuit shown. Sketch $v_o(t)$.

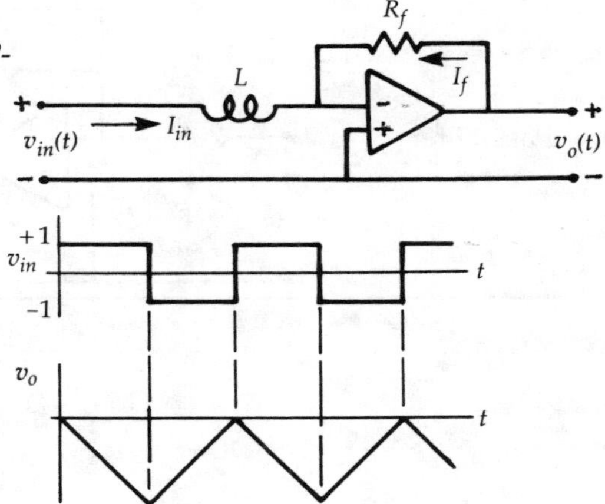

Solution. The KCL allows us to write

$$I_{in} + I_f = 0$$

But we know that

$$I_{in} = \frac{1}{L}\int v_{in}dt \qquad I_f = \frac{v_o}{R_f}$$

Hence,

$$\frac{1}{L}\int v_{in}dt + \frac{v_o}{R_f} = 0$$

or

$$v_o(t) = -\frac{R_f}{L}\int v_{in}dt$$

The circuit shown is an integrator so the output voltage is proportional to the integral of the input voltage.

Table 6.6 Operational Amplifier Functions

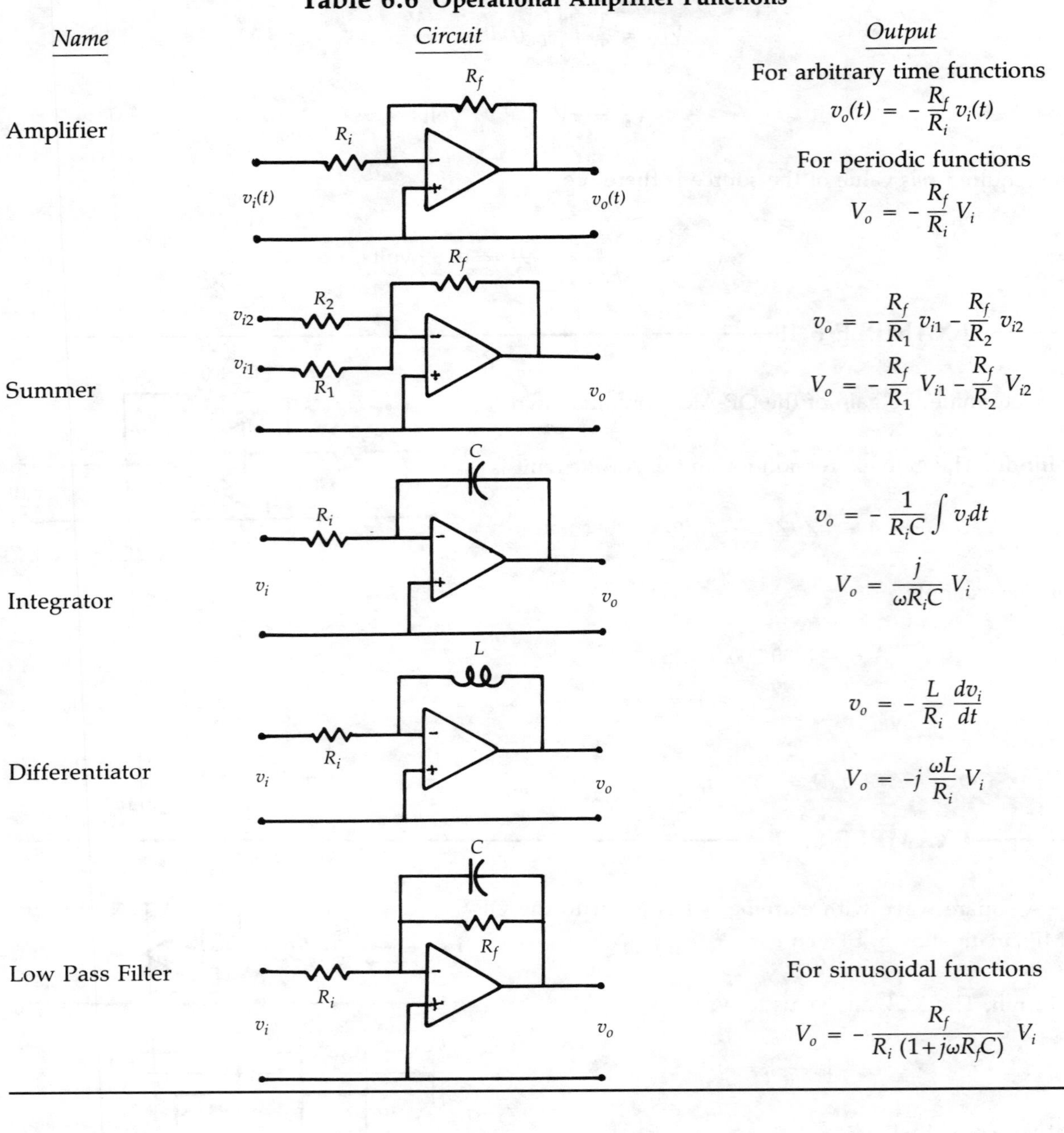

Name	*Circuit*	*Output*
Amplifier	R_f, R_i, $v_i(t)$, $v_o(t)$	For arbitrary time functions $v_o(t) = -\frac{R_f}{R_i} v_i(t)$ For periodic functions $V_o = -\frac{R_f}{R_i} V_i$
Summer	R_f, R_2, R_1, v_{i2}, v_{i1}, v_o	$v_o = -\frac{R_f}{R_1} v_{i1} - \frac{R_f}{R_2} v_{i2}$ $V_o = -\frac{R_f}{R_1} V_{i1} - \frac{R_f}{R_2} V_{i2}$
Integrator	C, R_i, v_i, v_o	$v_o = -\frac{1}{R_i C} \int v_i dt$ $V_o = \frac{j}{\omega R_i C} V_i$
Differentiator	L, R_i, v_i, v_o	$v_o = -\frac{L}{R_i} \frac{dv_i}{dt}$ $V_o = -j \frac{\omega L}{R_i} V_i$
Low Pass Filter	C, R_f, R_i, v_i, v_o	For sinusoidal functions $V_o = -\frac{R_f}{R_i(1+j\omega R_f C)} V_i$

Practice Problems

(If only a few, selected problems are desired, choose those with a star.)

DC CIRCUITS

*6.1 For the circuit below, with the voltages' polarities as shown, KVL in equation form is

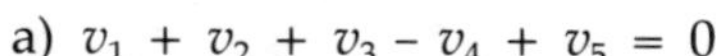

a) $v_1 + v_2 + v_3 - v_4 + v_5 = 0$
b) $-v_1 + v_2 + v_3 - v_4 + v_5 = 0$
c) $v_1 + v_2 - v_3 - v_4 + v_5 = 0$
d) $-v_1 - v_2 - v_3 + v_4 + v_5 = 0$
e) $v_1 - v_2 + v_3 + v_4 - v_5 = 0$

6.2 Find I_1 in amps.

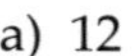

a) 12
b) 15
c) 18
d) 21
e) 27

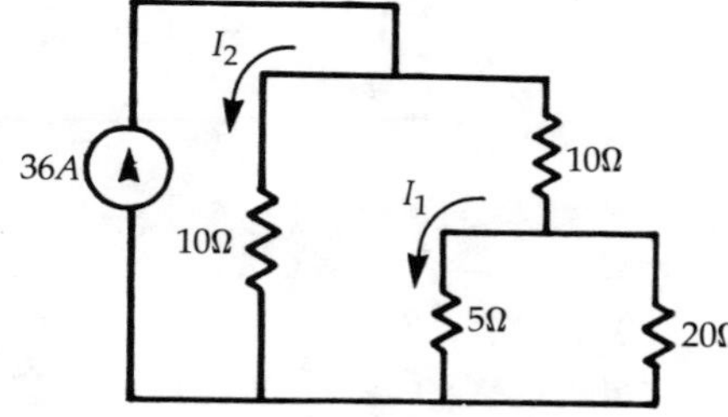

*6.3 Find the magnitude and sign of the power, in watts, absorbed by the circuit element in the box.

a) -20
b) -8
c) 8
d) 12
e) 20

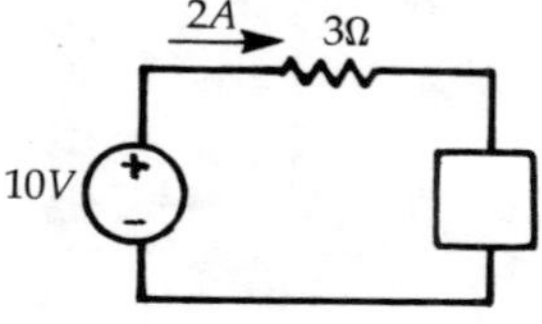

*6.4 For the circuit shown, the voltage across the 4 ohm resistor is

a) 1/4
b) 1/2
c) 2/3
d) 2
e) 4

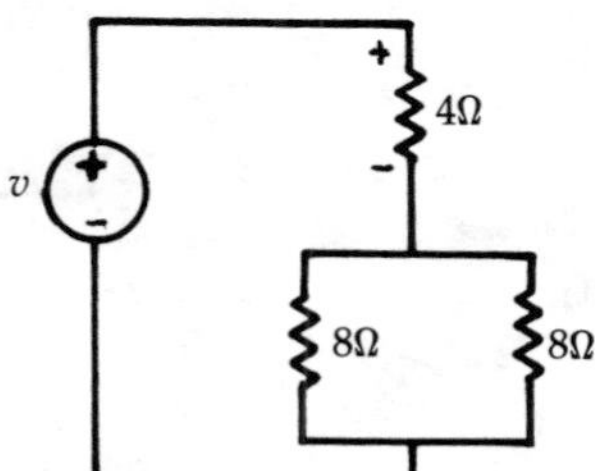

6.5 The total conductance, in mhos, in the circuit shown below is

a) 1/5
b) 1/2
c) 2
d) 5
e) 10

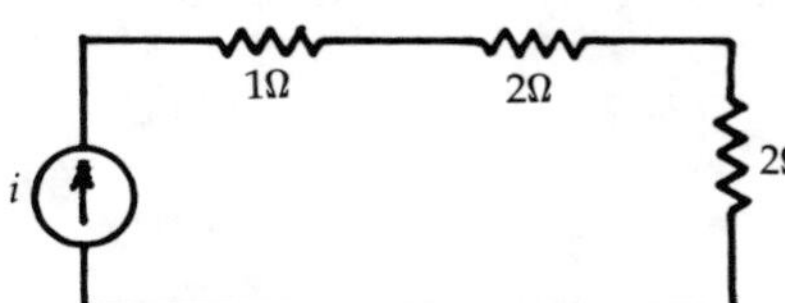

6.6 The power, in watts, absorbed by the 6 mho conductance in the circuit below is

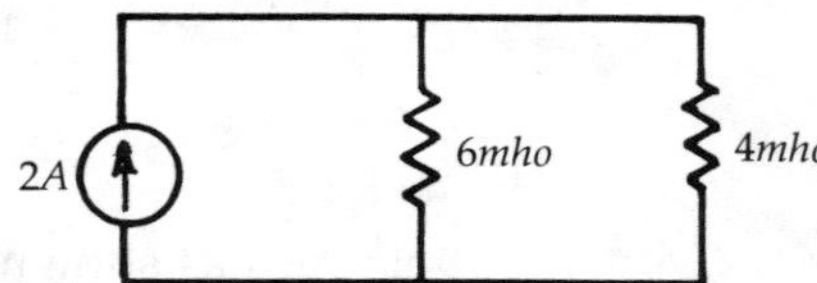

a) –.24
b) .2
c) .24
d) .48
e) 0.54

*6.7 The equivalent resistance, in ohms, between points *a* and *b* in the circuit below is

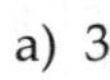

a) 3
b) 5
c) 7
d) 8
e) 10

6.8 The voltage V_2 is

a) 6.4
b) 4.0
c) 2.0
d) 5.6
e) 3.0

6.9 Find I_1 in amperes.

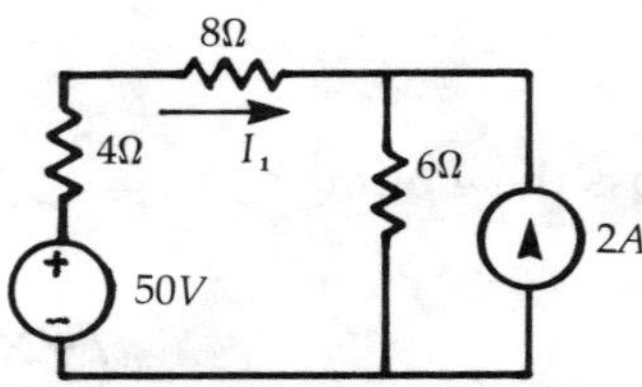

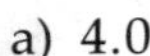

a) 4.0
b) 2.0
c) 4.11
d) 2.11
e) 3.0

AC CIRCUITS–SINGLE PHASE

*6.10 (2 + *j*2) (3 – *j*4) is most nearly

a) 6.0 $\angle -21.8°$
b) 14.1 $\angle -21.8°$
c) 14.1 $\angle -8.1°$
d) 28.0 $\angle -8.1°$
e) 46.0 $\angle -8.1°$

*6.11 The following sinuisoid is displayed on an oscilloscope. The RMS voltage and the radian frequency are most nearly

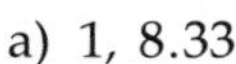

a) 1, 8.33
b) .7071, 52.36
c) 1.4142, 52.36
d) 2, 8.33
e) 2, 52.36

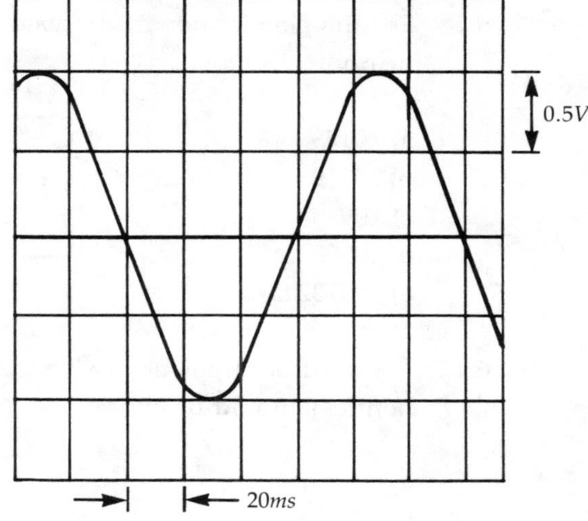

6.12 Find I_2 in amperes.

a) $0.29 + j0.68$
b) $-0.12 + j0.69$
c) $-0.82 - j0.37$
d) $1 - j2$
e) $-3.33 - j4.50$

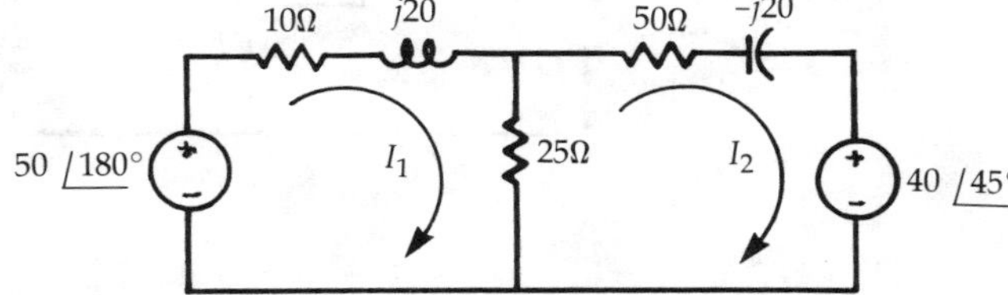

*6.13 Calculate the magnitude of the node voltage V_{AB}.

a) 85.1
b) 77.2
c) 68.8
d) 92.2
e) 102.2

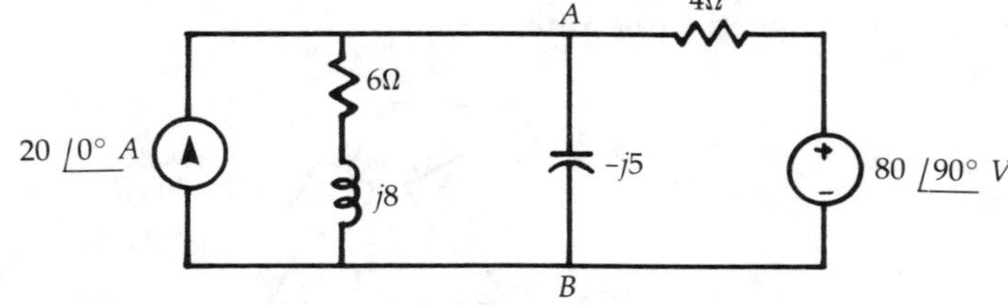

6.14 The peak value of $V(t)$ in the circuit shown is approximately

a) 2.0
b) 3.68
c) 25.9
d) 50.0
e) 71.6

$I(t) = 10 \cos 100t$

V(t)

5Ω

3300μF

*6.15 The power factor of the circuit shown is most nearly

a) 0.5
b) 0.6
c) 0.7
d) 0.8
e) 0.9

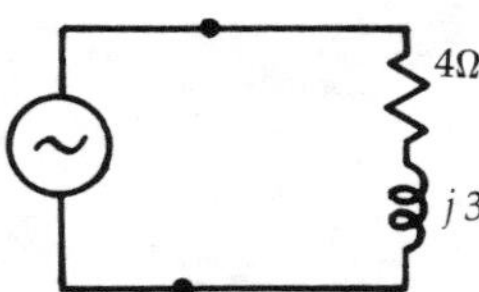

6.16 For the circuit shown, the value of capacitance C that will give a power factor of 1.0 is most nearly

$I = 10 \cos 50t$

C   0.01H 10Ω

a) .0173
b) .0519
c) .0938
d) .0393
e) 0.0732

6.17 For maximum power dissipation in the load of the circuit shown, R (in ohms) and L (in milli-henries) should be chosen as

a) 26, 50
b) 20, 100
c) 20, 50
d) 25, 100
e) 10, 25

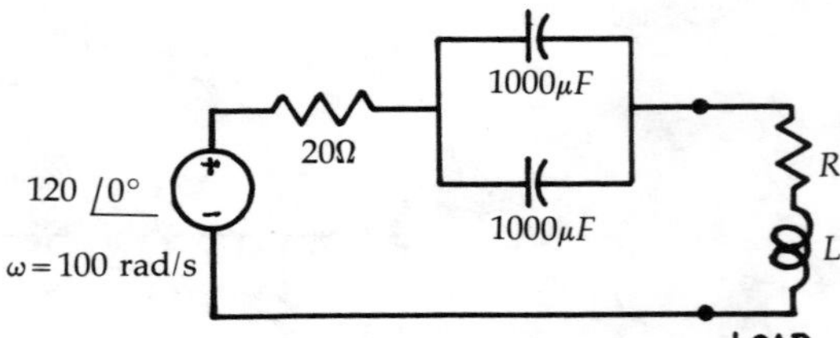

AC CIRCUITS—THREE PHASE

*6.18 Calculate the total average power, in watts, dissipated in the balanced three phase load.

a) 2507
b) 5276
c) 3456
d) 978
e) 1728

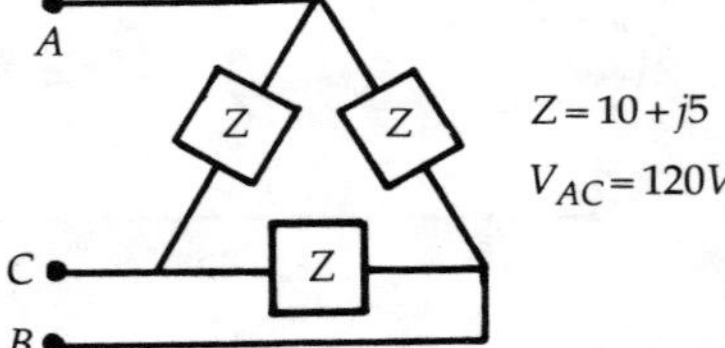

6.19 The value of the line current I_{aA} in the balanced Y-connected system shown is most nearly

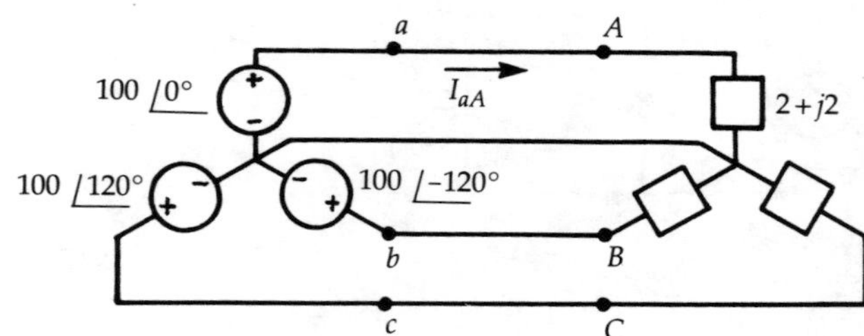

a) 20.6 $\angle 30°$
b) 35.3 $\angle -45°$
c) 35.3 $\angle 45°$
d) 15.1 $\angle -30°$
e) 15.1 $\angle 30°$

*6.20 For a balanced Y-connected system identify the incorrect statement.

a) $V_{PH} = \sqrt{3}\ V_L$
b) $I_L = I_{PH}$
c) $P_{total} = 3P_{PH}$
d) All phase impedances are equal.
e) The neutral current is zero.

6.21 A 100μF capacitor has $I_C(t)$. The capacitor voltage $V_C(t)$ at $t = 2.5$ seconds ($V(0) = 1.0V$) is most nearly

$I_C(t)$; 1.0 2.0 3.0 t; $-1mA$

a) −24
b) −25
c) 25
d) 26
e) .0025

6.22 The voltage across a 10μF capacitor is $50t^2V$. The time, in seconds, it will take to store 200 J of energy is most nearly

a) 0.15 b) .21 c) 1.38 d) 2.275 e) 11.25

*6.23 The value of the voltage across C at $t = 30 \times 10^{-6}$s, if the switch is closed at $t = 0$, is

S; 50Ω; 10V; $10^{-6}F$

a) 3.51 b) 4.51 c) 5.46 d) 6.32 e) 7.43

6.24 How long, in microseconds, does it take for the current to reach half its final value, if the switch is closed at $t = 0$?

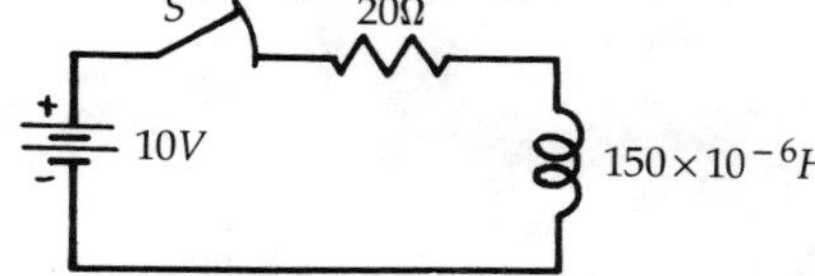

a) 3.1
b) 4.7
c) 5.2
d) 7.3
e) 8.4

ELECTRIC FIELDS

*6.25 The electric flux passing out through a closed surface is equal to:

a) the line integral of the current around the surface.
b) zero.
c) the flux density at the surface.
d) the total charge enclosed by the surface.
e) the net surface integral of the current.

*6.26 The direction of the force acting on a moving charge placed in a magnetic field is:

a) perpendicular to the magnetic field.
b) opposite to the direction of motion of the charge.
c) along the direction of the magnetic field.
d) along the direction of motion of the charge.
e) at an oblique angle to the magnetic field.

6.27 Two infinitely long lines of charge are parallel to the z-axis and located as shown. The force on an electron at (1,0,0) will be in the direction

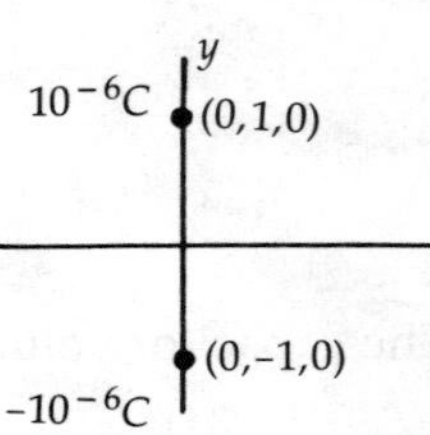

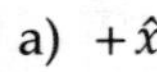
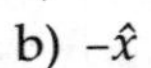

a) $+\hat{x}$
b) $-\hat{x}$
c) $+\hat{y}$
d) $-\hat{y}$
e) $\hat{x} + \hat{y}$

6.28 A point charge of 2×10^{-7} C is located at the origin of coordinates. A spherical shell with center at the origin and radius of 20 cm has a surface charge density 1×10^{-7} C/m^2. The electric flux density at $r = 50$ cm, in C/m^2, is

a) 3.18×10^{-8}
b) 7.96×10^{-8}
c) 9.55×10^{-8}
d) 11.14×10^{-8}
e) 14.22×10^{-8}

*6.29 A uniform line charge of $\rho_L = 30$ nC/m lies along the z axis. The flux density D at (3,–4,5) is:

a) 1.91×10^{-10}
b) $9.55\times10^{-}$
c) 11.94×10^{-10}
d) 15.92×10^{-10}
e) 30×10^{-10}

6.30 An electric field in rectangular coordinates is given by

$$\vec{E} = 4y\hat{x} + 4x\hat{y} \text{ V/m}$$

The voltage drop from (1,1,1) to (5,1,1) is

a) +12
b) –12
c) +16
d) –16
e) +25

6.31 Static electric field distributions refer to cases where

a) all time derivations of field quantities are zero.
b) the time derivatives of the displacement current are not zero.
c) the electric fields vary with time.
d) the electric scalar potential is two-dimensional.
e) the electric fields are uniform in space.

*6.32 A point change of 50×10^{-9} C is placed 10 cm above a perfectly conducting infinitely large flat ground plane. What is the voltage 5 cm above the ground with respect to zero volts on the ground?

a) 3000
b) 4000
c) 5000
d) 6000
e) 7000

MAGNETIC FIELDS

*6.33 Two long, straight conductors located at (0,3,0) and (0,–3,0) each carry 10 amperes in the same direction. Distances are in meters. The magnitude of magnetic field intensity at (4,0,0) is

a) $1/\pi$ b) $2/5\pi$ c) $3/5\pi$ d) $4/5\pi$ e) π

6.34 A solenoid has 1000 turns and carries a current of 5 amperes. If $L = 50$ cm and $r_c = 2.5$ cm, what is the magnetic field intensity on the solenoid axis at the center of the solenoid?

a) 10^4 b) 2×10^5 c) 5×10^4 d) 10^5

*6.35 The inductor shown has an inductance of 4 mH. In order to increase the inductance to 40 mH one should

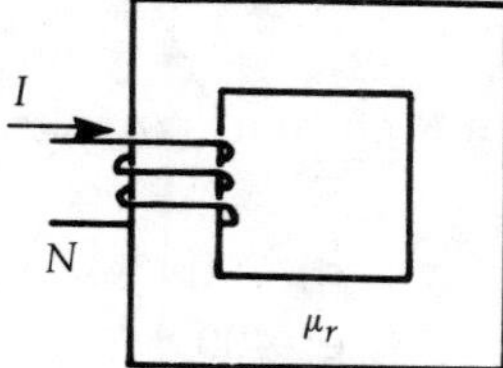

a) increase the current by 10.
b) increase the mean flux path length by 10.
c) increase the number of turns to 10 N.
d) increase the cross-sectional area of the iron by 10.
e) increase the number of turns to 100 N.

6.36 An iron ring with a mean diameter of 20 cm is wound with a coil of 200 turns. The permeability of the iron is 4×10^{-4} H/m. A current of 0.05 A is passed through the coil. The magnetic flux density in the iron, in W/m^2, is

a) 0.02 b) $0.01\pi^2$ c) $100/\pi$ d) π e) $10/\pi$

*6.37 An iron core is shown. The relative permittivity of the iron is 4000. The reluctance, in H^{-1}, of the magnetic circuit shown is

a) $1/(\pi\times10^{-8})$
b) $1/(4\pi\times10^{-5})$
c) $1/(4\pi\times10^{-8})$
d) $1/(16\pi\times10^{-8})$
e) $1/(32\pi\times10^{-8})$

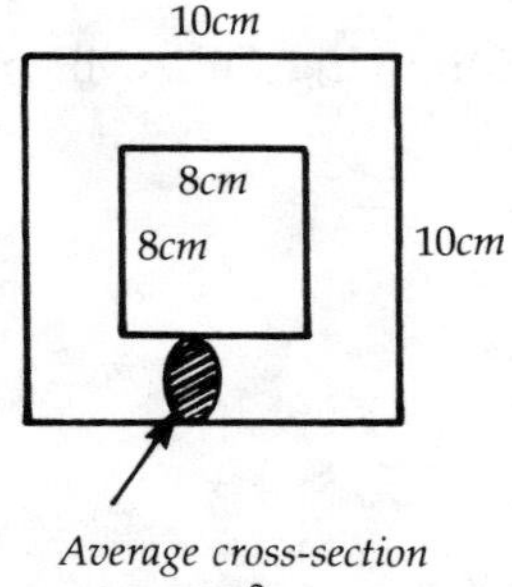

DIODES

*6.38 If the desired DC load voltage is 9 volts, what is the rms value of the source?

a) 4.1
b) 12.7
c) 20.0
d) 28.3
e) 32.5

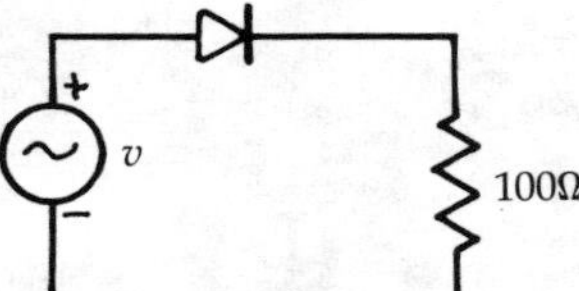

6.39 If the source voltage in the circuit of Prob. 6.38 is $v = 100 \sin 377t$, the peak reverse voltage applied to the diode would be

a) 2.5 b) 100 c) 141.4 d) 31.8 e) 87.2

6.40 If $R_L = 600\Omega$ what must be the rms value of the sinusoidal voltage v if $I = 150$ mA?

a) 60
b) 80
c) 100
d) 120
e) 140

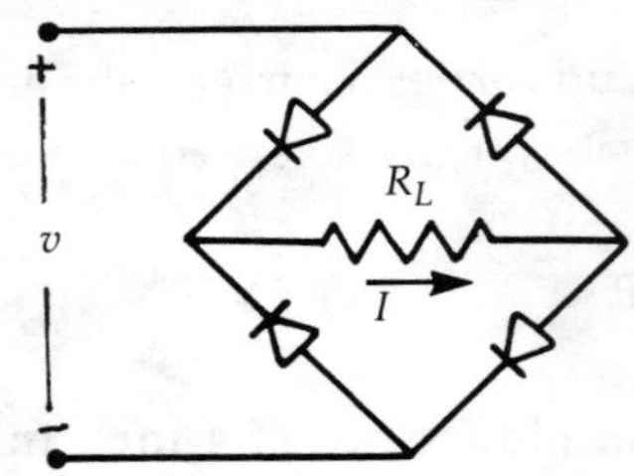

OPERATIONAL AMPLIFIERS

*6.41 Calculate R so that $v_o/v_{in} = 200$.

a) 50
b) 100
c) 200
d) 300
e) 400

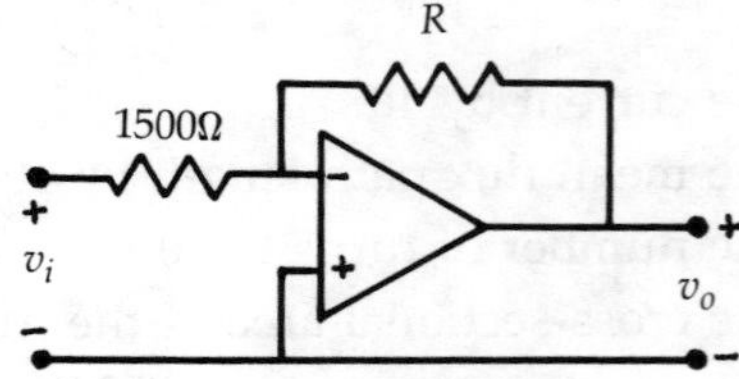

6.42 The gain of the following OP-AMP circuit is

a) −0.2
b) −1.2
c) −4
d) −5
e) −8

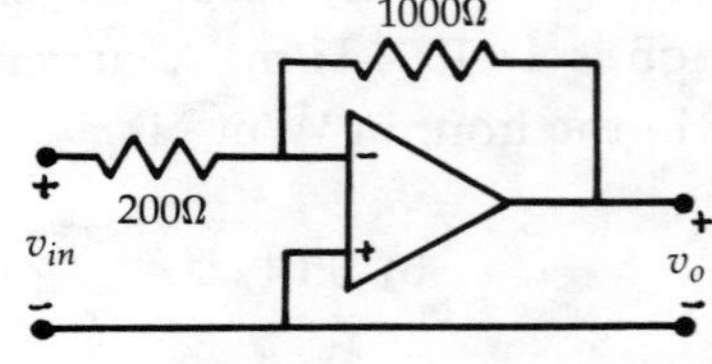

6.43 Given the voltages into the following OP-AMP network, the output voltage is

a) −2
b) −4
c) −7
d) −10
e) −15

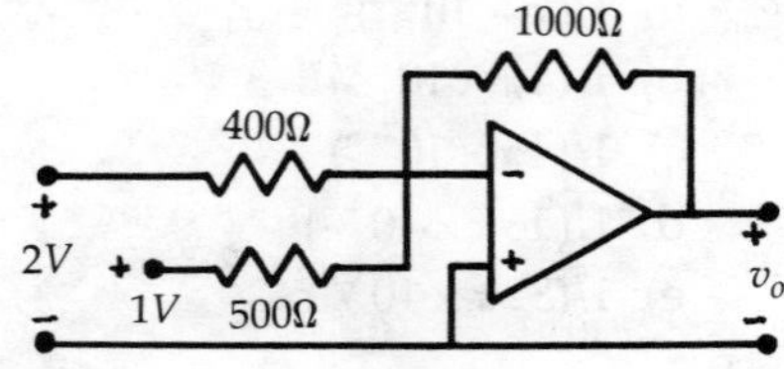

*6.44 The OP-AMP circuit below performs the function of

a) amplification.
b) integration.
c) differentiation.
d) summing.
e) attenuation.

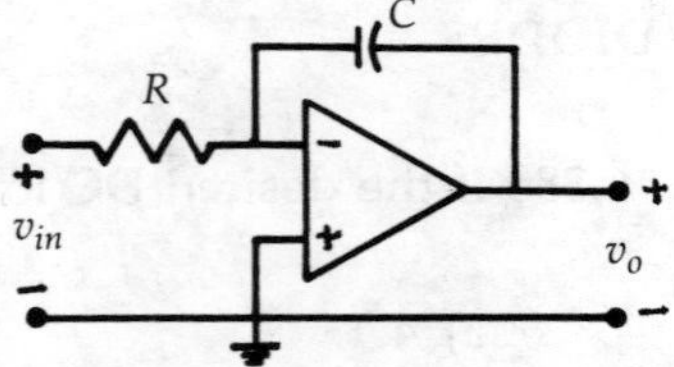

7. Thermodynamics

Thermodynamics involves the storage, transformation, and transfer of energy. It is stored as internal energy, kinetic energy, and potential energy; it is transformed between these various forms; and, it is transferred as work or heat transfer.

The *macroscopic* approach is used in this presentation, that is, we assume matter occupies all points in a region of interest. This is acceptable providing the density is sufficiently large, which it is in most engineering situations.

Both a *system*, a fixed quantity of matter, and a *control volume*, a volume into which and/or from which a substance flows, can be used in thermodynamics. (A control volume may also be referred to as an *open system*.) A system and its surroundings make up a *universe*. Some useful definitions follow:

phase — matter that has the same composition throughout; it is homogeneous.

mixture — a quantity of matter that has more than one phase.

property — a quantity which serves to describe a system.

simple system — a system composed of a single phase, free of magnetic, electrical, and surface effects. Only two properties are needed to fix a simple system.

state — the condition of a system described by giving values to its properties at a given instant.

intensive property — a property that does not depend on the mass.

extensive property — a property that depends on the mass of the system.

specific property — an extensive property divided by the mass.

thermodynamic equilibrium — when the properties do not vary from point to point and there is no tendency for additional change.

process — the path of successive states through which a system passes.

quasiequilibrium — if, in passing from one state to the next, the deviation from equilibrium is infinitesimal. It is also called a *quasistatic* process.

reversible process — a process which, when reversed, leaves no change in either the system or surroundings.

isothermal — temperature is constant.

isobaric — pressure is constant.

isometric — volume is constant.

isentropic — entropy is constant.

adiabatic — no heat transfer.

Experimental observations are organized into mathematical statements or *laws.* Some of those used in thermodynamics follow:

zeroith law of thermodynamics — if two bodies are equal in temperature to a third, they are equal in temperature to each other.

first law of thermodynamics — during a given process, the net heat transfer minus the net work output equals the change in energy.

second law of thermodynamics — a device cannot operate in a cycle and produce work while exchanging heat at a single temperature.

Boyle's law — the volume varies inversely with pressure for an ideal gas.

Charles' law — the volume varies inversely with temperature for an ideal gas.

Avogadro's law — equal volumes of different ideal gases with the same temperature and pressure contain equal molecules.

7.1 Density, Pressure, and Temperature

The density ρ is the mass divided by the volume,

$$\rho = \frac{M}{V}. \tag{7.1.1}$$

The specific volume is the reciprocal of the density,

$$v = \frac{1}{\rho} = \frac{V}{M}. \tag{7.1.2}$$

The pressure p is the normal force divided by the area upon which it acts. In thermodynamics, it is important to use *absolute pressure,* defined by

$$p_{abs} = p_{gauge} + p_{atmospheric} \tag{7.1.3}$$

where the atmospheric pressure is taken as 100 kPa (14.7 psi), unless otherwise stated. If the gauge pressure is negative, it is a *vacuum*.

The temperature scale is established by choosing a specified number of divisions, called degrees, between the ice point and the steam point, each at 101 kPa absolute. In the Celsius scale, the ice point is set to be 0°C and the steam point at 100°C. The absolute temperature in kelvins is

$$T = T_{celsius} + 273. \tag{7.1.4}$$

The temperature, pressure, and specific volume, for an ideal (perfect) gas are related by the ideal gas law

$$pv = RT \tag{7.1.5}$$

where R is the gas constant; for air it is 0.287 kJ/kg·K (53.3 ft-lb/lb$_m$-°R). Below moderate pressures this equation can be used if the temperature exceeds twice the *critical temperature*, the temperature above which the liquid and vapor phases do not coexist. Thus, if the temperature of steam is greater than 800°C (1500°F), Eq. 7.1.5 can be used with $R = 0.462$ kJ/kg·K (85.7 ft-lb/lb$_m$-°R) if the pressure is not excessive; for high pressures the temperature should be higher if Eq. 7.1.5 is to be used.

If the temperature is below twice the critical temperature, tables relating the three variables p, v, T must be used. For water such tables are called the "steam tables" and are presented as Table 7.3. Data is presented in Table 7.3.1 and 7.3.2 that is used to relate p, v, and T when water exists in the liquid phase, as a liquid/vapor mixture, or in the saturated vapor phase. These situations are best described referring to a

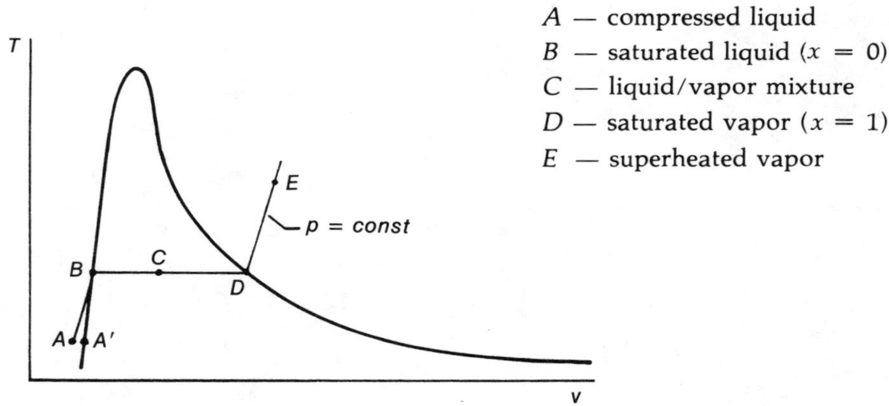

Figure 7.1 **A T-v diagram.**

T-v diagram, shown in Fig. 7.1. Suppose we start with a constant pressure container (a cylinder with a floating piston) containing water at room temperature; it would undergo the following changes if heat is transferred to the volume:

- the temperature would rise above T_A in the *compressed liquid* and $v_A \cong v_{A'}$.
- at state B *saturated liquid* results and vaporization (boiling) begins; $v_B = v_f$. A subscript f will always denote a saturated liquid.
- at state C the liquid phase and vapor phase are in equilibrium; a *liquid/vapor mixture* occurs and

$$v = v_f + x\,(v_g - v_f) \tag{7.1.6}$$

where x is the *quality*.

- at state D a *saturated vapor* exists and vaporization is complete; a subscript g will always denote a saturated vapor.

$$v_D = v_g.$$

- at state E the vapor is *superheated* and v is found in Table 7.3.3.

Eq. 7.1.6 above results from the definition of *quality*, the ratio of the vapor mass to the total mass:

$$x = \frac{m_{vapor}}{m_{total}} . \tag{7.1.7}$$

Note that the entries in the tables at the end of this chapter are in absolute pressure and degrees Celsius.

EXAMPLE 7.1

What mass of air is contained in a room 20 m × 40 m × 3 m at standard conditions?

Solution. Standard conditions are $T = 15°C$ and $p = 100$ kPa abs. Hence,

$$\rho = \frac{p}{RT}$$

$$= \frac{100}{0.287 \times 288} = 1.21 \text{ kg/m}^3.$$

$$\therefore m = \rho V$$

$$= 1.21 \times 20 \times 40 \times 3 = 2900 \text{ kg}.$$

EXAMPLE 7.2

What is the mass of water contained in 2 m³ at 1000 kPa abs and 200°C?

Solution. The specific volume of water is insensitive to pressure so we use v_f at 200°C. It is, using Table 7.3.1,

$$v_f = 0.001156 \text{ m}^3/\text{kg}.$$

The mass is then

$$m = V/v$$

$$= 2/0.001156 = 1730 \text{ kg}.$$

EXAMPLE 7.3

The volume occupied by 20 kg of a water-vapor mixture at 200°C is 2 m³. Calculate the quality.

Solution. The specific volume is

$$v = V/m$$

$$= 2/20 = 0.1 \text{ m}^3/\text{kg}.$$

Using Eq. 7.1.6 we have, using Table 7.3.1,

$$x = \frac{v - v_f}{v_g - v_f}$$

$$= \frac{0.1 - 0.001156}{0.1274 - 0.001156} = 0.783 \text{ or } 78.3\%.$$

EXAMPLE 7.4

Find the volume occupied by 20 kg of steam at 4 MPa and 400°C.

Solution. The specific volume is found in Table 7.3.3 to be

$$v = 0.07341 \text{ m}^3/\text{kg}.$$

the volume is then

$$V = mv$$

$$= 20 \times 0.07341 = 1.468 \text{ m}^3.$$

7.2 The First Law of Thermodynamics For A System

The first law of thermodynamics, referred to as the "first law" or the "energy equation," is expressed for a cycle as

$$Q_{net} = W_{net} \tag{7.2.1}$$

and for a process as

$$Q - W = \Delta E \tag{7.2.2}$$

where Q is the heat transfer, W is the work, and E represents the energy (kinetic, potential, and internal) of the system.* In thermodynamics attention is focused on internal energy with kinetic and potential energy changes neglected (unless otherwise stated) so that we have

$$Q - W = \Delta U \tag{7.2.3}$$

where the specific internal energy is

$$u = U/m. \tag{7.2.4}$$

Heat transfer may occur during any of the three following modes:

Conduction — heat transfer due to molecular activity. For steady-state through a constant wall area Fourier's law states

$$\dot{Q} = kA\Delta T/L = UA\Delta T = A\Delta T/R \tag{7.2.5}$$

where k is the conductivity, U is the heat transfer coefficient, R is the resistance factor, and L is normal to the heat flow.† For a three-layer composite wall we could use

$$\dot{Q} = A\Delta T/(R_1 + R_2 + R_3). \tag{7.2.6}$$

Convection — heat transfer due to fluid motion. The mathematical expression used is

$$\dot{Q} = hA\Delta T \tag{7.2.7}$$

where h is the convective heat transfer coefficient.

Radiation — heat transfer due to the transmission of waves. The heat transfer from body 1 is

$$\dot{Q} = \sigma\varepsilon A(T_1^4 - T_2^4)F_{1\text{-}2} \tag{7.2.8}$$

in which $\sigma = 5.67 \times 10^{-11}$ kJ/(s·m²·K⁴), ε is emissivity ($\varepsilon = 1$ for a black body), and $F_{1\text{-}2}$ is the shape factor ($F_{1\text{-}2} = 1$ if body 2 encloses body 1).

In thermodynamics, the heat transfer is usually specified, or calculated using Eq. 7.2.3, and is *not* calculated with the previous four equations. The equations in the one-dimensional form above, and in two-dimensions, are the focus of attention in a course in Heat Transfer and are not of particular interest in thermodynamics. They are presented here for the sake of completeness.

Work can be accomplished mechanically by moving a boundary, resulting in the quasiequilibrium work mode

$$W = \int p\,dV \tag{7.2.9}$$

or by electrical means; it can also be accomplished in nonquasiequilibrium modes such as with a paddlewheel. But then Eq. 7.2.9 cannot be used.

We introduce *enthalpy* for convenience, and define it to be

$$H = U + pV$$
$$h = u + pv. \tag{7.2.10}$$

*In this book heat transferred to the system is positive and work done by the system is positive. It is also conventional to define work done on the system as positive so that $Q + W = \Delta E$.

†A dot signifies a rate, so that $\dot{Q}$ has units of kJ/s.

For substances such as steam, the specific internal energy and specific enthalpy are found in the steam tables. For compressed liquids, u and h are insensitive to pressure and are found in Table 7.3.1. For the liquid/vapor mixture we use

$$u = u_f + x(u_g - u_f) \tag{7.2.11}$$

$$h = h_f + x\, h_{fg}$$

where $h_{fg} = h_g - h_f$.

For ideal gases we assume constant specific heats and use

$$\Delta u = c_v \, \Delta T \tag{7.2.12}$$

$$\Delta h = c_p \, \Delta T \tag{7.2.13}$$

where c_v is the *constant volume specific heat*, and c_p is the *constant pressure specific heat*. From the differential forms of the above we can find

$$c_p = c_v + R. \tag{7.2.14}$$

We also define the *ratio of specific heats* k to be

$$k = c_p / c_v. \tag{7.2.15}$$

For air $c_v = 0.716$ kJ/kg·K (0.171 BTU/lb$_m$-°F), $c_p = 1.00$ kJ/kg·K (0.24 BTU/lb$_m$-°F), $k = 1.4$. For most solids and liquids we can find the heat transfer using

$$Q = m\, c\, \Delta T. \tag{7.2.16}$$

For water $c = 4.18$ kJ/kg·°C (1.00 BTU/lb$_m$-°F), and for ice $c \cong 2.1$ kJ/kg·°C (0.50 BTU/lb$_m$-°F).

When a substance changes phase the *latent heat* is involved. The energy necessary to melt a unit mass of a solid is the *heat of fusion*; the energy necessary to vaporize a unit mass of liquid is the *heat of vaporization*, equal to $(h_g - h_f)$; the energy necessary to vaporize a unit of mass of solid is the *heat of sublimation*. For ice, the heat of fusion is approximately 320 kJ/kg (140 BTU/lb$_m$), and the heat of sublimation is about 2040 kJ/kg (877 BTU/lb$_m$); the heat of vaporization varies from 2500 kJ/kg at 0°C (1075 BTU/lb$_m$ at 32°F) to zero at the critical point.

For specific processes, we consider the preceding paragraphs and summarize as follows:

Isothermal Process

1st law: $$Q - W = m\Delta u \tag{7.2.17}$$

ideal gas: $$Q = W = mRT\ell n\frac{v_2}{v_1} = mRT\ell n\frac{p_1}{p_2} \tag{7.2.18}$$

$$p_2 = p_1 v_1 / v_2 \tag{7.2.19}$$

Constant Pressure

1st law: $$Q = m\Delta h \tag{7.2.20}$$

$$W = m\,p\Delta v \tag{7.2.21}$$

ideal gas: $$Q = mc_p\Delta T \tag{7.2.22}$$

$$T_2 = T_1 v_2 / v_1 \tag{7.2.23}$$

Constant Volume

1st law: $$Q = m\Delta u \tag{7.2.24}$$

$$W = 0 \tag{7.2.25}$$

ideal gas: $$Q = mc_v\Delta T \tag{7.2.26}$$

$$T_2 = T_1 p_2 / p_1 \tag{7.2.27}$$

Adiabatic Process (Isentropic)

1st law: $$-W = m\Delta u \tag{7.2.28}$$

$$Q = 0 \tag{7.2.29}$$

ideal gas: $$-W = mc_v\Delta T \tag{7.2.30}$$

$$T_2 = T_1(v_1/v_2)^{k-1} = T_1(p_2/p_1)^{k-1/k} \tag{7.2.31}$$

$$p_2 = p_1(v_1/v_2)^k \tag{7.2.32}$$

A *polytropic* process results if k in Eqs. 7.2.31 and 7.2.32 is replaced with n. Then n must be specified. Note that the adiabatic, quasiequilibrium process is often referred to as an adiabatic, reversible process.

EXAMPLE 7.5

How much heat must be added to 2 kg of steam contained in a rigid volume, if the initial pressure of 2 MPa abs is increased to 4 MPa abs? $T_1 = 250°C$.

Solution. The first law, with $\Delta KE = \Delta PE = 0$, is

$$Q - W = \Delta U.$$

For a rigid container $W = 0$ so that

$$Q = m(u_2 - u_1).$$

From the steam Table 7.3.3, we find $u_1 = 2679.6$ kJ/kg and $v_1 = 0.1114$ m³/kg. We can locate state 2 because the container is rigid, so that

$$v_2 = v_1 \cong 0.111 \text{ m}^3/\text{kg}.$$

The temperature T_2 that has $p_2 = 4$ MPa and $v_2 = 0.111$ m³/kg is 700°C. At that state $u_2 = 3462$ kJ/kg. Thus,

$$Q = m(u_2 - u_1)$$

$$= 2(3462 - 2680) = 1564 \text{ kJ}.$$

EXAMPLE 7.6

Calculate the heat transfer necessary to raise the temperature of 2 kg of saturated steam to 600°C if pressure is maintained constant at 2000 kPa abs.

Solution. The first law, for a constant pressure process, is

$$Q = m(h_2 - h_1).$$

Using Tables 7.3.2 and 7.3.3, we find $h_1 = 2799.5$ and $h_2 = 3690.1$ kJ/kg. Hence, we have

$$Q = 2(3690.1 - 2799.5) = 1781 \text{ kJ}.$$

EXAMPLE 7.7

How much heat is needed to completely vaporize 100 kg of ice at −10°C? The pressure is held constant at 200 kPa abs.

Solution. The heat transfer is related to the enthalpy by

$$Q = m\Delta h$$

$$= m(c\,\Delta T_{ice} + \text{heat of fusion} + c\,\Delta T_{water} + \text{heat of vaporization}).$$

Using values given in Article 7.2 and Table 7.3.2,

$$Q = 100(2.1 \times 10 + 320 + 4.18 \times 120.2 + 2201.9)$$

$$= 304\,500 \text{ kJ} \quad \text{or} \quad 304.5 \text{ MJ}.$$

EXAMPLE 7.8

Estimate the heat transfer necessary to increase the pressure of 2 kg of 50% quality steam from 200 kPa abs to 800 kPa abs if the volume is kept constant.

Solution. The first law, with $W = 0$ for the constant volume process, is

$$Q = m(u_2 - u_1).$$

To find state 2 we must use $v_2 = v_1$. At state 1 we have

$$v_1 = v_f + x(v_g - v_f)$$

$$= 0.00106 + 0.5(0.8857 - 0.00106) = 0.4434 \text{ m}^3/\text{kg},$$

$$u_1 = u_f + x(u_g - u_f)$$

$$= 504.5 + 0.5(2529.5 - 504.5) = 1517 \text{ kJ/kg}.$$

At state 2, $p_2 = 0.8$ MPa and $v_2 = 0.4434$ m³/kg so that $u_2 = 3126$ kJ/kg. Hence,

$$Q = 2(3126 - 1517) = 3218 \text{ kJ}.$$

EXAMPLE 7.9

Calculate the work done by a piston if the 2 m³ volume of air is tripled while the temperature is maintained at 40°C. The initial pressure is 400 kPa abs.

Solution. The mass is needed in order to use Eq. 7.2.18 to find the work; it is

$$m = \frac{pV}{RT}$$

$$= \frac{400 \times 2}{0.287 \times 313} = 8.91 \text{ kg}.$$

The work is then found to be

$$W = m\,RT\,\ell n\, v_2/v_1$$

$$= 8.91 \times 0.287 \times 313\ \ell n\ 3 = 879 \text{ kJ}.$$

Note: The temperature is expressed as 40 + 273 = 313 K.

EXAMPLE 7.10

How much work is necessary to compress air in an insulated cylinder from 0.2 m³ to 0.01 m³? Use $T_1 = 20°C$ and $p_1 = 100$ kPa abs.

Solution. For the adiabatic process $Q = 0$ so that the first law is

$$-W = m(u_2 - u_1)$$

$$= m\,c_v\ (T_2 - T_1).$$

To find the mass m we use the ideal gas equation as

$$m = \frac{pV}{RT}$$

$$= \frac{100 \times 0.2}{0.287 \times 293} = 0.2378 \text{ kg}.$$

The temperature T_2 is found to be

$$T_2 = T_1(v_1/v_2)^{k-1}$$

$$= 293(0.2/0.01)^{0.4} = 971.1 \text{ K}.$$

The work is then

$$W = 0.2378 \times 0.716(971.1 - 293) = 115.5 \text{ kJ}.$$

7.3 The First Law of Thermodynamics for a Control Volume.

The continuity equation, which accounts for the conservation of mass, may be used in certain situations involving control volumes. It is stated as

$$\dot{m} = \rho_1 A_1 V_1 = \rho_2 A_2 V_2 \tag{7.3.1}$$

where, in control volume formulations, V is the velocity; $\dot{m}$ is called the *mass flux*. In the above continuity equation, we assume *steady flow*, that is, the variables are independent of time. For such steady flow situations, the first law takes the form

$$\frac{\dot{Q} - \dot{W}_S}{\dot{m}} = \frac{V_2^2 - V_1^2}{2} + h_2 - h_1 + g(z_2 - z_1) \tag{7.3.2}$$

where the dot signifies a rate so that $\dot{Q}$ and $\dot{W}_S$ have units of kJ/s. In most devices the potential energy change is negligible. Also, the kinetic energy change can often be ignored (but if sufficient information is given, it should be included) so that the first law simplifies to

$$\dot{Q} - \dot{W}_S = \dot{m}\,(h_2 - h_1). \tag{7.3.3}$$

Particular devices are of special interest. The energy equation for a *valve* or a *throttle plate* is simply

$$h_2 = h_1 \tag{7.3.4}$$

providing kinetic energy can be neglected.

For a *turbine* expanding a gas, the heat transfer is negligible so that

$$\dot{W}_T = \dot{m}\,(h_1 - h_2). \tag{7.3.5}$$

The work input to a *gas compressor* with negligible heat transfer is

$$\dot{W}_C = \dot{m}\,(h_2 - h_1). \tag{7.3.6}$$

A *boiler* and a *condenser* are simply heat transfer devices. The first law then simplifies to

$$\dot{Q} = \dot{m}(h_2 - h_1). \tag{7.3.7}$$

For a *nozzle* or a *diffuser* there is no work or heat transfer; we must include, however, the kinetic energy change resulting in

$$0 = \frac{V_2^2 - V_1^2}{2} + h_2 - h_1. \tag{7.3.8}$$

For a *pump* or a *hydroturbine* we take a slightly different approach. We return to Eq. 7.3.2 and write it, using $v = 1/\rho$, and with Eq. 7.2.10, as

$$\frac{\dot{Q} - \dot{W}_S}{\dot{m}} = \frac{V_2^2 - V_1^2}{2} + u_2 - u_1 + \frac{p_2 - p_1}{\rho} + g(z_2 - z_1). \tag{7.3.9}$$

For an ideal situation we do not transfer heat. Neglecting kinetic and potential energy changes we find that

$$-\dot{W}_S = \dot{m}\,\frac{p_2 - p_1}{\rho}. \tag{7.3.10}$$

This would provide the minimum pump power requirement or the maximum turbine power output. The inclusion of an efficiency would increase the pump power requirement or decrease the turbine output.

A gas turbine or compressor efficiency is based on an isentropic process ($s_2 = s_1$) as the ideal process, that is, for a gas turbine

$$\eta_T = \frac{\dot{W}_a}{\dot{W}_s} \tag{7.3.11}$$

where $\dot{W}_a$ is the actual output and $\dot{W}_s$ is the work output assuming an isentropic process.

EXAMPLE 7.11

Refrigerant-12 expands through a valve from a state of saturated liquid at 800 kPa abs to a pressure of 100 kPa abs. What is the final quality?

Solution. The first law states that

$$h_1 = h_2.$$

Using Table 7.4.2 for Refrigerant-12 we find, using $h_1 = h_f$,

$$h_1 = 67.3 \text{ kJ/kg}.$$

There follows at $p_2 = 100$ kPa,

$$67.3 = 8.78 + x_2(165.37).$$

$$\therefore x_2 = 0.354 \quad \text{or} \quad 35.4\%.$$

EXAMPLE 7.12

Steam expands through a turbine from 6 MPa abs, 600°C to 2 kPa abs with $x_2 = 1.0$. For a flow rate of 10 kg/s, find the work output.

Solution. The first law gives

$$\dot{W}_T = \dot{m}\,(h_1 - h_2)$$

$$= 10(3658.4 - 2533.5) = 11\ 250 \text{ kW},$$

where h_2 is h_g at $p_2 = 0.002$ MPa, as given in Table 7.3.2.

EXAMPLE 7.13

What is the turbine efficiency in Example 7.12?

Solution. The turbine efficiency is based on an isentropic process. Let state 2′ be at 2 kPa abs with

$$s_{2'} = s_1 = 7.1685 \quad \text{kJ/kg·K}.$$

At 0.002 MPa abs we find, from Table 7.3.2,

$$s_{2'} = s_f + x_{2'}\, s_{fg}.$$

$$7.1685 = 0.2606 + x_{2'}(8.4639).$$

$$\therefore x_{2'} = 0.816.$$

At this ideal state we find

$$h_{2'} = h_f + x_{2'} h_{fg}$$

$$= 73.5 + 0.816 \times 2460 = 2080 \text{ kJ/kg}.$$

Finally, we use $\dot{W}_a$ as the answer in Example 7.12 and obtain

$$\eta_T = \frac{\dot{W}_a}{\dot{W}_s} = \frac{\dot{W}_a}{\dot{m}(h_1 - h_2)}$$

$$= \frac{11\,250}{10(3658.4 - 2080)} = 0.713 \quad \text{or} \quad 71.3\%.$$

EXAMPLE 7.14

What is the minimum power requirement of a pump that is to increase the pressure from 2 kPa abs to 6 MPa abs for a mass flux of 10 kg/s of water?

Solution. The first law is simplified to

$$\dot{W}_P = \dot{m}\,\frac{p_2 - p_1}{\rho}$$

$$= 10\,\frac{6000 - 2}{1000} = 59.98 \text{ kW}.$$

Note how small this is relative to the power output of the turbine of Example 7.12 operating between the same pressures. Because this is less than 1% of the turbine output, the pump work may usually be neglected in the analysis of a cycle involving a turbine and a pump.

EXAMPLE 7.15

A nozzle accelerates air from 100 m/s, 400°C and 400 kPa abs to a receiver where p = 20 kPa abs. Assuming an isentropic process find V_2.

Solution. The energy equation takes the form

$$0 = \frac{V_2^2 - V_1^2}{2} + h_2 - h_1.$$

Assuming air to be an ideal gas with constant c_p we have

$$c_p(T_1 - T_2) = \frac{V_2^2 - V_1^2}{2}.$$

We can find T_2 from Eq. 7.2.31 to be

$$T_2 = T_1(p_2/p_1)^{k-1/k}$$

$$= 673(20/400)^{0.4/1.4} = 286 \text{ K}.$$

Thus, the exiting velocity is found as follows:

$$1000(673 - 286) = \frac{V_2^2 - 100^2}{2}.$$

$$\therefore V_2 = 874 \text{ m/s}.$$

Note: c_p must be used as 1000 J/kg·K so that the units are consistent.

7.4 The Second Law of Thermodynamics

The two scientific statements of the Second Law of Thermodynamics can be shown to be equivalent. They are stated and shown schematically in Fig. 7.2.

Kelvin-Planck statement — A device, which operates in a cycle, cannot produce work while exchanging heat at a single temperature.

Clasius statement — A device, which operates in a cycle, cannot transfer heat from a cooler body to a hotter body without a work input.

Figure 7.2 Violations of the second law.

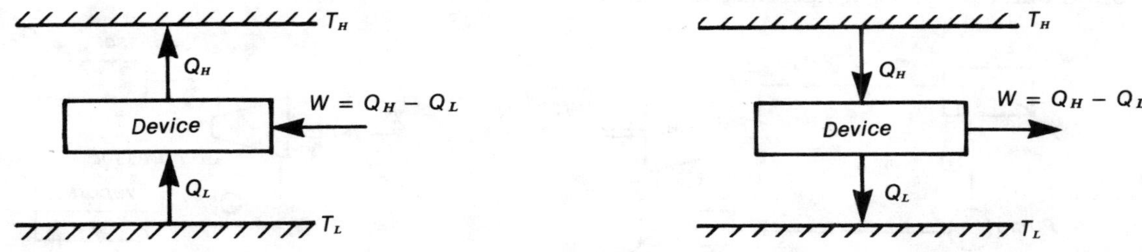

Figure 7.3 Devices that satisfy the second law.

Figure 7.3 shows an engine (it produces work) and a refrigerator (it transfers heat from a body at low temperature) that satisfy the second law. The devices of Fig. 7.2 do not violate the first law — energy is conserved — however, they represent impossibilities, violations of the second law.

To write a mathematical statement of the second law, we use entropy. It is written as

$$\Delta S_{universe} = \Delta S_{system} + \Delta S_{surroundings} \geqslant 0. \tag{7.4.1}$$

The equal sign applies to a reversible process; losses, friction, unrestrained expansion, and heat transfer across a finite temperature difference all lead to irreversibilities. Entropy changes can be found using tables, or for an ideal gas with constant specific heats, we can use

$$\Delta s = c_p \,\ell n \frac{T_2}{T_1} - R \,\ell n \frac{p_2}{p_1} \tag{7.4.2}$$

For constant temperature processes, such as heat transfer to a reservoir, we use

$$\Delta S = \frac{Q}{T} \, . \tag{7.4.3}$$

For a solid or a liquid we use

$$\Delta S = m \, c \, \ln \frac{T_2}{T_1} \, . \tag{7.4.4}$$

A Carnot engine or refrigerator is a fictitious device that operates with reversible processes. It provides us with the maximum possible efficiency of an engine:

$$\eta = \frac{W_{out}}{Q_{in}} = 1 - \frac{T_L}{T_H} \tag{7.4.5}$$

It also provides the maximum possible *coefficient of performance* COP of a refrigerator,

$$\text{COP} = \frac{Q_L}{W_{in}} = \frac{1}{T_H/T_L - 1} \tag{7.4.6}$$

or, the upper limit for the COP of a heat pump,

$$\text{COP} = \frac{Q_H}{W_{in}} = \frac{1}{1 - T_L/T_H} \, . \tag{7.4.7}$$

The temperature T_H is the temperature of the heat source and T_L is the temperature of the heat sink.

The vapor power cycle that is the basic cycle for most power plants, the Rankine cycle, is sketched in Fig. 7.4a. The vapor refrigeration cycle is shown in Fig. 7.4b.

The Carnot cycle is sketched in Fig. 7.5 (on the next page), along with some other common gas cycles. The efficiency of each of the other cycles is less than that of the Carnot cycle, primarily due to the transfer of heat across a finite temperature difference.

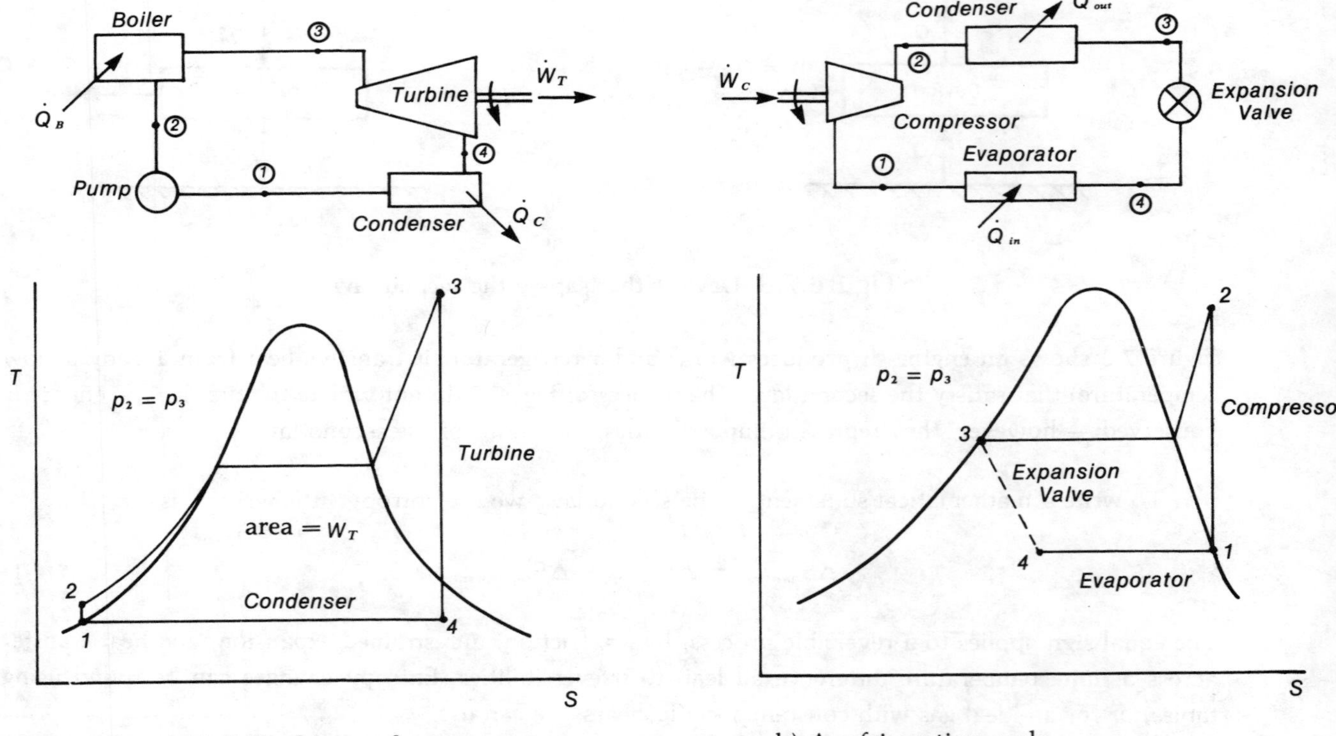

a) A Rankine cycle.

b) A refrigeration cycle.

Figure 7.4. Vapor cycles.

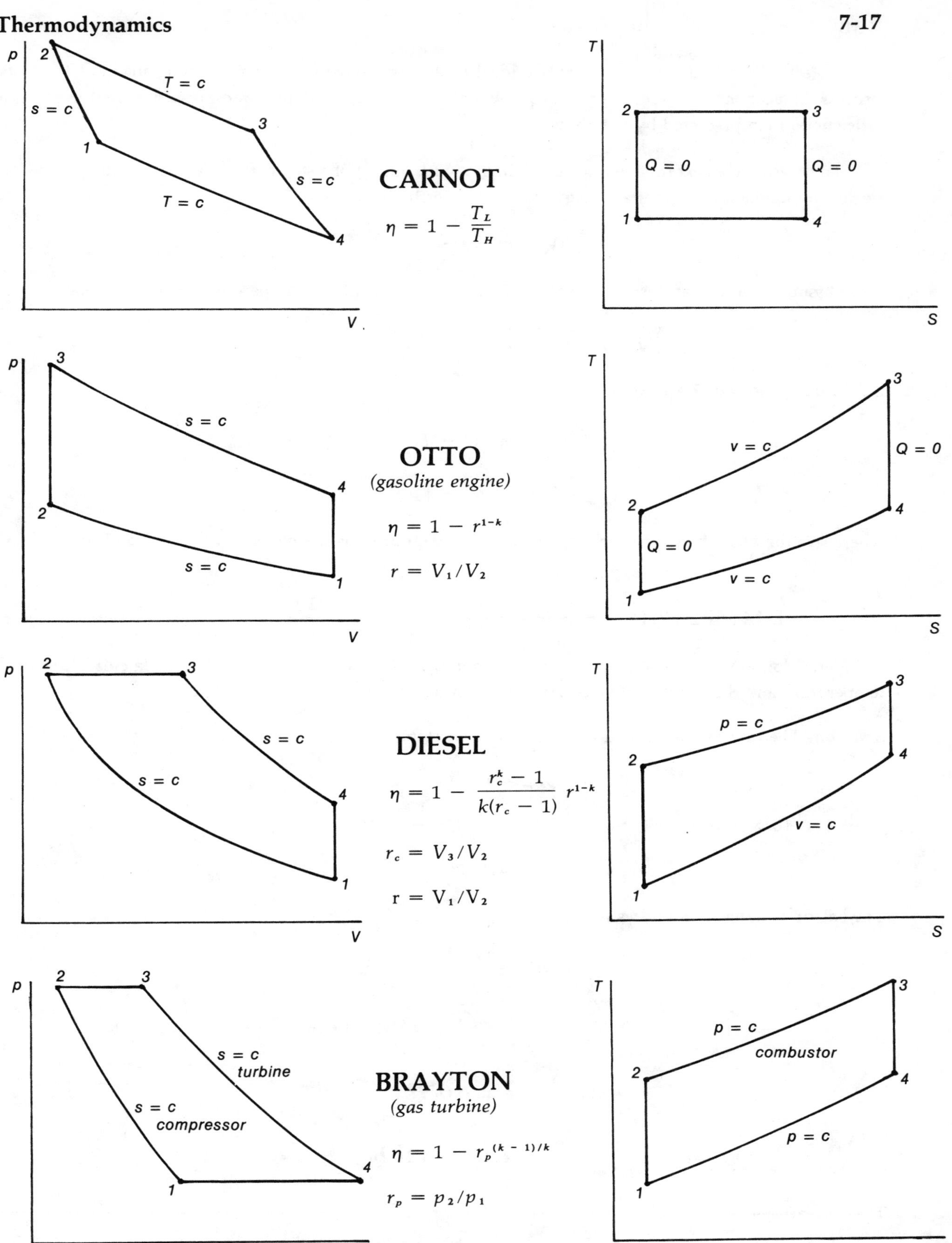

Figure 7.5. Common thermodynamic gas cycles.

Note that all of the above cycles are ideal cycles; the entropy is assumed constant in at least one process in each cycle. Actual processes deviate from these ideal processes resulting in lower cycle efficiencies than predicted by the above.

To determine an expression for the maximum work output of a steady flow device, use the first law — neglecting kinetic and potential energy changes — in the form:

$$\dot{Q} - \dot{W}_S = \dot{m}\,(h_2 - h_1). \tag{7.4.8}$$

If we assume heat is transferred to the surroundings at atmospheric temperature T_o we can relate

$$\dot{m}\,(s_2 - s_1) = \dot{Q}/T_o. \tag{7.4.9}$$

Substituting into Eq. 7.4.8 we have

$$\dot{W}_{max} = \dot{m}\,(h_1 - T_o s_1) - \dot{m}\,(h_2 - T_o s_2) \tag{7.4.10}$$

$$= \dot{m}\,(\phi_1 - \phi_2)$$

where ϕ is the *availability*. Hence, the maximum work output is the change in the availability.

EXAMPLE 7.16

Ten kilograms of ice at 0°C are melted in 100 kg of water initially at 25°C. Calculate the final temperature and the entropy change. Assume no heat transfer to the surroundings.

Solution. The first law is applied to the ice-water system:

$$Q_{gain} = Q_{lost}$$

$$10 \times 320 + 10(T - 0) \times 4.18 = 100(25 - T) \times 4.18.$$

$$\therefore T = 15.8°C.$$

The entropy change is found as follows:

$$\text{ice: } \Delta S = \frac{Q}{T_1} + m\,c\,\ell n \frac{T_2}{T_1}$$

$$= \frac{3200}{273} + 10 \times 4.18\,\ell n \frac{288.8}{273} = 14.073 \text{ kJ/K}.$$

$$\text{water: } \Delta S = m\,c\,\ell n \frac{T_2}{T_1}$$

$$= 100 \times 4.18\,\ell n \frac{288.8}{298} = -13.107 \text{ kJ/K}.$$

The net entropy change is

$$\Delta S_{net} = 14.073 - 13.107 = 0.966 \text{ kJ/K}.$$

This is positive, as required by the second law.

EXAMPLE 7.17

An inventor proposed to have invented an engine, using a 160°C geothermal heat source, which operates with an efficiency of 30%. If it exhausts to the 20°C atmosphere, is it a possibility?

Solution. The maximum possible efficiency, as limited by the second law, is given by

$$\eta_{max} = 1 - \frac{T_L}{T_H}$$

$$= 1 - \frac{293}{433} = 0.323 \quad \text{or} \quad 32.3\%.$$

The proposal is a possibility, however, the proposed effiency is quite close to the maximum efficiency. It would be extremely difficult to obtain the 30% because of the losses due to heat transfer across finite temperature differences and friction.

EXAMPLE 7.18

A heat pump delivers 20 000 kJ/hr of heat with a 1.39 kW input. Calculate the COP.

Solution. Using the definition of the COP we find

$$\text{COP} = \frac{\dot{Q}_H}{\dot{W}_{in}}$$

$$= \frac{20\,000/3600}{1.39} = 4.00.$$

The factor 3600 converts hours to seconds.

EXAMPLE 7.19

Compare the efficiency of an Otto cycle operating on an 8 to 1 compression ratio ($r = 8$) with a Diesel cycle that has a 20 to 1 compression ratio and a cut-off ratio of 2 to 1 ($r_c = 2$). Use air.

Solution. The efficiency of the Otto cycle (see Fig. 7.4) is

$$\eta = 1 - r^{1-k}$$

$$= 1 - 8^{-0.4} = 0.565 \quad \text{or} \quad 56.5\%$$

where air is assumed to be the working fluid.

The efficiency of the Diesel cycle is

$$\eta = 1 - \frac{r_c^{\,k} - 1}{k(r_c - 1)} r^{1-k}$$

$$= 1 - \frac{2^{1.4} - 1}{1.4(2 - 1)} 20^{-0.4} = 0.647 \quad \text{or} \quad 64.7\%.$$

The efficiency of the Diesel cycle is higher than that of the Otto cycle because it operates at a higher compression ratio. If the Otto cycle could operate at $r = 20$, its efficiency would be greater than that of the Diesel.

EXAMPLE 7.20

If a power plant operates on a simple Rankine cycle using water between 600°C, 6 MPa abs and a low pressure of 10 kPa, calculate η_{max}.

Solution. Referring to Fig. 7.5a, we define the efficiency to be

$$\eta_{max} = \frac{\dot{W}_T}{\dot{Q}_B}.$$

The pump work is neglected (see Example 7.14). To find $\dot{W}_T$ we must find h_4. This is accomplished, using $p_3 = 6$ MPa and $T_3 = 600$°C, as follows:

$$s_4 = s_3 = 7.1685 \text{ kJ/kg}\cdot\text{K}.$$

$$\text{at } p = 10 \text{ kPa:} \quad 7.1685 = 0.6491 + x_4\,(7.5019).$$

$$\therefore x_4 = 0.869.$$

$$\therefore h_4 = 191.8 + 0.869 \times 2392.8 = 2271 \text{ kJ/kg}.$$

The turbine output is then, assuming $\dot{m} = 1$ kg/s since it is not given,

$$\dot{W}_T = \dot{m}(h_3 - h_4)$$

$$= 1 \times (3658.4 - 2271) = 1387 \text{ kW}.$$

The energy input occurs in the boiler. It is

$$\dot{Q}_B = \dot{m}(h_3 - h_2)$$

$$= 1 \times (3658.4 - 191.8) = 3467 \text{ kW}.$$

Note: Be careful finding h_2. Remember, in a liquid use h_f at the temperature of the liquid, ignoring the pressure; hence, we use h_f at $T_2 \cong T_1 = 45.8$°C (find it in Table 7.3.2).

Finally, the efficiency of this idealized cycle is

$$\eta_{max} = \frac{1387}{3467} = 0.400 \quad \text{or} \quad 40.0\%.$$

EXAMPLE 7.21

A refrigeration system, using Refrigerant-12, operates between −20°C and 41.64°C. What is the maximum possible COP?

Solution. The refrigeration effect takes place in the evaporator. Hence, referring to Fig. 7.4, the COP is defined as

$$\text{COP} = \frac{\dot{Q}_{in}}{\dot{W}_c}.$$

To find $\dot{W}_c$ we must locate state 2. This is done as follows (recognizing that $T_3 = 41.64°C$):

$$s_2 = s_1 = 0.7082 \text{ kJ/kg·K.}$$

$$\text{state 2:} \quad p_2 = 1.0 \text{ MPa}, \qquad s_2 = 0.7082 \text{ kJ/kg·K.}$$

$$\therefore h_2 = \frac{0.7082 - 0.7021}{0.7254 - 0.7021}(217.8 - 210.9) + 210.9 = 212.7 \text{ kJ/kg.}$$

Note: A linear interpolation was used to find h_2. The compressor work is then, assuming $\dot{m} = 1$ kg/s,

$$\dot{W}_c = \dot{m}\,(h_2 - h_1)$$

$$= 1 \times (212.7 - 178.61) = 34.1 \text{ kW.}$$

To find $\dot{Q}_{in}$ we recognize that $h_4 = h_3 = 76.26$ kJ/kg, using Table 7.4.2. Thus we find,

$$\dot{Q}_{in} = \dot{m}\,(h_1 - h_4)$$

$$= 1 \times (178.61 - 76.26) = 102.4 \text{ kW.}$$

The maximum COP for this idealized cycle is

$$\text{COP} = \frac{102.4}{34.1} = 3.00.$$

EXAMPLE 7.22

Steam, at 200°C and 200 kPa abs, is available to produce work by expanding it to the atmosphere at 20°C and 100 kPa abs. What is $\dot{W}_{max}$ if $\dot{m} = 2$ kg/s?

Solution. We will use the equation

$$\dot{W}_{max} = \dot{m}\,(\phi_1 - \phi_2).$$

The availabilities are found to be (use Table 7.3)

$$\phi_1 = h_1 - T_o s_1$$

$$= 2870.5 - 293 \times 7.5074 = 670.8 \text{ kJ/kg.}$$

$$\phi_2 = h_2 - T_o s_2$$

$$= 83.9 - 293 \times 0.2965 = -3.0 \text{ kJ/kg.}$$

Note: State 2 is at 20°C and 100 kPa abs. This is the liquid state so we use $h_2 = h_f$ and $s_2 = s_f$ at 20°C (see Table 7.3.1); simply ignore the pressure. Finally,

$$\dot{W}_{max} = 2[670.8 - (-3.0)] = 1348 \text{ kW.}$$

Thermodynamics

Water

$\rho = 1000 \text{ kg/m}^3 = 62.4 \text{ lb}_m/\text{ft}^3$

$c_p = 4.18 \text{ kJ/kg}\cdot°\text{C}$

$c_p = 1.00 \text{ BTU/lb}_m\text{-}°\text{R}$

Air (standard conditions)

$k = 1.4$

$\rho = 1.23 \text{ kg/m}^3 = 0.0024 \text{ lb}_m/\text{ft}^3$

$c_p = 1.00 \text{ kJ/kg}\cdot°\text{C} = 0.24 \text{ BTU/lb}_m\text{-}°\text{F}$

$c_v = 0.716 \text{ kJ/kg}\cdot°\text{C} = 0.171 \text{ BTU/lb}_m\text{-}°\text{F}$

$R = 287 \text{ J/kg}\cdot°\text{C} = 53.3 \text{ BTU/lb}_m\text{-}°\text{F}$

Ice

$h_{fusion} = 320 \text{ kJ/kg} = 140 \text{ BTU/lb}_m$

$h_{sublimination} = 2040 \text{ kJ/kg} = 877 \text{ BTU/lb}_m$

Vapor relations

$$v = v_f + x(v_g - v_f) = v_f - xv_{fg}$$

$$x = \frac{\textit{mass of vapor}}{\textit{total mass}}$$

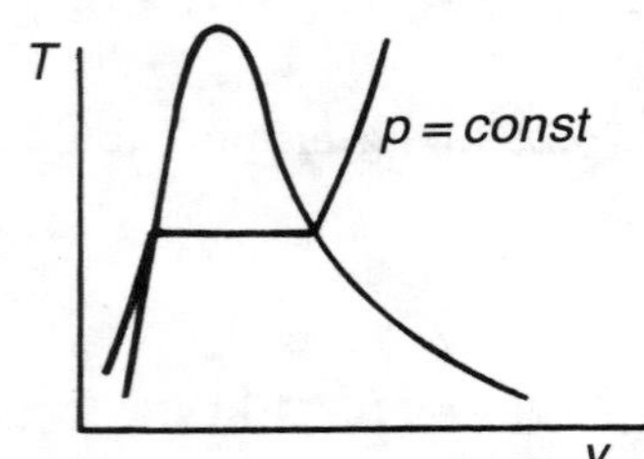

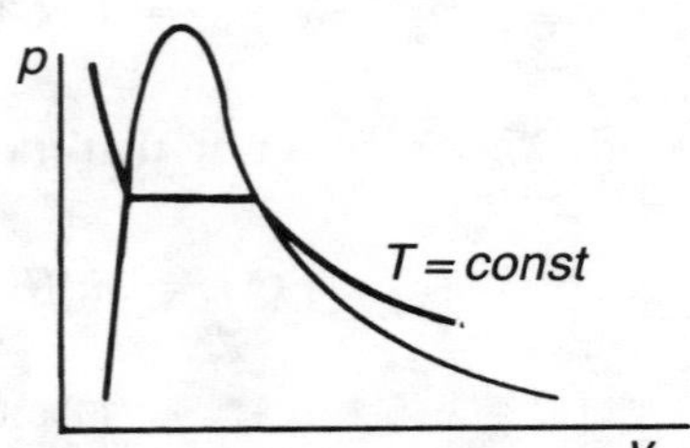

1st law – system

$$Q - W = \Delta U + \Delta KE + \Delta PE \qquad KE = \frac{1}{2}mV^2 \qquad PE = mgh$$

1st law – control volume

$$\frac{\dot{Q}-\dot{W}_S}{\dot{m}} = h_2-h_1 + \frac{V_2^2-V_1^2}{2} + g\,(z_2-z_1) \qquad \textit{(gas or vapor)}$$

$$\frac{\dot{Q}-\dot{W}_S}{\dot{m}} = \frac{p_2-p_1}{\rho} + u_2-u_1 + \frac{V_2^2-V_1^2}{2} + g\,(z_2-z_1) \qquad \textit{(liquid)}$$

General equations

$$\dot{m} = \rho AV \qquad \Delta h = c_p\Delta T \qquad \Delta u = c_v\Delta T \qquad pV = mRT \qquad v = \frac{1}{\rho} \qquad c_p - c_v = R$$

$$W = \int pdV \qquad v = \frac{V}{m} \qquad h = u + pv$$

$R_{universal} = 8.315 \text{ kJ/kg mol}\cdot\text{L}$ $\qquad R = R_{universal}/\text{Molar mass}$

Second law equations

$$\Delta s = c_p \ln \frac{T_2}{T_1} - R \ln \frac{p_2}{p_1} \qquad \eta_{max} = 1 - \frac{T_L}{T_H} \qquad COP_{ref} = \frac{T_L}{T_H - T_L}$$

$$\Delta s = c_v \ln \frac{T_2}{T_1} + R \ln \frac{v_2}{v_1} \qquad COP_{heat\ pump} = \frac{T_L}{T_L - T_H}$$

$$\Delta S \le \int \frac{đQ}{T}$$

TABLE 7.1. Properties of Ideal Gases — Metric Units

Gas	Chemical Formula	Molar Mass	$R \frac{kJ}{kg \cdot K}$	$c_p \frac{kJ}{kg \cdot K}$	$c_v \frac{kJ}{kg \cdot K}$	k
Air	—	28.97	0.287 00	1.0035	0.7165	1.400
Argon	Ar	39.948	0.208 13	0.5203	0.3122	1.667
Butane	C_4H_{10}	58.124	0.143 04	1.7164	1.5734	1.091
Carbon Dioxide	CO_2	44.01	0.188 92	0.8418	0.6529	1.289
Carbon Monoxide	CO	28.01	0.296 83	1.0413	0.7445	1.400
Ethane	C_2H_6	30.07	0.276 50	1.7662	1.4897	1.186
Ethylene	C_2H_4	28.054	0.296 37	1.5482	1.2518	1.237
Helium	He	4.003	2.077 03	5.1926	3.1156	1.667
Hydrogen	H_2	2.016	4.124 18	14.2091	10.0849	1.409
Methane	CH_4	16.04	0.518 35	2.2537	1.7354	1.299
Neon	Ne	20.183	0.411 95	1.0299	0.6179	1.667
Nitrogen	N_2	28.013	0.296 80	1.0416	0.7448	1.400
Octane	C_8H_{18}	114.23	0.072 79	1.7113	1.6385	1.044
Oxygen	O_2	31.999	0.259 83	0.9216	0.6618	1.393
Propane	C_3H_8	44.097	0.188 55	1.6794	1.4909	1.126
Steam	H_2O	18.015	0.461 52	1.8723	1.4108	1.327

TABLE 7.2. Specific Heats of Liquids and Solids — Metric Units

c_p kJ/(kg · °C)

A. LIQUIDS

Substance	State	c_p	Substance	State	c_p
Water	1 atm, 25°C	4.177	Glycerin	1 atm, 10°C	2.32
Ammonia	sat., −20°C	4.52	Bismuth	1 atm, 425°C	0.144
	sat., 50°C	5.10	Mercury	1 atm, 10°C	0.138
Refrigerant 12	sat., −20°C	0.908	Sodium	1 atm, 95°C	1.38
	sat., 50°C	1.02	Propane	1 atm, 0°C	2.41
Benzene	1 atm, 15°C	1.80	Ethyl Alcohol	1 atm, 25°C	2.43

B. SOLIDS

Substance	T, °C	c_p	Substance	T, °C	c_p
Ice	− 11	2.033	Lead	−100	0.118
	− 2.2	2.10		0	0.124
Aluminum	−100	0.699		100	0.134
	0	0.870	Copper	−100	0.328
	100	0.941		0	0.381
Iron	20	0.448		100	0.393
Silver	20	0.233			

TABLE 7.3. Thermodynamic Properties of Water (Steam Tables) — Metric Units

7.3.1 Saturated H_2O — Temperature Table

		Volume, m^3/kg		Energy, kJ/kg		Enthalpy, kJ/kg			Entropy, kJ/(kg·K)		
T,°C	p,MPa	v_f	v_g	u_f	u_g	h_f	h_{fg}	h_g	s_f	s_{fg}	s_g
0.010	0.0006113	0.001000	206.1	0.0	2375.3	0.0	2501.3	2501.3	0.0000	9.1571	9.1571
5	0.0008721	0.001000	147.1	21.0	2382.2	21.0	2489.5	2510.5	0.0761	8.9505	9.0266
10	0.001228	0.001000	106.4	42.0	2389.2	42.0	2477.7	2519.7	0.1510	8.7506	8.9016
20	0.002338	0.001002	57.79	83.9	2402.9	83.9	2454.2	2538.1	0.2965	8.3715	8.6680
30	0.004246	0.001004	32.90	125.8	2416.6	125.8	2430.4	2556.2	0.4367	8.0174	8.4541
40	0.007383	0.001008	19.52	167.5	2430.1	167.5	2406.8	2574.3	0.5723	7.6855	8.2578
50	0.01235	0.001012	12.03	209.3	2443.5	209.3	2382.8	2592.1	0.7036	7.3735	8.0771
60	0.01994	0.001017	7.671	251.1	2456.6	251.1	2358.5	2609.6	0.8310	7.0794	7.9104
70	0.03119	0.001023	5.042	292.9	2469.5	293.0	2333.8	2626.8	0.9549	6.8012	7.7561
80	0.04739	0.001029	3.407	334.8	2482.2	334.9	2308.8	2643.7	1.0754	6.5376	7.6130
90	0.07013	0.001036	2.361	376.8	2494.5	376.9	2283.2	2660.1	1.1927	6.2872	7.4799
100	0.1013	0.001044	1.673	418.9	2506.5	419.0	2257.0	2676.0	1.3071	6.0486	7.3557
120	0.1985	0.001060	0.8919	503.5	2529.2	503.7	2202.6	2706.3	1.5280	5.6024	7.1304
140	0.3613	0.001080	0.5089	588.7	2550.0	589.1	2144.8	2733.9	1.7395	5.1912	6.9307
160	0.6178	0.001102	0.3071	674.9	2568.4	675.5	2082.6	2758.1	1.9431	4.8079	6.7510
180	1.002	0.001127	0.1941	762.1	2583.7	763.2	2015.0	2778.2	2.1400	4.4466	6.5866
200	1.554	0.001156	0.1274	850.6	2595.3	852.4	1940.8	2793.2	2.3313	4.1018	6.4331
220	2.318	0.001190	0.08620	940.9	2602.4	943.6	1858.5	2802.1	2.5183	3.7686	6.2869
240	3.344	0.001229	0.5977	1033.2	2604.0	1037.3	1766.5	2803.8	2.7021	3.4425	6.1446
260	4.688	0.001276	0.04221	1128.4	2599.0	1134.4	1662.5	2796.9	2.8844	3.1184	6.0028
280	6.411	0.001332	0.03017	1227.4	2586.1	1236.0	1543.6	2779.6	3.0674	2.7905	5.8579
300	8.580	0.001404	0.02168	1332.0	2563.0	1344.0	1405.0	2749.0	3.2540	2.4513	5.7053
320	11.27	0.001499	0.01549	1444.6	2525.5	1461.4	1238.7	2700.1	3.4487	2.0883	5.5370
340	14.59	0.001638	0.01080	1570.3	2464.6	1594.2	1027.9	2622.1	3.6601	1.6765	5.3366
360	18.65	0.001892	0.006947	1725.2	2351.6	1760.5	720.7	2481.2	3.9154	1.1382	5.0536
374.136	22.088	0.003155	0.003155	2029.6	2029.6	2099.3	0.0	2099.3	4.4305	0.0000	4.4305

7.3.2 Saturated H_2O — Pressure Table

		Volume, m^3/kg		Energy, kJ/kg		Enthalpy, kJ/kg			Entropy, kJ/(kg·K)		
p,MPa	T,°C	v_f	v_g	u_f	u_g	h_f	h_{fg}	h_g	s_f	s_{fg}	s_g
0.000611	0.01	0.001000	206.1	0.0	2375.3	0.0	2501.3	2501.3	0.0000	9.1571	9.1571
0.001	7.0	0.001000	129.2	29.3	2385.0	29.3	2484.9	2514.2	0.1059	8.8706	8.9765
0.002	17.5	0.001001	67.00	73.5	2399.5	73.5	2460.0	2533.5	0.2606	8.4639	8.7245
0.01	45.8	0.001010	14.67	191.8	2437.9	191.8	2392.8	2584.6	0.6491	7.5019	8.1510
0.02	60.1	0.001017	7.649	251.4	2456.7	251.4	2358.3	2609.7	0.8319	7.0774	7.9093
0.04	75.9	0.001026	3.993	317.5	2477.0	317.6	2319.1	2636.7	1.0260	6.6449	7.6709
0.06	85.9	0.001033	2.732	359.8	2489.6	359.8	2293.7	2653.5	1.1455	6.3873	7.5328
0.08	93.5	0.001039	2.087	391.6	2498.8	391.6	2274.1	2665.7	1.2331	6.2023	7.4354
0.1(atm)	99.6	0.001043	1.694	417.3	2506.1	417.4	2258.1	2675.5	1.3029	6.0573	7.3602
0.12	104.8	0.001047	1.428	439.2	2512.1	439.3	2244.2	2683.5	1.3611	5.9378	7.2989
0.16	113.3	0.001054	1.091	475.2	2521.8	475.3	2221.2	2696.5	1.4553	5.7472	7.2025
0.2	120.2	0.001061	0.8857	504.5	2529.5	504.7	2201.9	2706.6	1.5305	5.5975	7.1280
0.4	143.6	0.001084	0.4625	604.3	2553.6	604.7	2133.8	2738.5	1.7770	5.1197	6.8967
0.6	158.9	0.001101	0.3157	669.9	2567.4	670.6	2086.2	2756.8	1.9316	4.8293	6.7609
0.8	170.4	0.001115	0.2404	720.2	2576.8	721.1	2048.0	2769.1	2.0466	4.6170	6.6636
1	179.9	0.001127	0.1944	761.7	2583.6	762.8	2015.3	2778.1	2.1391	4.4482	6.5873
1.2	188.0	0.001139	0.1633	797.3	2588.8	798.6	1986.2	2784.8	2.2170	4.3072	6.5242
1.6	201.4	0.001159	0.1238	856.9	2596.0	858.8	1935.2	2794.0	2.3446	4.0780	6.4226
2	212.4	0.001177	0.09963	906.4	2600.3	908.8	1890.7	2799.5	2.4478	3.8939	6.3417
4	250.4	0.001252	0.04978	1082.3	2602.3	1087.3	1714.1	2801.4	2.7970	3.2739	6.0709
6	275.6	0.001319	0.03244	1205.4	2589.7	1213.3	1571.0	2784.3	3.0273	2.8627	5.8900
8	295.1	0.001384	0.02352	1305.6	2569.8	1316.6	1441.4	2758.0	3.2075	2.5365	5.7440
12	324.8	0.001527	0.01426	1472.9	2513.7	1491.3	1193.6	2684.9	3.4970	1.9963	5.4933
16	347.4	0.001711	0.009307	1622.7	2431.8	1650.0	930.7	2580.7	3.7468	1.4996	5.2464
20	365.8	0.002036	0.005836	1785.6	2293.2	1826.3	583.7	2410.0	4.0146	0.9135	4.9281
22.088	374.136	0.003155	0.003155	2029.6	2029.6	2099.3	0.0	2099.3	4.4305	0.0000	4.4305

7.3.3 Superheated Steam

p, MPa (T_{sat} °C)		Temperature, °C											
		50	100	150	200	250	300	350	400	500	600	700	800
0.01	v	14.87	17.20	19.51	21.83	24.14	26.45	28.75	31.06	35.68	40.29	44.91	49.53
(45.8)	u	2443.9	2515.5	2587.9	2661.3	2736.0	2812.1	2889.7	2968.9	3132.3	3302.5	3479.6	3663.8
	h	2592.6	2687.5	2783.0	2879.5	2977.3	3076.5	3177.2	3279.5	3489.0	3705.4	3928.7	4159.1
	s	8.1757	8.4487	8.6890	8.9046	9.1010	9.2821	9.4506	9.6084	9.8985	10.1616	10.4037	10.6290
0.05	v		3.418	3.889	4.356	4.820	5.284	5.747	6.209	7.134	8.057	8.981	9.904
(81.3)	u		2511.6	2585.6	2659.8	2735.0	2811.3	2889.1	2968.4	3131.9	3302.2	3479.5	3663.7
	h		2682.5	2780.1	2877.6	2976.0	3075.5	3176.4	3278.9	3488.6	3705.1	3928.5	4158.9
	s		7.6955	7.9409	8.1588	8.3564	8.5380	8.7069	8.8650	9.1554	9.4186	9.6608	9.8861
0.1 (one atm)	v		1.696	1.936	2.172	2.406	2.639	2.871	3.103	3.565	4.028	4.490	4.952
(99.6)	u		2506.6	2582.7	2658.0	2733.7	2810.4	2888.4	2967.8	3131.5	3301.9	3479.2	3663.5
	h		2676.2	2776.4	2875.3	2974.3	3074.3	3175.5	3278.1	3488.1	3704.7	3928.2	4158.7
	s		7.3622	7.6142	7.8351	8.0341	8.2165	8.3858	8.5442	8.8350	9.0984	9.3406	9.5660
0.2	v			0.9596	1.080	1.199	1.316	1.433	1.549	1.781	2.013	2.244	2.475
(120.2)	u			2576.9	2654.4	2731.2	2808.6	2886.9	2966.7	3130.7	3301.4	3478.8	3663.2
	h			2768.8	2870.5	2971.0	3071.8	3173.5	3276.5	3487.0	3704.0	3927.7	4158.3
	s			7.2803	7.5074	7.7094	7.8934	8.0636	8.2226	8.5140	8.7778	9.0203	9.2458
0.4	v			0.4708	0.5342	0.5951	0.6548	0.7139	0.7726	0.8893	1.006	1.121	1.237
(143.6)	u			2564.5	2646.8	2726.1	2804.8	2884.0	2964.4	3129.2	3300.2	3477.9	3662.5
	h			2752.8	2860.5	2964.2	3066.7	3169.6	3273.4	3484.9	3702.4	3926.5	4157.4
	s			6.9307	7.1714	7.3797	7.5670	7.7390	7.8992	8.1921	8.4566	8.6995	8.9253
0.6	v				0.3520	0.3938	0.4344	0.4742	0.5137	0.5920	0.6697	0.7472	0.8245
(158.9)	u				2638.9	2720.9	2801.0	2881.1	2962.0	3127.6	3299.1	3477.1	3661.8
	h				2850.1	2957.2	3061.6	3165.7	3270.2	3482.7	3700.9	3925.4	4156.5
	s				6.9673	7.1824	7.3732	7.5472	7.7086	8.0029	8.2682	8.5115	8.7375
1	v				0.2060	0.2327	0.2579	0.2825	0.3066	0.3541	0.4011	0.4478	0.4943
(179.9)	u				2621.9	2709.9	2793.2	2875.2	2957.3	3124.3	3296.8	3475.4	3660.5
	h				2827.9	2942.6	3051.2	3157.7	3263.9	3478.4	3697.9	3923.1	4154.8
	s				6.6948	6.9255	7.1237	7.3019	7.4658	7.7630	8.0298	8.2740	8.5005
2	v					0.1114	0.1255	0.1386	0.1512	0.1757	0.1996	0.2232	0.2467
(212.4)	u					2679.6	2772.6	2859.8	2945.2	3116.2	3290.9	3471.0	3657.0
	h					2902.5	3023.5	3137.0	3247.6	3467.6	3690.1	3917.5	4150.4
	s					6.5461	6.7672	6.9571	7.1279	7.4325	7.7032	7.9496	8.1774
4	v						0.05884	0.06645	0.07341	0.08643	0.09885	0.1109	0.1229
(250.4)	u						2725.3	2826.6	2919.9	3099.5	3279.1	3462.1	3650.1
	h						2960.7	3092.4	3213.5	3445.2	3674.4	3905.9	4141.6
	s						6.3622	6.5828	6.7698	7.0908	7.3696	7.6206	7.8511
6	v						0.03616	0.04223	0.04739	0.05665	0.06525	0.07352	0.08160
(275.6)	u						2667.2	2789.6	2892.8	3082.2	3266.9	3453.2	3643.1
	h						2884.2	3043.0	3177.2	3422.1	3658.4	3894.3	4132.7
	s						6.0682	6.3342	6.5415	6.8811	7.1685	7.4242	7.6575
10	v							0.02242	0.02641	0.03279	0.03837	0.04358	0.04859
(311.1)	u							2699.2	2832.4	3045.8	3241.7	3434.7	3629.0
	h							2923.4	3096.5	3373.6	3625.3	3870.5	4114.9
	s							5.9451	6.2127	6.5974	6.9037	7.1696	7.4086
20	v								0.00994	0.01477	0.01818	0.02113	0.02385
(365.8)	u								2619.2	2942.8	3174.0	3386.5	3592.7
	h								2818.1	3238.2	3537.6	3809.1	4069.8
	s								5.5548	6.1409	6.5056	6.8002	7.0553
40	v								0.00191	0.00562	0.00809	0.00994	0.01152
	u								1854.5	2678.4	3022.6	3283.6	3517.9
	h								1930.8	2903.3	3346.4	3681.3	3978.8
	s								4.1143	5.4707	6.0122	6.3759	6.6671

TABLE 7.4 Refrigerant-12 — Metric Units

7.4.1 Saturated Refrigerant-12 — Temperature Table

		Specific Volume m³/kg		Enthalpy kJ/kg			Entropy kJ/kg · K	
T, °C	p, MPa	v_f	v_g	h_f	h_{fg}	h_g	s_f	s_g
−90	0.0028	0.000 608	4.415 545	−43.243	189.618	146.375	−0.2084	0.8268
−80	0.0062	0.000 617	2.138 345	−34.688	185.612	150.924	−0.1630	0.7979
−70	0.0123	0.000 627	1.127 280	−26.103	181.640	155.536	−0.1197	0.7744
−60	0.0226	0.000 637	0.637 910	−17.469	177.653	160.184	−0.0782	0.7552
−50	0.0391	0.000 648	0.383 105	−8.772	173.611	164.840	−0.0384	0.7396
−40	0.0642	0.000 659	0.241 910	−0.000	169.479	169.479	−0.0000	0.7269
−30	0.1004	0.000 672	0.159 375	8.854	165.222	174.076	0.0371	0.7165
−20	0.1509	0.000 685	0.108 847	17.800	160.810	178.610	0.0730	0.7082
−10	0.2191	0.000 700	0.076 646	26.851	156.207	183.058	0.1079	0.7014
0	0.3086	0.000 716	0.055 389	36.022	151.376	187.397	0.1418	0.6960
10	0.4233	0.000 733	0.040 914	45.337	146.265	191.602	0.1750	0.6916
20	0.5673	0.000 752	0.030 780	54.828	140.812	195.641	0.2076	0.6879
30	0.7449	0.000 774	0.023 508	64.539	134.936	199.475	0.2397	0.6848
40	0.9607	0.000 798	0.018 171	74.527	128.525	203.051	0.2716	0.6820
50	1.2193	0.000 826	0.014 170	84.868	121.430	206.298	0.3034	0.6792
60	1.5259	0.000 858	0.011 111	95.665	113.443	209.109	0.3355	0.6760
70	1.8858	0.000 897	0.008 725	107.067	104.255	211.321	0.3683	0.6721
80	2.3046	0.000 946	0.006 821	119.291	93.373	212.665	0.4023	0.6667
90	2.7885	0.001 012	0.005 258	132.708	79.907	212.614	0.4385	0.6585
100	3.3440	0.001 113	0.003 903	148.076	61.768	209.843	0.4788	0.6444
110	3.9784	0.001 364	0.002 462	168.059	28.425	196.484	0.5322	0.6064

7.4.2 Saturated Refrigerant-12 — Pressure Table

		Specific Volume m³/kg		Enthalpy kJ/kg			Entropy kJ/kg · K	
p, MPa	T, °C	v_f	v_g	h_f	h_{fg}	h_g	s_f	s_g
0.06	−41.42	.000 6578	0.257 5	−1.25	170.19	168.94	−0.0054	0.7290
0.10	−30.10	.000 6719	0.160 0	8.78	165.37	174.15	0.0368	.7171
0.12	−25.74	.000 6776	0.134 9	12.66	163.48	176.14	.0526	.7133
0.14	−21.91	.000 6828	0.116 8	16.09	161.78	177.87	.0663	.7102
0.16	−18.49	.000 6876	0.103 1	19.18	160.23	179.41	.0784	.7076
0.18	−15.38	.000 6921	0.092 25	21.98	158.82	180.80	.0893	.7054
0.20	−12.53	.000 6962	0.083 54	24.57	157.50	182.07	.0992	.7035
0.24	−7.42	.000 7040	0.070 33	29.23	155.09	184.32	.1168	.7004
0.28	−2.93	.000 7111	0.060 76	33.35	152.92	186.27	.1321	.6980
0.32	1.11	.000 7177	0.053 51	37.08	150.92	188.00	.1457	.6960
0.40	8.15	.000 7299	0.043 21	43.64	147.33	190.97	.1691	.6928
0.50	15.60	.000 7438	0.034 82	50.67	143.35	194.02	.1935	.6899
0.60	22.00	.000 7566	0.029 13	56.80	139.77	196.57	.2142	.6878
0.70	27.65	.000 7686	0.025 01	62.29	136.45	198.74	.2324	6860
0.80	32.74	.000 7802	0.021 88	67.30	133.33	200.63	.2487	.6845
0.90	37.37	.000 7914	0.019 42	71.93	130.36	202.29	.2634	.6832
1.00	41.64	.000 8023	0.017 44	76.26	127.50	203.76	.2770	.6820
1.20	49.31	.000 8237	0.014 41	84.21	122.03	206.24	.3015	.6799
1.40	56.09	.000 8448	0.012 22	91.46	116.76	208.22	.3232	.6778
1.60	62.19	.000 8660	0.010 54	98.19	111.62	209.81	.3329	.6758

7.4.3 Superheated Refrigerant-12

T, °C	v	h	s	v	h	s	v	h	s
	0.05 MPa			**0.10 MPa**			**0.20 MPa**		
−20.0	0.341 9	181.0	0.7912	0.167 7	179.9	0.7401			
−10.0	0.356 2	186.8	0.8133	0.175 2	185.7	0.7628			
0.0	0.370 5	192.6	0.8350	0.182 6	191.6	0.7849	0.088 61	189.7	0.7320
10.0	0.384 7	198.5	0.8562	0.190 0	197.6	0.8064	0.092 55	195.9	0.7543
20.0	0.398 9	204.5	0.8770	0.197 3	203.7	0.8275	0.096 42	202.1	0.7760
30.0	0.413 0	210.6	0.8974	0.204 5	209.9	0.8482	0.100 23	208.4	0.7972
40.0	0.427 0	216.7	0.9175	0.211 7	216.1	0.8684	0.103 99	214.8	0.8178
50.0	0.441 0	223.0	0.9372	0.218 8	222.4	0.8883	0.107 71	221.2	0.8381
60.0	0.455 0	229.3	0.9565	0.226 0	228.8	0.9078	0.111 40	227.7	0.8578
70.0	0.469 0	235.8	0.9755	0.233 0	235.3	0.9269	0.115 06	234.3	0.8772
80.0	0.482 9	242.3	0.9942	0.240 1	241.8	0.9457	0.118 69	240.9	0.8962
90.0	0.496 9	248.9	1.0126	0.247 2	248.4	0.9642	0.122 30	247.6	0.9149
	0.30 MPa			**0.40 MPa**			**0.60 MPa**		
20.0	0.062 73	200.5	0.7440	0.045 84	198.8	0.199			
30.0	0.065 41	207.0	0.7658	0.047 97	205.4	0.7423	0.030 42	202.1	0.7068
40.0	0.068 05	213.5	0.7869	0.050 05	212.1	0.7639	0.031 97	209.2	0.7291
50.0	0.070 64	220.0	0.8075	0.052 07	218.8	0.7849	0.033 45	216.1	0.7511
60.0	0.073 18	226.6	0.8276	0.054 06	225.5	0.8054	0.034 89	223.1	0.7723
70.0	0.075 70	233.3	0.8473	0.056 01	232.2	0.8253	0.036 29	230.1	0.7929
80.0	0.078 20	240.0	0.8665	0.057 94	239.0	0.8448	0.037 65	237.0	0.8129
90.0	0.080 67	246.7	0.8853	0.059 85	245.8	0.8638	0.039 00	244.0	0.8324
100.0	0.083 13	253.5	0.9038	0.061 73	252.7	0.8825	0.040 32	251.0	0.8514
110.0	0.085 57	260.4	0.9220	0.063 60	259.6	0.9008	0.041 62	258.1	0.8700
120.0				0.065 46	266.6	0.9187	0.042 91	265.1	0.8882
130.0				0.067 30	273.6	0.9364	0.044 18	272.2	0.9061
	0.80 MPa			**1.00 MPa**			**1.20 MPa**		
50.0	0.024 07	213.3	0.7248	0.018 37	210.9	0.7021	0.014 48	206.7	0.6812
60.0	0.025 25	220.6	0.7469	0.019 41	217.8	0.7254	0.015 46	214.8	0.7060
70.0	0.026 38	227.8	0.7682	0.020 40	225.3	0.7476	0.016 37	222.7	0.7293
80.0	0.027 48	234.9	0.7888	0.021 34	232.7	0.7689	0.017 22	230.4	0.7514
90.0	0.028 55	242.1	0.8088	0.022 25	240.1	0.7895	0.018 03	238.0	0.7727
100.0	0.029 59	249.3	0.8283	0.023 13	247.4	0.8094	0.018 81	245.5	0.7931
110.0	0.030 61	256.4	0.8472	0.023 99	254.7	0.8287	0.019 57	253.0	0.8129
120.0	0.031 62	263.6	0.8657	0.024 84	262.7	0.8475	0.020 30	260.4	0.8320
130.0	0.032 61	270.8	0.8838	0.025 66	269.7	0.8659	0.021 02	267.9	0.8507
140.0	0.033 59	278.1	0.9016	0.026 47	276.7	0.8839	0.021 72	275.3	0.8689
150.0	0.034 56	285.3	0.9189	0.027 28	284.0	0.9015	0.022 41	282.7	0.8867
160.0				0.028 07	291.4	0.9187	0.023 09	290.2	0.9041
	1.60 MPa			**2.00 MPa**			**3.00 MPa**		
70.0	0.011 21	216.6	0.6959						
80.0	0.011 98	225.2	0.7204	0.008 704	218.9	0.6909			
90.0	0.012 70	233.4	0.7433	0.009 406	228.1	0.7166			
100.0	0.013 37	241.4	0.7651	0.010 085	236.8	0.7482	0.005 231	220.5	0.6770
110.0	0.014 00	249.3	0.7859	0.010 615	245.2	0.7624	0.005 886	232.1	0.7075
120.0	0.014 61	257.0	0.8059	0.011 159	253.3	0.7835	0.006 419	242.2	0.7336
130.0	0.015 20	264.7	0.8253	0.011 676	261.4	0.8037	0.006 887	251.6	0.7573
140.0	0.015 76	272.4	0.8440	0.012 172	269.3	0.8232	0.007 313	260.6	0.7793
150.0	0.016 32	280.0	0.8623	0.012 651	277.2	0.8420	0.007 709	269.3	0.8001
160.0	0.016 86	287.7	0.8801	0.013 116	285.0	0.8603	0.008 083	277.8	0.8200
170.0	0.017 40	295.3	0.8975	0.013 570	292.8	0.8781	0.008 439	286.2	0.8391
180.0	0.017 92	302.9	0.9145	0.014 010	300.6	0.8955	0.008 782	294.4	0.8575

TABLE 7.1E. Properties of Ideal Gases — English Units

Gas	Chemical Formula	Molecular Weight	$R\ \frac{\text{ft-lb}}{\text{lbm}\cdot\text{R}}$	$c_p\ \frac{\text{Btu}}{\text{lbm}\cdot\text{R}}$	$c_v\ \frac{\text{Btu}}{\text{lbm}\cdot\text{R}}$	k
Air	. . .	28.97	53.34	0.240	0.171	1.400
Argon	Ar	39.94	38.68	0.1253	0.0756	1.667
Butane	C_4H_{10}	58.124	26.58	0.415	0.381	1.09
Carbon Dioxide	CO_2	44.01	35.10	0.203	0.158	1.285
Carbon Monoxide	CO	28.01	55.16	0.249	0.178	1.399
Ethane	C_2H_6	30.07	51.38	0.427	0.361	1.183
Ethylene	C_2H_4	28.054	55.07	0.411	0.340	1.208
Helium	He	4.003	386.0	1.25	0.753	1.667
Hydrogen	H_2	2.016	766.4	3.43	2.44	1.404
Methane	CH_4	16.04	96.35	0.532	0.403	1.32
Neon	Ne	20.183	76.55	0.246	0.1477	1.667
Nitrogen	N_2	28.016	55.15	0.248	0.177	1.400
Octane	C_8H_{18}	114.22	13.53	0.409	0.392	1.044
Oxygen	O_2	32.000	48.28	0.219	0.157	1.395
Propane	C_3H_8	44.097	35.04	0.407	0.362	1.124
Steam	H_2O	18.016	85.76	0.445	0.335	1.329

TABLE 7.2E. Specific Heats of Liquids and Solids — English Units

c_p Btu/(lbm-°F)

A. LIQUIDS

Substance	State	c_p	Substance	State	c_p
Water	1 atm, 77°F	1.00	Glycerin	1 atm, 50°F	0.555
Ammonia	sat., −4°F	1.08	Bismuth	1 atm, 800°F	0.0344
	sat., 120°F	1.22	Mercury	1 atm, 50°F	0.0330
Refrigerant-12	sat., −4°F	0.217	Sodium	1 atm, 200°F	0.330
	sat., 120°F	0.244	Propane	1 atm, 32°F	0.577
Benzene	1 atm, 60°F	0.431	Ethyl Alcohol	1 atm, 77°F	0.581

B. SOLIDS

Substance	T, °F	c_p	Substance	T, °F	c_p
Ice	−76	0.392	Silver	−4	0.0557
	12	0.486	Lead	−150	0.0282
	28	0.402		30	0.0297
Aluminum	−150	0.167		210	0.0321
	30	0.208	Copper	−150	0.0785
	210	0.225		30	0.0911
Iron	−4	0.107		210	0.0940

TABLE 7.3E. Thermodynamic Properties of Water (Steam Tables) — English Units

7.3.1E Saturated H_2O — Temperature Table

		Volume, ft³/lbm		Energy, Btu/lbm		Enthalpy, Btu/lbm			Entropy, Btu/lbm-°R		
T, °F	p, psia	v_f	v_g	u_f	u_g	h_f	h_{fg}	h_g	s_f	s_{fg}	s_g
32.018	0.08866	0.016022	3302	0.00	1021.2	0.01	1075.4	1075.4	0.00000	2.1869	2.1869
40	0.12166	0.016020	2445	8.02	1023.9	8.02	1070.9	1078.9	0.01617	2.1430	2.1592
60	0.2563	0.016035	1206.9	28.08	1030.4	28.08	1059.6	1087.7	0.05555	2.0388	2.0943
80	0.5073	0.016073	632.8	48.08	1037.0	48.09	1048.3	1096.4	0.09332	1.9423	2.0356
100	0.9503	0.016130	350.0	68.04	1043.5	68.05	1037.0	1105.0	0.12963	1.8526	1.9822
120	1.6945	0.016205	203.0	87.99	1049.9	88.00	1025.5	1113.5	0.16465	1.7690	1.9336
140	2.892	0.016293	122.88	107.95	1056.2	107.96	1014.0	1121.9	0.19851	1.6907	1.8892
160	4.745	0.016395	77.23	127.94	1062.3	127.96	1002.2	1130.1	0.23130	1.6171	1.8484
180	7.515	0.016509	50.20	147.97	1068.3	147.99	990.2	1138.2	0.26311	1.5478	1.8109
200	11.529	0.016634	33.63	168.04	1074.2	168.07	977.9	1145.9	0.29400	1.4822	1.7762
212	14.698	0.016716	26.80	180.11	1077.6	180.16	970.3	1150.5	0.31213	1.4446	1.7567
220	17.188	0.016772	23.15	188.17	1079.8	188.22	965.3	1153.5	0.32406	1.4201	1.7441
240	24.97	0.016922	16.327	208.36	1085.3	208.44	952.3	1160.7	0.35335	1.3609	1.7143
260	35.42	0.017084	11.768	228.64	1090.5	228.76	938.8	1167.6	0.38193	1.3044	1.6864
280	49.18	0.017259	8.650	249.02	1095.4	249.18	924.9	1174.1	0.40986	1.2504	1.6602
300	66.98	0.017448	6.472	269.52	1100.0	269.73	910.4	1180.2	0.43720	1.1984	1.6356
340	117.93	0.017872	3.792	310.91	1108.0	311.30	879.5	1190.8	0.49031	1.0997	1.5901
380	195.60	0.018363	2.339	352.95	1114.3	353.62	845.4	1199.0	0.54163	1.0067	1.5483
420	308.5	0.018936	1.5024	395.81	1118.3	396.89	807.2	1204.1	0.59152	0.9175	1.5091
460	466.3	0.019614	0.9961	439.7	1119.6	441.4	764.1	1205.5	0.6404	0.8308	1.4712
500	680.0	0.02043	0.6761	485.1	1117.4	487.7	714.8	1202.5	0.6888	0.7448	1.4335
540	961.5	0.02145	0.4658	532.6	1111.0	536.4	657.5	1193.8	0.7374	0.6576	1.3950
580	1324.3	0.02278	0.3225	583.1	1098.9	588.6	589.3	1178.0	0.7872	0.5668	1.3540
620	1784.4	0.02465	0.2209	638.3	1078.5	646.4	505.0	1151.4	0.8398	0.4677	1.3075
660	2362	0.02767	0.14459	702.3	1042.3	714.4	391.1	1105.5	0.8990	0.3493	1.2483
700	3090	0.03666	0.07438	801.7	947.7	822.7	167.5	990.2	0.9902	0.1444	1.1346
705.44	3204	0.05053	0.05053	872.6	872.6	902.5	0	902.5	1.0580	0	1.0580

7.3.2E Saturated H_2O — Pressure Table

		Volume, ft³/lbm		Energy, Btu/lbm		Enthalpy, Btu/lbm			Entropy, Btu/lbm-°R		
p, psia	T, °F	v_f	v_g	u_f	u_g	h_f	h_{fg}	h_g	s_f	s_{fg}	s_g
1.0	101.70	0.016136	333.6	69.74	1044.0	69.74	1036.0	1105.8	0.13266	1.8453	1.9779
2.0	126.04	0.016230	173.75	94.02	1051.8	94.02	1022.1	1116.1	0.17499	1.7448	1.9198
4.0	152.93	0.016358	90.64	120.88	1060.2	120.89	1006.4	1127.3	0.21983	1.6426	1.8624
6.0	170.03	0.016451	61.98	137.98	1065.4	138.00	996.2	1134.2	0.24736	1.5819	1.8292
10	193.19	0.016590	38.42	161.20	1072.2	161.23	982.1	1143.3	0.28358	1.5041	1.7877
14.696	211.99	0.016715	26.80	180.10	1077.6	180.15	970.4	1150.5	0.31212	1.4446	1.7567
20	227.96	0.016830	20.09	196.19	1082.0	196.26	960.1	1156.4	0.33580	1.3962	1.7320
30	250.34	0.017004	13.748	218.84	1088.0	218.93	945.4	1164.3	0.36821	1.3314	1.6996
40	267.26	0.017146	10.501	236.03	1092.3	236.16	933.8	1170.0	0.39214	1.2845	1.6767
50	281.03	0.017269	8.518	250.08	1095.6	250.24	924.2	1174.4	0.41129	1.2476	1.6589
60	292.73	0.017378	7.177	262.06	1098.3	262.25	915.8	1178.0	0.42733	1.2170	1.6444
70	302.96	0.017478	6.209	272.56	1100.6	272.79	908.3	1181.0	0.44120	1.1909	1.6321
80	312.07	0.017570	5.474	281.95	1102.6	282.21	901.4	1183.6	0.45344	1.1679	1.6214
90	320.31	0.017655	4.898	290.46	1104.3	290.76	895.1	1185.9	0.46442	1.1475	1.6119
100	327.86	0.017736	4.434	298.28	1105.8	298.61	889.2	1187.8	0.47439	1.1290	1.6034
120	341.30	0.017886	3.730	312.27	1108.3	312.67	878.5	1191.1	0.49201	1.0966	1.5886
140	353.08	0.018024	3.221	324.58	1110.3	325.05	868.7	1193.8	0.50727	1.0688	1.5761
160	363.60	0.018152	2.836	335.63	1112.0	336.16	859.8	1196.0	0.52078	1.0443	1.5651
180	373.13	0.018273	2.533	345.68	1113.4	346.29	851.5	1197.8	0.53292	1.0223	1.5553
200	381.86	0.018387	2.289	354.9	1114.6	355.6	843.7	1199.3	0.5440	1.0025	1.5464
300	417.43	0.018896	1.5442	393.0	1118.2	394.1	809.8	1203.9	0.5883	0.9232	1.5115
400	444.70	0.019340	1.1620	422.8	1119.5	424.2	781.2	1205.5	0.6218	0.8638	1.4856
500	467.13	0.019748	0.9283	447.7	1119.4	449.5	755.8	1205.3	0.6490	0.8154	1.4645
600	486.33	0.02013	0.7702	469.4	1118.6	471.7	732.4	1204.1	0.6723	0.7742	1.4464
800	518.36	0.02087	0.5691	506.6	1115.0	509.7	689.6	1199.3	0.7110	0.7050	1.4160
1000	544.75	0.02159	0.4459	538.4	1109.9	542.4	650.0	1192.4	0.7432	0.6471	1.3903
1400	587.25	0.02307	0.3016	592.7	1096.0	598.6	575.5	1174.1	0.7964	0.5497	1.3461
2000	636.00	0.02565	0.18813	662.4	1066.6	671.9	464.4	1136.3	0.8623	0.4238	1.2861
3000	695.52	0.03431	0.08404	783.4	968.8	802.5	213.0	1015.5	0.9732	0.1843	1.1575
3203.6	705.44	0.05053	0.05053	872.6	872.6	902.5	0	902.5	1.0580	0	1.0580

7.3.3E Superheated Vapor

°F	*v*	*u*	*h*	*s*	*v*	*u*	*h*	*s*	*v*	*u*	*h*	*s*
	1 psia				**10 psia**				**14.7 psia**			
200	392.5	1077.5	1150.1	2.0508	38.85	1074.7	1146.6	1.7927	...	...	...	...
300	452.3	1112.1	1195.8	2.1153	45.00	1110.6	1193.9	1.8595	30.53	1109.7	1192.8	1.8160
400	511.9	1147.0	1241.8	2.1720	51.03	1146.1	1240.5	1.9171	34.67	1145.6	1239.9	1.8741
500	571.5	1182.8	1288.5	2.2235	57.04	1182.2	1287.7	1.9690	38.77	1181.8	1287.3	1.9263
600	631.1	1219.3	1336.1	2.2706	63.03	1218.9	1335.5	2.0164	42.86	1218.6	1335.2	1.9737
700	690.7	1256.7	1384.5	2.3142	69.01	1256.3	1384.0	2.0601	46.93	1256.1	1383.8	2.0175
800	750.3	1294.9	1433.7	2.3550	74.98	1294.6	1433.3	2.1009	51.00	1294.4	1433.1	2.0584
1000	869.5	1373.9	1534.8	2.4294	86.91	1373.8	1534.6	2.1755	59.13	1373.7	1534.5	2.1330
1200	988.6	1456.7	1639.6	2.4967	98.84	1456.5	1639.4	2.2428	67.25	1456.5	1639.3	2.2003
1400	1107.7	1543.1	1748.1	2.5584	110.76	1543.0	1748.0	2.3045	75.36	1543.0	1747.9	2.2621
	20 psia				**60 psia**				**100 psia**			
300	22.36	1108.8	1191.6	1.7808	7.259	1101.0	1181.6	1.6492	...	...	...	...
400	25.43	1145.1	1239.2	1.8395	8.353	1140.8	1233.5	1.7134	4.934	1136.2	1227.5	1.6517
500	28.46	1181.5	1286.8	1.8919	9.399	1178.6	1283.0	1.7678	5.587	1175.7	1279.1	1.7085
600	31.47	1218.4	1334.8	1.9395	10.425	1216.3	1332.1	1.8165	6.216	1214.2	1329.3	1.7582
700	34.47	1255.9	1383.5	1.9834	11.440	1254.4	1381.4	1.8609	6.834	1252.8	1379.2	1.8033
800	37.46	1294.3	1432.9	2.0243	12.448	1293.0	1431.2	1.9022	7.445	1291.8	1429.6	1.8449
1000	43.44	1373.5	1534.3	2.0989	14.454	1372.7	1533.2	1.9773	8.647	1371.9	1532.1	1.9204
1200	49.41	1456.4	1639.2	2.1663	16.452	1455.8	1638.5	2.0448	9.861	1455.2	1637.7	1.9882
1400	55.37	1542.9	1747.9	2.2281	18.445	1542.5	1747.3	2.1067	11.060	1542.0	1746.7	2.0502
1600	61.33	1633.2	1860.1	2.2854	20.44	1632.8	1859.7	2.1641	12.257	1632.4	1859.3	2.1076
	120 psia				**160 psia**				**200 psia**			
400	4.079	1133.8	1224.4	1.6288	3.007	1128.8	1217.8	1.5911	2.361	1123.5	1210.8	1.5600
500	4.633	1174.2	1277.1	1.6868	3.440	1171.2	1273.0	1.6518	2.724	1168.0	1268.8	1.6239
600	5.164	1213.2	1327.8	1.7371	3.848	1211.1	1325.0	1.7034	3.058	1208.9	1322.1	1.6767
700	5.682	1252.0	1378.2	1.7825	4.243	1250.4	1376.0	1.7494	3.379	1248.8	1373.8	1.7234
800	6.195	1291.2	1428.7	1.8243	4.631	1289.9	1427.0	1.7916	3.693	1288.6	1425.3	1.7660
1000	7.208	1371.5	1531.5	1.9000	5.397	1370.6	1530.4	1.8677	4.310	1369.8	1529.3	1.8425
1200	8.213	1454.9	1637.3	1.9679	6.154	1454.3	1636.5	1.9358	4.918	1453.7	1635.7	1.9109
1400	9.214	1541.8	1746.4	2.0300	6.906	1541.4	1745.9	1.9980	5.521	1540.9	1745.3	1.9732
1600	10.212	1632.3	1859.0	2.0875	7.656	1631.9	1858.6	2.0556	6.123	1631.6	1858.2	2.0308
1800	11.209	1726.2	1975.1	2.1413	8.405	1725.9	1974.8	2.1094	6.722	1725.6	1974.4	2.0847
	250 psia				**300 psia**				**400 psia**			
500	2.150	1163.8	1263.3	1.5948	1.7662	1159.5	1257.5	1.5701	1.2843	1150.1	1245.2	1.5282
600	2.426	1206.1	1318.3	1.6494	2.004	1203.2	1314.5	1.6266	1.4760	1197.3	1306.6	1.5892
700	2.688	1246.7	1371.1	1.6970	2.227	1244.6	1368.3	1.6751	1.6503	1240.4	1362.5	1.6397
800	2.943	1287.0	1423.2	1.7401	2.442	1285.4	1421.0	1.7187	1.8163	1282.1	1416.6	1.6844
900	3.193	1327.6	1475.3	1.7799	2.653	1326.3	1473.6	1.7589	1.9776	1323.7	1470.1	1.7252
1000	3.440	1368.7	1527.9	1.8172	2.860	1367.7	1526.5	1.7964	2.136	1365.5	1523.6	1.7632
1200	3.929	1453.0	1634.8	1.8858	3.270	1452.2	1633.8	1.8653	2.446	1450.7	1631.8	1.8327
1400	4.414	1540.4	1744.6	1.9483	3.675	1539.8	1743.8	1.9279	2.752	1538.7	1742.4	1.8956
1600	4.896	1631.1	1857.6	2.0060	4.078	1630.7	1857.0	1.9857	3.055	1629.8	1855.9	1.9535
1800	5.376	1725.2	1974.0	2.0599	4.479	1724.9	1973.5	2.0396	3.357	1724.1	1972.6	2.0076
	600 psia				**800 psia**				**1000 psia**			
500	0.7947	1128.0	1216.2	1.4592	...	...	...	...	...	...	...	...
600	0.9456	1184.5	1289.5	1.5320	0.6776	1170.1	1270.4	1.4861	0.5140	1153.7	1248.8	1.4450
700	1.0727	1231.5	1350.6	1.5872	0.7829	1222.1	1338.0	1.5471	0.6080	1212.0	1324.6	1.5135
800	1.1900	1275.4	1407.6	1.6343	0.8764	1268.5	1398.2	1.5969	0.6878	1261.2	1388.5	1.5664
900	1.3021	1318.4	1462.9	1.6766	0.9640	1312.9	1455.6	1.6408	0.7610	1307.3	1448.1	1.6120
1000	1.4108	1361.2	1517.8	1.7155	1.0482	1356.7	1511.9	1.6807	0.8305	1352.2	1505.9	1.6530
1200	1.6222	1447.7	1627.8	1.7861	1.2102	1444.6	1623.8	1.7526	0.9630	1441.5	1619.7	1.7261
1400	1.8289	1536.5	1739.5	1.8497	1.3674	1534.2	1736.6	1.8167	1.0905	1531.9	1733.7	1.7909
1600	2.033	1628.0	1853.7	1.9080	1.5218	1626.2	1851.5	1.8754	1.2152	1624.4	1849.3	1.8499
1800	2.236	1722.6	1970.8	1.9622	1.6749	1721.0	1969.0	1.9298	1.3384	1719.5	1967.2	1.9046
	2000 psia				**3000 psia**				**4000 psia**			
700	0.2487	1147.7	1239.8	1.3782	0.09771	1003.9	1058.1	1.1944	0.02867	742.1	763.4	0.9345
800	0.3071	1220.1	1333.8	1.4562	0.17572	1167.6	1265.2	1.3675	0.10522	1095.0	1172.9	1.2740
900	0.3534	1276.8	1407.6	1.5126	0.2160	1241.8	1361.7	1.4414	0.14622	1201.5	1309.7	1.3789
1000	0.3945	1328.1	1474.1	1.5598	0.2485	1301.7	1439.6	1.4967	0.17520	1272.9	1402.6	1.4449
1200	0.4685	1425.2	1598.6	1.6398	0.3036	1408.0	1576.6	1.5848	0.2213	1390.1	1553.9	1.5423
1400	0.5368	1520.2	1718.8	1.7082	0.3524	1508.1	1703.7	1.6571	0.2603	1495.7	1688.4	1.6188
1600	0.6020	1615.4	1838.2	1.7692	0.3978	1606.3	1827.1	1.7201	0.2959	1597.1	1816.1	1.6841
1800	0.6656	1712.0	1958.3	1.8249	0.4416	1704.5	1949.6	1.7769	0.3296	1697.1	1941.1	1.7420
2000	0.7284	1810.6	2080.2	1.8765	0.4844	1803.9	2072.8	1.8291	0.3625	1797.3	2065.6	1.7948

TABLE 7.4 Refrigerant-12 — English Units

7.4.1E Saturated Refrigerant-12 — Temperature Table

		Specific Volume ft³/lbm		Enthalpy Btu/lbm			Entropy Btu/lbm · R	
T, °F	p, psia	v_f	v_g	h_f	h_{fg}	h_g	s_f	s_g
−120	0.64190	0.009816	46.741	−16.565	80.617	64.052	−0.04372	0.19359
−100	1.4280	0.009985	22.164	−12.466	78.714	66.248	−0.03200	0.18683
−80	2.8807	0.010164	11.533	−8.3451	76.812	68.467	−0.02086	0.18143
−60	5.3575	0.010357	6.4774	−4.1919	74.885	70.693	−0.01021	0.17714
−40	9.3076	0.010564	3.8750	0	72.913	72.913	0	0.17373
−20	15.267	0.010788	2.4429	4.2357	70.874	75.110	0.00983	0.17102
0	23.849	0.011030	1.6089	8.5207	68.750	77.271	0.01932	0.16888
20	35.736	0.011296	1.0988	12.863	66.522	79.385	0.02852	0.16719
40	51.667	0.011588	0.77357	17.273	64.163	81.436	0.03745	0.16586
60	72.433	0.011913	0.55839	21.766	61.643	83.409	0.04618	0.16479
80	98.870	0.012277	0.41135	26.365	58.917	85.282	0.05475	0.16392
100	131.86	0.012693	0.30794	31.100	55.929	87.029	0.06323	0.16315
120	172.35	0.013174	0.23326	36.013	52.597	88.610	0.07168	0.16241
140	221.32	0.013746	0.17799	41.162	48.805	89.967	0.08021	0.16159
160	279.82	0.014449	0.13604	46.633	44.373	91.006	0.08893	0.16053
180	349.00	0.015360	0.10330	52.562	38.999	91.561	0.09804	0.15900
200	430.09	0.016659	0.076728	59.203	32.075	91.278	0.10789	0.15651
220	524.43	0.018986	0.053140	67.246	21.790	89.036	0.11943	0.15149
233.6 (critical)	596.9	0.02870	0.02870	78.86	0	78.86	0.1359	0.1359

7.4.2E Superheated Refrigerant-12

T, °F	v	h	s	v	h	s	v	h	s
		5 psia			**10 psia**			**20 psia**	
0	8.0611	78.582	0.19663	3.9809	78.246	0.18471	...	...	...
20	8.4265	81.309	0.20244	4.1691	81.014	0.19061	2.0391	80.403	0.17829
40	8.7903	84.090	0.20812	4.3556	83.828	0.19635	2.1373	83.289	0.18419
60	9.1528	86.922	0.21367	4.5408	86.689	0.20197	2.2340	86.210	0.18992
80	9.5142	89.806	0.21912	4.7248	89.596	0.20746	2.3295	89.168	0.19550
100	9.8747	92.738	0.22445	4.9079	92.548	0.21283	2.4241	92.164	0.20095
120	10.234	95.717	0.22968	5.0903	95.546	0.21809	2.5179	95.198	0.20628
140	10.594	98.743	0.23481	5.2720	98.586	0.22325	2.6110	98.270	0.21149
160	10.952	101.812	0.23985	5.4533	101.669	0.22830	2.7036	101.380	0.21659
180	11.311	104.925	0.24479	5.6341	104.793	0.23326	2.7957	104.528	0.22159
200	11.668	108.079	0.24964	5.8145	107.957	0.23813	2.8874	107.712	0.22649
220	12.026	111.272	0.25441	5.9946	111.159	0.24291	2.9789	110.932	0.23130
		30 psia			**40 psia**			**60 psia**	
20	1.3278	79.765	0.17065	...	...	...	...	...	...
40	1.3969	82.730	0.17671	1.0258	82.148	0.17112	...	...	...
60	1.4644	85.716	0.18257	1.0789	85.206	0.17712	0.69210	84.126	0.16892
80	1.5306	88.729	0.18826	1.1306	88.277	0.18292	0.72964	87.330	0.17497
100	1.5957	91.770	0.19379	1.1812	91.367	0.18854	0.76588	90.528	0.18079
120	1.6600	94.843	0.19918	1.2309	94.480	0.19401	0.80110	93.731	0.18641
140	1.7237	97.948	0.20445	1.2798	97.620	0.19933	0.83551	96.945	0.19186
160	1.7868	101.086	0.20960	1.3282	100.788	0.20453	0.86928	100.776	0.19716
180	1.8494	104.258	0.21463	1.3761	103.985	0.20961	0.90252	103.427	0.20233
200	1.9116	107.464	0.21957	1.4236	107.212	0.21457	0.93531	106.700	0.20736
220	1.9735	110.702	0.22440	1.4707	110.469	0.21944	0.96775	109.997	0.21229
240	2.0351	113.973	0.22915	1.5176	113.757	0.22420	0.99988	113.319	0.21710
		80 psia			**100 psia**			**150 psia**	
100	0.52795	86.316	0.16885	0.43138	88.694	0.16996	...	...	...
120	0.55734	89.640	0.17489	0.45562	92.116	0.17597	0.28007	89.800	0.16629
140	0.58556	92.945	0.18070	0.47881	95.507	0.18172	0.29845	93.498	0.17256
160	0.61286	96.242	0.18629	0.50118	98.884	0.18726	0.31566	97.112	0.17849
180	0.63943	99.542	0.19170	0.52291	102.257	0.19262	0.33200	100.675	0.18415
200	0.66543	102.851	0.19696	0.54413	105.633	0.19782	0.34769	104.206	0.18958
220	0.69095	106.174	0.20207	0.56492	109.018	0.20287	0.36285	107.720	0.19483
240	0.71609	109.513	0.20706	0.58538	112.415	0.20780	0.37761	111.226	0.19992
260	0.74090	112.872	0.21193	0.60554	115.828	0.21261	0.39203	114.732	0.20485
280	0.76544	116.251	0.21669	0.62546	119.258	0.21731	0.40617	118.242	0.20967
300	0.78975	119.652	0.22135	0.64518	122.707	0.22191	0.42008	121.761	0.21436
		200 psia			**300 psia**			**400 psia**	
140	0.20579	91.137	0.16480	...	...	...	...	...	...
160	0.22121	95.100	0.17130	...	...	...	...	...	...
180	0.23535	98.921	0.17737	0.13182	94.556	0.16537	...	...	...
200	0.24860	102.652	0.18311	0.14697	98.975	0.17217	0.091005	93.718	0.16092
220	0.26117	106.325	0.18860	0.15774	103.136	0.17838	0.10316	99.046	0.16888
240	0.27323	109.962	0.19387	0.16761	107.140	0.18419	0.11300	103.735	0.17568
260	0.28489	113.576	0.19896	0.17685	111.043	0.18969	0.12163	108.105	0.18183
280	0.29623	117.178	0.20390	0.18562	114.879	0.19495	0.12949	112.286	0.18756
300	0.30730	120.775	0.20870	0.19402	118.670	0.20000	0.13680	116.343	0.19298
320	0.31815	124.373	0.21337	0.20214	122.430	0.20489	0.14372	120.318	0.19814
340	0.32881	127.974	0.21793	0.21002	126.171	0.20963	0.15032	124.235	0.20310

Practice Problems (Metric Units)

(If you choose to work only a few problems, select those with a star.)

GENERAL

*7.1 Which of the following would be considered a system rather than a control volume?

a) a pump b) a tire c) a pressure cooker d) a turbine e) a jet nozzle

*7.2 Which of the following is an extensive property?

a) Temperature b) Velocity c) Pressure d) Mass e) Stress

7.3 An automobile heats up while sitting in a parking lot on a sunny day. The process can be assumed to be

a) isothermal b) isobaric c) isometric d) isentropic e) adiabatic

*7.4 If a quasiequilibrium process exists we have assumed

a) the pressure at any instant to be everywhere constant.
b) an isothermal process.
c) the heat transfer to be small.
d) the boundary motion to be infinitesimally small.
e) that no friction exists.

*7.5 A scientific law is a statement that

a) we postulate to be true.
b) is generally observed to be true.
c) is derived from a mathematical theorem.
d) is agreed upon by the scientific community.
e) is a summary of experimental observations.

DENSITY, PRESSURE, AND TEMPERATURE

*7.6 The density of air at a vacuum of 40 kPa and $-40°C$ is, in kg/m^3,

a) 0.598 b) 0.638 c) 0.697 d) 0.753 e) 0.897

7.7 The specific volume of water is 0.5 m^3/kg at a pressure of 200 kPa abs. Find the quality.

a) 0.623 b) 0.564 c) 0.478 d) 0.423 e) 0.356

7.8 The specific volume of steam at 4 MPa abs and 1200°C, in m^3/kg, is

a) 0.20 b) 0.19 c) 0.18 d) 0.17 e) 0.16

7.9 There are 20 kg of steam contained in 2 m^3 at 4 MPa. What is the temperature, in °C?

a) 600 b) 610 c) 620 d) 630 e) 640

7.10 The pressure in a cylinder containing water is 200 kPa and the temperature is 115°C. What state is it in?

a) compressed liquid b) saturated liquid c) liquid/vapor mixture
d) saturated vapor e) superheated vapor

*7.11 A cold tire has a volume of 0.03 m^3 at −10°C and 180 kPa gage. If the pressure and temperature increase to 210 kPa gage and 30°C find the final volume, in m^3.

a) 0.0304 b) 0.0308 c) 0.0312 d) 0.0316 e) 0.0320

THE FIRST LAW FOR A SYSTEM

*7.12 A 300-watt light bulb provides energy in a 10-m-dia spherical space. If the outside temperature is 20°C, find the inside steady-state temperature, in °C, if $R = 1.5$ hr·m²·°C/kJ for the wall.

a) 40.5 b) 35.5 c) 32.6 d) 29.4 e) 25.2

7.13 A 10 kg mass, which is attached to a pulley and a paddle wheel submerged in water, drops 3 m. Find the subsequent heat transfer, in joules, needed to return the temperature of the water to its original value.

a) −30 b) −95 c) −126 d) −195 e) −294

*7.14 A cycle undergoes the following processes. All units are kJ. Find E_{after} for process 1→2.

	Q	W	ΔE	E_{before}	E_{after}
1→2	20	5		10	
2→3		−5	5		10
3→1	30			20	

a) 10 b) 15 c) 20 d) 25 e) 30

7.15 For the cycle of Prob. 7.14 find W_{3-1}.

a) 50 b) 60 c) 70 d) 80 e) 90

7.16 A 2000 kg automobile, travelling at 25 m/s, strikes a plunger in 10 000 cm^3 of water, bringing the auto to a stop. What is the maximum temperature rise, in °C, in the water?

a) 5 b) 10 c) 15 d) 20 e) 25

7.17 Ten kilograms of −10°C ice is added to 100 kg of 20°C water. What is the eventual temperature, in °C, of the water? Assume an insulated container.

a) 9.2 b) 10.8 c) 11.4 d) 12.6 e) 13.9

*7.18 Of the following first law statements, choose the one that is wrong:

a) the net heat transfer equals the net work for a cycle.
b) the heat transfer cannot exceed the work done.
c) the heat transfer equals the work plus the energy change.
d) the heat transfer equals the energy change if no work is done.
e) the energy of an isolated system remains constant.

ISOTHERMAL PROCESS

*7.19 Determine the work, in kJ, necessary to compress 2 kg of air from 100 kPa abs to 4000 kPa abs if the temperature is held constant at 300°C.

a) −1210 b) −1105 c) −932 d) −812 e) −733

*7.20 Steam is compressed from 100 kPa abs to 4000 kPa abs holding the temperature constant at 300°C. What is the internal energy change, in kJ, for m = 2 kg?

a) −180 b) −170 c) −160 d) −150 e) −140

7.21 How much heat transfer, in kJ, is needed to convert 2 kg of saturated liquid water to saturated vapor if temperature is held constant at 200°C?

a) 2380 b) 2980 c) 3210 d) 3520 e) 3880

CONSTANT PRESSURE PROCESS

7.22 There are 200 people in a 2000 m^2 room, lighted with 30 W/m^2. Estimate the maximum temperature increase, in °C, if the ventilation system fails for 20 min. Each person generates 400 kJ/h. The room is 3 m high.

a) 75.6 b) 6.8 c) 8.6 d) 11.4 e) 13.4

*7.23 Estimate the average c_p-value in kJ/kg·K of a gas if 522 kJ of heat are necessary to raise the temperature of one kilogram from 300 K to 800 K holding the pressure constant.

a) 1.00 b) 1.026 c) 1.038 d) 1.044 e) 1.052

*7.24 How much heat, in kJ, must be transferred to 10 kg of air to increase the temperature from 10°C to 230°C if the pressure is maintained constant?

a) 1780 b) 1620 c) 1890 d) 2090 e) 2200

7.25 Ten kilograms of water, initially at 10°C, are heated until T = 300°C. Estimate the heat transfer, in MJ, if the pressure is held constant at 200 kPa abs.

a) 29.2 b) 29.7 c) 30.3 d) 30.9 e) 31.3

7.26 Calculate the work done, in MJ, in Prob. 7.25.

a) 2.63 b) 4.72 c) 8.96 d) 11.4 e) 15.6

CONSTANT VOLUME PROCESS

7.27 How much heat, in kJ, must be added to a rigid volume, containing 2 kg of a water/vapor mixture with $x = 0.5$, to increase the temperature from 200°C to 500°C?

a) 2730 b) 2620 c) 2510 d) 2390 e) 2250

*7.28 A tire is pressurized to 100 kPa gauge in Michigan where $T = 0$°C. In Arizona the tire is at 70°C; assuming a rigid tire estimate the pressure, in kPa gauge.

a) 120 b) 130 c) 140 d) 150 e) 160

7.29 A sealed, rigid 10 m³ air tank is heated by the sun from 20°C to 80°C. How much energy, in kJ, is transferred to the tank? Assume $p_1 = 100$ kPa abs.

a) 720 b) 680 c) 620 d) 560 e) 510

7.30 Ten kilograms of water are heated in a rigid container from 10°C to 200°C. What is the final quality if $Q = 9000$ kJ?

a) 0.258 b) 0.162 c) 0.093 d) 0.078 e) 0.052

ISENTROPIC PROCESS

7.31 Air expands from an insulated cylinder from 200°C and 400 kPa abs to 20 kPa abs. Find T_2, in °C.

a) 24 b) −28 c) −51 d) −72 e) −93

*7.32 During an isentropic expansion of air, the volume triples. If the initial temperature is 200°C, find T_2, in °C.

a) 32 b) 28 c) 16 d) 8 e) −12

7.33 Superheated steam expands isentropically from 600°C and 6 MPa abs to 10 kPa abs. Find the final quality.

a) 0.79 b) 0.83 c) 0.87 d) 0.91 e) 0.95

7.34 Superheated steam expands isentropically from 600°C and 6 MPa abs to 400kPa abs. Find T_2, in °C.

a) 220 b) 200 c) 190 d) 160 e) 140

7.35 Find the work, in kJ/kg, needed to compress air isentropically in a cylinder from 20°C and 100 kPa abs to 6 MPa abs.

a) −523 b) −466 c) −423 d) −392 e) −376

*7.36 During an isentropic process which one of the following is true?

a) the temperature increases as the pressure decreases.
b) the temperature increases as the volume increases.
c) the heat transfer equals the enthalpy change.
d) the heat transfer is zero.
e) the volume decreases as the pressure decreases.

POLYTROPIC PROCESS

*7.37 Find T_2, in °C, if the pressure triples and $T_1 = 10$°C. Let $n = 1.2$.

a) 179 b) 113 c) 93 d) 79 e) 67

THE FIRST LAW FOR A CONTROL VOLUME

7.38 Steam enters a turbine in a 20-cm-dia pipe at 600°C and 6MPa abs. It exits from a 5-cm-dia pipe at 20 kPa abs with $x = 1$. What is the velocity rate V_{out}/V_{in}?

a) 1880 b) 1640 c) 1210 d) 820 e) 420

*7.39 Water enters a boiler at 60°C and 4 MPa abs. How much energy, in kJ/kg, must be added to obtain 600°C at the exit if the pressure remains constant?

a) 2340 b) 2630 c) 2970 d) 3280 e) 3420

7.40 A condenser is cooled by heating water from 20°C to 30°C. If the condenser inlets 10 kg/s of saturated water vapor at 20 kPa abs and exits saturated liquid, what is the mass flux, in kg/s, of the cooling water?

a) 640 b) 560 c) 500 d) 410 e) 350

7.41 Steam at 400°C and 4 MPa abs expands isentropically through a turbine to 10 kPa abs. Estimate the maximum work output, in kJ/kg.

a) 1030 b) 1050 c) 1070 d) 1090 e) 1110

*7.42 What is the energy requirement, in kW, for a pump that is 75% efficient if it increases the pressure of 10 kg/s of water from 10 kPa to 6 MPa abs?

a) 40 b) 50 c) 60 d) 70 e) 80

7.43 A river 60 m wide and 2 m deep flows at 2 m/s. A hydro plant develops a pressure of 300 kPa gage just before the turbine. What maximum power, in MW, is possible?

a) 40 b) 48 c) 56 d) 64 e) 72

*7.44 A nozzle expands air isentropically from 400°C and 2 MPa abs to the atmosphere at 80 kPa abs. If the inlet velocity is small, what exit velocity, in m/s, can be expected?

a) 500 b) 600 c) 700 d) 800 e) 900

7.45 If the efficiency of a turbine, that expands steam at 400°C and 6 MPa abs to 20 kPa abs, is 87%, find the work output, in kJ/kg.

a) 723 b) 891 c) 933 d) 996 e) 1123

THE SECOND LAW

*7.46 An inventor proposes to take 10 kg/s of geothermal water at 120°C and generate 4180 kW of energy by exhausting very near to ambient temperature of 20°C. This proposal should not be supported because

a) it violates the first law.
b) it violates the second law.
c) it would be too expensive.
d) friction must be accounted for.
e) the geothermal water is not hot enough.

*7.47 The net entropy change in the universe during any real process

a) is equal to zero.
b) is positive.
c) is negative.
d) must be calculated to determine its sense.
e) is positive if $T_{system} > T_{surroundings}$.

*7.48 A Carnot engine

a) provides a fictitious model which is of little use.
b) can be experimentally modeled.
c) supplies us with the lower limit for engine efficiency.
d) operates between two constant temperature reservoirs.
e) has two reversible and two irreversible processes.

7.49 Ninety kilograms of ice at 0°C are completely melted. Find the entropy change, in kJ/K, if $T_2 = 0$°C.

a) 0 b) 45 c) 85 d) 105 e) 145

7.50 Forty kilograms of ice at 0°C are mixed with 100 kg of water at 20°C. What is the entropy change, in kJ/K?

a) 2.36 b) 2.15 c) 1.96 d) 1.53 e) 1.04

*7.51 An inventor proposes to have developed a small power plant that operates at 70% efficiency. It operates between temperature extremes of 600°C and 50°C. Your analysis shows that the maximum possible efficiency is

a) 56 b) 63 c) 67 d) 72 e) 81

POWER CYCLES

*7.52 A Carnot engine operates on air such that $p_4 = 160$ kPa abs and $v_4 = 0.5$ m^3/kg (see Fig. 7.5). If 30 kJ/kg of heat are added at $T_H = 200$°C find the work produced in kJ/kg.

a) 20.3 b) 18.2 c) 16.8 d) 14.2 e) 12.3

7.53 An Otto cycle operates with volumes of 40 cm³ and 400 cm³ at top dead center (TDC) and bottom dead center (BDC), respectively. If the power output is 100 kW, what is the heat input, in kJ/s? Assume $k = 1.4$.

a) 166 b) 145 c) 110 d) 93 e) 60

7.54 The volumes of states 1, 2, and 3 in Fig. 7.5 of the Diesel cycle are 450 cm³, 25 cm³, and 45 cm³, respectively. If the power produced is 120 kW, what is the required heat input, in kJ/s? Assume $k = 1.4$.

a) 187 b) 172 c) 157 d) 146 e) 132

7.55 A simple Rankine cycle operates between superheated steam at 600°C and 6 MPa abs entering the turbine, and 10 kPa abs entering the pump. What is the maximum possible efficiency, in percent?

a) 30 b) 35 c) 40 d) 45 e) 50

7.56 The water vapor that expands from 600°C and 6 MPa abs in the turbine of a Rankine cycle is intercepted at 200 kPa abs and reheated at constant pressure to 600°C in the boiler after which it is reinjected in the turbine and expanded to 10 kPa abs. (This is a *reheat cycle*.) Calculate the maximum possible efficiency of this cycle, in percent.

a) 32 b) 37 c) 41 d) 45 e) 49

7.57 The primary effect of reheating, as illustrated in Prob. 7.56 (compare with the result in Prob. 7.55), is to

a) increase the efficiency.
b) increase the work output.
c) decrease the heat requirement.
d) eliminate unnecessary piping.
e) decrease or eliminate moisture condensation in the turbine.

REFRIGERATION CYCLES

7.58 A refrigeration cycle operates with Refrigerant-12 between −30°C and 49.3°C. What is the maximum possible COP?

a) 1.5 b) 2 c) 2.5 d) 3 e) 3.5

7.59 The quality of the refrigerant immediately after the expansion valve of the cycle of Prob. 7.58 is

a) 0.254 b) 0.324 c) 0.387 d) 0.421 e) 0.456

AVAILABILITY

7.60 Determine the maximum possible power output, in kW, if 10 kg/s of air is expanded in a turbine from 100°C and 6 MPa abs to the surroundings at 20°C and 80 kPa abs.

a) 3720 b) 2610 c) 2030 d) 1890 e) 1620

(This page is intentionally blank.)

Practice Problems (English Units)

(If you choose to work only a few problems, select those with a star.)

GENERAL

*7.1 Which of the following would be considered a system rather than a control volume?

a) a pump b) a tire c) a pressure cooker d) a turbine e) a jet nozzle

*7.2 Which of the following is an extensive property?

a) Temperature b) Velocity c) Pressure d) Mass e) Stress

7.3 An automobile heats up while sitting in a parking lot on a sunny day. The process can be assumed to be

a) isothermal b) isobaric c) isometric d) isentropic e) adiabatic

*7.4 If a quasiequilibrium process exists we have assumed

a) the pressure at any instant to be everywhere constant.
b) an isothermal process.
c) the heat transfer to be small.
d) the boundary motion to be infinitesimally small.
e) that no friction exists.

*7.5 A scientific law is a statement that

a) we postulate to be true.
b) is generally observed to be true.
c) is derived from a mathematical theorem.
d) is agreed upon by the scientific community.
e) is a summary of experimental observations.

DENSITY, PRESSURE, AND TEMPERATURE

*7.6 The density of air at a vacuum of 6 psi and $-40°F$ is, in lbm/ft^3,

a) 0.076 b) 0.072 c) 0.068 d) 0.062 e) 0.056

7.7 The specific volume of water is 2 ft^3/lbm at a pressure of 30 psia. Find the quality.

a) 0.183 b) 0.144 c) 0.112 d) 0.092 e) 0.061

7.8 The specific volume in ft^3/lbm, of steam at 600 psia and 2200°F, is

a) 4.26 b) 3.90 c) 3.21 d) 2.64 e) 2.11

7.9 There are 40 lbm of steam contained in 60ft³ at 600 psia. What is the temperature, in °F?

a) 1121 b) 1084 c) 936 d) 902 e) 876

7.10 The pressure in a cylinder containing water is 30 psia and the temperature is 230°F. What state is it in?

a) compressed liquid b) saturated liquid c) liquid/vapor mixture
d) saturated vapor e) superheated vapor

*7.11 A cold tire has a volume of 0.9 ft³ at 14°F and 26 psia. If the pressure and temperature increase to 30 psia and 90°F, find the final volume, in ft³.

a) 0.907 b) 0.906 c) 0.905 d) 0.904 e) 0.903

THE FIRST LAW FOR A SYSTEM

*7.12 A 300-watt light bulb provides energy in a 30-ft-dia spherical space. If the outside temperature is 70°F, find the steady-state temperature, in °F, if R = 30 hr-ft²-°F/BTU for the wall.

a) 121.2 b) 115.6 c) 106.2 d) 90.9 e) 80.9

7.13 A 20 lb mass, which is attached to a paddle wheel submerged in water, drops 10 ft. Find the subsequent heat transfer, in BTU's, needed to return the temperature of the water to its original value.

a) −0.667 b) −.576 c) −0.415 d) −0.311 e) −0.257

*7.14 A cycle undergoes the following processes. All units are BTU. Find E_{after} for process 1→2.

	Q	W	ΔE	E_{before}	E_{after}
1→2	20	5		10	
2→3		−5	5		10
3→1	30			20	

a) 10 b) 15 c) 20 d) 25 e) 30

7.15 For the cycle of Prob. 7.14 find $W_{3\text{-}1}$.

a) 50 b) 45 c) 40 d) 35 e) 30

7.16 A 4000 lb automobile, travelling at 90 fps, strikes a plunger in 600 in³ of water, bringing the auto to a stop. What is the maximum temperature rise, in °F?

a) 50 b) 40 c) 30 d) 20 e) 10

7.17 Twenty pounds of 15°F ice is added to 200 lbs of 70°F water. What is the eventual temperature, in °F, of the water? Assume an insulated container.

a) 58 b) 53 c) 49 d) 44 e) 40

*7.18 Of the following first law statements, choose the wrong one:

a) the heat transfer equals the net work for a cycle.
b) the heat transfer cannot exceed the work done.
c) the heat transfer equals the work plus the energy change.
d) the heat transfer equals the energy change if no work is done.
e) the energy of an isolated system remains constant.

ISOTHERMAL PROCESS

*7.19 Determine the work, in ft-lb, necessary to compress 4 lbm of air from 15 psia to 600 psia if the temperature is held constant at 600°F.

a) −834,000 b) −726,000 c) −592,000 d) −421,000 e) −302,000

*7.20 Steam is compressed from 20 psia to 600 psia holding the temperature constant at 600°F. What is the internal energy change, in BTU, if $m = 4$ lbs?

a) −144 b) −136 c) −127 d) −112 e) −103

7.21 How much heat transfer, in BTU, is needed to convert 4 lbm of saturated liquid water to saturated vapor if $T = 400$°F?

a) 2320 b) 2560 c) 2810 d) 3020 e) 3310

CONSTANT PRESSURE PROCESS

7.22 There are 200 people in a 20,000 ft² room, lighted with 3 W/ft². Estimate the maximum temperature increase, in °F, if the ventilation system fails for 20 min. Each person generates 400 BTU/hr. The room is 10 ft high.

a) 18 b) 20 c) 22 d) 24 e) 26

*7.23 Estimate the average c_p-value, in BTU/lbm-°F, of a gas if 520 BTU of heat are necessary to raise the temperature of two pounds from 100°F to 1000°F, holding the pressure constant.

a) 0.202 b) 0.243 c) 0.261 d) 0.289 e) 0.310

*7.24 How much heat, in BTU, must be transferred to 20 lbm of air to increase the temperature from 50°F to 400°F if the pressure remains constant?

a) 2050 b) 1930 c) 1860 d) 1720 e) 1680

7.25 Twenty pounds of water, initially at 40°F, are heated until $T = 600$°F. Estimate the heat transfer, in BTU, if the pressure remains constant at 60 psia.

a) 36,500 b) 31,500 c) 26,500 d) 21,500 e) 16,500

7.26 Calculate the work done, in ft-lb, in Prob. 7.25.

a) 1.8×10^6 b) 1.8×10^5 c) 1.8×10^4 d) 1.8×10^3 e) 1.8×10^2

CONSTANT VOLUME PROCESS

7.27 How much heat, in BTU, must be added to a rigid volume, containing 4 lbm of water/vapor mixture with $x = 0.5$, to increase the temperature from 500°F to 900°F?

a) 1900 b) 2260 c) 2680 d) 2810 e) 2990

*7.28 A tire is pressurized to 28 psia gauge in Michigan where $T = 30$°F. In Arizona the tire is at 160°F; assuming a rigid tire estimate the pressure, in psi gauge.

a) 33 b) 35 c) 37 d) 39 e) 41

7.29 A sealed, rigid 300 ft^3 air tank is heated by the sun from 70°F to 150°F. How much energy, in BTU, is transferred to the tank if $p_1 = 14.7$ psia?

a) 401 b) 382 c) 356 d) 331 e) 308

7.30 Twenty pounds of water are heated in a rigid container from 40°F to 380°F. What is the final quality if $Q = 9000$ BTU?

a) 0.431 b) 0.374 c) 0.246 d) 0.193 e) 0.138

ISENTROPIC PROCESS

7.31 Air expands from an insulated cylinder from 400°F and 60 psia to 2 psia. Find T_2, in °F.

a) −63 b) −96 c) −110 d) −134 e) −178

*7.32 During an isentropic expansion, the volume triples. If the initial temperature is 400°F, find T_2, in °F.

a) 94 b) 106 c) 112 d) 126 e) 137

7.33 Superheated steam expands isentropically from 1000°F and 400 psia to 2 psia. Find the final quality.

a) 0.87 b) 0.89 c) 0.91 d) 0.93 e) 0.95

7.34 Superheated steam expands isentropically from 1000°F and 400 psia to 60 psia. Find T_2, in °F.

a) 500 b) 490 c) 480 d) 470 e) 460

7.35 Find the work, in ft-lb/lbm, needed to compress air isentropically in a cylinder from 70°F and 14.7 psia to 400 psia.

a) 121×10^3 b) 111×10^3 c) 101×10^3 d) 92.1×10^3 e) 85.1×10^3

*7.36 During an isentropic process, which one of the following is true?

a) the temperature increases as the pressure decreases.
b) the temperature increases as the volume increases.
c) the heat transfer equals the enthalpy change.
d) the heat transfer is zero.
e) the volume decreases as the pressure decreases.

POLYTROPIC PROCESS

*7.37 Find T_2, in °F, if the pressure triples and $T_1 = 50°F$. Let $n = 1.2$.

a) 194 b) 186 c) 179 d) 161 e) 152

THE FIRST LAW FOR A CONTROL VOLUME

7.38 Steam enters a turbine in an 8-in-dia pipe at 1000°F and 400 psia. It exits from a 2-in-dia pipe at 2 psia with $x = 1$. What is the velocity ratio V_{out}/V_{in}?

a) 1300 b) 1100 c) 900 d) 700 e) 500

*7.39 Water enters a boiler at 120°F and 300 psia. How much energy, in BTU/lbm, must be added to obtain 1000°F at the exit if the pressure remains constant?

a) 1080 b) 1110 c) 1260 d) 1390 e) 1440

7.40 A condenser is cooled by heating water from 70°F to 90°F. If the condenser inlets 20 lbm/sec of saturated water vapor at 2 psia and exits saturated liquid, what is the mass flux, in lbm/sec, of the cooling water?

a) 1130 b) 1022 c) 961 d) 902 e) 851

7.41 Steam at 700°F and 300 psia expands isentropically through a turbine to 2 psia. Estimate the maximum work output, in ft-lb/lbm.

a) 5.21×10^5 b) 4.12×10^5 c) 3.07×10^5 d) 2.13×10^5 e) 1.09×10^5

*7.42 What is the energy requirement, in horsepower, for a pump that is 75% efficient if it increases the pressure of 20 lbm/sec of water from 2 psia to 400 psia?

a) 22.4 b) 28.6 c) 31.6 d) 38.2 e) 44.5

7.43 A river 200 ft wide and 6 ft deep flows at 6 fps. A hydro plant develops a pressure of 50 psi gauge just before the turbine. What maximum power, in horsepower, is possible?

a) 21,300 b) 41,900 c) 65,200 d) 81,100 e) 94,300

*7.44 A nozzle expands air isentropically from 700°F and 100 psia to the atmosphere at 12 psia. If the inlet velocity is small, what exit velocity, in fps, is expected?

a) 222 b) 276 c) 333 d) 393 e) 444

7.45 If the efficiency of a turbine, that expands steam at 700°F and 400 psia to 2 psia, is 87%, find the work output, in ft-lb/lbm.

a) 302,000 b) 278,000 c) 233,000 d) 201,000 e) 187,000

THE SECOND LAW

*7.46 An inventor proposes to take 20 lbm/sec of geothermal water at 240°F and generate 2600 hp of energy by exhausting very near to ambient temperature of 70°F. This proposal should not be supported because

a) it violates the first law.
b) it violates the second law.
c) it would be too expensive.
d) friction must be accounted for.
e) the geothermal water is not hot enough.

*7.47 The net entropy change in the universe during any real process

a) is equal to zero.
b) is positive.
c) is negative.
d) must be calculated to determine its sense.
e) is positive if $T_{system} > T_{surroundings}$.

*7.48 A Carnot engine

a) provides a fictitious model which is of little use.
b) can be experimentally modeled.
c) supplies the lower limit for engine efficiency.
d) operates between two constant temperature reservoirs.
e) has two reversible and two irreversible processes.

7.49 Two hundred pounds of ice at 32°F are completely melted. Find the entropy change, in BTU/°R, if T_2 = 32°F.

a) 87 b) 77 c) 67 d) 57 e) 47

7.50 Eighty pounds of ice at 32°F are mixed with 100 lbm of water at 60°F. What is the entropy change, in BTU/°F?

a) 1.32 b) 0.926 c) 0.711 d) 0.526 e) 0.312

*7.51 An inventor proposes to have developed a small power plant that operates at 70% efficiency. It operates between temperature extremes of 1000°F and 100°F. Your analysis shows that the maximum possible efficiency is

a) 65 b) 62 c) 57 d) 51 e) 46

POWER CYCLES

*7.52 A Carnot engine operates on air such that p_4 = 24 psia and v_4 = 8 ft³/lbm (see Fig. 7.5). If 15 BTU/lbm of heat are added at T_H = 400°F find the work produced in ft-lb/lbm.

a) 23,400 b) 14,200 c) 11,600 d) 7820 e) 4630

7.53 An Otto cycle operates with volumes of 4 in^3 and 40 in^3 at top dead center (TDC) and bottom dead center (BDC), respectively. If the power output is 130 hp, what is the heat input, in BTU/sec? Assume $k = 1.4$.

a) 153 b) 172 c) 196 d) 210 e) 236

7.54 The volumes of states 1, 2, and 3 in Fig. 7.5 of the Diesel cycle are 45 in^3, 3 in^3, and 5 in^3, respectively. If the power produced is 160 hp, what is the required heat input, in BTU/sec? Assume $k = 1.4$.

a) 182 b) 171 c) 159 d) 142 e) 136

7.55 A simple Rankine cycle operates between superheated steam at 1000°F and 400 psia entering the turbine, and 2 psia entering the pump. What is the maximum possible efficiency, in percent?

a) 39 b) 37 c) 35 d) 33 e) 31

7.56 The water vapor that expands from 1000°F and 300 psia in the turbine of a Rankine cycle is intercepted at 20 psia and reheated at constant pressure to 1000°F in the boiler after which it is reinjected in the turbine and expanded to 1 psia. (This is a *reheat cycle*.) Calculate the maximum possible efficiency of this cycle, in percent.

a) 41 b) 39 c) 37 d) 35 e) 33

7.57 The primary effect of reheating, as illustrated in Prob. 7.56 (compare with the result in Prob. 7.55), is to

a) increase the efficiency.
b) increase the work output.
c) decrease the heat requirement.
d) eliminate unnecessary piping.
e) decrease or eliminate moisture condensation in the turbine.

REFRIGERATION CYCLES

7.58 A refrigeration cycle operates with Refrigerant-12 between −20°F and 114.3°F. What is the maximum possible COP?

a) 3.8 b) 3.4 c) 2.8 d) 2.4 e) 2.1

7.59 The quality of the refrigerant immediately after the expansion valve of the cycle of Prob. 7.58 is

a) 0.51 b) 0.49 c) 0.47 d) 0.45 e) 0.31

AVAILABILITY

7.60 Determine the maximum possible power output, in horsepower, if 20 lbm/sec of air is expanded in a turbine from 200°F and 400 psia to the surroundings at 70°F and 12 psia.

a) 3760 b) 3500 c) 3300 d) 3100 e) 2900

(This page is intentionally blank.)

8. Statics

The subject of mechanics is divided into two parts, *statics*, which deals with the equilibrium of bodies under the action of various forces, and *dynamics*, which considers the motion of bodies due to the action of various forces. Dynamics will be reviewed in the following chapter and statics in this chapter. To be in equilibrium a body may be either stationary or translating at constant velocity; it cannot be accelerating.

8.1 Forces and Moments

Statics is concerned primarily with forces, a *force* being the action of one body on another. A force is a *vector* quantity possessing both magnitude and direction; its point of application is also important. Forces are either body forces — due to gravitational, electrical, or magnetic fields — or surface forces, due to direct physical contact. The effect of a force acting on a body in equilibrium is both external and internal. The internal forces, which result in internal stresses and movement, are reviewed in Chapter 10 under the subject *The Mechanics of Materials*. Thus, in statics we are concerned only with the external effects.

If interest is in external effects only, we are allowed to locate a force at any point along its line of action. This short cut often allows us to simplify problems.

The tendency of a force to rotate the body upon which it acts about an axis is called the *moment* of the force about that axis. The magnitude of the moment is the magnitude of the force multiplied by the perpendicular distance between line of action of the force and the axis. Hence, the units on the moment are newton·meters, N·m. The expression used to calculate the moment $\vec{M}$ due to a force $\vec{F}$ is

$$\vec{M} = \vec{r} \times \vec{F} \tag{8.1.1}$$

where $\vec{r}$ is the vector connecting any point on the axis and the force; see Fig. 8.1. The components M_x, M_y, and M_z would represent the moments about the x-, y-, and z-axes, respectively. We should note that a positive moment can be either clockwise or counterclockwise; but we must be consistent in any given problem.

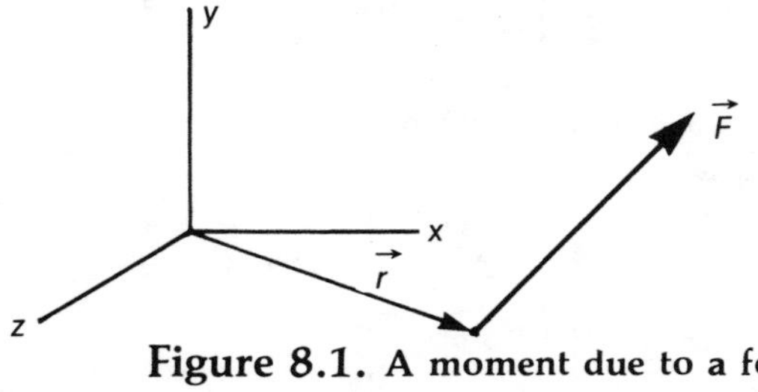

Figure 8.1. A moment due to a force.

EXAMPLE 8.1

Find the component of $\vec{A} = 15\hat{i} - 12\hat{j} - 9\hat{k}$ in the direction of the vector $\vec{B} = 2\hat{i} + \hat{j} + 2\hat{k}$.

Solution. First, let's find a unit vector in the direction of $\vec{B}$. It is

$$\hat{i}_B = (2\hat{i} + \hat{j} + 2\hat{k})/3.$$

To find the component of $\vec{A}$ in the direction of $\vec{B}$, we simply find the dot (scalar) product $\vec{A} \cdot \hat{i}_B$. This is

$$\begin{aligned} \vec{A} \cdot \hat{i}_B &= (15\hat{i} - 12\hat{j} + 9\hat{k}) \cdot (2\hat{i} + \hat{j} + 2\hat{k})/3 \\ &= (30 - 12 + 18)/3 = 12 \end{aligned}$$

where we have used $\hat{i} \cdot \hat{i} = 1$, $\hat{j} \cdot \hat{j} = 1$, $\hat{k} \cdot \hat{k} = 1$, and $\hat{i} \cdot \hat{j} = 0$, $\hat{i} \cdot \hat{k} = 0$, $\hat{j} \cdot \hat{k} = 0$.

EXAMPLE 8.2

Find the moment about the x-axis of the force $\vec{F} = 10\hat{i} + 8\hat{j} - 12\hat{k}$ acting at the point $(2, -3, 2)$.

Solution. The vector moment is found by taking the cross (vector) product $\vec{r} \times \vec{F}$. There results

$$\begin{aligned} \vec{M} &= \vec{r} \times \vec{F} \\ &= (2\hat{i} - 3\hat{j} + 2\hat{k}) \times (10\hat{i} + 8\hat{j} - 12\hat{k}) \\ &= (36 - 16)\hat{i} + (20 + 24)\hat{j} + (16 + 30)\hat{k} \\ &= 20\hat{i} + 44\hat{j} + 46\hat{k}. \end{aligned}$$

where we have used $\hat{i} \times \hat{j} = \hat{k}$, $\hat{j} \times \hat{k} = \hat{i}$, $\hat{k} \times \hat{i} = \hat{j}$, $\hat{i} \times \hat{k} = -\hat{j}$, $\hat{j} \times \hat{i} = -\hat{k}$, $\hat{k} \times \hat{j} = -\hat{i}$.

The moment about the x-axis is the x-component of $\vec{M}$; that is

$$M_x = 20.$$

If the units on the force are kilonewtons kN and the units on position are meters m, the units on the moment are kN·m.

8.2 Equilibrium

A body is in equilibrium when both the resultant force and the resultant moment acting on the body are zero. This is stated as

$$\Sigma \vec{F} = 0 \qquad \Sigma \vec{M} = 0 \tag{8.2.1}$$

These equations can be written as the six scalar equations

$$\begin{aligned} \Sigma F_x &= 0 \qquad & \Sigma M_x &= 0 \\ \Sigma F_y &= 0 \qquad & \Sigma M_y &= 0 \\ \Sigma F_z &= 0 \qquad & \Sigma M_z &= 0 \end{aligned} \tag{8.2.2}$$

For problems involving forces acting in only the x- and y-directions, equilibrium is assured if

$$\Sigma F_x = 0 \qquad \Sigma F_y = 0 \qquad \Sigma M_z = 0 \tag{8.2.3}$$

If only three forces act on a body, the forces must be planar and either have lines of action that pass through the same point (i.e., *concurrent*) or all act along parallel lines of action, as shown in Fig. 8.2.

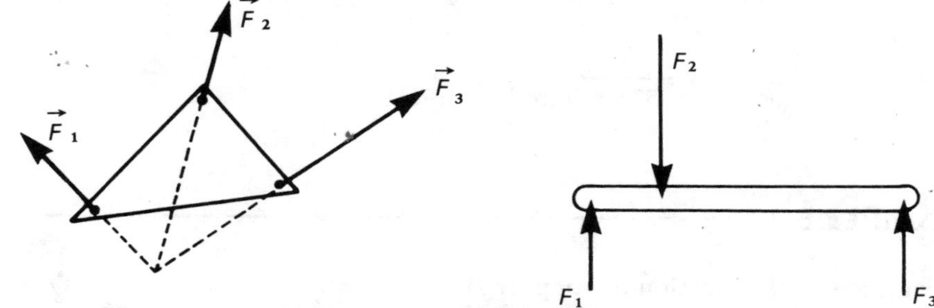

Figure 8.2. Three-force bodies.

If only two forces act on a body, the forces must be equal, opposite, and colinear, as shown in Fig. 8.3. Elements in truss sections are two-force bodies.

When summing forces on a body, it is very important that all forces are included. This is best assured by first drawing a *free-body diagram*, a sketch of an object showing all forces and moments acting on the object. If the object is in equilibrium, Eq. 8.1.1 can be applied and unknown forces or moments can be determined. It may also be necessary to draw free-body diagrams of parts of an object in order to find certain internal forces acting at locations of interest. The first step in solving a problem is usually the drawing of a free-body diagram.

Figure 8.3. Two-force body.

EXAMPLE 8.3

Find the magnitude of the resultant of the three vectors shown.

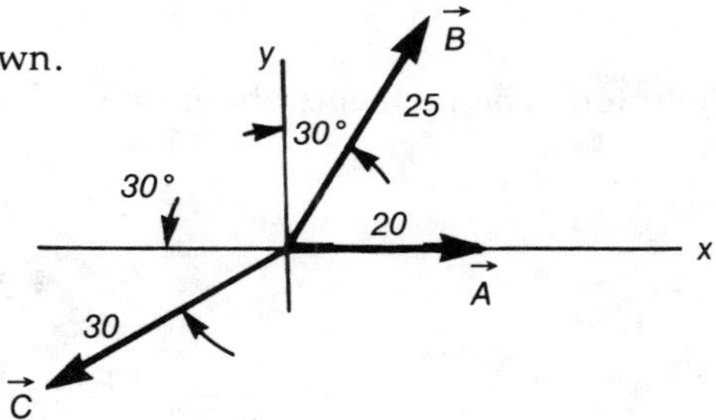

Solution. First, we express the three vectors as follows:

$$\vec{A} = 20\hat{i}$$

$$\vec{B} = 12.5\hat{i} + 21.65\hat{j}$$

$$\vec{C} = -25.98\hat{i} - 15\hat{j}.$$

Now we add the three vectors together to form the resultant; it is

$$\vec{R} = \vec{A} + \vec{B} + \vec{C}$$

$$= 20\hat{i} + (12.5\hat{i} + 21.65\hat{j}) + (-25.98\hat{i} - 15\hat{j})$$

$$= 6.52\hat{i} + 6.65\hat{j}.$$

Its magnitude would be

$$R = \sqrt{6.52^2 + 6.65^2} = 9.31.$$

EXAMPLE 8.4

Find the magnitude of the reaction at support A.

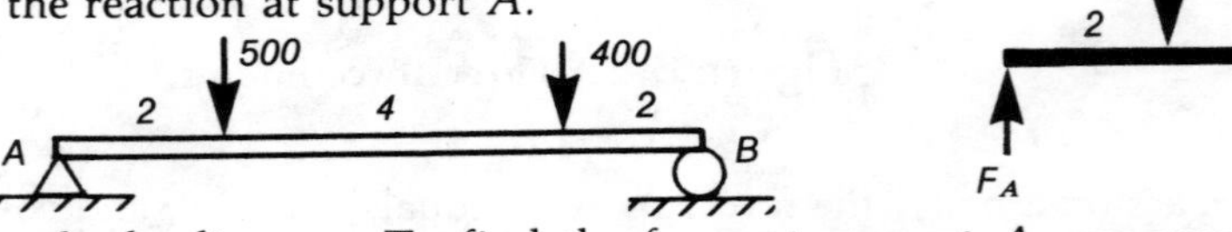

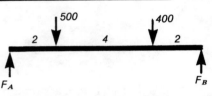

Solution. First draw the free-body diagram. To find the force at support A, we sum moments about support B. There results

$$\Sigma M_B = 0$$

$$= 8F_A - 6 \times 500 - 2 \times 400.$$

$$\therefore F_A = 475.$$

Note: A roller at the right end infers only a normal (vertical) force.

EXAMPLE 8.5

Three forces hold the beam shown in equilibrium. Find F_1 and the magnitude of the force which acts at point A.

Solution. The moments about point A must be zero. Thus,

$$\Sigma M_A = 0$$

$$= 2 \times 100 \cos 45° - 2 \times F_1 \cos 30°.$$

$$\therefore F_1 = 81.64.$$

To find F_A we determine its components. We find $(F_A)_x$ from

$$\Sigma F_x = 0$$

$$= 100 \sin 45° - 81.64 \sin 30° + (F_A)_x.$$

$$\therefore (F_A)_x = -29.88.$$

The y-component can be found similarly:

$$\Sigma F_y = 0$$

$$= -100 \cos 45^\circ - 81.64 \cos 30^\circ + (F_A)_y$$

$$\therefore (F_A)_y = 141.4.$$

The magnitude F_A is found to be

$$F_A = \sqrt{141.4^2 + 29.88^2} = 144.5$$

EXAMPLE 8.6

A beam supports the loading shown. Find the magnitude of the reaction at support A.

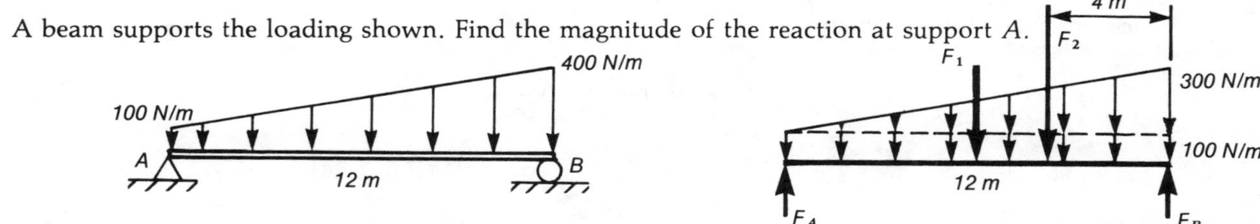

Solution. Let us consider the loading to be a linear load of 100 N/m and a triangular load, as shown. The resultant F_1 of the linear load acts through the middle, and the resultant F_2 of the triangular load acts 4 m from support B (see the section on centroids). Thus, summing moments about support B gives

$$\Sigma M_B = 0$$

$$= 4 \times (150 \times 12) + 6 \times (100 \times 12) - F_A \times 12.$$

$$\therefore F_A = 1200 \text{ N}.$$

8.3 Trusses and Frames

A truss is an assembly of two force members coupled by pins. The member of a frame may have forces acting at more than two points. The analysis of a frame usually involves drawing a free-body diagram of each member and then summing forces and moments on each member to determine the desired unknowns. This may require the simultaneous solution of several equations. To solve for an unknown force in the middle of a truss, we can proceed joint by joint until we reach the unknown force, or we can cut the truss judiciously and possibly find the unknown force with one cut. Examples illustrate.

EXAMPLE 8.7

Find the magnitude of the force acting at support A.

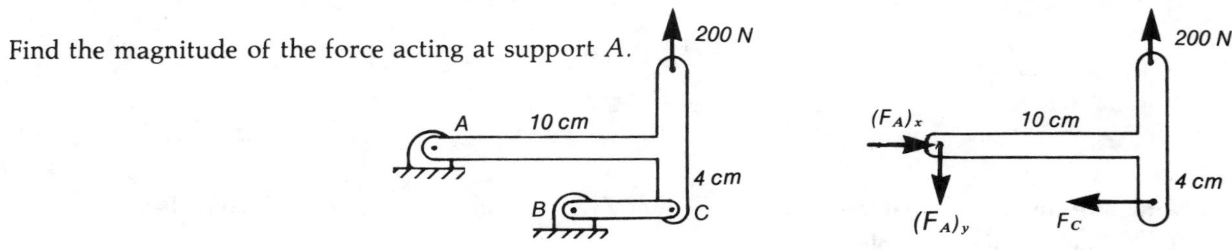

Solution. First, we note that link BC is a two-force member. Thus, F_C acts to the left. Then, let us take moments about point A. There results

$$\Sigma M_A = 0$$

$$= 10 \times 200 - 4 \times F_C.$$

$$\therefore F_C = 500 \text{ N}.$$

Summing forces in the x- and y-directions yields

$$(F_A)_x = 500 \text{ N} \qquad (F_A)_y = 200 \text{ N}.$$

Hence, the magnitude of the force is

$$F_A = \sqrt{500^2 + 200^2} = 539 \text{ N}.$$

EXAMPLE 8.8

Find the force in links DF and BE. Each link length is 4 m.

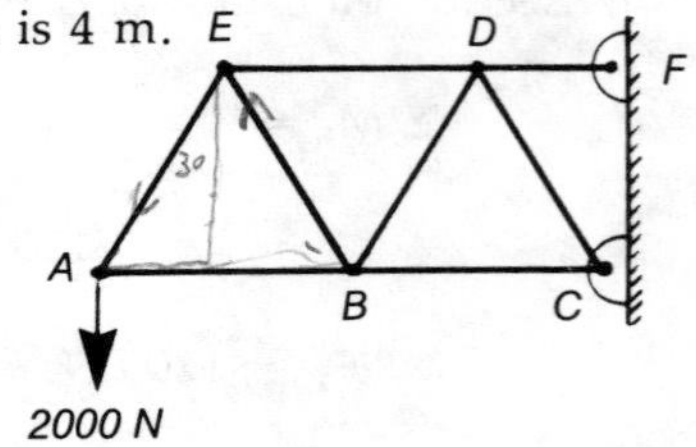

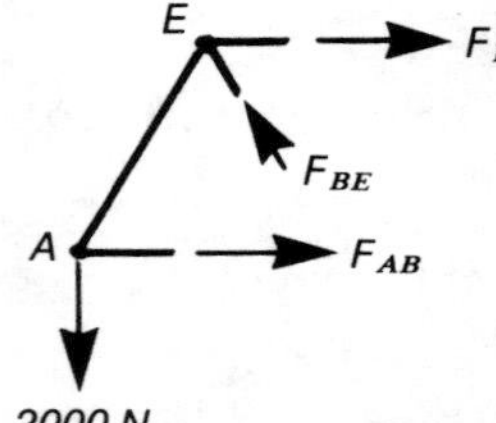

Solution. To find the force in link DF we simply take moments about point C:

$$\Sigma M_C = 0$$

$$= 8 \times 2000 - 4 \cos 30^\circ \times F_{DF}.$$

$$\therefore F_{DF} = 4619 \text{ N}.$$

The quickest way to find the force in link BE is to cut the section, as shown, cutting members ED, BE, and AB. Then, summing forces in the y-direction gives

$$\Sigma F_y = 0$$

$$= F_{BE} \cos 60^\circ - 2000.$$

$$\therefore F_{BE} = 2309 \text{ N}.$$

Note: We could have analyzed joint A and found F_{AE}. Then an analysis of joint E would have produced F_{BE}. The above cut simplified the analysis.

8.4 Friction

Frictional forces that exist between contacting surfaces of solid bodies in the absence of lubrication are due to dry friction (Coulomb friction). The maximum frictional force is given by the relationship

$$F = \mu_s N \tag{8.4.1}$$

where μ_s is the coefficient of static friction and N is the normal force between the two surfaces. If relative motion occurs between the two solid surfaces, the coefficient of dynamic friction μ_d is used; μ_d is always less than μ_s and it decreases slightly with increased relative speed. The frictional force always opposes the impending motion, or the actual motion.

For a belt, or rope, around a stationary circular member, the maximum increased tension T_2 is given by

$$T_2 = T_1 e^{\mu\beta} \tag{8.4.2}$$

where T_1 is the initial tension, μ is the coefficient of friction, and β is the total angle of contact.

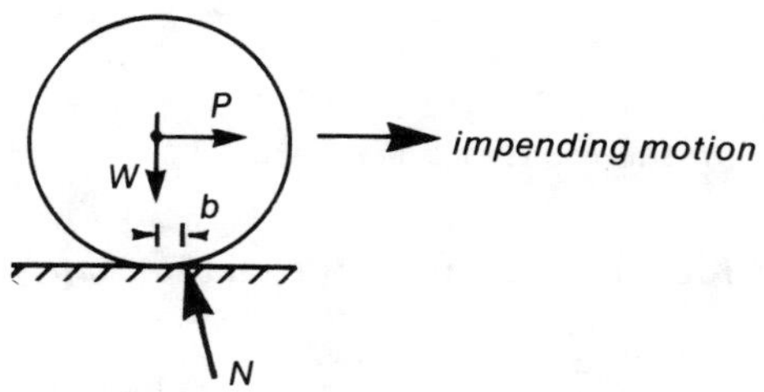

Figure 8.4. Rolling friction.

Rolling friction occurs due to the normal force acting a distance b ahead of the point where the line of action of the weight intersects the solid surface. The force P needed to cause motion is given by

$$P = \frac{b}{r} W = \mu W \tag{8.4.3}$$

where we assume $N \cong W$ and r is the radius of the wheel. Note that the coefficient of friction is the ratio of the two dimensions b and r.

EXAMPLE 8.9

Calculate the minimum and maximum weight needed to avoid motion.

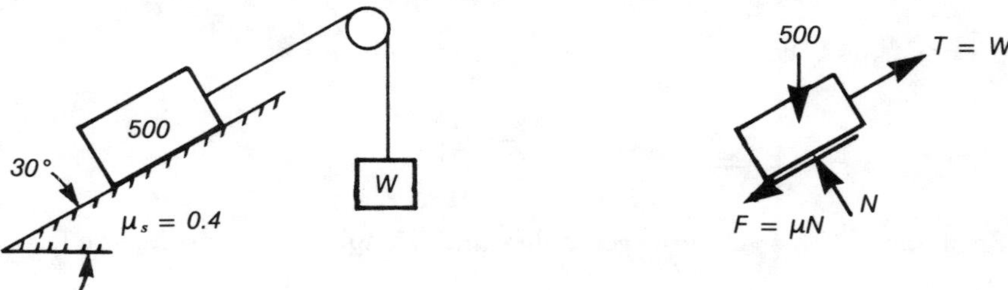

Solution. The free-body diagram is shown for impending motion up the plane; this would yield W_{max}. The normal force is found to be

$$N = 500 \cos 30° = 433 \text{ N}.$$

Then, the maximum weight is found to be

$$W_{max} = \mu N + 500 \sin 30°$$

$$= 0.4 \times 433 + 250 = 423 \text{ N}.$$

For the minimum weight the impending motion would be downward so that the friction force would act up the plane. The normal force would remain unchanged so that

$$W_{min} = 500 \sin 30° - \mu N$$

$$= 250 - 0.4 \times 433 = 76.8 \text{ N}.$$

EXAMPLE 8.10

A rope, wrapped around a post three times, is to resist a 10 000 N force. What minimum force is required by the holder on the loose end if $\mu = 0.3$?

Solution. The required force is found from the relationship

$$T_1 = T_2 e^{-\mu\beta}$$

$$= 10{,}000\, e^{-0.3 \times 6\pi} = 35.0 \text{ N}.$$

8.5 Centroids and Moments of Inertia

The *center of mass*, or *center of gravity*, is that point where all the mass could be concentrated producing no change in the external forces. If the density is constant, the center of mass is coincident with the *centroid*, a geometrical quantity whose coordinates are given by

$$\bar{x} = \frac{\int x dV}{V} \qquad \bar{y} = \frac{\int y dV}{V} \qquad \bar{z} = \frac{\int z dV}{V} \tag{8.5.1}$$

where V is the volume of the body.

An area also possesses a centroid; its coordinates are given by

$$\bar{x} = \frac{\int x dA}{A} \qquad \bar{y} = \frac{\int y dA}{A} . \tag{8.5.2}$$

For common shapes, such as circles, rectangles, and triangles the centroids are tabulated in Table 8.1.

If an area is composed of several common shapes (a composite area), we can locate the centroid using

$$\bar{x} = \frac{\Sigma \bar{x}_i A_i}{A} \qquad \bar{y} = \frac{\Sigma \bar{y}_i A_i}{A}. \tag{8.5.3}$$

The *moments of inertia* of an area about the x-axis and the y-axis are defined as

$$I_x = \int y^2 dA \qquad I_y = \int x^2 dA. \tag{8.5.4}$$

These are also referred to as the second moments of the area about the x- and y-axis, respectively.

The moment of inertia about an axis passing through the centroid is the centroidal moment of inertia I_c. The moment of inertia about an axis parallel to a centroidal axis can be found using the *parallel-axis theorem*:

$$I = I_c + Ad^2 \tag{8.5.5}$$

where d is the distance between the two axes. Moments of inertia for some common areas are given in Table 8.1.

TABLE 8.1. Properties of Areas.

Shape	Dimensions	Centroid	Inertia
Rectangle	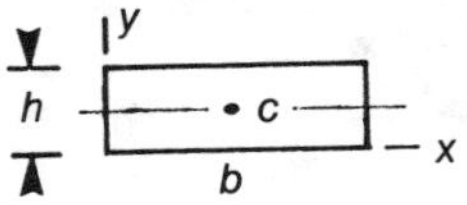	$\bar{x} = b/2$ $\bar{y} = h/2$	$I_c = bh^3/12$ $I_x = bh^3/3$ $I_y = hb^3/3$
Triangle	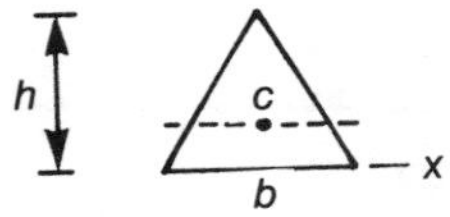	$\bar{y} = h/3$	$I_c = bh^3/36$ $I_x = bh^3/12$
Circle	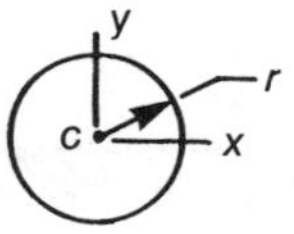	$\bar{x} = 0$ $\bar{y} = 0$	$I_x = \pi r^4/4$ $J = \pi d^4/32$
Quarter Circle	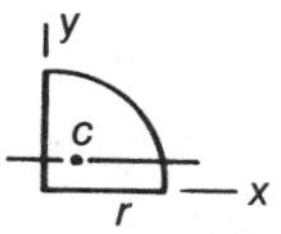	$\bar{y} = 4r/3\pi$	$I_x = \pi r^4/16$
Half Circle	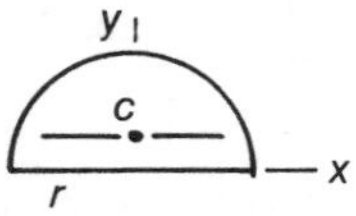	$\bar{y} = 4r/3\pi$	$I_x = \pi r^4/8$

Finally, the *mass moment of inertia* is defined as

$$\begin{aligned} I_x &= \int (y^2 + z^2)\, \rho dV \\ I_y &= \int (x^2 + z^2)\, \rho dV \\ I_z &= \int (x^2 + y^2)\, \rho dV \end{aligned} \tag{8.5.6}$$

It is a measure of resistance to rotational acceleration. The units on the mass moment of inertia are obviously $kg \cdot m^2$. Alternatively, the expressions for the mass moment of inertia could have been written as

$$I = \int r^2 dm \tag{8.5.7}$$

where r is measured perpendicular to the axis about which the mass moment of inertia is measured and dm is the mass element. The parallel axis theorem is also used for the mass moment of inertia.

Mass moments of inertia of some common objects are given in Table 8.2.

TABLE 8.2. Mass Moments of Inertia.

Shape	Dimensions	Moment of Inertia
Slender rod		$I_y = mL^2/12$ $I_{y'} = mL^2/3$
Circular cylinder		$I_x = mr^2/2$ $I_y = m(L^2 + 3r^2)/12$
Disk		$I_x = mr^2/2$ $I_y = mr^2/4$
Rectangular parallelepiped		$I_x = m(a^2 + b^2)/12$ $I_y = m(L^2 + b^2)/12$ $I_z = m(L^2 + a^2)/12$ $I_{y'} = m(4L^2 + b^2)/12$
Sphere		$I_x = 2mr^2/5$

EXAMPLE 8.11

Find the y-coordinate of the centroid of the area shown.

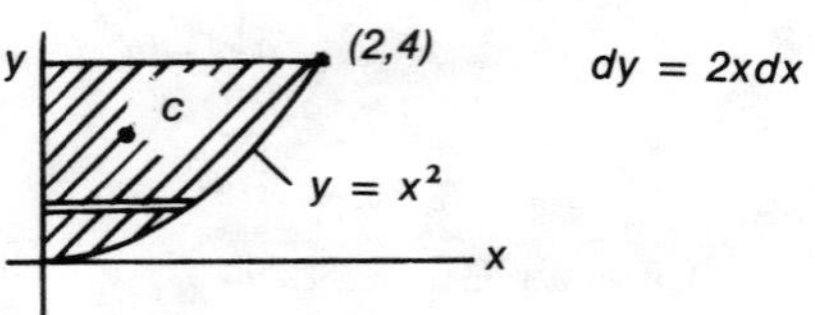

Solution. The y-coordinate of the centroid is found from

$$\bar{y} = \frac{\int y dA}{A} .$$

First, let's find the area A. It is

$$A = \int x dy$$

$$= \int_0^2 x\,(2xdx) = 2\int_0^2 x^2\,dx = \frac{16}{3}$$

Then, the y-coordinate of the centroid is given by

$$\bar{y} = \frac{1}{16/3}\int y dA = \frac{3}{16}\int yxdy$$

$$= \frac{3}{16}\int_0^2 x^2\,x(2xdx) = \frac{3}{8}\int_0^2 x^4 dx = 2.4.$$

EXAMPLE 8.12

Find the x-coordinate of the centroid of the composite area shown.

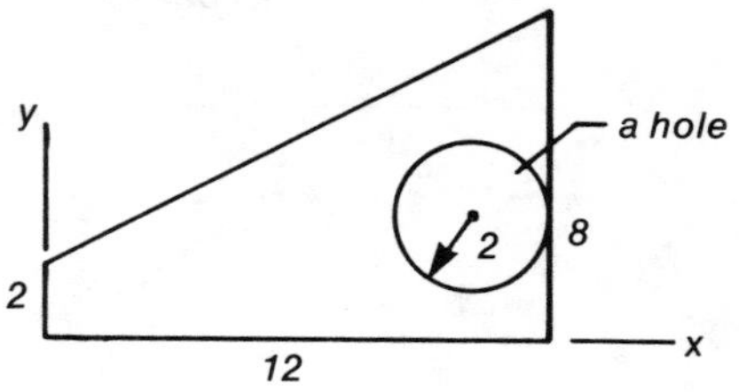

Solution. Divide the composite area into a rectangle, a triangle, and a circle. The x-coordinate is given by

$$\bar{x} = \frac{\bar{x}_1 A_1 + \bar{x}_2 A_2 - \bar{x}_3 A_3}{A_1 + A_2 - A_3}$$

$$= \frac{6 \times 24 + 8 \times 36 - 10 \times 4\pi}{24 + 36 - 4\pi} = 6.46.$$

EXAMPLE 8.13

Find the moment of inertia of the area of Example 8.11 about the x-axis.

Solution. Using the horizontal strip, which has all elements at a distance y from the x-axis, we have

$$I_x = \int y^2 dA$$

$$= \int_0^4 y^2 (x dy) = \int_0^4 y^2 y^{1/2}\, dy = \int_0^4 y^{5/2}\, dy = 36.57.$$

EXAMPLE 8.14

If the mass moment of inertia about the centroidal axis of a long slender bar is $mL^2/12$, find the moment of inertia about one end.

Solution. To do this we use the parallel-axis theorem. There results

$$I_{end} = I_c + md^2$$

$$= mL^2/12 + m\,(L/2)^2 = mL^2/3.$$

(This page is intentionally blank.)

Practice Problems

(If only a few select problems are desired, choose those with a star.)

GENERAL

*8.1 Find the component of the vector $\vec{A} = 15\hat{i} - 9\hat{j} + 15\hat{k}$ in the direction of $\vec{B} = \hat{i} - 2\hat{j} - 2\hat{k}$.

a) 1 b) 3 c) 5 d) 7 e) 9

8.2 Find the magnitude of the resultant of $\vec{A} = 2\hat{i} + 5\hat{j}$, $\vec{B} = 6\hat{i} - 7\hat{k}$, and $\vec{C} = 2\hat{i} - 6\hat{j} + 10\hat{k}$.

a) 8.2 b) 9.3 c) 10.5 d) 11.7 e) 12.8

*8.3 Determine the moment about the y-axis of the force $\vec{F} = 200\hat{i} + 400\hat{j}$ acting at (4, −6, 4).

a) 0 b) 200 c) 400 d) 600 e) 800

8.4 What moment do the two forces $\vec{F_1} = 50\hat{i} - 40\hat{k}$ and $\vec{F_2} = 60\hat{j} + 80\hat{k}$ acting at (2, 0, −4) and (−4, 2, 0), respectively, produce about the x-axis?

a) 0 b) 80 c) 160 d) 240 e) 320

*8.5 The force system shown may be referred to as being

a) concurrent b) coplanar c) parallel

d) two-dimensional e) three-dimensional

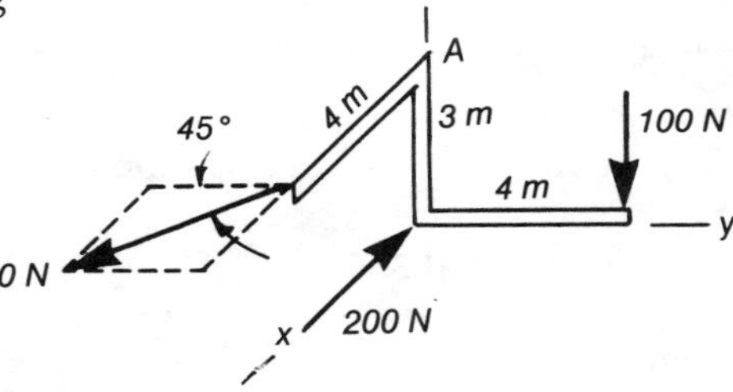

EQUILIBRIUM

8.6 If equilibrium exists due to a rigid support at A in the figure of Prob. 8.5, what reactive force must exist at A?

a) $-59\hat{i} - 141\hat{j} + 10\hat{k}$
b) $59\hat{i} + 141\hat{j} + 100\hat{k}$
c) $341\hat{i} - 141\hat{j} - 100\hat{k}$
d) $341\hat{i} + 141\hat{j} - 100\hat{k}$
e) $59\hat{i} - 141\hat{j} + 100\hat{k}$

*8.7. If equilibrium exists on the object in Prob. 8.5, what moment must exist at the rigid support A?

a) $600\hat{i} + 400\hat{j} + 564\hat{k}$
b) $400\hat{i} + 564\hat{k}$
c) $400\hat{i} - 600\hat{j} + 564\hat{k}$
d) $400\hat{i} - 600\hat{j}$
e) $-600\hat{j} + 564\hat{j}$

*8.8 If three nonparallel forces hold a rigid body in equilibrium, they must

a) be equal in magnitude.
b) be concurrent.
c) be nonconcurrent.
d) form an equilateral triangle.
e) be colinear.

*8.9 A truss element

a) is a two-force body.
b) is a three-force body.
c) resists forces in compression only.
d) may resist three concurrent forces.
e) resists lateral forces.

*8.10 Find the magnitude of the reaction at support B.

a) 400 b) 500 c) 600

d) 700 e) 800

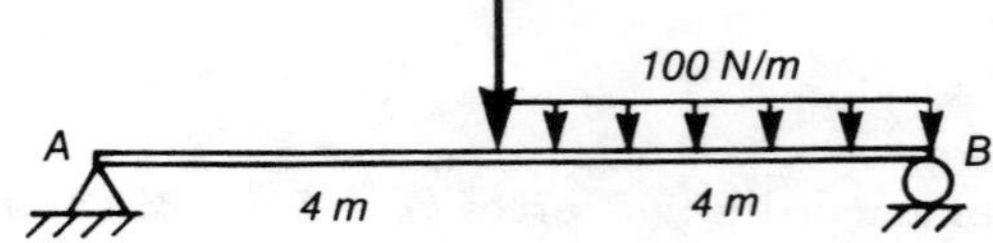

8.11 What moment exists at support A?

a) 5600 b) 5000 c) 4400

d) 4000 e) 3600

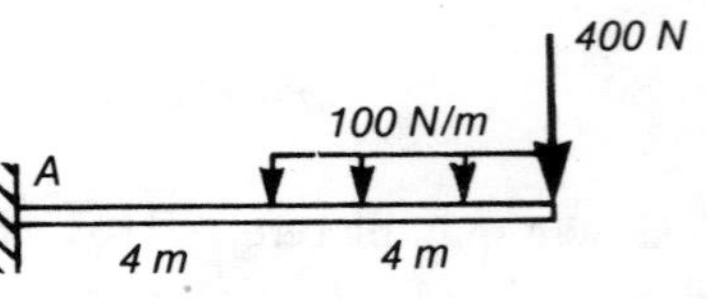

8.12 Calculate the reactive force at support A.

a) 250 b) 350 c) 450

d) 550 e) 650

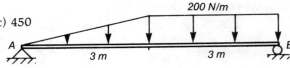

8.13 Find the support moment at A.

a) 66 b) 77 c) 88

d) 99 e) 111

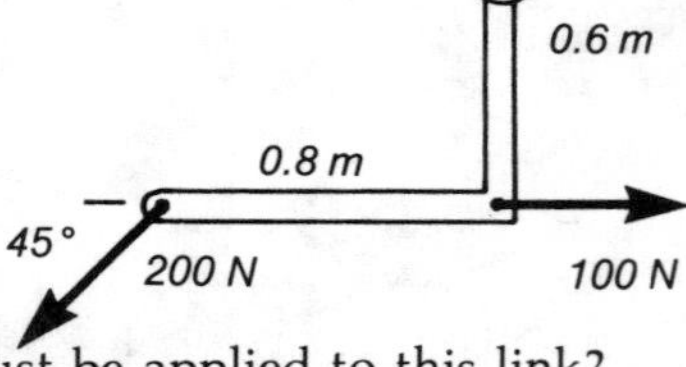

*8.14 To ensure equilibrium, what couple must be applied to this link?

a) 283 *cw* b) 283 *ccw* c) 400 *cw*

d) 400 *ccw* e) 0

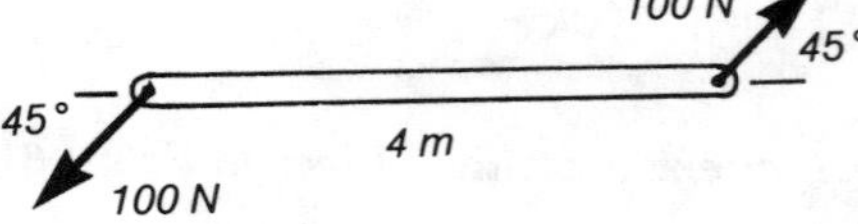

*8.15 Calculate the magnitude of the force at A for the three-force body shown.

a) 217 b) 287 c) 343

d) 385 e)492

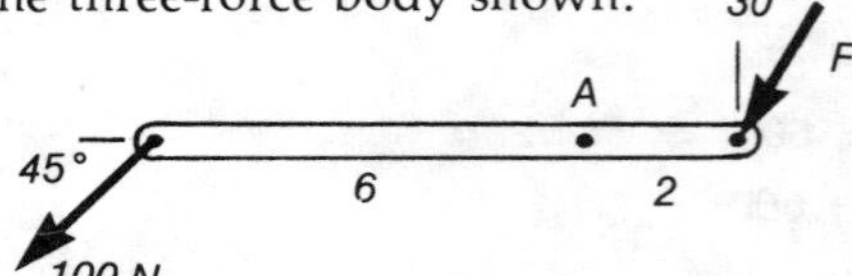

8.16 Find the magnitude of the force at point A.

a) 187 b) 142 c) 114

d) 99 e) 84

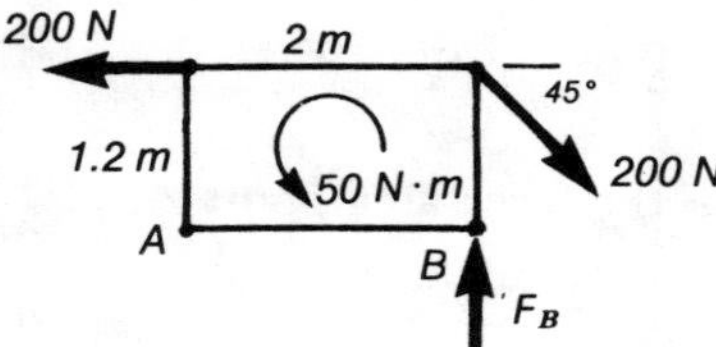

TRUSSES AND FRAMES

8.17 Find F_{DE} if all angles are equal.

a) 121 b) 142 c) 163

d) 176 e) 189

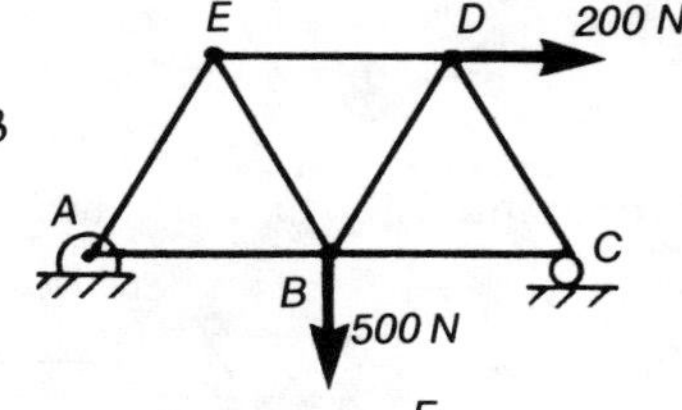

*8.18 Find F_{DE}.

a) 0 b) 1000 c) 2000

d) 2500 e) 5000

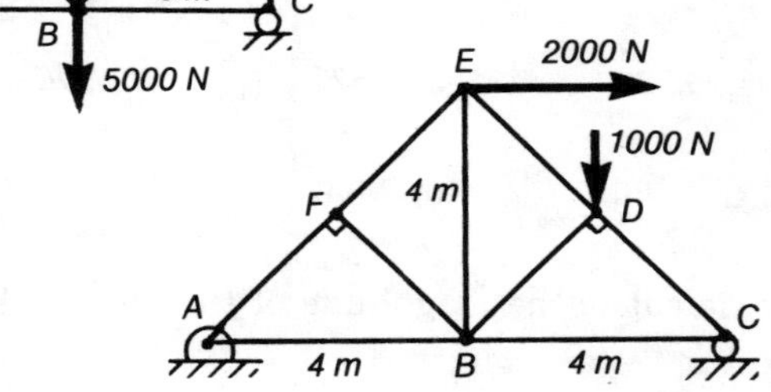

8.19 What is the force in link DE?

a) 1532 b) 1768 c) 1895

d) 1946 e) 2231

*8.20 Calculate F_{FB} in the truss of Problem 8.19.

a) 0 b) 932 c) 1561 d) 1732 e) 1887

8.21 Find the force in member IC.

a) 0 b) 1000 c) 1250

d) 1500 e) 2000

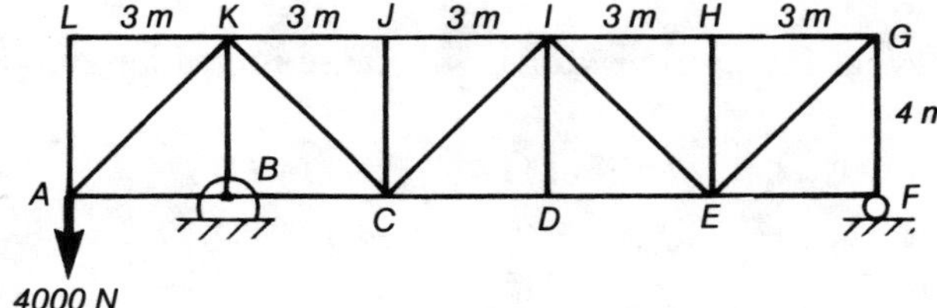

8.22 What force exists in member BC in the truss of Prob. 8.21?

a) 0 b) 1000 c) 2000 d) 3000 e) 4000

8.23 Find the force in link FC.

a) 5320 b) 3420 c) 2560

d) 936 e) 0

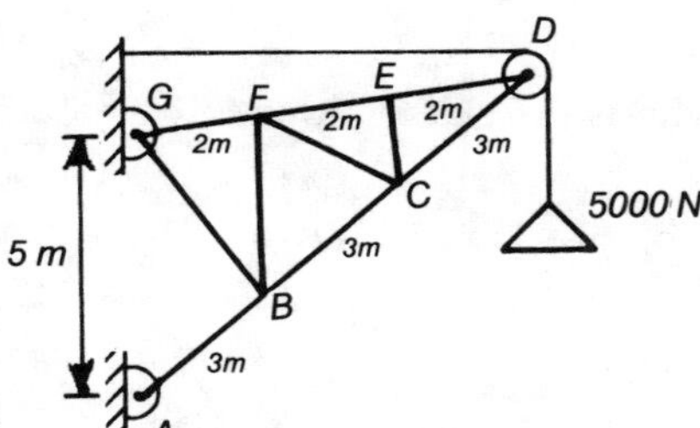

8.24 Determine the force in link *BC* in the truss of Prob. 8.23.

a) 3560 b) 4230 c) 4960 d) 5230 e) 5820

*8.25 Find the magnitude of the force at support *A*.

a) 1400 b) 1300 c) 1200

d) 1100 e) 1000

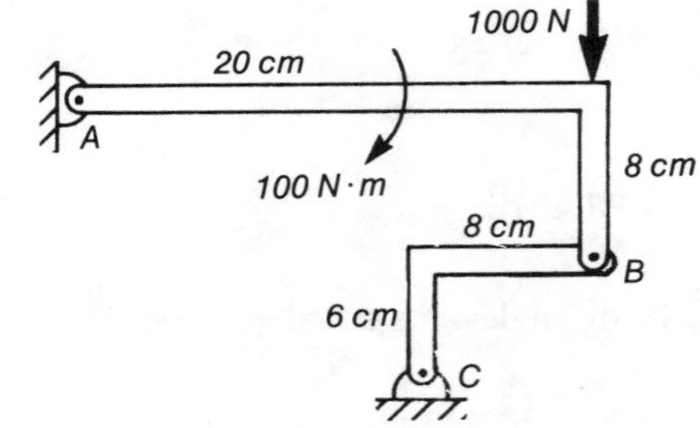

8.26 Determine the distributed force intensity w, in N/m.

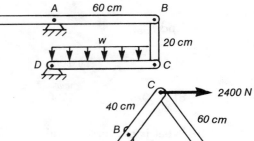

a) 2000 b) 4000 c) 6000

d) 8000 e) 10 000

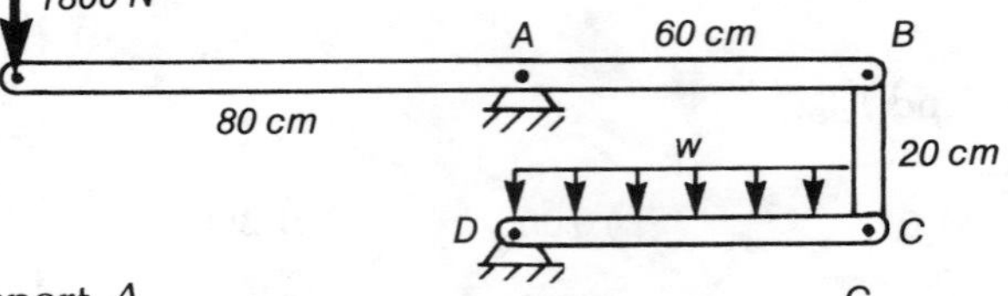

*8.27 Find the magnitude of the force acting at support *A*.

a) 2580 b) 2670 c) 2790

d) 2880 e) 2960

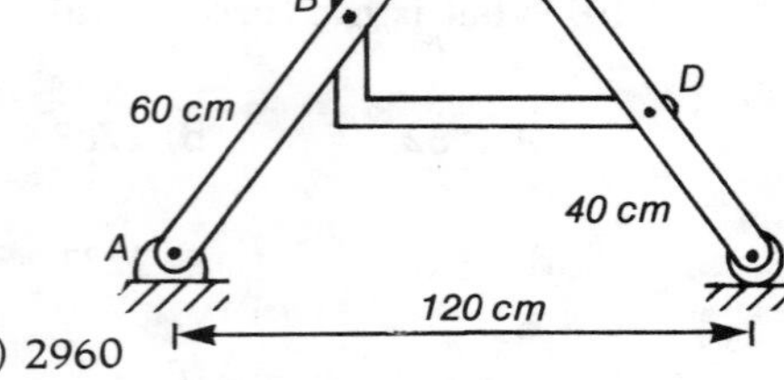

8.28 Calculate the magnitude of the force in link *BD* of Prob. 8.27.

a) 2590 b) 2670 c) 2790 d) 2880 e) 2960

FRICTION

*8.29. What force, in newtons, will cause impending motion?

a) 731 b) 821 c) 973

d) 1102 e) 1245

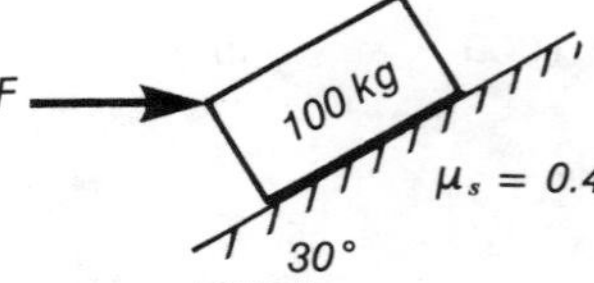

8.30 What is the maximum force *F*, in newtons, that will not cause motion?

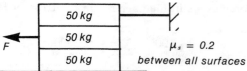

a) 184 b) 294 c) 316

d) 346 e) 392

*8.31 Only the rear wheels provide braking. At what angle will the car slide if $\mu_s = 0.6$?

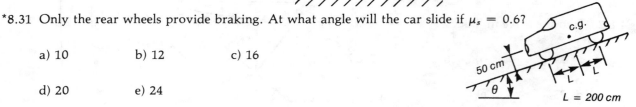

a) 10 b) 12 c) 16

d) 20 e) 24

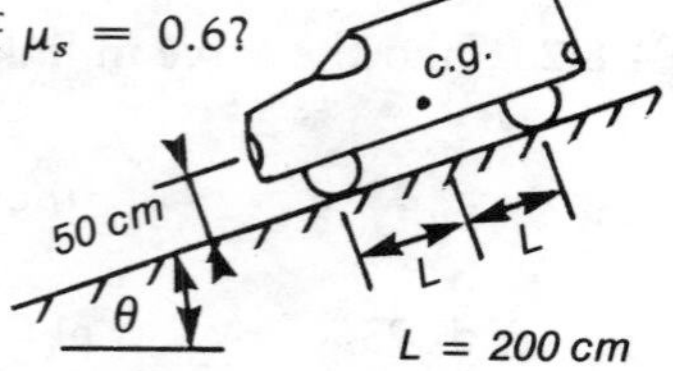

8.32 Find the minimum h value that will allow tipping.

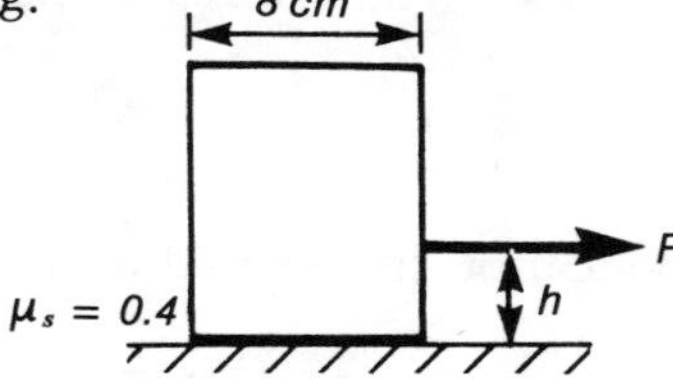

a) 8 b) 10 c) 12

d) 14 e) 16

*8.33 What force F, in newtons, will cause impending motion?

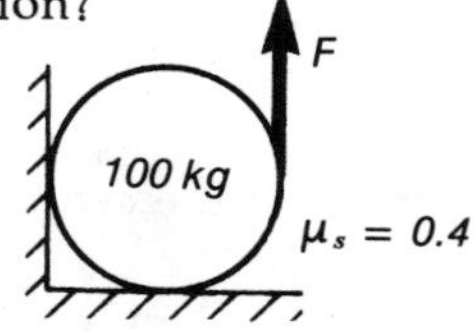

a) 240 b) 260 c) 280

d) 300 e) 320

8.34 The angle θ that will cause the ladder to slip is

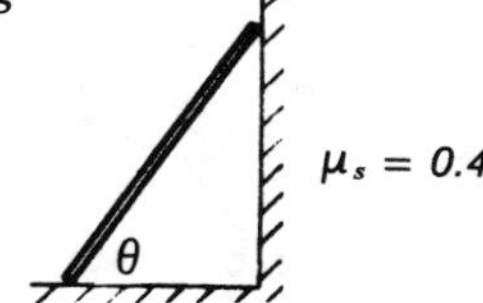

a) 50 b) 46 c) 42

d) 38 e) 34

*8.35 A boy and his dad put a rope around a tree and stand side by side. What force by the boy can resist a force of 800 N by the dad? Use $\mu = 0.5$.

a) 166 b) 192 c) 231 d) 246 e) 297

8.36 What moment, in N·m, will cause impending motion?

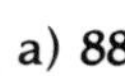

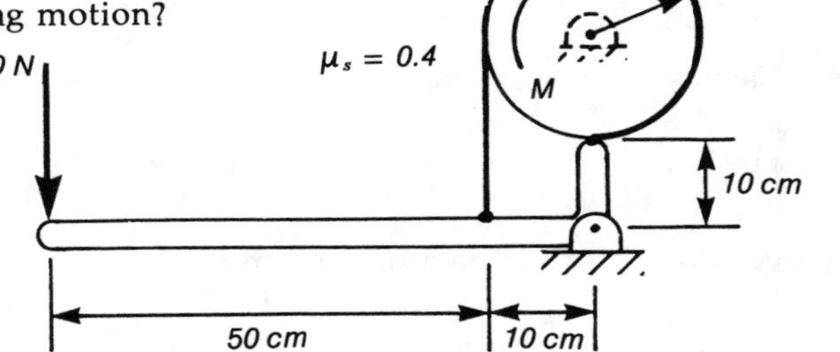

a) 88 b) 99 c) 110

d) 121 e) 146

8.37 A 12-m-long rope is draped over a horizontal cylinder 1.2-m-diameter so that both ends hang free. How long must an end be for impending motion? Use $\mu_s = 0.5$.

a) 6.98 b) 7.65 c) 7.92 d) 8.37 e) 8.83

CENTROIDS AND MOMENTS OF INERTIA

*8.38. Find the x-coordinate of the centroid of the area bounded by the x-axis, the line $x = 3$, and the parabola $y = x^2$.

a) 2.0 b) 2.15 c) 2.20 d) 2.25 e) 2.30

8.39 What is the y-coordinate of the centroid of the area of Prob. 8.38?

a) 2.70 b) 2.65 c) 2.60 d) 2.55 e) 2.50

8.40 Calculate the x-coordinate of the centroid of the area enclosed by the parabolas $y = x^2$ and $x = y^2$.

a) 0.43 b) 0.44 c) 0.45 d) 0.46 e) 0.47

8.41 Find the y-component of the centroid of the area shown.

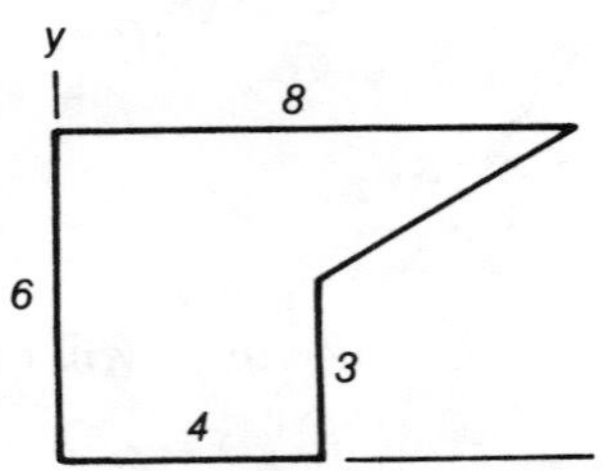

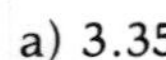

a) 3.35 b) 3.40 c) 3.45

d) 3.50 e) 3.55

*8.42 Calculate the y-component of the centroid of the area shown.

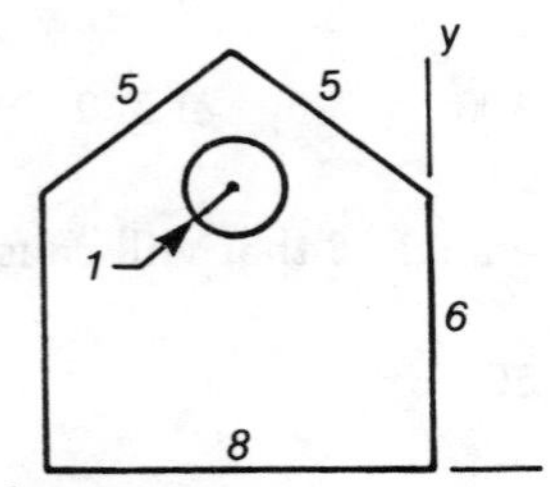

a) 3.52 b) 3.56 c) 3.60

d) 3.64 e) 3.68

8.43 Find the x-component of the center of gravity of the three objects shown.

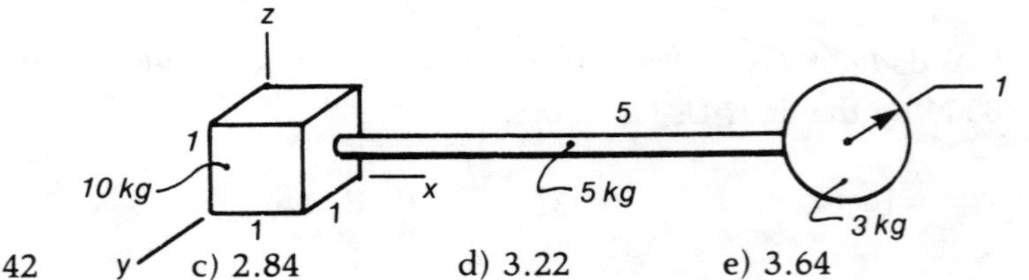

a) 2.33 b) 2.42 c) 2.84 d) 3.22 e) 3.64

*8.44 Calculate the moment of inertia about the x-axis of the area of Prob. 8.38.

a) 94 b) 104 c) 112 d) 124 e) 132

8.45 What is I_x for the area of Prob. 8.42?

a) 736 b) 842 c) 936 d) 982 e) 1056

*8.46 Find I_y for the symmetric area shown.

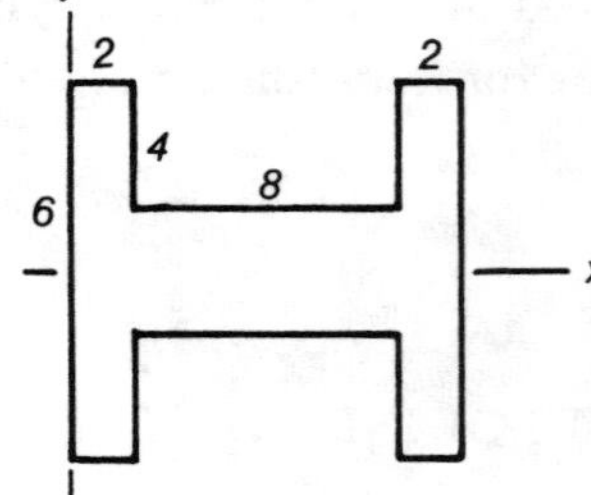

a) 4267 b) 4036 c) 3827

d) 3652 e) 3421

8.47 Determine the mass moment of inertia of a cube with edges of length b, about an axis passing through an edge.

a) $2mb^2/3$ b) $mb^2/6$ c) $3mb^2/2$ d) $mb^2/2$ e) $3mb^2/4$

*8.48 Find the mass moment of inertia about the x-axis if the mass of the rods per unit length is 1.0 kg/m.

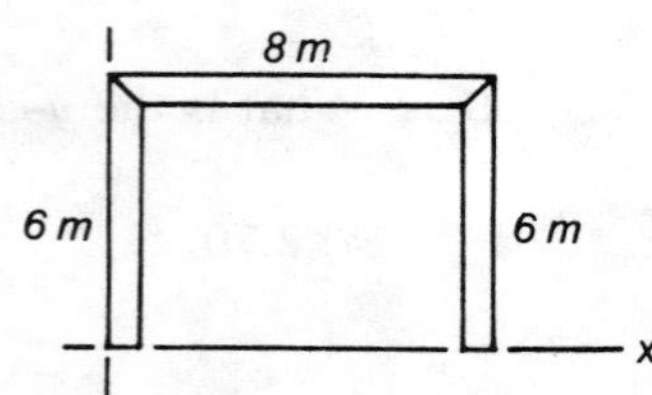

a) 224 b) 268 c) 336 d) 386 e) 432

9. Dynamics

Dynamics is separated into two major divisions: *kinematics,* which is a study of motion without reference to the forces causing the motion, and *kinetics,* which relates the forces on bodies to their resulting motions. Newton's laws of motion are necessary in relating forces to motions; they are:

1st law: A particle remains at rest or continues to move in a straight line with a constant velocity if no unbalanced force acts on it.

2nd law: The acceleration of a particle is proportional to the force acting on it and inversely proportional to the particle mass; the direction of acceleration is the same as the force direction.

3rd law: The forces of action and reaction between contacting bodies are equal in magnitude, opposite in direction, and colinear.

Law of gravitation: The force of attraction between two bodies is proportional to the product of their masses and inversely proportional to the square of the distance between their centers.

9.1 Kinematics

In kinematics we will consider three different kinds of particle motion: rectilinear motion, angular motion, and curvilinear motion. These will be followed by a review of the motion of rigid bodies.

9.1.1 Rectilinear Motion

In rectilinear motion of a particle, in which the particle moves in a straight line, the acceleration a, the velocity v, and the displacement s are related by

$$a = \frac{dv}{dt}, \qquad v = \frac{ds}{dt}, \qquad a = v\frac{dv}{ds} = \frac{d^2s}{dt^2} \tag{9.1.1}$$

If the acceleration is a known function of time, the above can be integrated to give $v(t)$ and $s(t)$. For the important case of constant acceleration, integration yields

$$v = v_o + at$$

$$s = v_o t + at^2/2 \tag{9.1.2}$$

$$v^2 = v_o^2 + 2\,as$$

where at $t = 0$, $v = v_o$ and $s_o = 0$.

9.1.2 Angular Motion

Angular displacement is the angle θ that a line makes with a fixed axis, usually the positive x-axis. Counterclockwise motion is assumed to be positive, as shown in Fig. 9.1. The angular acceleration α, the

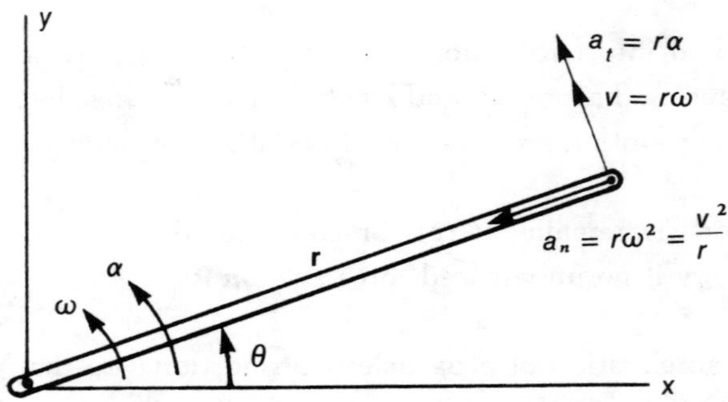

Figure 9.1. Angular motion.

angular velocity ω, and θ are related by

$$\alpha = \frac{d\omega}{dt}, \qquad \omega = \frac{d\theta}{dt}, \qquad \alpha = \omega\frac{d\omega}{d\theta} = \frac{d^2\theta}{dt^2} \tag{9.1.3}$$

If α is constant, integration of these equations gives

$$\begin{aligned} \omega &= \omega_o + \alpha t \\ \theta &= \omega_o t + \alpha t^2/2 \\ \omega^2 &= \omega_o^2 + 2\,\alpha\theta \end{aligned} \tag{9.1.4}$$

where we have assumed that $\omega = \omega_o$ and $\theta_o = 0$ at $t = 0$.

9.1.3 Curvilinear Motion

When a particle moves on a plane curve, as shown in Fig. 9.2, the motion may be described in terms of coordinates along the normal n and the tangent t to the curve at the instantaneous position of the particle.

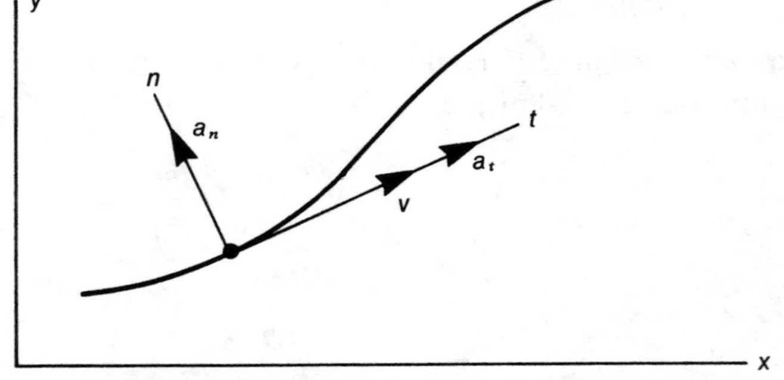

Figure 9.2. Motion on a plane curve.

The acceleration is the vector sum of the normal acceleration a_n and the tangential acceleration a_t. These components are given by

$$a_n = \frac{v^2}{r}, \qquad a_t = \frac{dv}{dt} \tag{9.1.5}$$

where r is the radius of curvature and v is the magnitude of the velocity. The velocity is always tangential to the curve so no subscript is necessary to identify the velocity.

It should be noted that a rigid body travelling without rotation can be treated as particle motion.

EXAMPLE 9.1

The velocity of a particle undergoing rectilinear motion is $v(t) = 3\ t_2 + 10\ t$ m/s. Find the acceleration and the displacement at $t = 10$ s, if $s_o = 0$, at $t = 0$.

Solution. The acceleration is found to be

$$a = \frac{dv}{dt}$$

$$= 6\ t + 10 = 6 \times 10 + 10 = 70 \text{ m/s}^2.$$

The displacement is found by integration

$$s = \int_0^{10} v\, dt$$

$$= \int_0^{10} (3\ t^2 + 10\ t)dt = t^3 + 5\ t^2\Big|_0^{10} = 1500 \text{ m}.$$

EXAMPLE 9.2

An automobile skids to a stop 60 m after its brakes are applied when it is travelling 25 m/s. What is its deceleration?

Solution. We use the relationship

$$v^2 = v_o^2 + 2\ as.$$

Letting $v = 0$, we find

$$a = -\frac{v_0^2}{2\ s}$$

$$= -\frac{25^2}{2 \times 60} = -5.21 \text{ m/s}^2.$$

EXAMPLE 9.3

A wheel, rotating at 100 rad/s ccw (counterclockwise), is subjected to an angular acceleration of 20 rad/s² cw. Find the total number of revolutions (cw plus ccw) through which the wheel rotates in 8 seconds.

Solution. The time at which the angular velocity is zero is found as follows:

$$\cancel{\omega}^{0} = \omega_o + \alpha t.$$

$$\therefore t = -\frac{\omega_o}{\alpha} = -\frac{100}{(-20)} = 5 \text{ s}.$$

After three additional seconds the angular velocity is

$$\omega = \cancel{\omega_o}^{0} + \alpha t$$

$$= -20 \times 3 = -60 \text{ rad/s}.$$

The angular displacement from 0 to 5 *s* is

$$\theta = \omega_o t + \alpha t^2/2$$

$$= 100 \times 5 - 20 \times 5^2/2 = 250 \text{ rad}.$$

During the next 3 *s* the angular displacement is

$$\theta = \alpha t^2/2$$

$$= -20 \times 3^2/2 = -90 \text{ rad}.$$

The total number of revolutions rotated is

$$\theta = (250 + 90)/2\pi = 54.1 \text{ rev}.$$

EXAMPLE 9.4

Consider idealized (no air drag) projectile motion in which $a_x = 0$ and $a_y = -g$. Find expressions for the range R and the maximum height H in terms of v_o and θ.

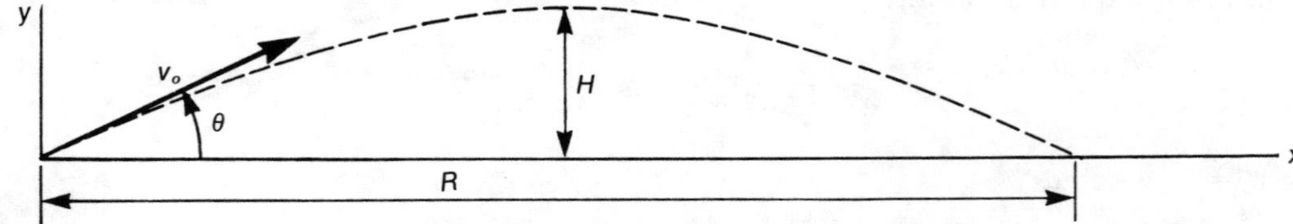

Solution. Using Eq. (9.1.2) for constant acceleration we have the point $(R,0)$:

$$\cancel{y}^{0} = (v_o \sin\theta)t - gt^2/2.$$

$$\therefore t = 2\, v_o \sin\theta/g.$$

From the x-component equation:

$$x = (v_o \cos\theta)t + a_x^{\,0}\, t^2/2.$$

$$\therefore R = (v_o \cos\theta)\, 2\, v_o \sin\theta/g = v_o^2 \sin 2\theta/g.$$

Obviously, the maximum height occurs when the time is one-half that which yields the range R. Hence,

$$H = (v_o \sin\theta)\frac{v_o \sin\theta}{g} - g\left(\frac{v_o \sin\theta}{g}\right)^2/2$$

$$= \frac{v_o^2}{2g}\sin^2\theta.$$

Note: The maximum R for a given v_o occurs when $\sin 2\theta = 1$ which means $\theta = 45°$ for R_{max}.

EXAMPLE 9.5

It is desired that the normal acceleration of a satellite be 9.6 m/s² at an elevation of 200 km. What should the velocity be for a circular orbit? The radius of the earth is 6400 km.

Solution. The normal acceleration, which points toward the center of the earth, is

$$a_n = \frac{v^2}{r}.$$

$$\therefore v = \sqrt{a_n r}$$

$$= \sqrt{9.6 \times (6400 + 200) \times 1000} = 7960 \text{ m/s}$$

9.1.4 Rigid Body Motion

The motion of a rigid body can be described using the relative velocity and relative acceleration equations

$$\vec{v}_A = \vec{v}_B + \vec{v}_{A/B}$$

$$\vec{a}_A = \vec{a}_B + \vec{a}_{A/B} \tag{9.1.6}$$

where the velocity $\vec{v}_{A/B}$ is the velocity of point A with respect to point B and the acceleration $\vec{a}_{A/B}$ is the acceleration of point A with respect to point B. If points A and B are on the same rigid body then point A must move perpendicular to the line AB and

$$v_{A/B} = r\omega$$

$$(a_{A/B})_n = r\omega^2 \tag{9.1.7}$$

$$(a_{A/B})_t = r\alpha$$

where ω is the angular velocity, α is the angular acceleration of the body, and r is the length of $\overline{AB}$.

If point A is located on a body which moves with a constant velocity v relative to a coincident point B which is located on a second body rotating with an angular velocity ω (see Fig. 9.3), the acceleration of A

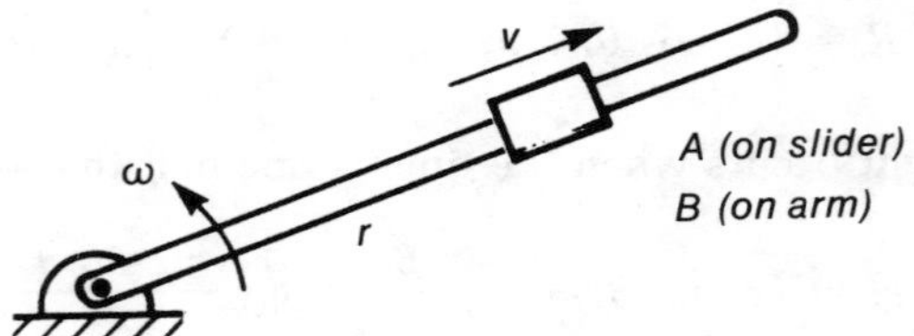

Figure 9.3. Coriolis acceleration.

with respect to B is called the Coriolis acceleration, given by

$$\vec{a}_{A/B} = 2\,\vec{\omega} \times \vec{v}. \tag{9.1.8}$$

We note that the Coriolis acceleration acts normal to both vectors $\vec{\omega}$ (use the right-hand rule) and $\vec{v}$, it acts normal to the arm.

A final note regards the use of the instant center of zero velocity, a point, which is often off the body, that has zero velocity. If such a point B can be located then the magnitude of the velocity of point A is simply $r\omega$.

EXAMPLE 9.6

Find the magnitudes of the velocity and acceleration of point A.

Solution. The velocity is found to be

$$v_A = r\omega$$

$$= 0.2 \times 10 = 2 \text{ m/s}.$$

The acceleration components are

$$a_n = r\omega^2 = 0.2 \times 10^2 = 20 \text{ m/s}^2.$$

$$a_t = r\alpha = 0.2 \times 40 = 8 \text{ m/s}^2.$$

Thus,

$$a = \sqrt{a_n^2 + a_t^2} = \sqrt{20^2 + 8^2} = 21.5 \text{ m/s}^2.$$

EXAMPLE 9.7

Find the velocity of C and ω_{BC}.

Solution. The velocity of B is normal to $\overline{AB}$ and equal to

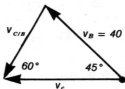

$$v_B = r\omega$$
$$= 0.4 \times 100 = 40 \text{ m/s}.$$

To find V_C we use the relative motion equation

$$\vec{v}_C = \vec{v}_B + \vec{v}_{C/B}$$

The velocity of C must be horizontal; $\vec{v}_{C/B}$ must be normal to $\overline{BC}$. This can be displayed in a velocity polygon as follows.

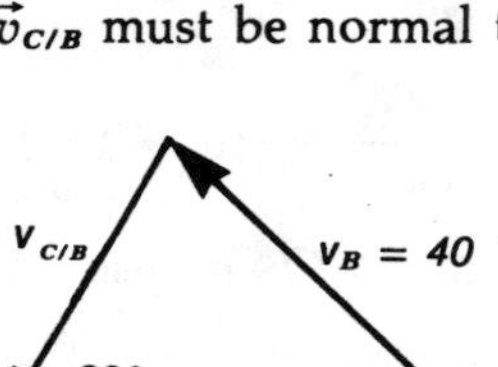

From the velocity polygon we use some simple trig and find v_c to be to the left with magnitude

$$v_c = 44.6 \text{ m/s}.$$

The angular velocity of $\overline{BC}$ is found to be

$$\omega_{BC} = v_{C/B}/r_{BC}$$
$$= 32.7/0.5656 = 57.8 \text{ rad/s cw}.$$

EXAMPLE 9.8

Find the acceleration of C in Ex. 9.7, assuming $\alpha_{AB} = 0$.

Solution. The acceleration of B is

$$(a_B)_n = r\omega^2$$
$$= 0.4 \times 100^2 = 4000 \text{ m/s}^2.$$

The relative acceleration equation

$$\vec{a}_C = \vec{a}_B + \vec{a}_{C/B}$$

can be displayed in an acceleration polygon, realizing that $\vec{a}_C$ must be horizontal, and $\vec{a}_{C/B}$ has both normal and tangential components. We find $(a_{C/B})_n$ to be

$$(a_{C/B})_n = r_{BC}\omega_{BC}^2$$
$$= 0.5656 \times 57.8^2 = 1890 \text{ m/s}^2.$$

The acceleration polygon follows.

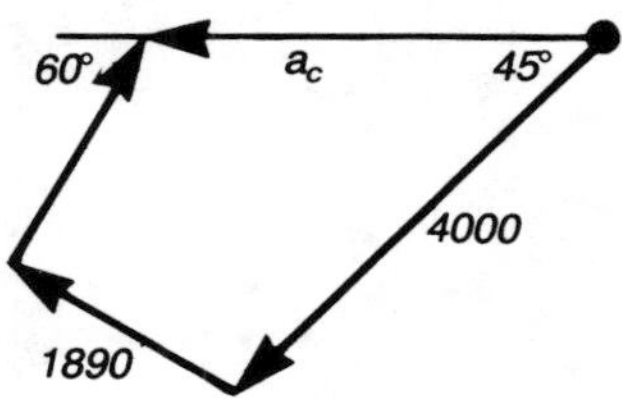

From the polygon we can find a_c to be to the left with magnitude

$$a_c = 3380 \text{ m/s}^2.$$

EXAMPLE 9.9

Find v_A and a_A.

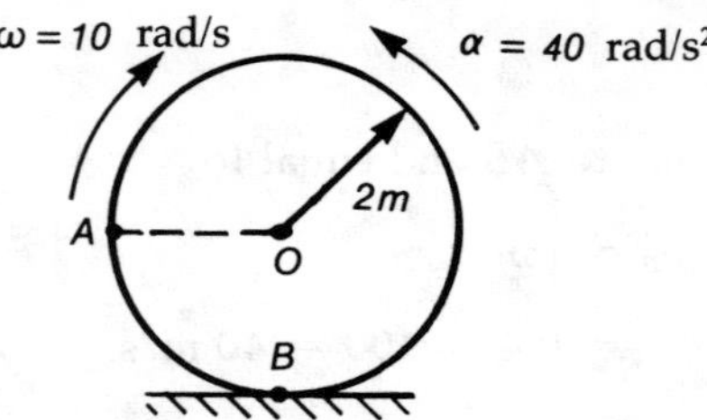

Solution. At this instant the wheel rotates about B. Hence, we can find v_A by using

$$v_A = r_{AB}\omega$$

$$= \frac{2}{\sin 45^\circ} \times 10 = 28.3 \text{ m/s}.$$

To find the acceleration we relate A to the center O and use

$$\vec{a}_A = \vec{a}_O + \vec{a}_{A/O}.$$

The acceleration polygon follows using

$$a_O = r\alpha = 2 \times 40 = 80 \text{ m/s}^2,$$

$$(a_{A/O})_t = r\alpha = 2 \times 40 = 80 \text{ m/s}^2,$$

$$(a_{A/O})_n = r\omega^2 = 2 \times 10^2 = 200 \text{ m/s}^2.$$

From the polygon we find a_A to be

$$a_A = 144 \text{ m/s}^2.$$

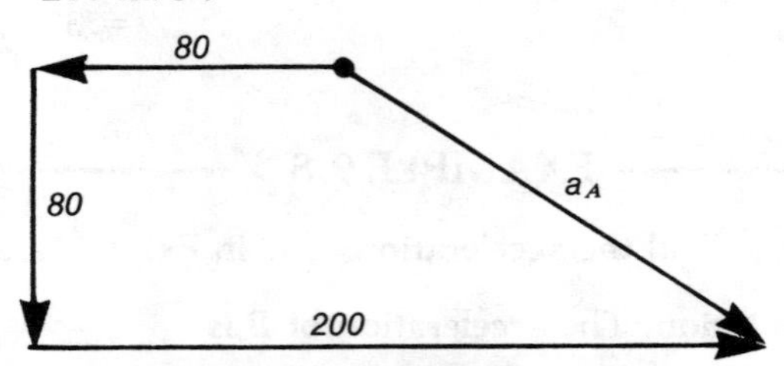

EXAMPLE 9.10

Calculate a_A where A is on the slider.

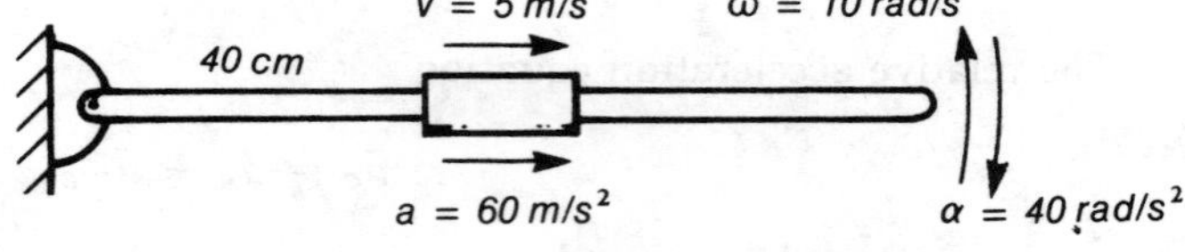

Solution. The relative acceleration equation is

$$\vec{a}_A = \vec{a}_B + \vec{a}_{A/B}$$

where point B is on the arm, coincident with point A. We know that

$$(a_B)_n = r\omega^2 = 0.4 \times 10^2 = 40 \text{ m/s}^2$$

$$(a_B)_t = r\alpha = 0.4 \times 40 = 16 \text{ m/s}^2$$

$$(a_{A/B})_t = 60 \text{ m/s}^2$$

$$(a_{A/B})_c = 2\omega v = 2 \times 10 \times 5 = 100 \text{ m/s}^2.$$

The acceleration polygon appears as follows.

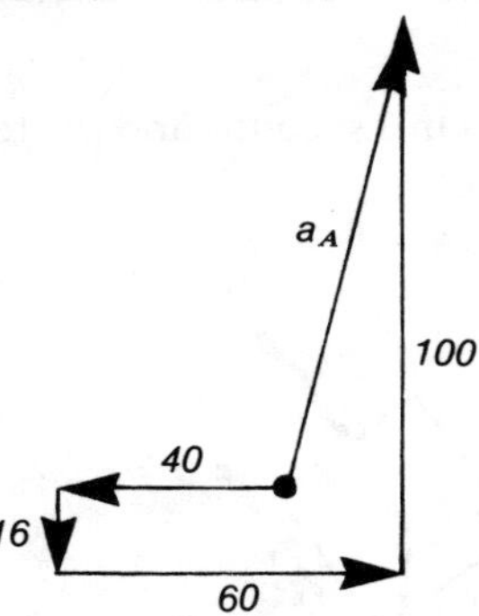

The acceleration a_A is found to be

$$a_A = 86.4 \text{ m/s}^2.$$

9.2 Kinetics

To relate the forces acting on a body to the motion of that body we use Newton's laws of motion. Newton's 2nd law is used in the form

$$\Sigma\vec{F} = m\vec{a} \tag{9.2.1}$$

where the mass of the body is assumed to be constant and $\vec{a}$ is the acceleration of the center of mass (center of gravity) if the body is rotating. We also require

$$\Sigma\vec{M} = I\vec{\alpha} \tag{9.2.2}$$

where the moments must be summed about an axis passing through the center of mass. The mass moment of inertia I is often found by using the radius of gyration k and the relation $I = mk^2$.

The gravitational attractive force between one body and another is given by

$$F = k\frac{m_1m_2}{r^2} \tag{9.2.3}$$

where $k = 6.67 \times 10^{-11}$ N·m²/kg².

Note: Since metric units are used in the above relations, the mass must be measured in kilograms; the weight is related to the mass by

$$W = mg \tag{9.2.4}$$

where we will use $g = 9.8$ m/s², unless otherwise stated.

EXAMPLE 9.11

Find the distance the 600 kg mass moves in 3 seconds and the tension in the rope. The mass starts from rest and the mass of the pulleys is negligible.

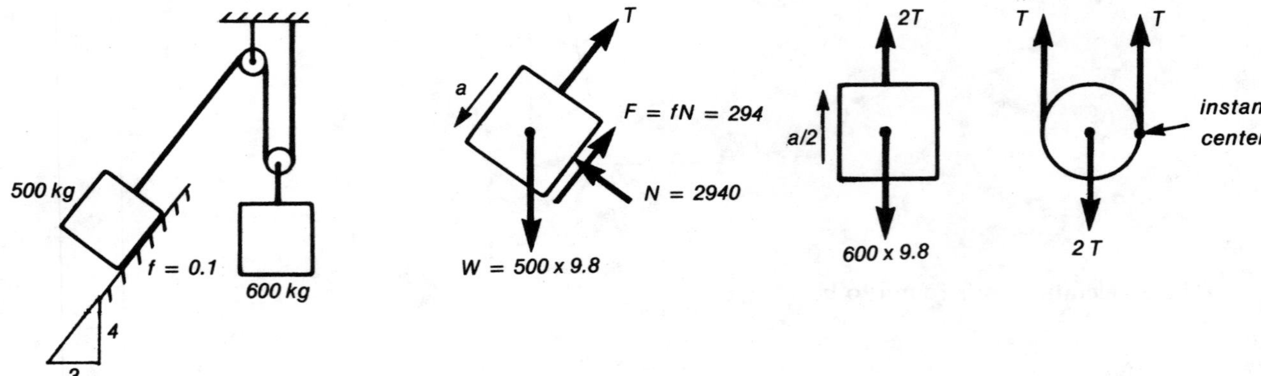

Solution. Applying Newton's 2nd law to the 500 kg mass gives

$$\Sigma F = ma,$$

$$0.8 \times 500 \times 9.8 - 294 - T = 500\,a.$$

By studying the lower pulley we observe the 600 kg mass to be accelerating at $a/2$. Hence, we have

$$\Sigma F = ma,$$

$$2\,T - 600 \times 9.8 = 600 \times a/2.$$

Solving the above equations simultaneously results in

$$a = 1.055 \text{ m/s}^2, \qquad T = 3100 \text{ N}.$$

The distance the 600 kg mass moves is

$$s = \frac{1}{2}at^2$$

$$= \frac{1}{2} \times 1.055 \times 3^2 = 4.75 \text{ m}.$$

EXAMPLE 9.12

Find the tension in the string if at the position shown $v = 4$ m/s. Also, calculate the angular acceleration.

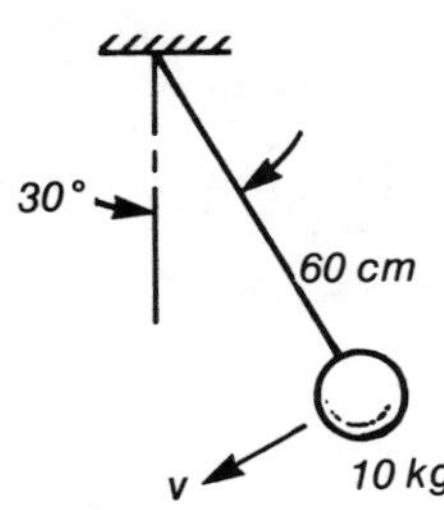

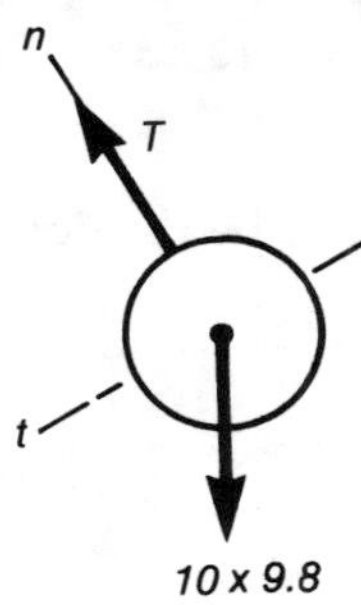

Solution. Sum forces in the normal direction and obtain

$$\Sigma F_n = ma_n = mv^2/r$$

$$T - 10 \times 9.8 \cos 30° = 10 \times 4^2/0.6.$$

$$\therefore T = 352 \text{ N}.$$

Sum forces in the tangential direction and find

$$\Sigma F_t = ma_t = mr\alpha$$

$$10 \times 9.8 \sin 30° = 10 \times 0.6\ \alpha$$

$$\therefore \alpha = 8.17 \text{ rad/s}^2.$$

EXAMPLE 9.13

What is the angular acceleration of the 60 kg cylindrical pulley and the tension in the rope?

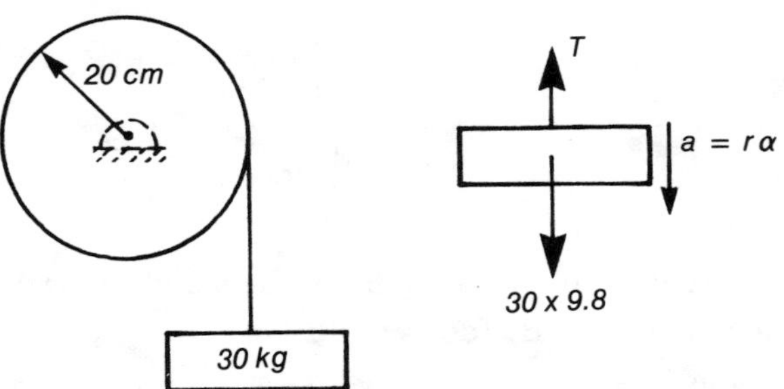

Solution. Summing forces on the 30 kg mass gives

$$\Sigma F = ma$$

$$-T + 30 \times 9.8 = 30 \times 0.2\ \alpha.$$

Summing moments about the center of the pulley yields

$$\Sigma M = I\alpha$$

$$T \times 0.2 = \frac{1}{2} \times 60 \times 0.2^2\ \alpha$$

where $I = mr_2/2$ for a cylinder (see Table 8.2). A simultaneous solution results in

$$\alpha = 24.5 \text{ rad/s}^2, \qquad T = 147 \text{ N}.$$

9.3 Work and Energy

Work is defined to be the dot product between a force and the distance it moves, that is,

$$W = \int \vec{F} \cdot \vec{d}s, \tag{9.3.1}$$

or, if the force is constant,

$$W = \vec{F} \cdot \Delta\vec{s}. \tag{9.3.2}$$

For a rotating body the work is

$$W = M\,\Delta\theta. \tag{9.3.3}$$

The *work-energy equation*, which results from integrating Newton's 2nd law, states that the net work done on a body (or several connected bodies) equals the change in energy of the body (or the several bodies). This is expressed as

$$W_{net} = \Delta E \tag{9.3.4}$$

where E represents the kinetic energy, given by

$$E = \frac{1}{2}mv^2 \tag{9.3.5}$$

for a translating body, and

$$E = \frac{1}{2}I\omega^2 \tag{9.3.6}$$

for a rotating body. For a translating and rotating body use v and I referred to the mass center. Potential energy can be realized by allowing the body forces to do work; or they can be incorporated in the ΔE-term by using the potential energy as

$$E_p = mgh \tag{9.3.7}$$

where h is the distance above a selected datum.

By applying Eq. 9.3.1 to a spring, the work necessary to compress that spring a distance x is

$$W = \frac{1}{2}Kx^2. \tag{9.3.8}$$

This quantity $Kx^2/2$ can be considered the potential energy stored in the spring.

EXAMPLE 9.14

Neglecting friction find v of the slider when it hits B if it starts from rest at A.

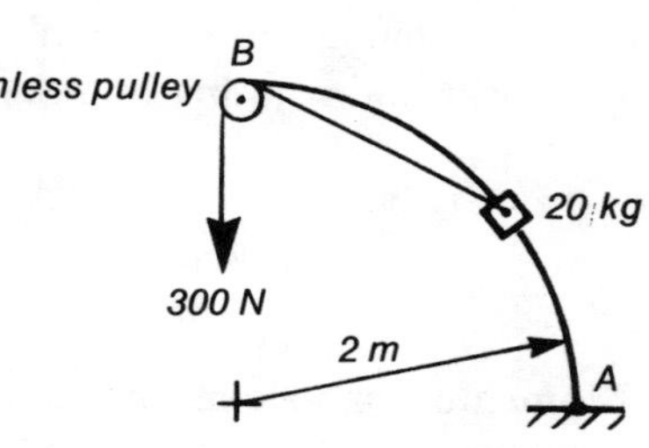

Solution. The distance the force F moves is

$$s = \frac{2}{\sin 45°} = 2.828 \text{ m}.$$

The work-energy equation is written as

$$W_{net} = \frac{1}{2} mv^2$$

$$300 \times 2.828 - (20 \times 9.8) \times 2 = \frac{1}{2} \times 20 \, v^2.$$

$$\therefore v = 6.76 \text{ m/s}.$$

Note that the body force does negative work since the motion is up and the body force acts down.

EXAMPLE 9.15

Neglect friction and estimate the angular velocity of the cylinder after the mass falls 2 m from rest.

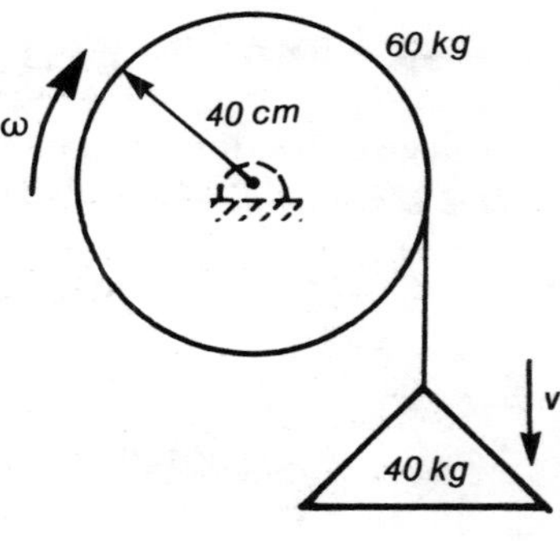

Solution. The work-energy equation provides

$$W_{net} = \frac{1}{2} mv^2 + \frac{1}{2} I \omega^2$$

$$(40 \times 9.8) \times 2 = \frac{1}{2} \times 40 \times (0.4 \, \omega)^2 + \frac{1}{2} \left(\frac{1}{2} \times 60 \times 0.4^2\right) \omega^2.$$

$$\therefore \omega = 11.83 \text{ rad/s}$$

where $I = \frac{1}{2} m r^2$.

EXAMPLE 9.16

What is the velocity of the 40 kg mass after it falls 20 cm from rest. The spring is initially stretched 10 cm.

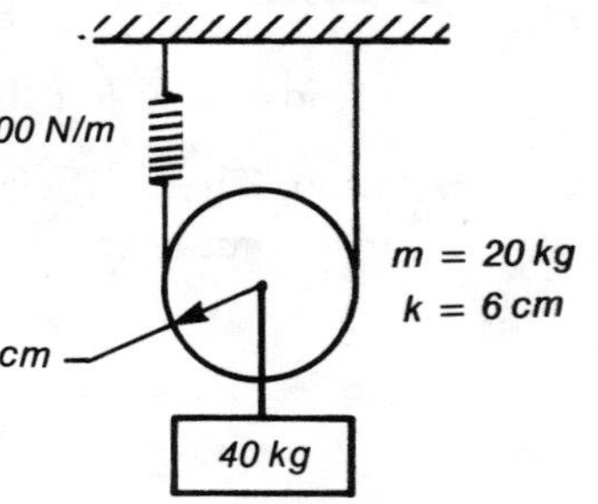

Solution. The spring will stretch an additional 40 cm. Thus, the work-energy equation results in

$$W_{net} = \frac{1}{2} mv^2 + \frac{1}{2} I\omega^2,$$

$$(20 + 40) \times 9.8 \times 0.2 - \frac{1}{2} 800 \, (0.5^2 - 0.1^2) =$$

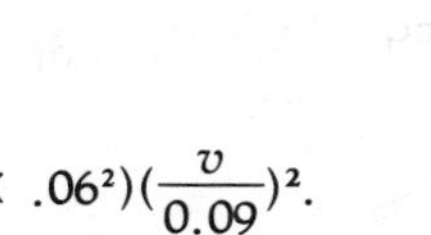

$$\frac{1}{2}(20 + 40) \, v^2 + \frac{1}{2} (20 \times .06^2)\left(\frac{v}{0.09}\right)^2.$$

$$\therefore v = 0.792 \text{ m/s}.$$

Note: The spring stretches an additional 40 cm.

9.4 Impulse and Momentum

The impulse-momentum equations also result from integrating Newton's 2nd law. Impulse is defined for linear and rotating bodies, respectively, as

$$i_\ell = \int F dt$$
$$i_r = \int M dt. \tag{9.4.1}$$

Momentum is velocity multiplied by mass. The impulse-momentum equations, for a constant force and moment, take the form

$$F\Delta t = m\Delta v$$
$$M\Delta t = I\Delta\omega. \tag{9.4.2}$$

Objects impacting each other, with no external forces acting, experience a conservation of momentum. The *coefficient of restitution* e is used in such problems. It is defined as

$$e = \frac{\text{relative separation velocity}}{\text{relative approach velocity}} \tag{9.4.3}$$

If $e = 1$, the collision is *elastic* with no energy loss; if $e = 0$, the collision is *plastic* with maximum energy loss.

EXAMPLE 9.17

Neglect friction and estimate the angular velocity of the cylinder after 2 seconds if motion starts from rest.

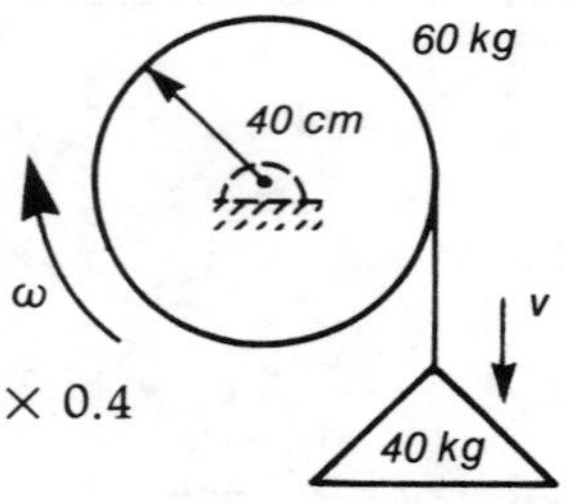

Solution. The impulse-momentum equation is used as follows:

$$M\Delta t = I\Delta\omega + m\Delta v \times r$$

$$0.4 \times (40 \times 9.8) \times 2 = \frac{1}{2} \times 60 \times 0.4^2\, \omega + 40 \times 0.4\, \omega \times 0.4$$

$$\therefore \omega = 28.0 \text{ rad/s}.$$

60 kg
40 cm
ω
V
40 kg

EXAMPLE 9.18

Find v' and θ if the coefficient of restitution is 0.8.

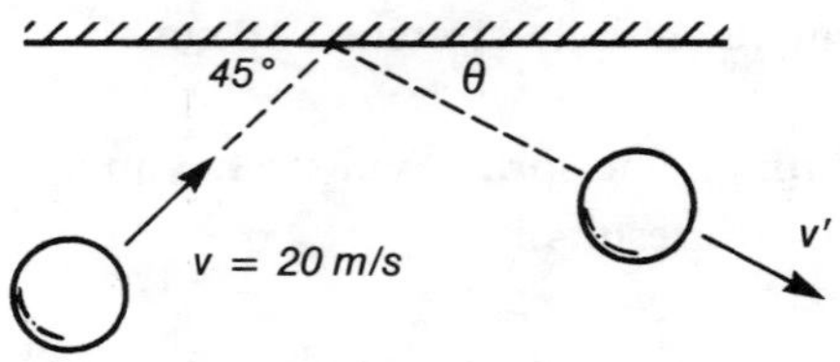

Solution. The coefficient of restitution is based on the normal components of velocity. Thus,

$$e = \frac{v' \sin\theta}{v \sin 45°}.$$

45°
θ
V = 20 m/s
V'

$$\therefore v' \sin\theta = 0.8 \times 20 \times 0.707 = 11.31.$$

The tangential velocity component remains unchanged so that

$$v' \cos\theta = 20 \cos 45° = 14.14.$$

Simultaneous solution of the above results in

$$v' = 18.11 \text{ m/s}, \qquad \theta = 38.65°.$$

Practice Problems

(If only a few selected problems are desired, choose those with a star.)

RECTILINEAR MOTION

*9.1 An object is moving with an initial velocity of 20 m/s. If it is decelerating at 5 m/s², how far, in meters, does it travel before it stops?

a) 10 b) 20 c) 30 d) 40 e) 50

9.2 If the particle starts from rest, what is its velocity, in m/s, at $t = 4$ s?

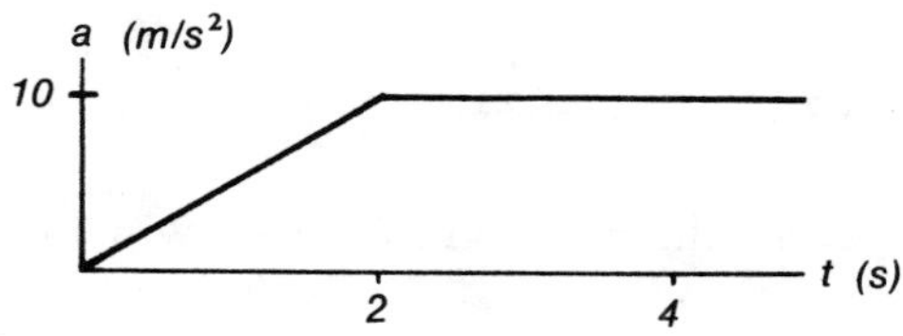

a) 10 b) 20 c) 30

d) 40 e) 50

*9.3 A projectile is shot straight up with $v_o = 40$ m/s. After how many seconds will it return if drag is neglected?

a) 4 b) 6 c) 8 d) 10 e) 12

9.4 An automobile is travelling at 25 m/s. It takes 0.3 s to apply the brakes after which the deceleration is 6 m/s². How far, in meters, does the automobile travel before it stops?

a) 40 b) 45 c) 50 d) 55 e) 60

ANGULAR MOTION

9.5 A wheel accelerates from rest with $\alpha = 6$ rad/s². How many revolutions are experienced in 4 s?

a) 7.64 b) 9.82 c) 12.36 d) 25.6 e) 38.4

*9.6 A 2-m-long shaft rotates about one end at 20 rad/s. It begins to accelerate with $\alpha = 10$ rad/s². After how long, in seconds, will the velocity of the free end reach 100 m/s?

a) 7 b) 6 c) 5 d) 4 e) 3

CURVILINEAR MOTION

*9.7 A roller-coaster reaches a velocity of 20 m/s at a location where the radius of curvature is 40 m. Calculate the acceleration, in m/s².

a) 8 b) 9 c) 10 d) 11 e) 12

9.8 A bucket full of water is to be rotated in the vertical plane. What minimum angular velocity, in rad/s, is necessary to keep the water inside it if the rotating arm is 120 cm?

a) 5.31 b) 4.26 c) 3.86 d) 3.15 e) 2.86

PROJECTILE MOTION

*9.9 Find the maximum height, in meters.

y x V_o = 100 m/s 45° H 10 m (L, – 10)

a) 295 b) 275 c) 255

d) 235 e) 215

9.10 Calculate the time to reach the low point, for the projectile of Prob. 9.9.

a) 14.6 b) 12.2 c) 11.0 d) 10.2 e) 8.31

9.11 What is the distance L, in meters, in Prob. 9.9?

a) 530 b) 730 c) 930 d) 1030 e) 1330

RIGID BODY MOTION

*9.12 The acceleration of the center O is given by

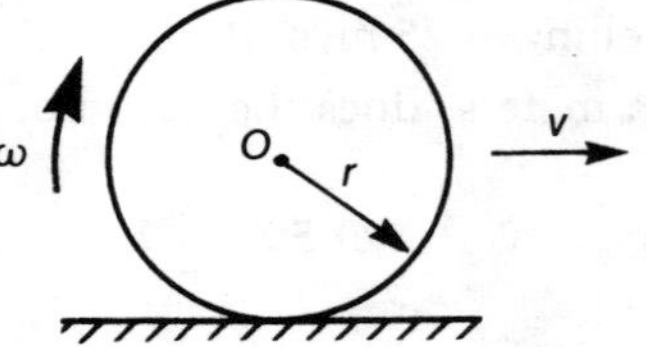

a) $r\omega^2$ b) v^2/r c) 0 d) ω^2/r e) rv^2

9.13 The spool rolls without slipping. Find v_o in m/s.

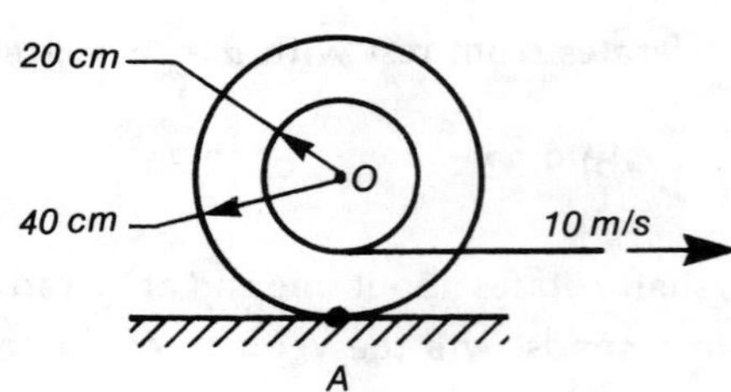

a) 30 b) 25 c) 20

d) 10 e) 5

*9.14 The acceleration a_A in m/s² in Prob. 9.13 is

a) 1000 b) 800 c) 600 d) 400 e) 200

*9.15 If the acceleration of B is $60\hat{i} - 20\hat{j}$ m/s², find $\vec{a}_A$, in m/s².

y α = 100 rad/s² 40 cm B A x ω = 20 rad/s

a) $220\hat{i} + 60\hat{j}$ b) $100\hat{i} + 20j$ c) $-100\hat{i} + 20\hat{j}$

d) $-100\hat{i} - 20j$ e) $200\hat{i} + 20j$

*9.16 Find ω_{AB}, in rad/s.

a) 56.6 ccw b) 56.6 cw

c) 34.1 ccw d) 28.3 ccw

e) 28.3 cw

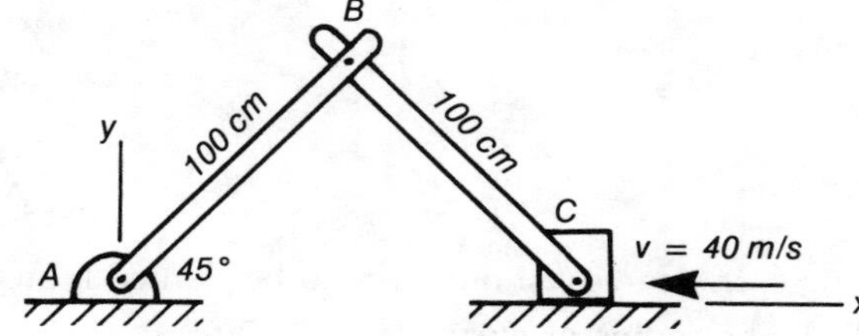

*9.17 What is ω_{BC}, in rad/s, for the linkage of Prob. 9.16?

a) 56.6 ccw b) 56.6 cw c) 34.1 ccw d) 28.3 ccw e) 28.3 cw

9.18 Determine α_{AB}, in rad/s², if $a_c = 0$ in Prob. 9.16.

a) 800 cw b) 800 ccw c) 1160 cw d) 3200 cw e) 3200 ccw

9.19 What is $\vec{a}_B$, in m/s², if $a_c = 0$, in Prob. 9.16?

a) 1130 $\hat{i}$ b) $-1130\ \hat{i}$ c) 1130 $\hat{j}$ d) $-1130\ \hat{j}$ e) 1130 $\hat{k}$

9.20 Find ω_{AB}, in rad/s.

a) 20 ccw b) 20 cw c) 10 ccw

d) 10 cw e) 0

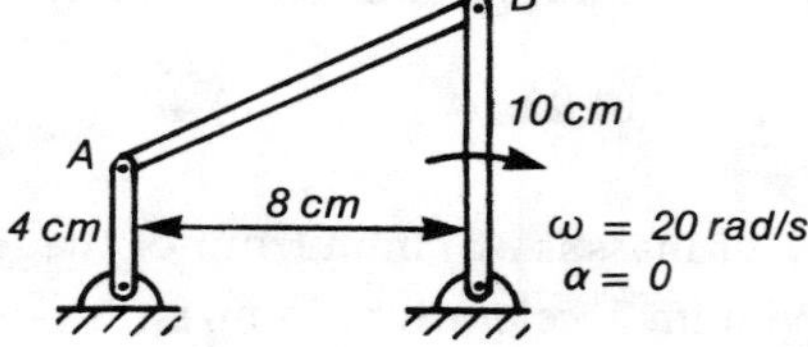

9.21 Find α_{AB}, in rad/s², for the linkage of Prob. 9.20.

a) 750 cw b) 750 ccw c) 400 cw d) 1000 cw e) 1000 ccw

*9.22 What is the acceleration in m/s², of the bead if it is 10 cm from the center?

a) $20\ \hat{i} + 40\ \hat{j}$ b) $-40\ \hat{i} - 20\ \hat{j}$

c) $-40\ \hat{i} + 40\ \hat{j}$ d) $20\ \hat{i} - 40\ \hat{j}$

e) $40\ \hat{i} + 20\ \hat{j}$

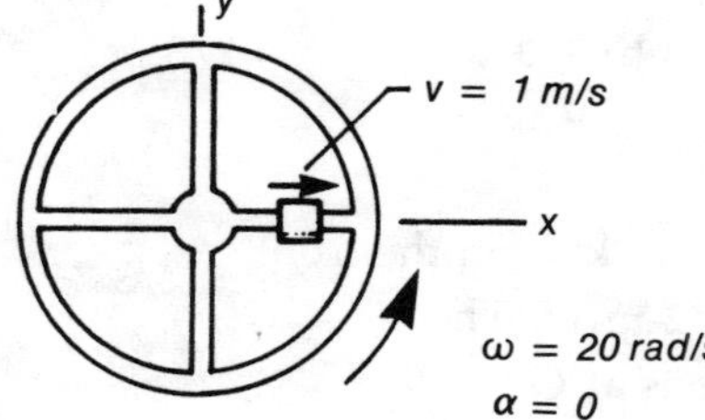

KINETICS

*9.23 What is a_A, in m/s²?

a) 2.09 b) 1.85

c) 1.63 d) 1.47

e) 1.22

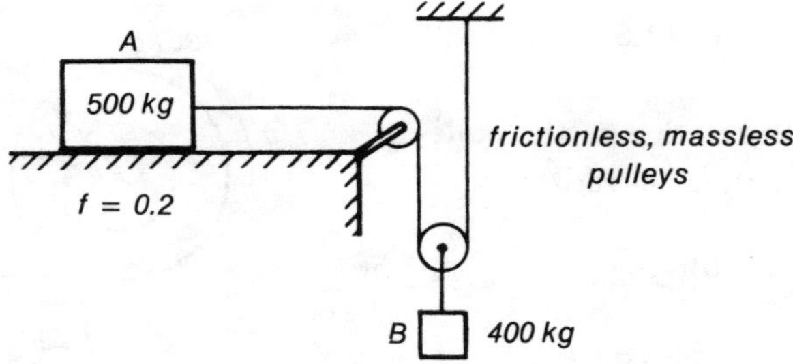

*9.24 How far, in meters, will the weight move in 10 s if released from rest?

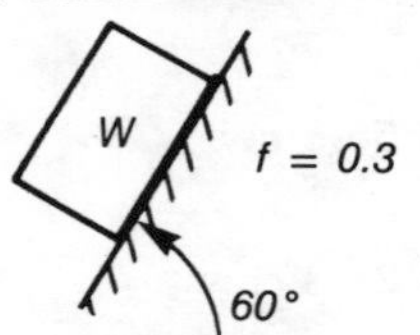

a) 350 b) 300 c) 250

d) 200 e) 150

*9.25 At what angle, in degrees, should a road be slanted if an automobile travelling at 25 m/s, does not tend to slip? The radius of curvature is 200 m.

a) 22 b) 20 c) 18 d) 16 e) 14

9.26 A satellite orbits the earth 200 km above the surface. What speed, in m/s, is necessary for a circular orbit? The radius of the earth is 6400 km and $g = 9.2$ m/s².

a) 7800 b) 7200 c) 6600 d) 6000 e) 5400

9.27 Determine the mass, in kg, of the earth if its radius is 6400 km.

a) 6×10^{22} b) 6×10^{23} c) 6×10^{24} d) 6×10^{25} e) 6×10^{26}

9.28 The coefficient of sliding friction between rubber and asphalt is about 0.6. What minimum distance, in meters, can an automobile slide on a horizontal surface if it is travelling at 25 m/s?

a) 38 b) 43 c) 48 d) 53 e) 58

*9.29 The center of mass is 30 cm in front of the rear wheel of a motorcycle and 80 cm above the roadway. What maximum acceleration, in m/s², is possible?

a) 4.5 b) 4.3 c) 4.1 d) 3.9 e) 3.7

9.30 Find the ratio of the tension in the wire before and immediately after the string is cut.

30° wire string

a) $2/\sqrt{3}$ b) 4/3 c) 2/3
d) 3/4 e) $\sqrt{3}/2$

*9.31 Find the force, in kN, on the front wheels if $a = 2$ m/s². The center of mass is at G.

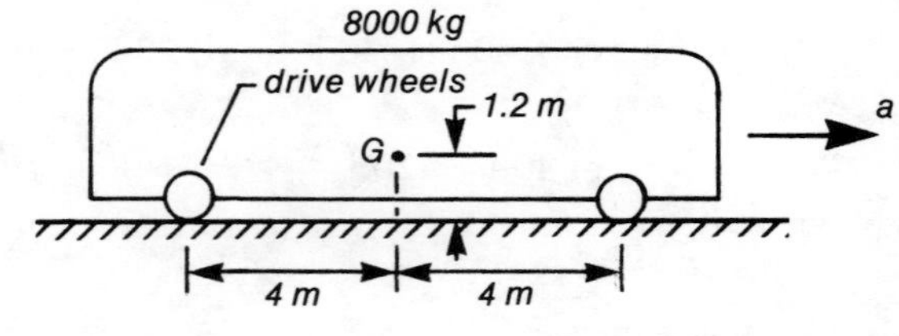

a) 58.2 b) 47.3 c) 41.6

d) 36.8 e) 22.8

*9.32 The radius of gyration of the pulley is 10 cm. Calculate α, in rad/s².

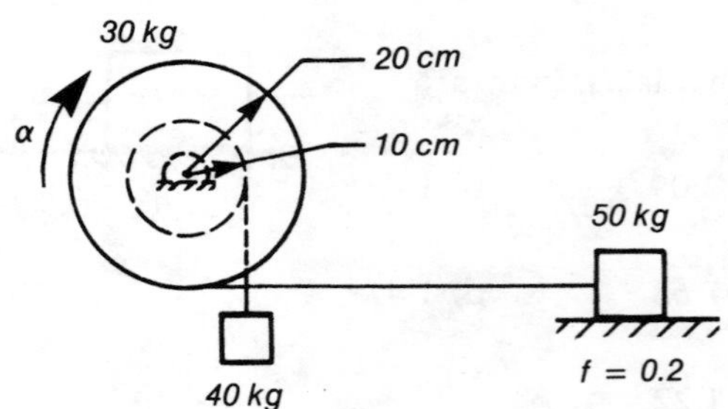

a) 8.52 b) 7.26 c) 6.58

d) 5.32 e) 4.69

9.33 What is the force at 0 immediately after the string is cut?

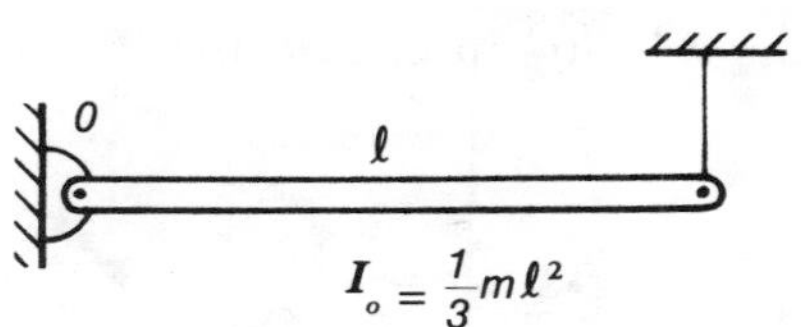

a) $\frac{mg}{2}$ b) $\frac{mg}{3}$ c) $\frac{mg}{4}$

d) $\frac{mg}{5}$ e) 0

WORK ENERGY

9.34 Find the velocity, in m/s, after the mass moves 10 m if it starts from rest.

a) 3 b) 4 c) 5 d) 6 e) 7

*9.35 If the force acts through 4 m, what is the angular velocity, in rad/s, of the solid cylinder? Assume no slip and the cylinder starts from rest.

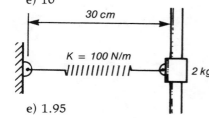

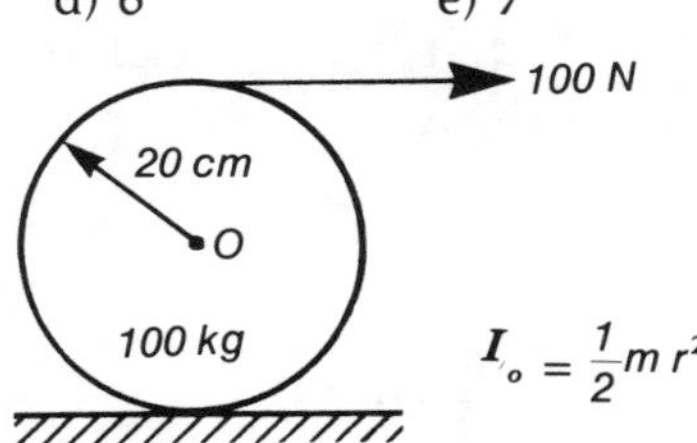

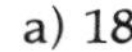

a) 18 b) 16 c) 14 d) 12 e) 10

*9.36 The spring is initially free. Calculate the velocity, in m/s, of the 2 kg mass after it falls 40 cm. It starts from rest.

a) 4.62 b) 3.84 c) 2.96 d) 2.42 e) 1.95

9.37 Find the velocity, in m/s, of the end of the 10 kg bar as it passes *A*. The free spring length is 30 cm. The moment of inertia of a bar about its mass center is $m\ell^2/12$.

a) 5.2 b) 4.6 c) 3.5

d) 2.4 e) 1.2

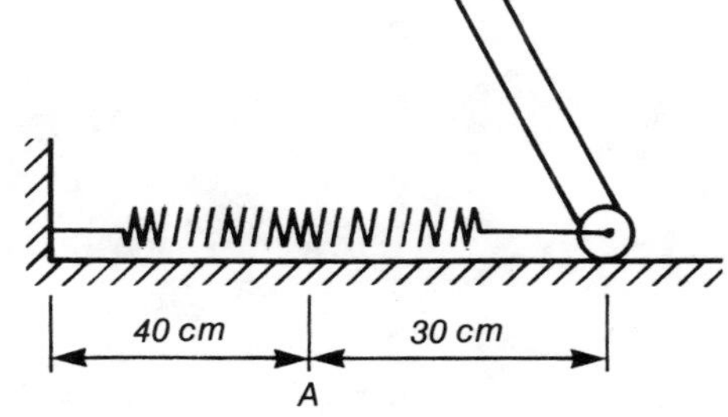

IMPULSE-MOMENTUM

*9.38 If the force in Prob. 9.35 acts for 4 seconds, find the angular velocity, in rad/s, assuming no slip. The cylinder starts from rest.

a) 2.4 b) 5.2 c) 8.6 d) 10.2 e) 26.7

*9.39 Find the velocity, in m/s, of the 100 kg mass at $t = 2$ s.

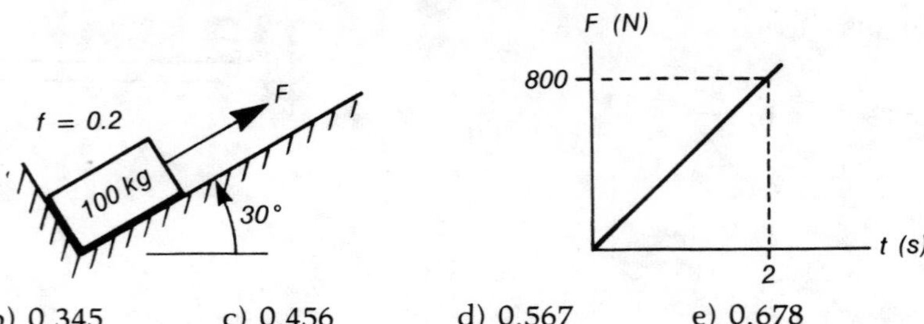

a) .245 b) 0.345 c) 0.456 d) 0.567 e) 0.678

*9.40 If the coefficient of restitution is 0.8 find v_B', in m/s.

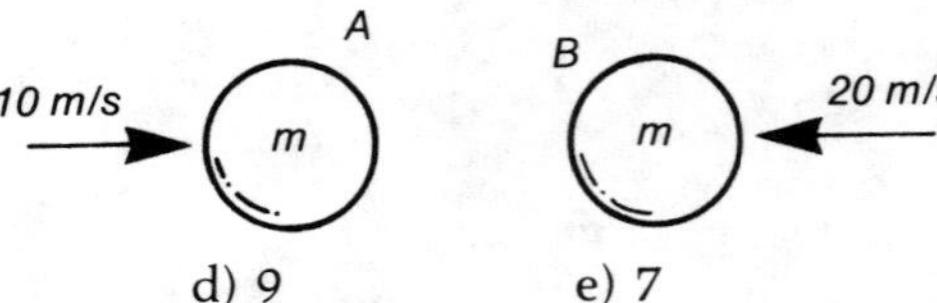

a) 16 b) 13 c) 11 d) 9 e) 7

*9.41 Calculate the energy lost, in joules, in the collision of Prob. 9.40 if $m = 2$ kg.

a) 200 b) 180 c) 160 d) 140 e) 120

Dynamics

Rectilinear motion

$$a = \frac{dv}{dt} \qquad v = \frac{ds}{dt} \qquad a = v\,\frac{dv}{ds} = \frac{d^2s}{dt^2}$$

$$v = v_o + at$$
$$s = v_o t + at^2/2$$
$$v^2 = v_o^2 = 2as$$

Rotational motion

$$\alpha = \frac{d\omega}{dt} \qquad \omega = \frac{d\theta}{dt} \qquad \alpha = \omega\,\frac{d\omega}{d\theta} = \frac{d^2\theta}{dt^2}$$

$$\omega = \omega_o + \alpha t$$
$$\theta = \omega_o t + \alpha t^2/2$$
$$\omega^2 = \omega_o^2 + 2\alpha\theta$$

Projectile motion

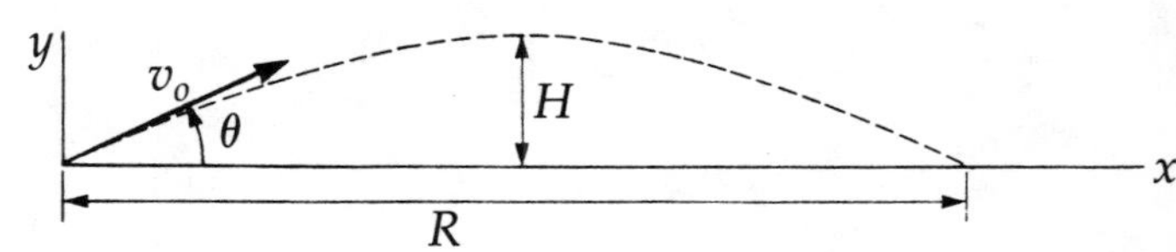

$$H = \frac{v_o^2}{2g}\sin^2\theta \qquad R = \frac{v_o^2}{g}\sin 2\theta$$

$$R_{max} = \frac{v_o^2}{g}, \qquad \theta = 45°$$

General equations

$$a_n = r\omega^2 = \frac{v^2}{r} \qquad a_t = \frac{dv}{dt} \qquad v = r\omega$$

$$\vec{v}_A = \vec{v}_B = \vec{v}_{A/B} \qquad \vec{a}_{coriolis} = 2\,\vec{\omega} x \vec{v}$$

$$\Sigma\vec{F} = m\vec{a} \qquad W = mg \qquad F = k\,\frac{m_1 m_2}{r^2} \qquad k = 6.67\times 10^{-11}\ \text{N·m}^2/\text{kg}^2$$

$\Sigma\vec{M} = I\vec{\alpha}$ *where* I = *mass moment of inertia (see Chapter 8)*
or $I = mk^2$ *and* k = *radius of gyration*

Work-Energy

$$Work = \vec{F}\cdot\Delta\vec{s} = \vec{M}\cdot\Delta\vec{\theta} \qquad Work_{net} = \Delta E$$

$E = \frac{1}{2}mv^2$ *(translation)*
$= \frac{1}{2}I\omega^2$ *(rotation)*
$= mgh$ *(potential)*
$= \frac{1}{2}Kx^2$ *(spring)*

Impulse-Momentum

$$F\Delta t = m\Delta v$$
$$M\Delta t = I\Delta\omega$$

$$e = \frac{\Delta v_{approach}}{\Delta v_{separation}} = \begin{cases} 1 & \textit{elastic} \\ 0 & \textit{plastic} \end{cases}$$

(This page is intentionally blank.)

10. The Mechanics of Materials

The mechanics of materials is concerned with the internal stresses and deformations that occur in a body due to externally applied forces. To simplify the analysis we assume:

— the material is homogeneous and isotropic (material properties are independent of position and angular orientation).

— the weights of structural members are negligible.

— the temperature is uniform throughout.

— the material possesses a linear stress-strain relation (the material is Hookean).

Since fracture is a distinct possibility in solids subjected to loads, we must be able to predict the maximum stresses that exist in a structural member when it is being stressed. Ductile materials fail most easily due to shearing stresses, while brittle materials fail most easily under tensile stresses. So we must be able to determine both maximum tensile and shearing stresses.

Beams (slender members subjected to transverse loads), columns (slender members subjected to compressive loads), and shafts in torsion are of primary interest. But composite sections will also be considered.

10.1. Stress and Strain

Stress results from a force F acting on an area A and is given by

$$\sigma = \frac{F}{A} \,. \tag{10.1.1}$$

We use σ to denote a normal stress and τ to denote a shearing stress. If an element is subjected to both normal and shearing stresses, as in Fig. 10.1, the maximum shearing stress is given by

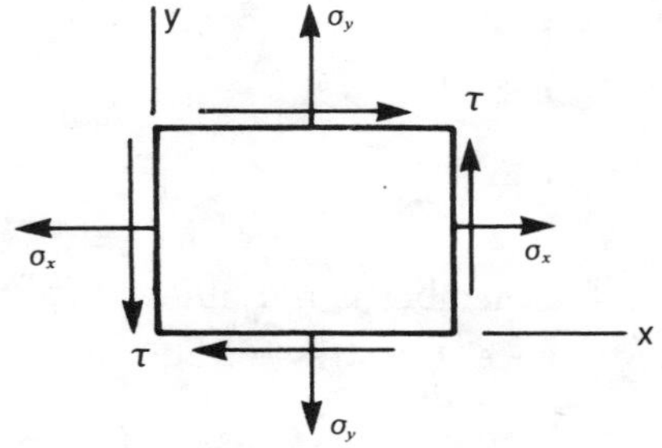

Note: A positive stress acts in the negative direction on a negative face (a face whose outward normal vector points in a negative direction).

Figure 10.1. Positive stresses acting on a material element.

$$\tau_{max} = \frac{1}{2}\sqrt{(\sigma_x - \sigma_y)^2 + 4\,\tau^2}\ . \tag{10.1.2}$$

It acts at an angle to the x-axis given by

$$\theta = \frac{1}{2}\text{arc tan}\,\frac{\sigma_x - \sigma_y}{2\tau} \tag{10.1.3}$$

where counterclockwise is positive. The maximum and minimum normal stresses, called principal stresses, are given by

$$\sigma_{\substack{max\\min}} = \frac{1}{2}(\sigma_x + \sigma_y) \pm \tau_{max}. \tag{10.1.4}$$

They act at the angles to the x-axis given by

$$\theta = \frac{1}{2}\text{arc tan}\,\frac{-2\tau}{\sigma_x - \sigma_y}\ . \tag{10.1.5}$$

Note: The principal stresses act on mutually perpendicular planes.

Stress is related to strain by *Hooke's Law* which, for a member loaded axially, is

$$\sigma = E\varepsilon \tag{10.1.6}$$

where E is the modulus of elasticity and ε is the normal strain defined by

$$\varepsilon = \frac{\Delta L}{L} \tag{10.1.7}$$

where ΔL is the change in original length L. For a member undergoing shear deformation, Hooke's law takes the form

$$\tau = G\,\gamma \tag{10.1.8}$$

where G is the shear modulus and γ is the shear strain. As a member is elongated, it will decrease in its lateral dimension according to *Poisson's ratio* μ, which is the ratio of the lateral strain to the axial strain.

Strain may also result from a temperature change ΔT. If a member is unrestrained, it will expand according to

$$\Delta L = \alpha L\,\Delta T \tag{10.1.9}$$

where α is the coefficient of thermal expansion. If the member is not allowed to expand, the ΔL of Eq. 10.1.9 will give rise to the strain of Eq. 10.1.7 and, hence, the stress of Eq. 10.1.6.

Material properties depend on several factors; average values for these factors are listed in Table 10.1.

TABLE 10.1. Average Material Properties.

	Modulus of Elasticity E		Shear Modulus G		Poisson's Ratio μ	Density ρ		Coefficient of Thermal Expansion α	
	$\times 10^6$ kPa	$\times 10^6$ psi	$\times 10^6$ kPa	$\times 10^6$ psi		kg/m³	lb/ft³	$\times 10^{-6}$ °C⁻¹	$\times 10^{-6}$ °F⁻¹
steel	210	30	83	12	.30	7850	490	11.7	6.5
aluminum	70	10	27	3.9	.33	2770	173	23.0	12.8
magnesium	45	6.5	17	2.4	.35	1790	112	26.1	14.5
cast iron	140	20	55	8	.27	7080	442	10.1	5.6
titanium	106	15.4	40	6	.34	4520	282	8.8	4.9
brass	100	15	40	6	.33	8410	525	21.2	11.8
concrete	20	3	—	—	—	2400	150	11.2	6.2

EXAMPLE 10.1

Find the maxium normal and shearing stresses at the point shown.

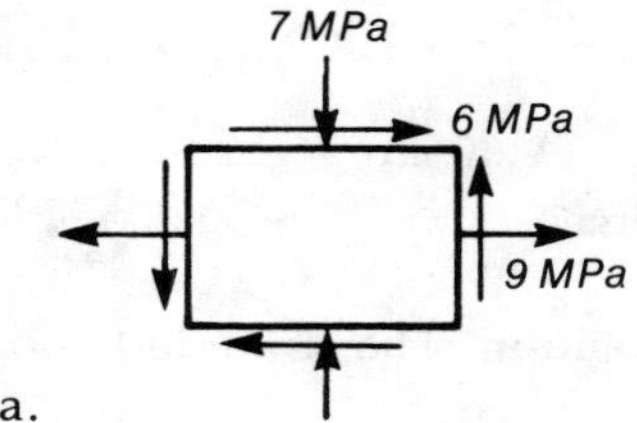

Solution. The maximum shearing stress, using Eq. 10.1.2, is

$$\tau_{max} = \frac{1}{2}\sqrt{(\sigma_x - \sigma_y)^2 + 4\tau^2}$$

$$= \frac{1}{2}\sqrt{[9 - (-7)]^2 + 4 \times 6^2} = 10 \text{ MPa}.$$

The maximum normal stress is

$$\sigma_{max} = \frac{1}{2}(\sigma_x + \sigma_y) + \tau_{max}$$

$$= \frac{1}{2}(9 - 7) + 10 = 11 \text{ MPa}.$$

EXAMPLE 10.2

How far will point A move to the right?

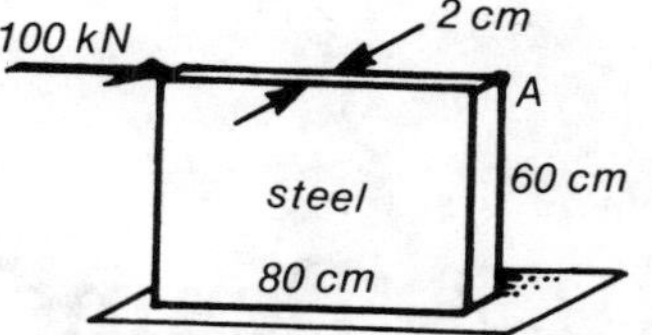

Solution. The distance point A will move to the right depends on the shear modulus G. The force is producing shearing stresses in the member resulting in a shearing strain γ. To use Eq. 10.1.8, we must know the shearing stress; it is

$$\tau = \frac{F}{A} = \frac{100}{0.8 \times 0.02} = 6250 \text{ kPa}.$$

The shearing strain is given by

$$\gamma = \frac{\tau}{G} = \frac{6250}{83 \times 10^6} = 7.53 \times 10^{-5}.$$

Hence, the deflection is

$$\Delta L = \gamma L = 7.53 \times 10^{-5} \times 600 = 0.0452 \text{ mm}.$$

EXAMPLE 10.3

A 5-cm-dia, 80-cm-long steel bar is restrained from moving. If its temperature is increased 100°C what compressive stress is induced?

Solution. The strain can be found from Eq. 10.1.9 to be

$$\varepsilon = \frac{\Delta L}{L} = \alpha \, \Delta T$$

$$= 11.7 \times 10^{-6} \times 100 = 11.7 \times 10^{-4}.$$

Hence, the induced stress is

$$\sigma = \varepsilon E$$

$$= 11.7 \times 10^{-4} \times 210 \times 10^{6} = 246\ 000 \text{ kPa}$$

EXAMPLE 10.4

A 60 cm × 120 cm × 5 mm rectangular sheet of aluminum is loaded in the direction of its long dimension with a 80 000 N load. How much does it shrink in the direction of its 60-cm dimension?

Solution. The strain in the direction of the load is

$$\varepsilon = \frac{\sigma}{E} = \frac{F}{AE}$$

$$= \frac{80\ 000}{1.2 \times 0.005 \times 70 \times 10^{9}} = 1.90 \times 10^{-4}.$$

The lateral deformation is then

$$\Delta L_{lateral} = \varepsilon \mu \times L$$

$$= 1.90 \times 10^{-4} \times 0.33 \times 600 = 0.038 \text{ mm}.$$

Note: The units on E must be Pa (i.e., N/m^2) in the above expression so that strain is dimensionless; hence, $E = 69 \times 10^{9}$ Pa.

10.2. Beams

The primary quantities of interest in the study of beams are the maximum bending stresses (both tensile and compressive), the maximum shearing stress, and the maximum deflection. These quantities depend on the geometry of the beam, the loads, and the maximum shear force and moment. The expression for the maximum bending stress, which is based on the assumption that plane sections remain planes, is

$$\sigma = \frac{Mc}{I} \tag{10.2.1}$$

where M is the moment acting on the section,
I is the second moment about the neutral axis,
c is the distance from the neutral axis.

The neutral axis passes through the centroid of the cross-sectional area as shown in Fig. 10.2. An example will illustrate the calculations.

The shearing stress, sketched in Fig. 10.3, is given by

$$\tau = \frac{VQ}{I\,b} \tag{10.2.2}$$

where V is the vertical shear force acting on the section,
Q is the first moment of the area outside the point of interest,
b is the width at the point of interest.

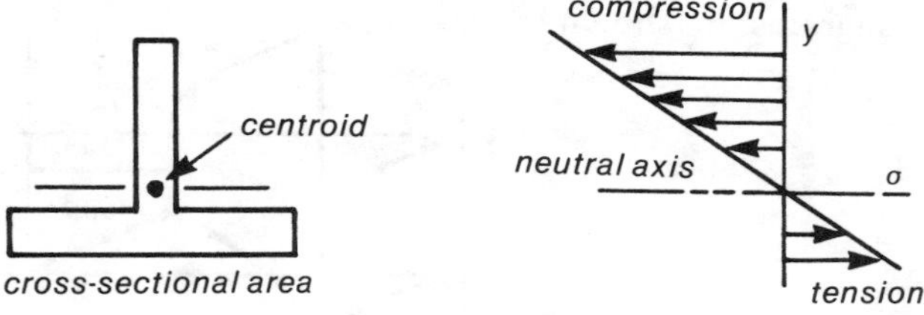

Figure 10.2. Bending stress distribution in a beam.

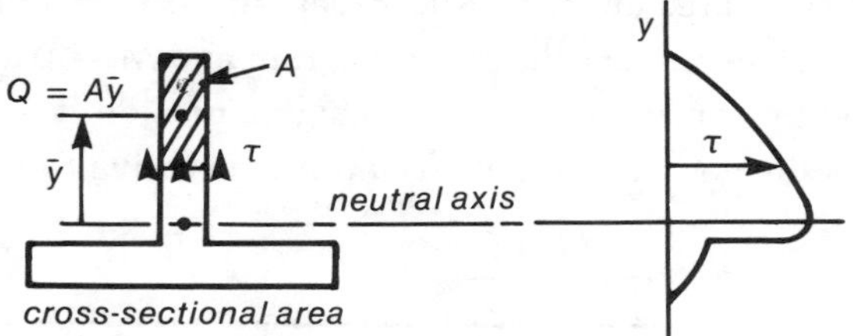

Figure 10.3. Shearing stress distribution in a beam.

Before bending and shearing stresses can be found, we must find the bending moment M and the vertical shear force V. These are usually presented as a shear diagram and a moment diagram which are related by the expression

$$V = \frac{dM}{dx}\,. \tag{10.2.3}$$

An example, to be presented shortly, will illustrate.

To calculate the deflection of a beam, we must solve the differential equation

$$EIy'' = M. \tag{10.2.4}$$

This is solved for a number of beams and the maximum deflection δ is included in Table 10.2.

TABLE 10.2. Beam Formulas

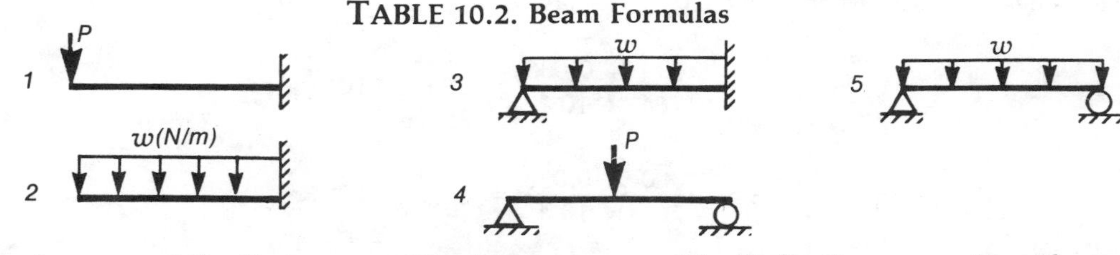

Number	Max Shear	Max Moment	Max Deflection	Max Slope
1	P	PL	$PL^3/3EI$	$PL^2/2EI$
2	wL	$wL^2/2$	$wL^4/8EI$	$wL^3/6EI$
3	$5wL/8$	$wL^2/8$	$wL^4/185EI$	
4	$P/2$	$PL/4$	$PL^3/48EI$	$PL^2/16EI$
5	$wL/2$	$wL^2/8$	$5wL^4/384EI$	$wL^3/24EI$

EXAMPLE 10.5

Draw the shear and moment diagrams for the beam shown.

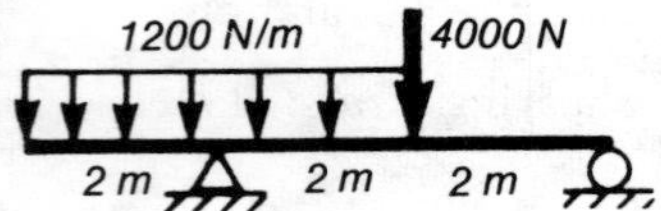

Solution. Summing moments about the right end gives

$$F_1 \times 4 = 4000 \times 2 + 1200 \times 4 \times 4.$$

$$\therefore F_1 = 6800 \text{ N} \qquad F_2 = 2000 \text{ N}$$

The shear and moment diagrams are shown.

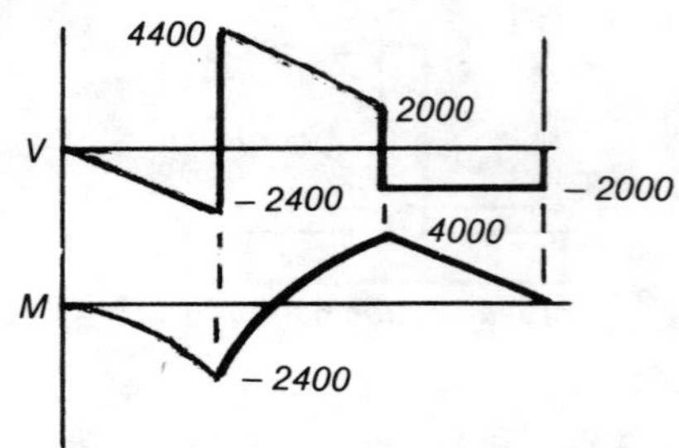

Notes: Change in the moment is the area under the shear diagram, $\Delta M = \int V dx$. Hence, the value of 4000 is obvious since that is the area under the shear diagram and $M = 0$ at the right end. The value of -2400 is quickly found by considering the overhanging portion. We know 4000 is the maximum because the shear (slope) is positive to the left and it is negative to the right.

EXAMPLE 10.6

If the beam of Example 10.5 has the cross-section area shown, find the maximum tensile, compressive, and shear stresses.

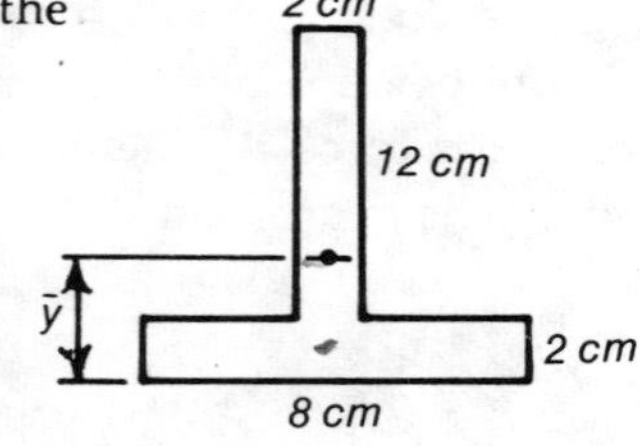

Solution. First, we must locate the centroid. The distance $\bar{y}$ is

$$\bar{y} = \frac{16 \times 1 + 24 \times 8}{16 + 24} = 5.2 \text{ cm}.$$

The second moment about the centroidal axis is then

$$I = \frac{8 \times 2^3}{3} + \frac{2 \times 12^3}{3} - 40 \times 3.2^2 = 764 \text{ cm}^4.$$

Using the values of V and M from Ex. 10.5, we have

$$\sigma_{tension} = \frac{Mc}{I} = \frac{2400 \times 0.088}{764 \times 10^{-8}} = 27.6 \times 10^6 \text{ Pa}$$

$$\sigma_{compression} = \frac{Mc}{I} = \frac{4000 \times 0.088}{764 \times 10^{-8}} = 46.1 \times 10^6 \text{ Pa}$$

$$\tau_{max} = \frac{VQ}{I\,b} = \frac{4400(0.02 \times .088 \times .044)}{764 \times 10^{-8} \times 0.02} = 2.23 \times 10^6 \text{ Pa}.$$

Note: At the left support, tension occurs in the top fibers, whereas, at the center it occurs in the bottom fibers. It is important to check both positions.

EXAMPLE 10.7

Find the maximum deflection of a 3 cm × 24 cm aluminum cantilever beam if a load of 800 N is located at its midpoint. It is 5 m long.

Solution. The 2.5 m beyond the load is simply a straight section at the slope given in Table 10.2. Hence, the deflection of the free end is, using a total length of $2L$,

$$\delta = \frac{PL^3}{3EI} + L \times \frac{PL^2}{2EI}$$

$$= \frac{5PL^3}{6EI} = \frac{5 \times 800 \times 2.5^3}{6 \times 70 \times 10^9 \times 0.24 \times 0.03^3/12} = 0.276 \text{ m}.$$

10.3 Torsion

The maximum shearing stress which exists at the outside surface of a circular shaft is given by the expression

$$\tau = \frac{Tc}{J} \tag{10.3.1}$$

where T is the torque experienced by the section,
c is the shaft radius,
J is the polar moment of inertia.

The angle of twist between two sections subjected to a torque T is found from

$$\theta = \frac{TL}{JG}. \tag{10.3.2}$$

EXAMPLE 10.8

Find the maximum shearing stress and the twist angle of a hollow 3-m-long, steel shaft if the outside diameter is 10 cm and the thickness is 5 mm. It is subjected to a torque of 1600 N·m.

Solution. The maximum shear stress is

$$\tau = \frac{Tc}{J}$$

$$= \frac{1600 \times .05}{\pi(.1^4 - .09^4)/32} = 23.7 \times 10^6 \text{ Pa}.$$

The twist angle is

$$\theta = \frac{TL}{JG}$$

$$= \frac{1600 \times 3}{[\pi(.1^4 - .09^4)/32]\ 210 \times 10^9} = 0.0068 \text{ rad or } 0.39°.$$

10.4 Columns

Long, slender members subjected to compressive loads are referred to as columns. If the slenderness ratio L/k (k is the least radius of gyration of the cross-sectional area A, $k = \sqrt{I/A}$) is greater than 80 to 120, the compressive member can usually be considered a long column. For a long column, the maximum load that can be applied without buckling, the *critical load* P_{cr}, is determined by *Euler's equation* which depends on the end conditions as follows:

$$\begin{aligned} \text{round or pinned ends:} \quad P_{cr} &= \pi^2 EI/L^2 \\ \text{fixed ends:} \quad P_{cr} &= 4\pi^2 EI/L^2 \\ \text{one free end, one fixed end:} \quad P_{cr} &= \pi^2 EI/4L^2 \\ \text{one pinned end, one fixed end:} \quad P_{cr} &= 9\pi^2 EI/4L^2 \end{aligned} \tag{10.4.1}$$

In general, the above equations can be used if the *Euler allowable stress* (P_{cr}/A) does not exceed one-half of the compressive yield stress. For intermediate length columns, other empirical formulas must be used.

EXAMPLE 10.9

A hollow 8-m-long steel member has inside and outside diameters of 8 cm and 10 cm, respectively. Calculate the maximum allowable compressive load using a safety factor of 2 and assuming round ends.

Solution. First, let's check the slenderness ratio. It is

$$\frac{L}{k} = \frac{L}{\sqrt{I/A}}$$

$$= \frac{8}{\sqrt{\dfrac{\pi(.1^4 - .08^4)/64}{\pi(.1^2 - .08^2)/4}}} = 250.$$

This is sufficiently large to use Euler's equation. It gives

$$P_{cr} = \frac{\pi^2 EI}{L^2}$$

$$= \frac{\pi^2 \times 210 \times 10^9 \times \pi(.1^4 - .08^4)/64}{8^2} = 93\,900 \text{ N}.$$

For a safety factor of 2, the maximum allowable force is

$$P_{max} = \frac{P_{cr}}{2} = 46\,900 \text{ N}.$$

10.5 Combined Stresses

The stress situations described in the preceding sections can be superimposed to form a combined stress condition. It is then necessary to determine the maximum shearing stress and maximum normal stresses due to the combined stresses. For example, the situation shown in Fig. 10.4 combines a normal stress σ_1 due to the axial force P; a shearing stress τ due to the torque PL on the circular shaft; and a tensile stress σ_2 due to a bending moment of $PL/2$, assuming the element shown is in the center of the circular beam. Let's work a specific example.

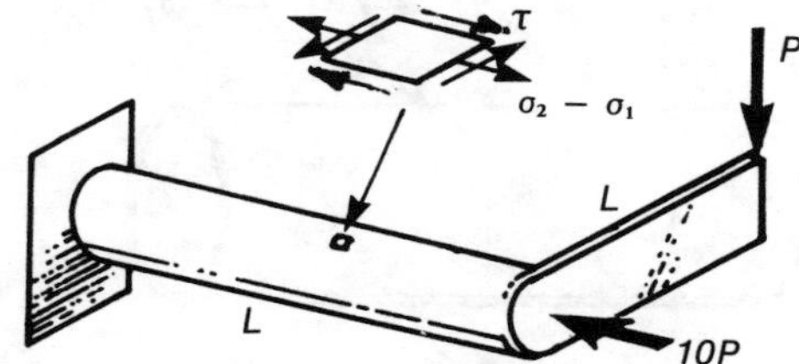

Figure 10.4. **A combined stress situation.**

EXAMPLE 10.10

For the situation of Fig. 10.4, let $L = 2$ m, $P = 8000$ N, and the shaft diameter be 10 cm. Determine the maximum shearing stress and the maximum tensile stress.

Solution. The torque is

$$T = PL = 8000 \times 2 = 16\,000 \text{ N}\cdot\text{m}.$$

The shearing stress τ is then

$$\tau = \frac{Tc}{J} = \frac{16\,000 \times .05}{\pi \times .1^4/32} = 81.5 \times 10^6 \text{ Pa}$$

The axial stress is

$$\sigma_1 = \frac{10P}{A} = \frac{10 \times 8000}{\pi \times .1^2/4} = 10.2 \times 10^6 \text{ Pa}.$$

The bending stress, on an element at the very top in the center of the beam, is

$$\sigma_2 = \frac{Mc}{I} = \frac{(8000 \times 1) \times .05}{\pi \times .1^4/64} = 81.5 \times 10^6 \text{ Pa}.$$

Using Eq. 10.1.2, we find the maximum shearing stress to be, using $\sigma_x = 81.5 - 10.2 = 70.3$ MPa:

$$\tau_{max} = \frac{1}{2}\sqrt{(\sigma_x - \overset{0}{\not{\sigma}_y})^2 + 4\tau^2} =$$

$$= \frac{1}{2}\sqrt{70.3^2 + 4 \times 81.5^2} = 88.8 \text{ MPa}.$$

The maximum tensile stress is found using Eq. 10.1.4 to be

$$\sigma_{max} = \frac{1}{2}(\sigma_x + \overset{0}{\not{\sigma}_y}) + \tau_{max}$$

$$= \frac{1}{2} \times 70.3 + 88.8 = 124 \text{ MPa}.$$

10.6 Thin-Walled Pressure Vessels

A cylinder whose ratio of wall thickness t to radius r is less than 0.1 can be considered to be thin-walled. The circumferential stress (hoop-stress) can be found (draw a free-body of half of the cylinder) to be

$$\sigma_c = \frac{pr}{t} \tag{10.6.1}$$

where the inside radius is used.

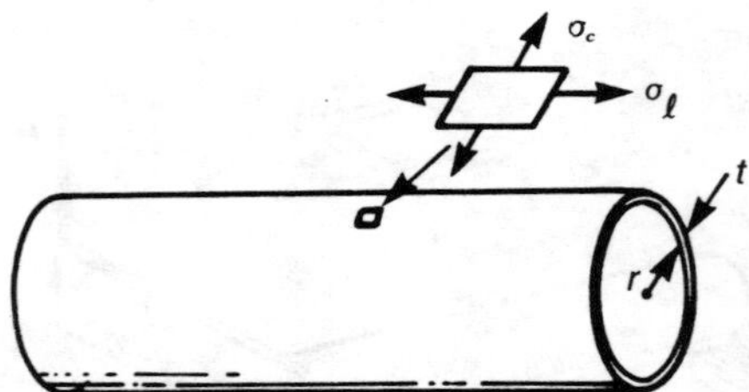

Figure 10.5. Circumferential and longitudinal stresses in a thin-walled cylinder.

Likewise, the longitudinal stress is given by the expression

$$\sigma_\ell = \frac{pr}{2t}\,. \tag{10.6.2}$$

For a spherical thin-walled tank the normal stresses on an element, similar to the one displayed in Fig. 10.5, would be equal and given by

$$\sigma = \frac{pr}{2t}\,. \tag{10.6.3}$$

Because the normal stresses on two perpendicular faces are equal, Eq. 10.1.2 predicts that no shearing stress exists and Eq. 10.1.4 predicts that the maximum normal stress is the same as the applied normal stresses.

EXAMPLE 10.11

A pressure of 2 MPa exists in a 80-cm-dia cylindrical tank with wall thickness of 3 mm. Determine the maximum shearing stress in the tank.

Solution. The normal stresses are the circumferential and longitudinal stresses,

$$\sigma_c = \frac{pr}{t}$$

$$= \frac{2 \times 0.4}{0.003} = 267 \text{ MPa}.$$

$$\sigma_\ell = \frac{pr}{2t}$$

$$= \frac{2 \times 0.4}{2 \times 0.003} = 133 \text{ MPa}.$$

The maximum shearing stress is (see Eq. 10.1.2),

$$\tau_{max} = \frac{1}{2}\sqrt{(\sigma_x - \sigma_y)^2 + \cancel{4\tau^2}^{\,0}}$$

$$= \frac{1}{2}\sqrt{(267 - 133)^2} = 66.7 \text{ MPa}.$$

This stress would act on a plane oriented 45° to the longitudinal axis, as predicted by Eq. 10.1.3.

10.7 Composite Sections

A composite section is composed of two or more different materials, each of which carries part of the load. The method used to approximate the stress in each component material is outlined as follows:

1. For each material find the ratio

$$n = \frac{E}{E_{min}} \tag{10.7.1}$$

where E_{min} is the least modulus of elasticity of the materials used. For the material with E_{min} we use $n = 1$.

2. Multiply the area of each material by its n-value. Then assume each enlarged area A_t to have the modulus of elasticity E_{min}.

3. For simple tension or compression members, the normal stress is calculated from

$$\sigma = \frac{nP}{A_t} \tag{10.7.2}$$

4. For beams we continue through these remaining steps. Each area is enlarged by increasing the width; the depth must remain unchanged.

5. Find the centroid of the transformed composite areas, then find I_t of the transformed area.

6. Calculate the respective stresses using

$$\sigma = \frac{nMc}{I_t}, \qquad \tau = \frac{nVQ_t}{I_t b_t}$$

EXAMPLE 10.12

A 10-cm-wide rectangular composite section of a cantilevered beam has a 6 mm plate of steel on the top and bottom, and a middle of 5-cm magnesium. If a force of 3000 N acts on the end of the 3-m-long beam, find the maximum bending stress in each material.

Solution. The value of n for the steel is

$$n = \frac{E_s}{E_m} = \frac{210 \times 10^6}{45 \times 10^6} = 4.67.$$

The transformed beam is as shown.

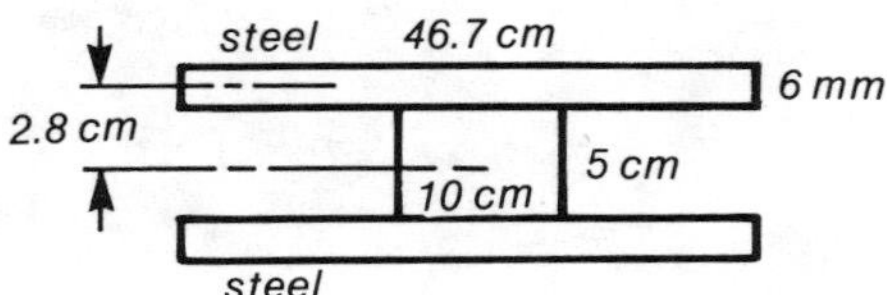

10-12

The moment of inertia of the transformed area is

$$I_t = 2\left[\frac{0.467 \times .006^3}{12} + (0.467 \times .006) \times 0.028^2\right] + \frac{.1 \times .05^3}{12} = 5.45 \times 10^{-6}\ \text{m}^4.$$

The largest moment occurs at the fixed end and is equal to

$$M = PL = 3000 \times 3 = 9000\ \text{N}\cdot\text{m}.$$

The respective maximum bending stresses are

$$\sigma_{steel} = \frac{nMc}{I_t}$$

$$= \frac{4.67 \times 9000 \times 0.031}{5.45 \times 10^{-6}} = 239 \times 10^6\ \text{Pa}$$

$$\sigma_{mag} = \frac{nMc}{I_t}$$

$$= \frac{1 \times 9000 \times 0.025}{5.45 \times 10^{-6}} = 41.3 \times 10^6\ \text{Pa}.$$

Practice Problems (Metric Units)

(If only a few select problems are desired, choose those with a star.)

STRESS AND STRAIN

*10.1 A structural member with the same material properties in all directions at any particular point is

a) homogeneous b) isotropic c) isentropic d) holomorphic e) orthotropic

*10.2 The amount of lateral deformation in a tension member can be calculated using

a) the bulk modulus.
b) the moment of inertia.
c) the yield stress.
d) Hooke's law.
e) Poisson's ratio.

*10.3 Find the allowable load, in kN, on a 2-cm-dia, 1-m-long, steel rod if its maximum deflection cannot exceed 0.1 cm.

a) 35 b) 45 c) 55 d) 66 e) 76

10.4 An elevator is suspended by a 2-cm-dia, 30-m-long steel cable. Twenty people, with a total weight of 14 000 N, enter. How far, in milimeters, does the elevator drop?

a) 3.5 b) 4.5 c) 5.5 d) 6.4 e) 7.4

10.5 A hole, one meter from the end of a structural steel member, is 0.8 mm shy of matching another hole for possible connection. What force, in kN, is necessary to stretch it for connection? The cross section is 25 mm × 3 mm.

a) 12.6 b) 13.6 c) 14.7 d) 15.8 e) 17.2

10.6 As the load is applied, pt. *A* moves 0.03 mm to the right. Determine the shear modulus, in MPa.

a) 50 300 b) 41 700 c) 38 600 d) 32 500 e) 26 300

10.7 A 5-cm-dia steel shaft is subjected to a force of 600 kN. What is the diameter, in cm, after the force is applied?

a) 4.999 b) 4.998 c) 4.997 d) 4.996 e) 4.995

*10.8 An aluminum cylinder carries a compressive load of 1500 kN. Its diameter measures exactly 12.015 cm and its height 19.311 cm. What was its original diameter, in cm?

a) 12.010 b) 12.008 c) 12.006 d) 12.004 e) 12.002

*10.9 Find the maximum shearing stress, in MPa.

a) 80 b) 70 c) 60

d) 50 e) 40

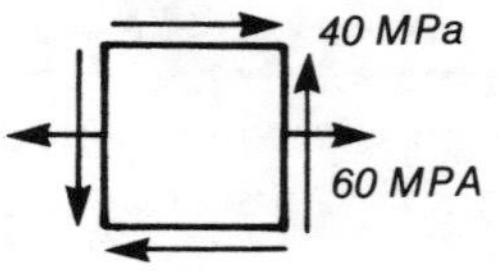

40 MPa

60 MPA

10.10 What is the maximum tensile stress, in MPa?

a) 40 b) 30 c) 20

d) 10 e) 0

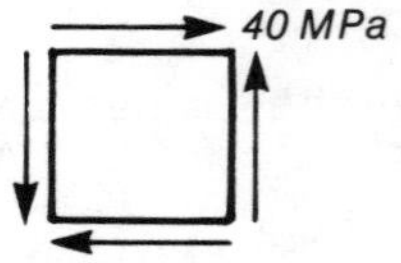

40 MPa

10.11 Determine the maximum shearing stress, in MPa.

a) 80 b) 60 c) 50

d) 40 e) 30

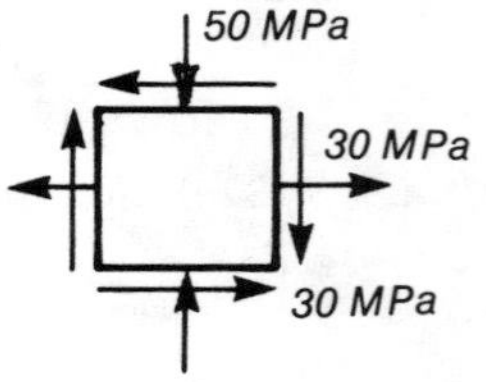

50 MPa

30 MPa

30 MPa

THERMAL STRESS

10.12 A tensile stress of 100 MPa exists in a 2-cm-dia steel rod which connects two rigid members. If the temperature increases by 30°C, find the final stress, in MPa.

a) 46.7 b) 41.2 c) 36.9 d) 31.2 e) 26.2

*10.13 A steel bridge span is normally 300 m long. What is the difference in length, in cm, between January (−35°C) and August (40°C)?

a) 26 b) 28 c) 30 d) 32 e) 34

10.14 An aluminum shaft at 30°C is inserted between two stationary members by inducing a compressive stress of 70 MPa. At what temperature, in °C, will the shaft drop out?

a) 10 b) 0 c) −8 d) −14 e) −20

*10.15 Brass could not be used to reinforce concrete because

a) its density is too large.
b) its density is too low.
c) it is too expensive.
d) its coefficient of thermal expansion is not right.
e) it does not adhere well to concrete.

BENDING MOMENTS IN BEAMS

*10.16 The maximum bending stress in an I-beam occurs

a) where the shearing stress is maximum.
b) at the outermost fiber.
c) at the joint of the web and the flange.
d) at the neutral axis.
e) just below the joint of the web and the flange.

*10.17 The moment diagram for a simply supported beam with a load in the midpoint is

a) a triangle. b) a parabola. c) a trapezoid. d) a rectangle. e) a semicircle.

10.18 Find the bending moment, in N·m, at A.

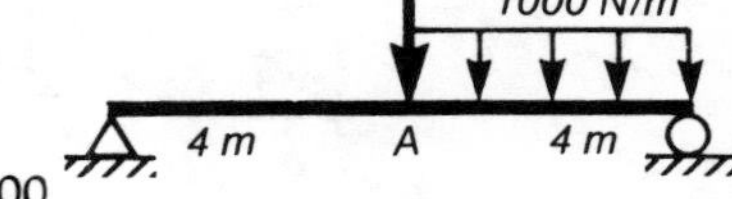

a) 12 000 b) 14 000 c) 16 000 d) 18 000 e) 20 000

*10.19 The bending moment, in N·m, at A is

1000 N/m
4000 N
A
4 m

a) 26 000 b) 24 000 c) 22 000 d) 20 000 e) 18 000

DEFLECTION OF BEAMS

*10.20 What is the maximum deflection, in cm, of a simply supported, 6-m-long steel beam with a 5 cm × 5 cm cross-section if it has a 2000 N load at the midpoint?

a) 6.35 b) 7.02 c) 7.63 d) 8.23 e) 8.92

10.21 Find the maximum deflection, in cm, if the 10-cm-wide, steel beam is 5 cm deep.

2000 N
1000 N/m
6 m

a) 3.97 b) 3.24 c) 2.83 d) 1.92 e) 1.18

10.22 If the deflection of the right end of the 5-cm-dia steel beam is 10 cm, what is the load P, in N?

a) 403 b) 523 c) 768 d) 872 e) 935

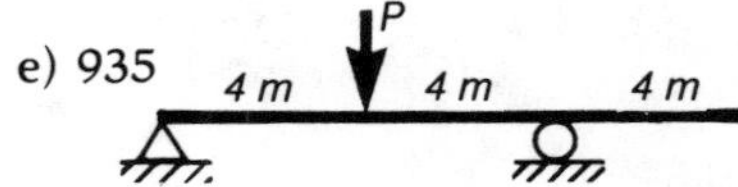

STRESSES IN BEAMS

*10.23 Find the maximum tensile stress, in MPa.

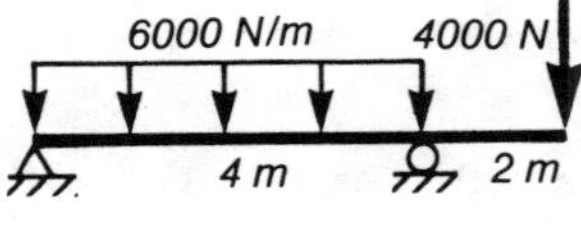

a) 94 b) 88 c) 82

d) 76 e) 72

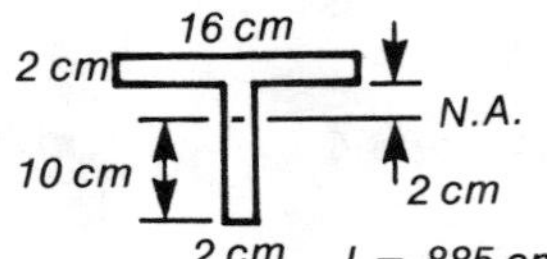

*10.24 What is the maximum compressive stress, in MPa, in the beam of Prob. 10.23?

a) 96 b) 90 c) 84 d) 76 e) 72

*10.25 What is the maximum shearing stress, in MPa, in the beam of Prob. 10.23?

a) 13.8 b) 11.3 c) 9.6 d) 8.2 e) 7.9

STRESSES IN BEAMS

*10.26 The shearing stress VQ/Ib on the cross-section of Prob. 10.23 most resembles which sketch?

a) 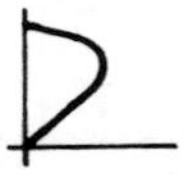 b) 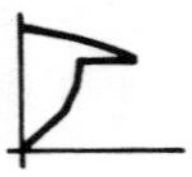 c) 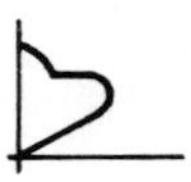 d)  e)

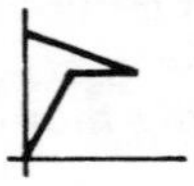

10.27 If the allowable bending stress is 140 MPa in the beam of Prob. 10.19, calculate the *section modulus* I/c, in cm³.

a) 196 b) 184 c) 171 d) 162 e) 153

10.28 Find the maximum bending stress, in MPa, in the beam of Prob. 10.21.

a) 200 b) 180 c) 160 d) 140 e) 120

10.29 If the beam of Prob. 10.21 were rotated 90°, find the maximum bending stress in MPa.

a) 130 b) 120 c) 110 d) 100 e) 90

10.30 Find the maximum shearing stress, in MPa, in the beam of Prob. 10.20.

a) 0.6 b) 0.9 c) 1.2 d) 1.6 e) 2.4

TORSION

*10.31 The maximum shearing stress, in MPa, that exists in a 6-cm-dia shaft subjected to a 200 N·m torque is

a) 4.72 b) 5.83 c) 7.29 d) 8.91 e) 9.97

10.32 The shaft of Prob. 10.31 is replaced with a 6-cm-outside diameter, 5-cm-inside diameter hollow shaft. What is the maximum shearing stress, in MPa?

a) 5.5 b) 6.4 c) 7.3 d) 8.2 e) 9.1

10.33 The maximum allowable shear stress in a 10-cm-dia shaft is 140 MPa. What maximum torque, in N·m, can be applied?

a) 27 500 b) 21 400 c) 19 300 d) 17 100 e) 15 300

10.34 A builder uses a 50-cm-long, 1-cm-dia steel drill. If two opposite forces of 200 N are applied normal to the shaft, each with a moment arm of 15 cm, what angle of twist, in degrees, occurs?

a) 29.3 b) 24.6 c) 22.8 d) 21.1 e) 19.2

COLUMNS

10.35 What is the minimum length, in meters, for which a 10 cm × 10 cm wooden post can be considered a long column? Assume a maximum slenderness ratio of 60.

a) 4.03 b) 3.12 c) 2.24 d) 1.73 e) 1.12

*10.36 A free-standing platform, holding 2000 N, is to be supported by a 10-cm-dia aluminum shaft. How long, in meters, can it be if a safety factor of 2 is used?

a) 18.3 b) 16.6 c) 14.6 d) 12.2 e) 9.32

10.37 What increase in temperature, in °C, is necessary to cause a 2-cm-dia, 4-m-long, steel rod with fixed ends to buckle? There is no initial stress.

a) 9.38 b) 8.03 c) 7.12 d) 6.34 e) 5.27

10.38 A column with both ends fixed buckles when subjected to a force of 30 000 N. One end is then allowed to be free. At what force, in newtons, will it buckle?

a) 2025 b) 1875 c) 1725 d) 1650 e) 1575

COMBINED STRESSES

600 N·m
5 cm dia
40 000 N

*10.39 Find the maximum shearing stress, in MPa, in the shaft.

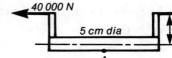

a) 29.5 b) 28.5 c) 27.5 d) 26.5 e) 25.5

10.40 The maximum normal stress, in MPa, in the shaft of Prob. 10.39 is

a) 52.8 b) 41.7 c) 36.7 d) 30.1 e) 25.3

10.41 The normal stress, in MPa, at pt. *A* is

5 cm dia
8 cm
A

a) 263 b) 241 c) 228

d) 213 e) 201

10.42 The maximum shearing stress, in MPa, at pt. *A* in Prob. 10.41 is

a) 140 b) 130 c) 120 d) 110 e) 100

10.43 The maximum shearing stress, in MPa, in the circular shaft is

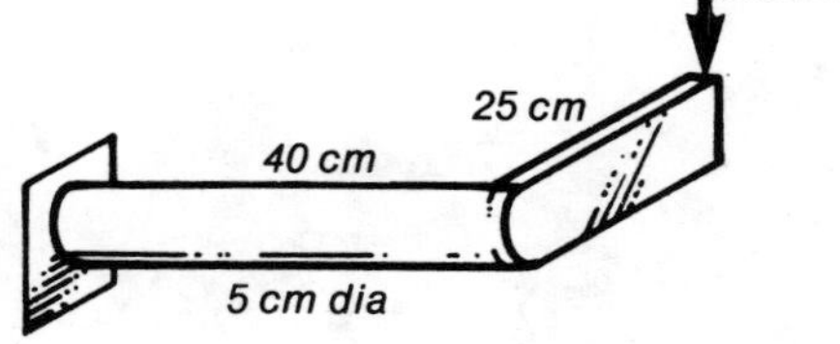

a) 171 b) 167 c) 154

d) 142 e) 133

10.44 The maximum tensile stress, in MPa, in the circular shaft of Prob. 10.43 is

a) 284 b) 248 c) 223 d) 212 e) 197

THIN-WALLED PRESSURE VESSELS

*10.45 The allowable tensile stress for a pressurized cylinder is 180 MPa. What maximum pressure, in kPa, is allowed if the 80-cm-dia cylinder is made of 0.5 cm thick material?

a) 2400 b) 2250 c) 2150 d) 2050 e) 1950

10.46 The maximum normal stress that can occur in a 120-cm-dia steel sphere is 200 MPa. If it is to contain a pressure of 8000 kPa, what must be the minimum thickness, in cm?

a) 1.6 b) 1.4 c) 1.2 d) 1.0 e) 0.8

10.47 What is the maximum shearing stress, in MPa, in the sphere of Prob. 10.46?

a) 200 b) 150 c) 100 d) 50 e) 0

COMPOSITE SECTIONS

*10.48 A compression member, composed of a 1.2-cm-thick steel pipe with a 25-cm-inside diameter, is filled with concrete. Find the stress, in MPa, in the steel if the load is 2000 kN.

a) 137 b) 145 c) 155 d) 165 e) 175

10.49 If the flanges are aluminum and the rib is steel, find the maximum tensile stress, in MPa, in the beam.

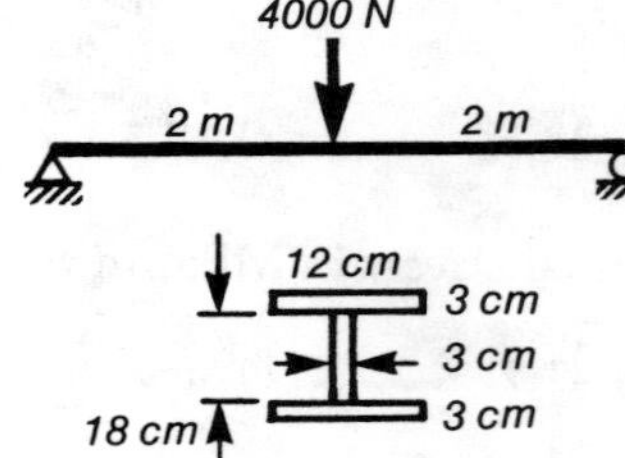

a) 7.89 b) 6.31 c) 5.72

d) 4.91 e) 3.88

10.50 If the flanges of the I-beam of Prob. 10.49 are steel and the rib is aluminum, what is the maximum tensile stress, in MPa, in the beam?

a) 7.89 b) 6.31 c) 5.67 d) 4.91 e) 3.88

Practice Problems (English Units)

(If only a few select problems are desired, choose those with a star.)

STRESS AND STRAIN

*10.1 A structural member with the same material properties in all directions at any particular point is

a) homogeneous b) isotropic c) isentropic d) holomorphic e) orthotropic

*10.2 The amount of lateral deformation in a tension member can be calculated using

a) the bulk modulus.
b) the moment of inertia.
c) the yield stress.
d) Hooke's law.
e) Poisson's ratio.

*10.3 Find the allowable load, in pounds, on a 1/2″ dia, 4-ft-long, steel rod if its maximum deflection cannot exceed 0.04 in.

a) 9290 b) 6990 c) 5630 d) 4910 e) 3220

10.4 An elevator is suspended by a 1/2″ dia, 100-ft-long steel cable. Twenty people, with a total weight of 3500 lbs, enter. How far, in inches, does the elevator drop?

a) 1.3 b) 1.1 c) 0.9 d) 0.7 e) 0.5

10.5 A hole in a 3-ft-long structural steel member is 1/32″ shy of matching another hole for possible connection. What force, in pounds, is necessary to stretch it for connection? The cross section is 1/8″ × 1″.

a) 3300 b) 3000 c) 2700 d) 2400 e) 2100

10.6 As the load is applied, pt. *A* moves 0.0012″ to the right. Determine the shear modulus, in psi.

a) 7.5×10^6 b) 6.2×10^6 c) 5.7×10^6
d) 5.2×10^6 e) 4.5×10^6

10.7 A 2″ dia steel shaft is subjected to a force of 150,000 lbs. What is the diameter, in inches, after the force is applied?

a) 1.999 b) 1.998 c) 1.997 d) 1.996 e) 1.995

*10.8 An aluminum cylinder carries a compressive load of 400,000 lbs. Its diameter measures exactly 5.923″ and its height 8.314″. What was its original diameter, in inches?

a) 5.922 b) 5.920 c) 5.918 d) 5.916 e) 5.914

*10.9 Find the maximum shearing stress, in psi.

4000 psi

6000 psi

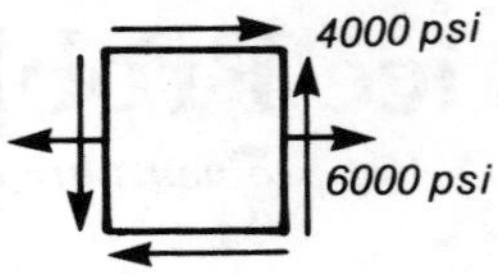

a) 8000 b) 7000 c) 6000

d) 5000 e) 4000

10.10 What is the maximum tensile stress, in psi?

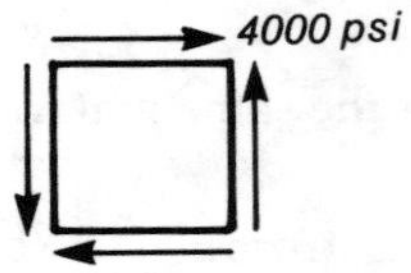

4000 psi

a) 4000 b) 3000 c) 2000

d) 1000 e) 0

10.11 Determine the maximum shearing stress, in psi.

5000 psi

3000 psi

3000 psi

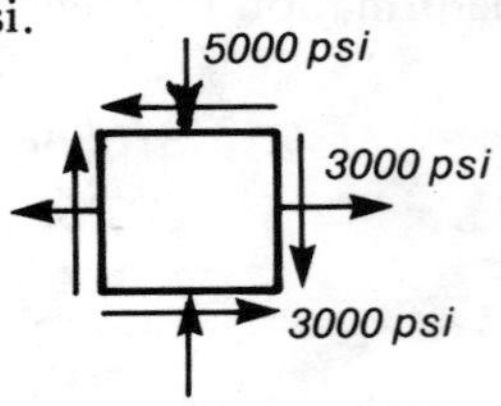

a) 8000 b) 6000 c) 5000

d) 4000 e) 3000

THERMAL STRESS

10.12 A tensile stress of 16,000 psi exists in a 1" dia steel rod which connects two rigid members. If the temperature increases by 50°F, determine the final stress, in psi.

a) 12,200 b) 9400 c) 8600 d) 7400 e) 6200

*10.13 A steel bridge span is normally 1000 ft long. What is the difference in length, in inches, between January (−30°F) and August (100°F)?

a) 10 b) 9 c) 8 d) 7 e) 6

10.14 An aluminum shaft at 80°F is inserted between two stationary members by inducing a compressive stress of 10,000 psi. At what temperature, in °F, will the shaft drop out?

a) 36 b) 24 c) 10 d) 2 e) −10

*10.15 Brass could not be used to reinforce concrete because

a) it is not sufficiently strong.
b) its density is too large.
c) it is too expensive.
d) its coefficient of thermal expansion is not right.
e) it does not adhere well to concrete.

BENDING MOMENTS IN BEAMS

*10.16 The maximum bending stress in an I-beam occurs

a) where the shearing stress is maximum.
b) at the outermost fiber.
c) at the joint of the web and the flange.
d) at the neutral axis.
e) just below the joint of the web and the flange.

10.17 The moment diagram for a simply supported beam with a load in the midpoint is

a) a triangle.
b) a parabola.
c) a trapezoid.
d) a rectangle.
e) a semicircle.

10.18 Find the bending moment, in ft-lb, at *A*.

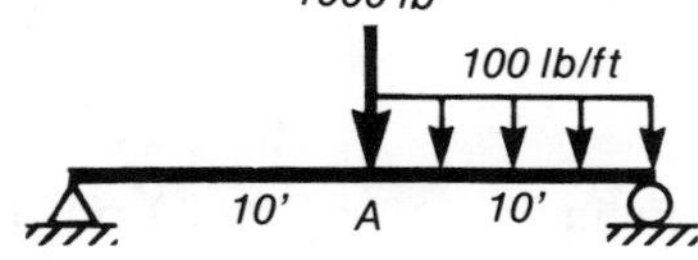

a) 7500 b) 7000 c) 6500

d) 6000 e) 5000

*10.19 What is the bending moment, in ft-lb, at *A*?

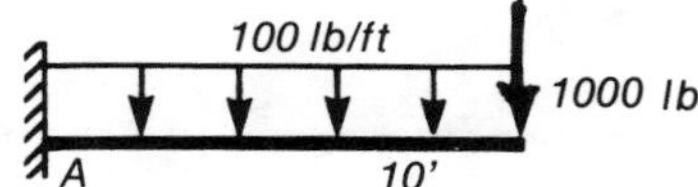

a) 18,000 b) 15,000 c) 12,000

d) 10,000 e) 8000

DEFLECTION OF BEAMS

*10.20 What is the maximum deflection, in inches, of a simply supported, 20-ft-long steel beam with a 2″ x 2″ cross-section if it has a 500-lb load at the midpoint?

a) 7.2 b) 6.0 c) 4.8 d) 3.6 e) 2.4

10.21 Find the maximum deflection, in inches, if the 4″ wide steel beam is 2″ deep.

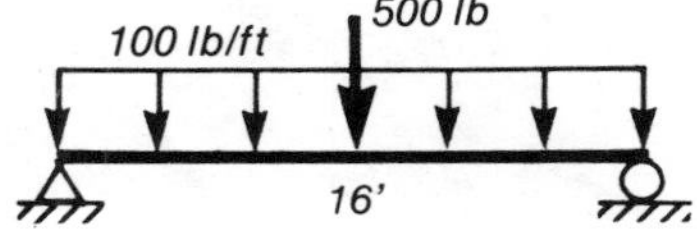

a) 4.22 b) 4.86 c) 3.52 d) 2.98 e) 2.76

10.22 If the deflection of the right end of the 2″ dia steel beam is 4″, what is the load *P*, in pounds?

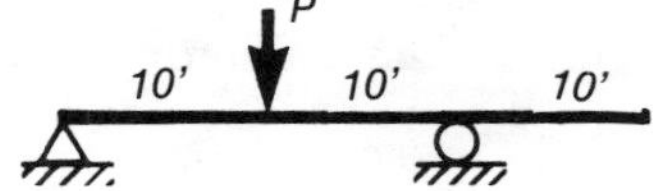

a) 220 b) 330 c) 440 d) 550 e) 660

STRESSES IN BEAMS

*10.23 What is the maximum tensile stress, in psi?

a) 4360 b) 3960 c) 3240

d) 2860 e) 2110

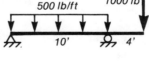

*10.24 What is the maximum compressive stress, in psi, in the beam of Prob. 10.23?

a) 4360 b) 3960 c) 3240 d) 2860 e) 2110

*10.25 What is the maximum shearing stress, in psi, in the beam of Prob. 10.23?

a) 1000 b) 900 c) 800 d) 700 e) 600

*10.26 The shearing stress VQ/Ib on the cross-section of Prob. 10.23 most resembles what sketch?

a)  b)  c) 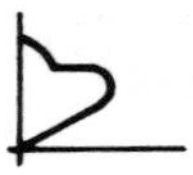 d)  e)

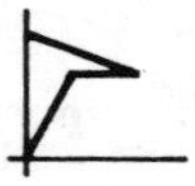

10.27 If the allowable bending stress is 20,000 psi in the beam of Prob. 10.19, calculate the *section modulus* I/c, in in^3.

a) 11 b) 10 c) 9 d) 8 e) 7

10.28 Find the maximum bending stress, in psi, in the beam of Prob. 10.21.

a) 29,200 b) 23,400 c) 18,600 d) 15,600 e) 11,700

10.29 If the beam of Prob. 10.21 were rotated 90°, find the maximum bending stress in psi.

a) 29,200 b) 23,400 c) 18,600 d) 15,600 e) 11,700

10.30 Find the maximum shearing stress, in psi, in the beam of Prob. 10.20.

a) 188 b) 152 c) 131 d) 109 e) 94

TORSION

*10.31 The maximum shearing stress, in psi, that exists in a 2″ dia shaft subjected to a 2000 in-lb torque is

a) 1270 b) 1630 c) 1950 d) 2610 e) 3080

10.32 The shaft of Prob. 10.31 is replaced with a 2″ outside diameter, 1.75″ inside diameter hollow shaft. What is the maximum shearing stress, in psi?

a) 1270 b) 1630 c) 1950 d) 2610 e) 3080

10.33 The maximum allowable shear stress in a 4″ dia shaft is 20,000 psi. What maximum torque, in ft-lb, can be applied?

a) 20,900 b) 15,600 c) 11,200 d) 8,600 e) 4,210

10.34 A mechanic fits an 18″ long, 7/8″ dia steel wrench to a nut. If two opposite forces of 160 lbs are applied normal to the shaft, each with a moment arm of 12″, what angle of twist, in degrees, occurs?

a) 10.3 b) 8.29 c) 6.95 d) 5.73 e) 4.68

COLUMNS

10.35 What is the minimum length, in ft, for which a 4″ × 4″ wooden post can be considered a long column? Assume a maximum slenderness ratio of 60.

a) 9.2 b) 7.6 c) 6.8 d) 5.8 e) 4.2

*10.36 A free-standing platform, holding 500 lb, is to be supported by a 4″ dia aluminum shaft. How long, in ft, can it be if a safety factor of 2 is used?

a) 66 b) 56 c) 46 d) 36 e) 26

10.37 What increase in temperature, in °F, is necessary to cause a 1″ dia, 10 ft long, steel rod with fixed ends to buckle? There is no initial stress.

a) 66 b) 56 c) 46 d) 36 e) 26

10.38 A column with both ends fixed buckles when subjected to a force of 8000 lbs. One end is then allowed to be free. At what force, in pounds, will it buckle?

a) 100 b) 500 c) 2000 d) 8000 e) 32,000

COMBINED STRESSES

*10.39 Find the maximum shearing stress, in psi, in the shaft.

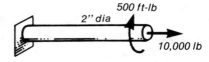

500 ft-lb
2″ dia
10,000 lb

a) 7340 b) 6520 c) 5730 d) 4140 e) 3160

10.40 The maximum normal stress, in psi, in the shaft of Prob. 10.39 is

a) 7340 b) 6520 c) 5730 d) 4140 e) 3160

10.41 The normal stress, in psi, at pt. A is

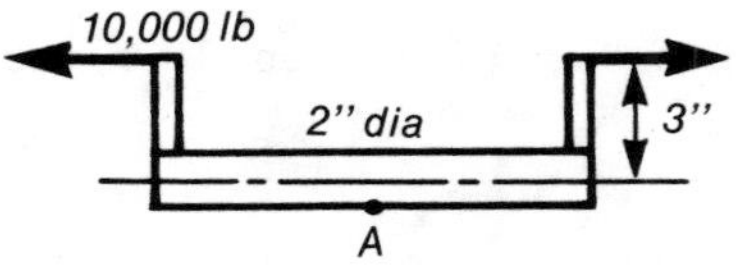

a) 25,000 b) 35,000 c) 41,000

d) 46,000 e) 55,000

10.42 The maximum shearing stress, in psi, at pt. A in Prob. 10.41 is

a) 12,500 b) 15,000 c) 17,500 d) 20,500 e) 23,000

10.43 The maximum shearing stress, in psi, in the circular shaft is

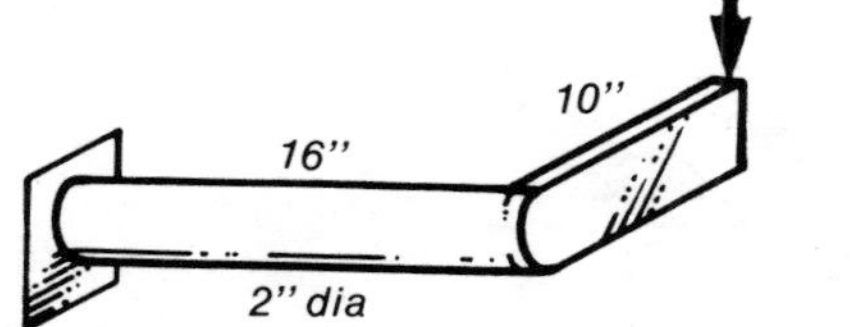

a) 12,000 b) 18,000 c) 24,000

d) 30,000 e) 36,000

10.44 The maximum tensile stress, in psi, in the circular shaft of Prob. 10.43 is

a) 11,100 b) 22,200 c) 33,300 d) 44,400 e) 55,500

THIN-WALLED PRESSURE VESSELS

*10.45 The allowable tensile stress for a pressurized cylinder is 24,000 psi. What maximum pressure, in psi, is allowed if the 2-ft-dia cylinder is made of 1/4" thick material?

a) 250 b) 500 c) 1000 d) 1500 e) 2000

10.46 The maximum normal stress that can occur in a 4-ft-dia steel sphere is 30,000 psi. If it is to contain a pressure of 2000 psi, what must be the minimum thickness, in inches?

a) 1.0 b) 0.9 c) 0.8 d) 0.7 e) 0.6

10.47 What is the maximum shearing stress, in psi, in the sphere of Prob. 10.46?

a) 30,000 b) 20,000 c) 15,000 d) 10,000 e) 0

COMPOSITE SECTIONS

*10.48 A compression member, composed of 1/2" thick steel pipe with a 10" inside diameter, is filled with concrete. Find the stress, in psi, in the steel if the load is 400,000 lbs.

a) 16,400 b) 14,200 c) 12,600 d) 10,100 e) 8,200

10.49 If the flanges are aluminum and the rib is steel, find the maximum tensile stress, in psi, in the beam.

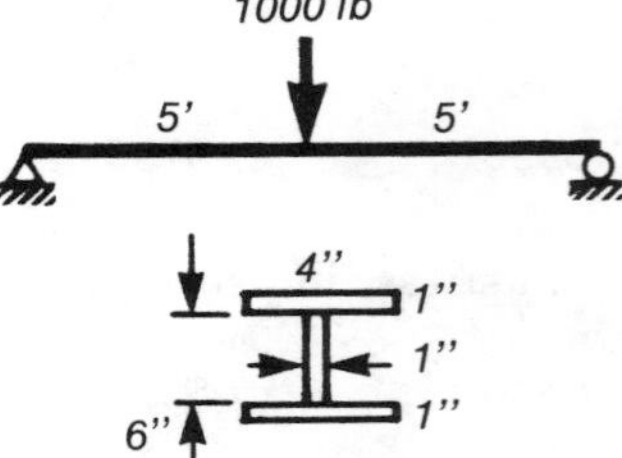

a) 2170 b) 1650 c) 1320

d) 1160 e) 1110

10.50 If the flanges of the I-beam of Prob. 10.49 are steel and the rib is aluminum, what is the maximum tensile stress?

a) 2170 b) 1650 c) 1320 d) 1150 e) 1110

Mechanics of Materials

Steel

$E = 210\times10^6$ kPa $= 30\times10^6$ psi $G = 83\times10^6$ kPa $= 12\times10^6$ psi $\mu = 0.3$ $\alpha = 11.7\times10^{-6}\ °C^{-1} = 6.6\times10^{-6}\ °F^{-1}$

Aluminum

$E = 70\times10^6$ kPa $= 10\times10^6$ psi $G = 27\times10^6$ kPa $= 3.9\times10^6$ psi $\mu = 0.33$ $\alpha = 23.0\times10^{-6}\ °C^{-1} = 12.8\times10^{-6}\ °F^{-1}$

Concrete

$E = 20\times10^6$ kPa $= 3\times10^6$ psi $\alpha = 11.2\times10^{-6}\ °C^{-1} = 6.2\times10^{-6}\ °F^{-1}$

Maximum stresses

$$\tau_{max} = \frac{1}{2}\sqrt{(\sigma_x-\sigma_y)^2 + 4\tau^2}$$

$$\sigma_{\substack{max\\min}} = \frac{1}{2}(\sigma_x-\sigma_y) \pm \tau_{max}$$

$$\theta = \frac{1}{2}\text{ arc tan }\frac{(\sigma_x-\sigma_y)}{2\tau}$$

$$\theta = \frac{1}{2}\text{ arc tan }\frac{-2\tau}{(\sigma_x-\sigma_y)}$$

One-Dimensional relations

$$\sigma = \frac{F}{A} \qquad \sigma = E\epsilon \qquad \epsilon = \frac{\Delta L}{L} \qquad \tau = G\gamma \qquad \Delta L = \alpha L \Delta T \qquad \Delta L_{lateral} = \mu \Delta L_{axial}$$

Beams

$$\sigma = \frac{Mc}{I} \qquad \tau = \frac{VQ}{Ib}$$

$$EIy'' = M \qquad V = \frac{dM}{dx}$$

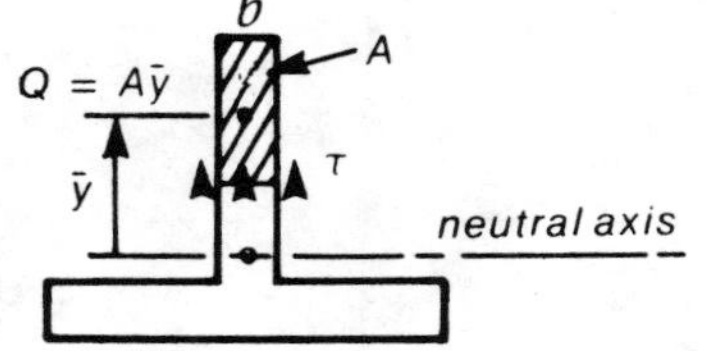

$$I = \frac{bh^3}{12}\ (rectangle)$$

$$I = \frac{\pi d^4}{64}\ (circle)$$

Torsion

$$\tau = \frac{Tc}{J} \qquad \theta = \frac{TL}{JG} \qquad J = \frac{\pi d^4}{32}\ (circle)$$

Columns

pinned ends $P_{cr} = \pi^2 EI/L^2$

fixed ends $P_{cr} = 4\pi^2 EI/L^2$

one free, one fixed $P_{cr} = \pi^2 EI/4L^2$

Pressure vessels

$$\sigma_c = \frac{pr}{t} \qquad \sigma_l = \frac{pr}{2t}\ (cylinder) \qquad \sigma = \frac{pr}{2t}\ (sphere)$$

(This page is intentionally blank.)

11. Fluid Mechanics

Fluid mechanics deals with the statics, kinematics and dynamics of fluids, including both gases and liquids. Most fluid flows can be assumed to be incompressible (constant density); such flows include liquid flows as well as low speed gas flows (with velocities less than about 100 m/s). In addition, particular flows are either viscous or inviscid. Viscous effects dominate internal flows — such as flow in a pipe — and must be included near the boundaries of external flows (flow near the surface of an airfoil). Viscous flows are laminar if well-behaved, or turbulent if chaotic and highly fluctuating. This review will focus on Newtonian fluids, that is, fluids which exhibit linear stress-strain-rate relationships; Newtonian fluids include air, water, oil, gasoline and tar.

11.1 Fluid Properties

Some of the more common fluid properties are defined below and listed in Tables 11.1 and 11.2 for water and air at standard conditions.

density $$\rho = \frac{M}{\forall} \quad (11.1.1)$$

specific weight $$\gamma = \rho g \quad (11.1.2)$$

viscosity $$\mu = \frac{\tau}{du/dy} \quad (11.1.3)$$

kinematic viscosity $$\nu = \frac{\mu}{\rho} \quad (11.1.4)$$

specific gravity $$S = \frac{\rho_x}{\rho_{H_2O}} \quad (11.1.5)$$

bulk modulus $$K = -\forall \frac{\Delta p}{\Delta \forall} \quad (11.1.6)$$

speed of sound $$c_{liquid} = \sqrt{K/\rho} \qquad c_{gas} = \sqrt{kRT} \qquad (k_{air} = 1.4) \quad (11.1.7)$$

Notes:
- kinematic viscosity is used because the ratio μ/ρ occurs frequently.
- surface tension is used primarily for calculating capillary rise.
- vapor pressure is used to predict cavitation which exists whenever the local pressure falls below the vapor pressure.

EXAMPLE 11.1

A velocity difference of 2.4 m/s is measured between radial points 2 mm apart in a pipe in which 20°C water is flowing. What is the shear stress?

Solution. Using Eq. 11.1.3 we find, with $\mu = 10^{-3}$ N·s/m² from Table 11.1,

$$\tau = \mu \frac{du}{dy} \cong \mu \frac{\Delta u}{\Delta y}$$

$$= 10^{-3} \frac{2.4}{0.002} = 1.2 \text{ Pa}.$$

TABLE 11.1 Properties

Property	Symbol	Definition	Water (20°C, 68°F)	Air (STP)
density	ρ	$\frac{\text{mass}}{\text{volume}}$	1000 kg/m³ 1.94 slug/ft³	1.23 kg/m³ 0.0023 slug/ft³
viscosity	μ	$\frac{\text{shear stress}}{\text{velocity gradient}}$	10^{-3} N·s/m² 2×10^{-5}lb-sec/ft²	2.0×10^{-5} N·s/m² 3.7×10^{-7}lb-sec/ft²
kinematic viscosity	ν	$\frac{\text{viscosity}}{\text{density}}$	10^{-6} m²/s 10^{-5} ft²/sec	1.6×10^{-5} m²/s 1.6×10^{-4} ft²/sec
speed of sound	c	velocity of propagation of a small wave	1480 m/s 4900 ft/sec	343 m/s 1130 ft/sec
specific weight	γ	$\frac{\text{weight}}{\text{volume}}$	9800 N/m³ 62.4 lb/ft³	12 N/m³ 0.077 lb/ft³
surface tension	σ	stored energy per unit area	0.073 J/m² 0.005 lb/ft	
bulk modulus	K	$-\text{volume}\,\frac{\Delta(\text{pressure})}{\Delta(\text{volume})}$	220×10^4 kPa 323,000 psi	
vapor pressure	p_v	pressure at which liquid and vapor are in equilibrium	2.45 kPa 0.34 psia	

EXAMPLE 11.2

Find the speed of sound of air at an elevation of 1000 m. From Table 11.2 we find $T = 281.7$ K. Using Eq. 11.1.7, with $R = 287$ J/kg·K, there results

$$\begin{aligned} c &= \sqrt{kRT} \\ &= \sqrt{1.4 \times 287 \times 281.7} = 336.4 \text{ m/s.} \end{aligned}$$

Note: Temperature must be absolute.

11.2 Fluid Statics

Typical problems in fluid statics involve manometers, forces on plane and curved surfaces, and buoyancy. All of these problems are solved by using the pressure distribution derived from summing forces on an infinitesimal element of fluid; in differential form with h positive downward, it is

$$dp = \gamma dh \tag{11.2.1}$$

For constant specific weight, assuming $p = 0$ at $h = 0$, we have

$$p = \gamma h. \tag{11.2.2}$$

TABLE 11.2 Properties of Water and Air

Properties of Water (Metric)

Temperature °C	Density kg/m³	Viscosity N·s/m²	Kinematic Viscosity m²/s	Bulk Modulus kPa	Surface Tension N/m	Vapor Pressure kPa
0	999.9	1.792×10^{-3}	1.792×10^{-6}	204×10^{4}	7.62×10^{-2}	0.588
5	1000.0	1.519	1.519	206	7.54	0.882
10	999.7	1.308	1.308	211	7.48	1.176
15	999.1	1.140	1.141	214	7.41	1.666
20	998.2	1.005	1.007	220	7.36	2.447
30	995.7	0.801	0.804	223	7.18	4.297
40	992.2	0.656	0.661	227	7.01	7.400
50	988.1	0.549	0.556	230	6.82	12.220
60	983.2	0.469	0.477	228	6.68	19.600
70	977.8	0.406	0.415	225	6.50	30.700
80	971.8	0.357	0.367	221	6.30	46.400
90	965.3	0.317	0.328	216	6.12	68.200
100	958.4	0.284×10^{-3}	0.296×10^{-6}	207×10^{4}	5.94×10^{-2}	97.500

Properties of Air at Standard Pressure (Metric)

Temperature	Density kg/m³	Specific Weight N/m³	Viscosity N·s/m²	Kinematic Viscosity m²/s
−20°C	1.39	13.6	1.56×10^{-5}	1.13×10^{-5}
−10°C	1.34	13.1	1.62×10^{-5}	1.21×10^{-5}
0°C	1.29	12.6	1.68×10^{-5}	1.30×10^{-5}
10°C	1.25	12.2	1.73×10^{-5}	1.39×10^{-5}
20°C	1.20	11.8	1.80×10^{-5}	1.49×10^{-5}
40°C	1.12	11.0	1.91×10^{-5}	1.70×10^{-5}
60°C	1.06	10.4	2.03×10^{-5}	1.92×10^{-5}
80°C	0.99	9.71	2.15×10^{-5}	2.17×10^{-5}
100°C	0.94	9.24	2.28×10^{-5}	2.45×10^{-5}

Properties of the Atmosphere (Metric)

Altitude m	Temperature K	p/p_0 (p_0 = 101 kPa)	ρ/ρ_0 (ρ_0 = 1.23 kg/m³)
0	288.2	1.000	1.000
1 000	281.7	0.8870	0.9075
2 000	275.2	0.7846	0.8217
4 000	262.2	0.6085	0.6689
6 000	249.2	0.4660	0.5389
8 000	236.2	0.3519	0.4292
10 000	223.3	0.2615	0.3376
12 000	216.7	0.1915	0.2546
14 000	216.7	0.1399	0.1860
16 000	216.7	0.1022	0.1359
18 000	216.7	0.07466	0.09930
20 000	216.7	0.05457	0.07258
22 000	218.6	0.03995	0.05266
26 000	222.5	0.02160	0.02797
30 000	226.5	0.01181	0.01503
40 000	250.4	0.2834×10^{-2}	0.3262×10^{-2}
50 000	270.7	0.7874×10^{-3}	0.8383×10^{-3}
60 000	255.8	0.2217×10^{-3}	0.2497×10^{-3}
70 000	219.7	0.5448×10^{-4}	0.7146×10^{-4}
80 000	180.7	0.1023×10^{-4}	0.1632×10^{-4}
90 000	180.7	0.1622×10^{-5}	0.2588×10^{-5}

TABLE 11.2E Properties of Water and Air

Properties of Water (English)

Temperature °F	Density slugs/ft³	Viscosity lb-sec/ft²	Surface Tension lb/ft	Vapor Pressure lb/in.²	Bulk Modulus lb/in.²
32	1.94	3.75×10^{-5}	0.518×10^{-2}	0.089	293,000
40	1.94	3.23	0.514	0.122	294,000
50	1.94	2.74	0.509	0.178	305,000
60	1.94	2.36	0.504	0.256	311,000
70	1.94	2.05	0.500	0.340	320,000
80	1.93	1.80	0.492	0.507	322,000
90	1.93	1.60	0.486	0.698	323,000
100	1.93	1.42	0.480	0.949	327,000
120	1.92	1.17	0.465	1.69	333,000
140	1.91	0.98	0.454	2.89	330,000
160	1.90	0.84	0.441	4.74	326,000
180	1.88	0.73	0.426	7.51	318,000
200	1.87	0.64	0.412	11.53	308,000
212	1.86	0.59×10^{-5}	0.404×10^{-2}	14.7	300,000

Properties of Air at Standard Pressure (English)

Temperature °F	Density slugs/ft³	Viscosity lb-sec/ft²	Kinematic Viscosity ft²/sec
0	0.00268	3.28×10^{-7}	12.6×10^{-5}
20	0.00257	3.50	13.6
40	0.00247	3.62	14.6
60	0.00237	3.74	15.8
68	0.00233	3.81	16.0
80	0.00228	3.85	16.9
100	0.00220	3.96	18.0
120	0.00215	4.07×10^{-7}	18.9×10^{-5}

Properties of the Atmosphere (English)

Altitude ft	Temperature °F	Pressure lb/ft²	Density slugs/ft³	Kinematic Viscosity ft²/sec	Velocity of Sound ft/sec
0	59.0	2116	0.00237	1.56×10^{-4}	1117
1,000	55.4	2041	0.00231	1.60	1113
2,000	51.9	1968	0.00224	1.64	1109
5,000	41.2	1760	0.00205	1.77	1098
10,000	23.4	1455	0.00176	2.00	1078
15,000	5.54	1194	0.00150	2.28	1058
20,000	−12.3	973	0.00127	2.61	1037
25,000	−30.1	785	0.00107	3.00	1016
30,000	−48.0	628	0.000890	3.47	995
35,000	−65.8	498	0.000737	4.04	973
36,000	−67.6	475	0.000709	4.18	971
40,000	−67.6	392	0.000586	5.06	971
50,000	−67.6	242	0.000362	8.18	971
100,000	−67.6	22.4	3.31×10^{-5}	89.5×10^{-4}	971
110,000	−47.4	13.9	1.97×10^{-5}	1.57×10^{-6}	996
150,000	113.5	3.00	3.05×10^{-6}	13.2	1174
200,000	160.0	0.665	6.20×10^{-7}	68.4	1220
260,000	−28	0.0742	1.0×10^{-7}	321×10^{-6}	1019

This relationship can be used to interpret manometer readings directly. By summing forces on elements of a plane surface, we would find the magnitude and location of a force acting on one side (refer to Fig. 11.1) to be

$$F = \gamma \bar{h} A \tag{11.2.3}$$

$$y_p = \bar{y} + \frac{\bar{I}}{\bar{y}A} \tag{11.2.4}$$

where $\bar{y}$ locates the centroid and $\bar{I}$ is the second moment* of the area about the centroidal axis.

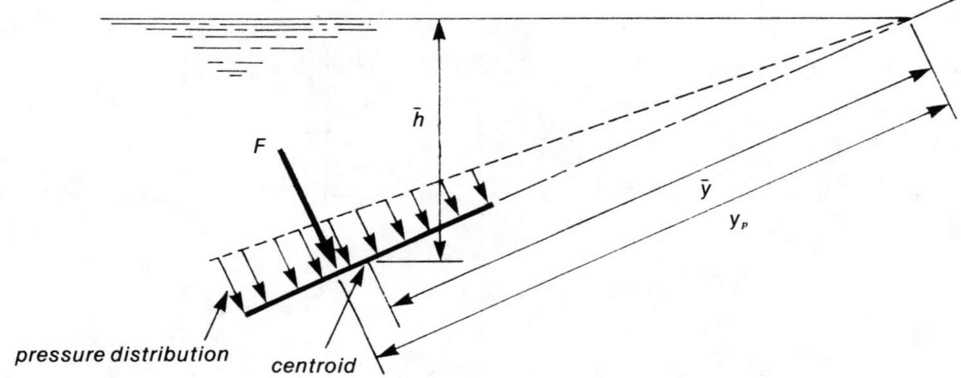

Figure 11.1. **Force on a plane surface.**

To solve problems involving curved surfaces, we simply draw a free-body diagram of the liquid contained above the curved surface, and using the above formulas, solve the problem.

To solve buoyancy-related problems, we use Archimedes' ancient principle, which states: the buoyant force on a submerged object is equal to the weight of displaced liquid; that is,

$$F_b = \gamma \forall_{displaced} \tag{11.2.5}$$

EXAMPLE 11.3

Find the pressure difference between the air pipe and the water pipe.

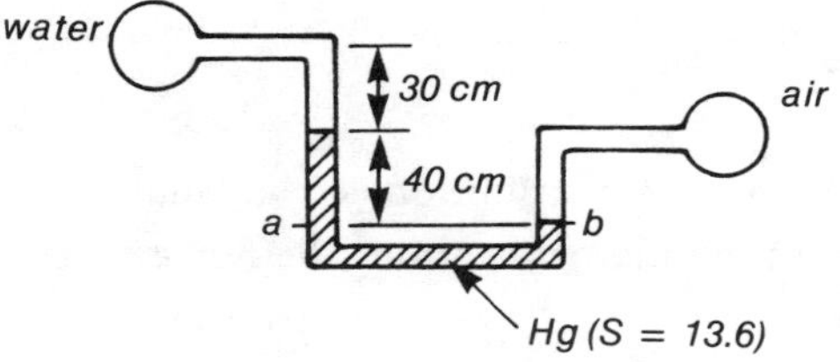

---

*The second moment $\bar{I}$ of three common areas:

b h $\bar{I} = \frac{bh^3}{12}$ h/3 h b $\bar{I} = \frac{bh^3}{36}$ r $\bar{I} = \frac{\pi r^4}{4}$

Solution. We first locate points "a" and "b" in the same fluid where $p_a = p_b$; then using Eq. 11.2.2 we find

$$p_{H_2O} + 9800 \times 0.3 + (9800 \times 13.6) \times 0.4 = p_{air} + \underbrace{\gamma_{air}}_{neglect} \times 0.4.$$

$$\therefore p_{air} - p_{H_2O} = 56\ 300 \text{ Pa} \quad \text{or} \quad 56.3 \text{ kPa}.$$

EXAMPLE 11.4

What is the pressure in pipe A?

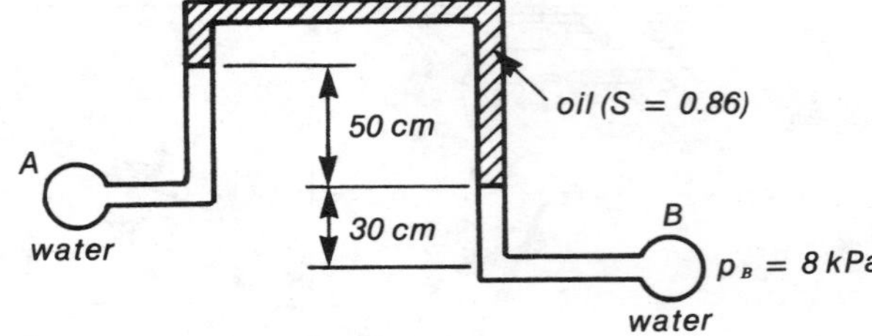

Solution. Locate points "a" and "b" so that $p_a = p_b$. Then, using Eq. 11.2.2 there results

$$p_A - 9800 \times 0.5 = 8000 - 9800 \times 0.3 - (9800 \times 0.86) \times 0.5.$$

$$\therefore p_A = 5750 \text{ Pa} \quad \text{or} \quad 5.75 \text{ kPa}.$$

EXAMPLE 11.5

Find the force P needed to hold the 5-m-wide gate closed.

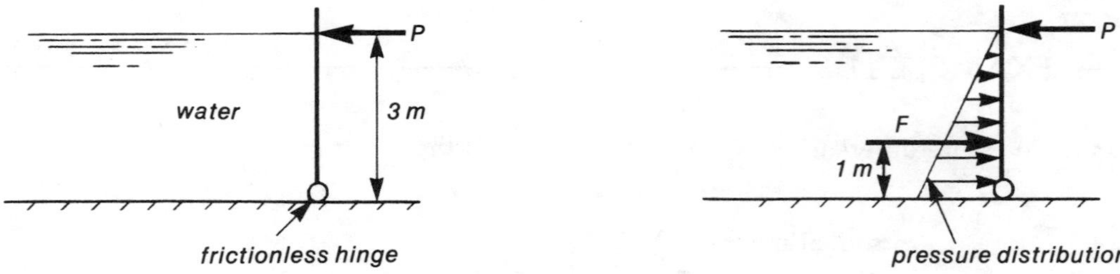

Solution. First, we note that the pressure distribution is triangular, as shown. Hence, the resultant force F acts 1/3 up from the hinge. Summing moments about the hinge gives

$$F \times 1 = P \times 3.$$

$$\therefore P = F/3 = \gamma \bar{h} A/3$$

$$= 9800 \times \frac{3}{2} \times (5 \times 3)/3 = 73\ 500 \text{ N}.$$

Note: If the top of the gate were not at the free surface, we would have to find y_p using Eq. 11.2.4.

EXAMPLE 11.6

A rectangular 4 m × 20 m vessel has a mass of 40 000 kg. How far will it sink in water when carrying a load of 100 000 kg?

Solution. The total weight of the loaded vessel must equal the weight of the displaced water. This is expressed as

$$W = \gamma \forall$$

$$(40\,000 + 100\,000) \times 9.8 = 9800 \times 4 \times 20 \times h.$$

$$\therefore h = 1.75 \text{ m}.$$

11.3 Dimensionless Parameters and Similitude

Information involving phenomena encountered in fluid mechanics is often presented in terms of dimensionless parameters. For example, the lift force F_L on a streamlined body will be represented by a lift coefficient C_L, a dimensionless parameter. Rather than plotting the lift force as a function of velocity, the lift coefficient could be plotted as a function of the Reynolds number, or the Mach number — two other dimensionless parameters.

To form dimensionless parameters, we first list the various quantities encountered in fluid mechanics in Table 11.3. A dimensionless parameter, involving several quantities is then formed by combining the quantities so that the combination of quantities is dimensionless. If all units are present in the quantities to be combined, this usually requires four quantities. For example, the four quantities power $\dot{W}$, flow rate Q, specific weight γ, and head H can be arranged as the dimensionless parameter $\dot{W}/\gamma QH$. Many dimensionless parameters have special significance; they are identified as follows:

$$\text{Reynolds number} = \frac{\text{inertial force}}{\text{viscous force}} \qquad Re = \frac{V\ell\rho}{\mu}$$

$$\text{Froude number} = \frac{\text{inertial force}}{\text{gravity force}} \qquad Fr = \frac{V^2}{\ell g}$$

$$\text{Mach number} = \frac{\text{inertial force}}{\text{compressibility force}} \qquad M = \frac{V}{c}$$

$$\text{Weber number} = \frac{\text{inertial force}}{\text{surface tension force}} \qquad We = \frac{V^2\ell\rho}{\sigma}$$

$$\text{Strouhal number} = \frac{\text{centrifugal force}}{\text{inertial force}} \qquad St = \frac{\ell\omega}{V}$$

$$\text{Pressure coefficient} = \frac{\text{pressure force}}{\text{inertial force}} \qquad C_p = \frac{\Delta p}{\rho V^2}$$

$$\text{Drag coefficient} = \frac{\text{drag force}}{\text{inertial force}} \qquad C_D = \frac{\text{drag}}{\frac{1}{2}\rho V^2 A}$$

So, rather than writing the drag force on a cylinder as a function of length ℓ, diameter d, velocity V, viscosity μ, and density ρ, i.e.,

$$F_D = f(\ell, d, V, \mu, \rho) \tag{11.3.2}$$

we express the relationship using dimensionless parameters as

$$C_D = f\left(\frac{V\rho d}{\mu}, \frac{\ell}{d}\right). \tag{11.3.3}$$

TABLE 11.3. Symbols and Dimensions of Quantities Used in Fluid Mechanics

Quantity	Symbol	Dimensions
Length	ℓ	L
Time	t	T
Mass	m	M
Force	F	ML/T^2
Velocity	V	L/T
Acceleration	a	L/T^2
Frequency	ω	T^{-1}
Gravity	g	L/T^2
Area	A	L^2
Flow Rate	Q	L^3/T
Mass Flux	$\dot{m}$	M/T
Pressure	p	M/LT^2
Stress	τ	M/LT^2
Density	ρ	M/L^3
Specific Weight	γ	M/L^2T^2
Viscosity	μ	M/LT
Kinematic Viscosity	ν	L^2/T
Work	W	ML^2/T^2
Power	$\dot{W}$	ML^2/T^3
Heat Flux	$\dot{Q}$	ML^2/T^3
Surface Tension	σ	M/T^2
Bulk Modulus	K	M/LT^2

The subject of similarity is encountered when attempting to use the results of a model study in predicting the performance of a prototype. We always assume *geometric similarity*, that is, the model is constructed to scale with the prototype; the length scale $\ell_p/\ell_m = \lambda$ is usually designated. The primary notion is simply stated: *Dimensionless quantities associated with the model are equal to corresponding dimensionless quantities associated with the prototype.* For example, if viscous effects dominate we would require

$$Re_m = Re_p. \tag{11.3.4}$$

Then if we are interested in, for example, the drag force we would demand

$$(F_D)^*_m = (F_D)^*_p \tag{11.3.5}$$

where the asterisk "*" denotes a dimensionless quantity. The above equation can be expressed as

$$\frac{(F_D)_m}{\rho_m \ell_m^2 V_m^2} = \frac{(F_D)_p}{\rho_p \ell_p^2 V_p^2}. \tag{11.3.6}$$

This would allow us to predict the drag force to be expected on the prototype as

$$(F_D)_p = (F_D)_m \frac{\rho_p \ell_p^2 V_p^2}{\rho_m \ell_m^2 V_m^2} \tag{11.3.7}$$

The same strategy is used for other quantities of interest.

EXAMPLE 11.7

Combine $\dot{W}$, Q, γ , and H as a dimensionless parameter.

Solution. First, let us note the dimensions on each variable.

$$[\dot{W}] = \frac{ML^2}{T^3} \qquad [Q] = \frac{L^3}{T} \qquad [\gamma] = \frac{M}{L^2T^2} \qquad [H] = L$$

Now, by inspection we simply form the dimensionless parameter: note that to eliminate the mass unit, $\dot{W}$ and γ must appear as the ratio, $\dot{W}/\gamma$. This puts an extra time unit in the denominator; hence, Q must appear with γ, $\dot{W}/\gamma Q$. Now, we inspect the length unit and find one length unit still in the numerator. This requires H to go in the denominator giving the dimensionless parameter as

$$\frac{\dot{W}}{\gamma Q H} .$$

EXAMPLE 11.8

If a flow rate of 0.2 m³/s is measured over a 9-to-1 scale model of a weir, what flow rate can be expected on the prototype?

Solution. First, we recognize that gravity forces dominate (as they do in all problems involving weirs, dams, ships, and open channels) and demand that

$$Fr_p = Fr_m,$$

$$\frac{V_p^2}{\ell_p g_p} = \frac{V_m^2}{\ell_m g_m} .$$

$$\therefore \frac{V_p}{V_m} = \sqrt{\frac{\ell_p}{\ell_m}} = 3.$$

The dimensionless flow rates are now equated:

$$Q_p^* = Q_m^* ,$$

$$\frac{Q_p}{V_p \ell_p^2} = \frac{Q_m}{V_m \ell_m^2} .$$

Recognizing that velocity times area ($V \times \ell^2$) gives the flow rate, we have

$$Q_p = Q_m \frac{V_p \ell_p^2}{V_m \ell_m^2}$$

$$= 0.2 \times 3 \times 9^2 = 48.6 \text{ m}^3/\text{s}.$$

11.4 Control Volume Equations

When solving problems in fluid dynamics, we are most often interested in volumes into which and from which fluid flows; such volumes are called *control volumes*. The control volume equations include the conservation of mass (the continuity equation), Newton's second law (the momentum equation), and the first law of thermodynamics (the energy equation). We will not derive the equations but simply state them and then apply them to some situations of interest. We will assume *steady, imcompressible flow* with *uniform velocity profiles*. The equations take the following forms:

$$\text{continuity:} \quad A_1V_1 = A_2V_2 \tag{11.4.1}$$

$$\text{momentum:} \quad \Sigma\vec{F} = \rho Q(\vec{V}_2 - \vec{V}_1) \tag{11.4.2}$$

$$\text{energy:} \quad -\frac{\dot{W}_s}{\rho Q g} = \frac{V_2^2 - V_1^2}{2g} + \frac{p_2 - p_1}{\gamma} + z_2 - z_1 + h_L \tag{11.4.3}$$

where

$$\begin{aligned} Q &= AV = \text{flow rate} \\ \dot{W}_s &= \text{work output (positive for a turbine)} \\ h_L &= \text{head loss.} \end{aligned} \tag{11.4.4}$$

If there is no shaft work term $\dot{W}_s$ (due to a pump or turbine) between the two sections and the losses are zero, then the energy equation reduces to the Bernoulli equation, namely,

$$\frac{V_2^2}{2g} + \frac{p_2}{\gamma} + z_2 = \frac{V_1^2}{2g} + \frac{p_1}{\gamma} + z_1. \tag{11.4.5}$$

For flow in a pipe, the head loss can be related to the friction factor by the Darcy-Weisbach equation, namely,

$$h_L = f\,\frac{L}{D}\,\frac{V^2}{2g} \tag{11.4.6}$$

where the friction factor is related to the Reynolds number, $Re = VD/\nu$, and the relative roughness e/D by the Moody diagram, Fig. 11.2; the roughness e is given for various materials. Note that for completely turbulent flows, the friction factor is constant so that the head loss varies with the square of the velocity; for laminar flow the head loss is directly proportional to the velocity.

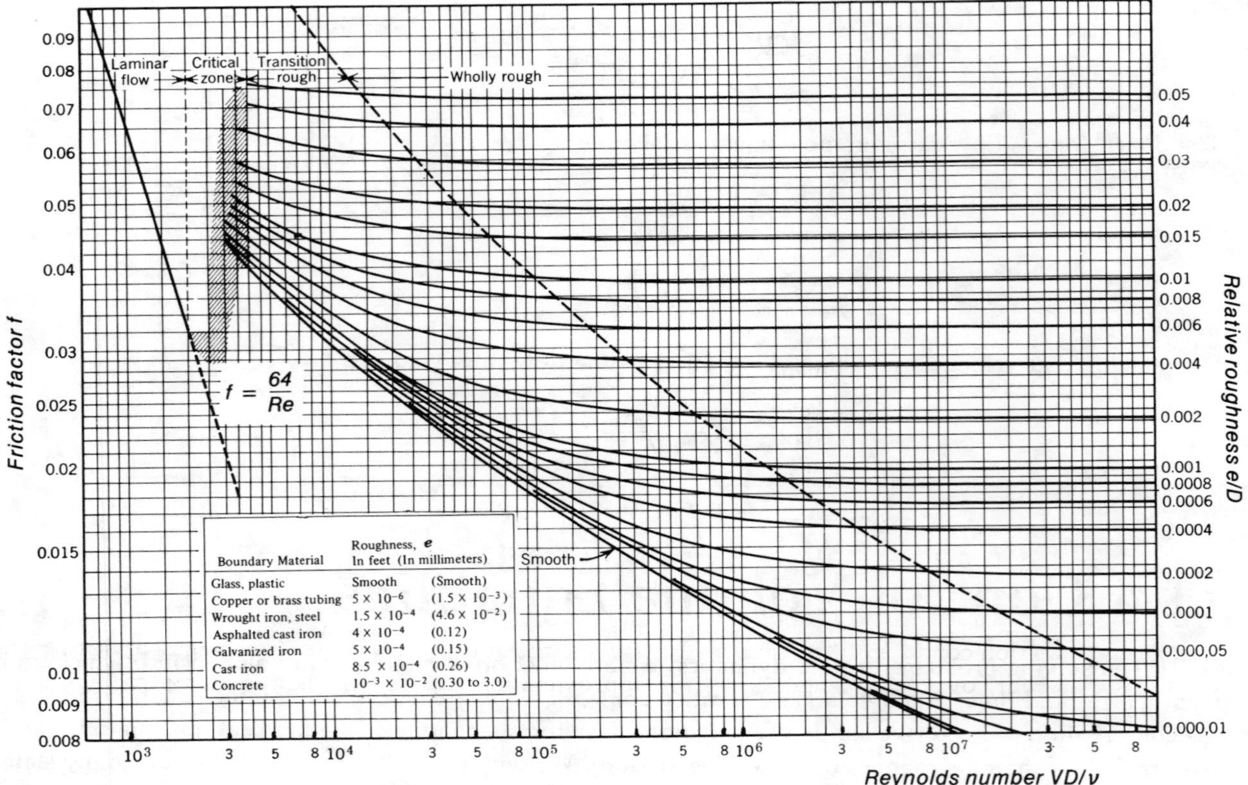

Boundary Material	Roughness, e In feet	(In millimeters)
Glass, plastic	Smooth	(Smooth)
Copper or brass tubing	5×10^{-6}	(1.5×10^{-3})
Wrought iron, steel	1.5×10^{-4}	(4.6×10^{-2})
Asphalted cast iron	4×10^{-4}	(0.12)
Galvanized iron	5×10^{-4}	(0.15)
Cast iron	8.5×10^{-4}	(0.26)
Concrete	$10^{-3} \times 10^{-2}$	(0.30 to 3.0)

Figure 11.2. The Moody diagram.

For sudden geometry changes, such as valves, elbows, and enlargements, the head loss (often called a minor loss) is determined by using a loss coefficient K; that is,

$$h_L = K\frac{V^2}{2g} \quad (11.4.7)$$

where the velocity V is the characteristic velocity associated with the device. Typical values are given in Table 11.4. In engineering practice, the loss coefficient is often expressed as an *equivalent length* L_e of pipe; if that is done, the equivalent length is expressed as

$$L_e = K\frac{D}{f}. \quad (11.4.8)$$

TABLE 11.4. Loss Coefficients

Geometry	K
Globe valve (fully open)	6.4
(half open)	9.5
Angle valve (fully open)	5.0
Swing check valve (fully open)	2.5
Gate valve (fully open)	0.2
(half open)	5.6
(one-quarter open)	24.0
Close return bend	2.2
Standard tee	1.8
Standard elbow	0.9
Medium sweep elbow	0.75
Long sweep elbow	0.6
45° elbow	0.4
Square-edged entrance	0.5
Reentrant entrance	0.8
Well-rounded entrance	0.03⁺⁺
Pipe exit	1.0
Sudden contraction (2 to 1)*	0.25⁺⁺
(5 to 1)*	0.41⁺⁺
(10 to 1)*	0.46⁺⁺
Orifice plate (1.5 to 1)*	0.85
(2 to 1)*	3.4
(4 to 1)*	29
Sudden enlargement⁺	$(1 - A_1/A_2)^2$
90° miter bend (without vanes)	1.1
(with vanes)	0.2
General contraction (30° included angle)	0.02
(70° included angle)	0.07

*Area ratio ⁺Based on V_1 ⁺⁺Based on V_2

The above analysis, using the Moody diagram and the loss coefficients, can be applied directly to only circular cross-section conduits; if the cross section is noncircular but fairly "open" (rectangular with aspect ratio less than four, oval, or triangular) a good approximation can be obtained by using the hydraulic radius defined by

$$R = A/P \quad (11.4.9)$$

where A is the cross-sectional area and P is the wetted perimeter (that perimeter where the fluid is in contact with the solid boundary). Using this formula the diameter of a pipe is $D = 4R$. The Reynolds number then takes the form

$$Re = \frac{4VR}{\nu} \tag{11.4.10}$$

If the shape is not "open," such as flow in an anulus, the error in using the above relationships will be quite significant.

A final note in this article defines the energy grade line (*EGL*) and the hydraulic grade line (*HGL*). The distance $(z + p/\gamma)$ above the datum (the zero elevation line) locates the *HGL*, and the distance $(z + p/\gamma + V^2/2g)$ above the datum locates the *EGL*; these are shown in Fig. 11.3. Note that the pump head H_p is given by

$$H_p = -\frac{\dot{W}_p}{\gamma Q}\,. \tag{11.4.11}$$

The negative sign is necessary since the pump power $\dot{W}_p$ is negative.

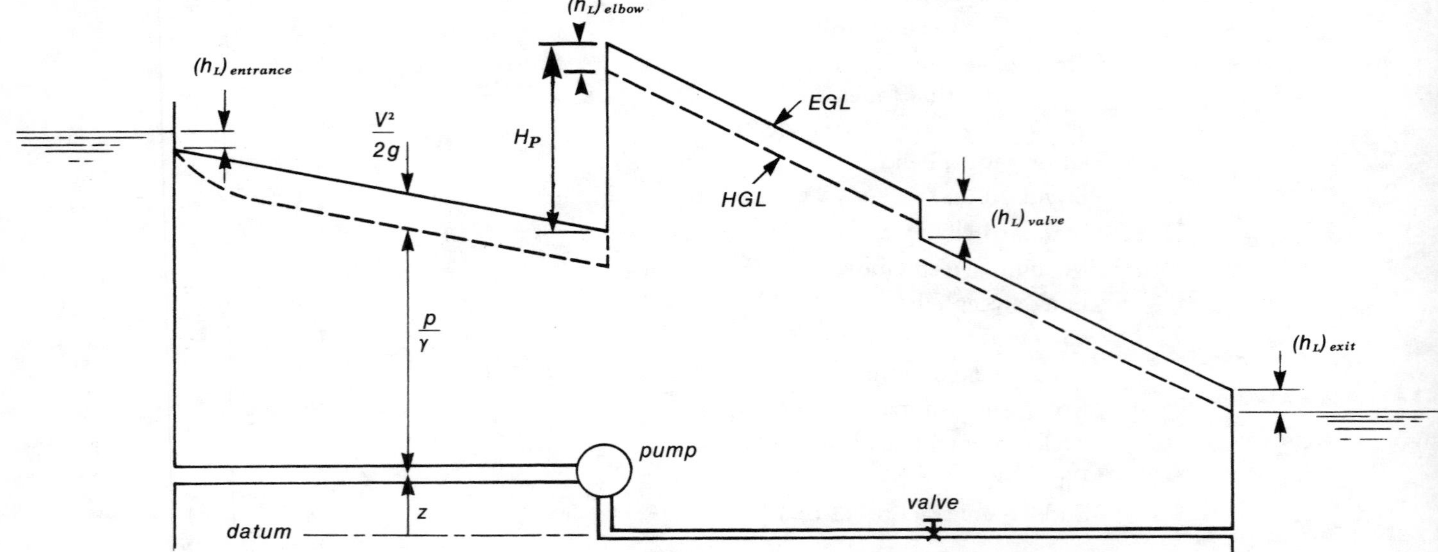

Figure 11.3. **The energy grade line (*EGL*) and the hydraulic grade line (*HGL*).**

EXAMPLE 11.9

The velocity in a 2-cm-dia pipe is 10 m/s. If the pipe enlarges to 4-cm-dia, find the velocity and the flow rate.

Solution. The continuity equation is used as follows:

$$A_1V_1 = A_2V_2$$

$$\frac{\pi D_1^2}{4}\,V_1 = \frac{\pi D_2^2}{4}\,V_2.$$

$$\therefore V_2 = V_1\,\frac{D_1^2}{D_2^2} = 10 \times \frac{2^2}{4^2} = 2.5 \text{ m/s}.$$

The flow rate is

$$Q = A_1V_1$$

$$= \pi \times 0.01^2 \times 10 = 0.00314 \text{ m}^3\text{/s}.$$

EXAMPLE 11.10

What is the force R needed to hold the plate as shown?

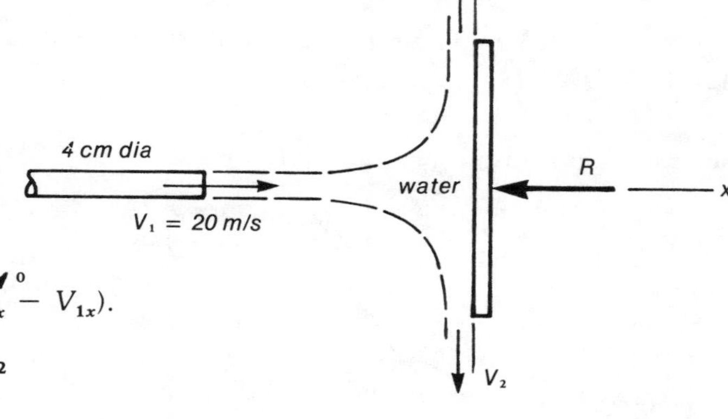

Solution. The momentum equation (11.4.2) is a vector equation; applying it in the x-direction results in

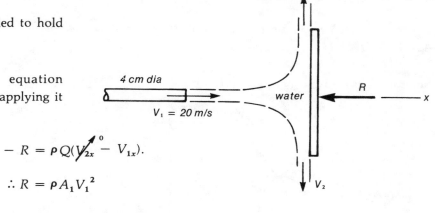

$$-R = \rho Q(\cancelto{0}{V_{2x}} - V_{1x}).$$

$$\therefore R = \rho A_1 V_1^2$$

$$= 1000 \times \pi \times 0.02^2 \times 20^2 = 503 \text{ N}.$$

Note: Since the water is open to the atmosphere $p_2 = p_1$, and if we neglect elevation changes, Bernoulli's equation requires $V_2 = V_1$. However, here $V_{2x} = 0$ so V_2 was not necessary.

EXAMPLE 11.11

What force is exerted on the joint if the flow rate of water is 0.01 m³/s?

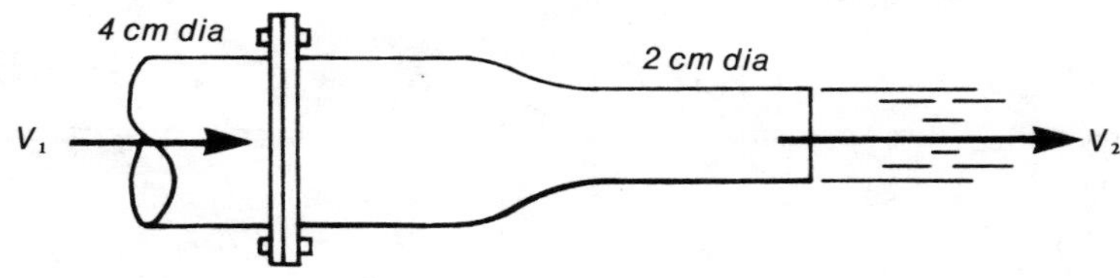

Solution. The velocities are found to be

$$V_1 = \frac{Q}{A_1} = \frac{0.01}{\pi \times 0.02^2} = 7.96 \text{ m/s}.$$

$$V_2 = \frac{Q}{A_2} = \frac{0.01}{\pi \times 0.01^2} = 31.8 \text{ m/s}.$$

Bernoulli's equation is used to find the pressure at section 1. There results, using $p_2 = 0$ (atmospheric pressure is zero gage),

$$\frac{V_1^2}{2g} + \frac{p_1}{\gamma} = \frac{V_2^2}{2g} + \cancelto{0}{\frac{p_2}{\gamma}}.$$

$$\frac{7.96^2}{2 \times 9.8} + \frac{p_1}{9800} = \frac{31.8^2}{2 \times 9.8} \qquad \therefore p_1 = 474\,000 \text{ Pa}.$$

Now, using the control volume shown, we can apply the momentum equation (11.4.2) in the x-direction:

p_1A_1 → F_j ← x

F_j = the force of the contraction on the water

$$p_1A_1 - F_j = \rho Q(V_2 - V_1)$$

$$474\,000 \times \pi \times 0.02^2 - F_j = 1000 \times 0.01\,(31.8 - 7.96).$$

$$\therefore F_j = 357 \text{ N}.$$

Note: Remember, all forces on the control volume must be included. Never forget the pressure force.

EXAMPLE 11.12

What is the pump power needed to increase the pressure by 600 kPa in a 8-cm-dia pipe transporting 0.04 m³/s of water?

Solution. The energy equation (11.4.3) is used:

$$-\frac{\dot{W}_s}{\gamma Q} = \frac{V_2^2 - V_1^2}{2g} + \frac{p_2 - p_1}{\gamma} + z_2 - z_1.$$

$$-\frac{-\dot{W}_p}{9800 \times 0.04} = \frac{600\,000}{9800} \quad \therefore \dot{W}_p = 24\,000 \text{ W} \quad \text{or} \quad 24 \text{ kW}.$$

EXAMPLE 11.13

A pitot tube is used to measure the velocity in the pipe. If $V = 15$ m/s, what is H?

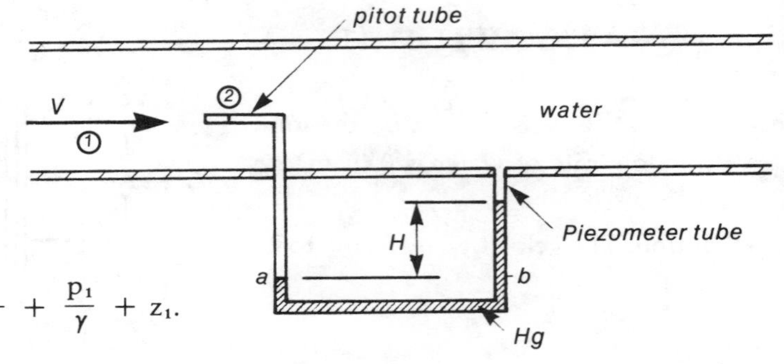

Solution. Bernoulli's equation can be used to relate the pressure at pt. 2 to the velocity V. It gives

$$\frac{V_2^2}{2g} + \frac{p_2}{\gamma} + z_2 = \frac{V_1^2}{2g} + \frac{p_1}{\gamma} + z_1.$$

$$\therefore p_2 = p_1 + \gamma \frac{V_1^2}{2g}.$$

The manometer allows us to write

$$p_a = p_b$$

$$\gamma H + p_2 = \gamma_{Hg} H + p_1.$$

Substituting for p_2 we have

$$\gamma H + p_1 + \gamma \frac{V_1^2}{2g} = \gamma_{Hg} H + p_1.$$

$$\therefore H = \frac{\gamma}{\gamma_{Hg} - \gamma} \frac{V_1^2}{2g}$$

$$= \frac{9800}{13.6 \times 9800 - 9800} \frac{15^2}{2 \times 9.8} = 0.91 \text{ m}.$$

Note: The piezometer tube on the right leg measures the pressure p_1 in the pipe.

EXAMPLE 11.14

For a flow rate of 0.02 m³/s, find the turbine output if it is 80% efficient.

①
100 m el
water
20°C
10 cm dia
60 m el
②
T
300 m of galvanized iron

Solution. The energy equation (11.4.3) takes the form:

$$-\frac{\dot{W}_T}{\gamma Q} = \frac{V_2^2 - V_1^2}{2g} + \frac{p_2 - p_1}{\gamma} + z_2 - z_1 + (K_{entrance} + K_{exit} + f\frac{L}{D})\frac{V^2}{2g}$$

where Eqs. (11.4.6) and (11.4.7) have been used for the head loss. To find f, using the Moody diagram, we need

$$V = \frac{0.02}{\pi \times 0.05^2} = 2.55 \text{ m/s} \qquad Re = \frac{2.55 \times 0.1}{10^{-6}} = 2.55 \times 10^5 \qquad \frac{e}{D} = \frac{0.15}{100} = 0.0015.$$

$$\therefore f = 0.022.$$

Using the loss coefficients from Table 11.4 we have

$$-\frac{\dot{W}_T/0.8}{9800 \times 0.02} = 60 - 100 + (0.5 + 1.0 + 0.022\frac{300}{0.1})\frac{2.55^2}{2 \times 9.8}.$$

$$\therefore \dot{W}_T = 2760 \text{ W} \quad \text{or} \quad 2.76 \text{ kW}.$$

EXAMPLE 11.15

The pressure drop over a 4-cm-dia, 300-m-long section of pipe is measured to be 120 kPa. If the elevation drops 25 m over that length of pipe and the flow rate is 0.003 m³/s, calculate the friction factor and the power loss.

Solution. The velocity is found to be

$$V = \frac{Q}{A} = \frac{0.003}{\pi \times 0.02^2} = 2.39 \text{ m/s}.$$

The energy equation (11.4.3) with Eq. (11.4.6) then gives

$$-\frac{\dot{W}_S}{\gamma Q} = \frac{V_2^2 - V_1^2}{2g} + \frac{p_1 - p_1}{\gamma} + z_2 - z_1 + f\frac{L}{D}\frac{V^2}{2g}.$$

$$0 = -\frac{120\,000}{9800} - 25 + f\frac{300}{0.04}\frac{2.39^2}{2 \times 9.8}.$$

$$\therefore f = 0.0170, \qquad h_L = 37.2.$$

The power loss is

$$\dot{W}_{friction} = h_L \gamma Q$$

$$= 37.2 \times 9800 \times 0.003 = 1095 \text{ W}.$$

EXAMPLE 11.16

Estimate the loss coefficient for an orifice place if $A_1/A_0 = 2$.

Solution. We approximate the flow situation shown as a gradual contraction up to A_c and a sudden enlargement forward from A_c to A_1. The loss coefficient for the contraction is very small so it will be neglected. For the enlargement, we need to know A_c; it is

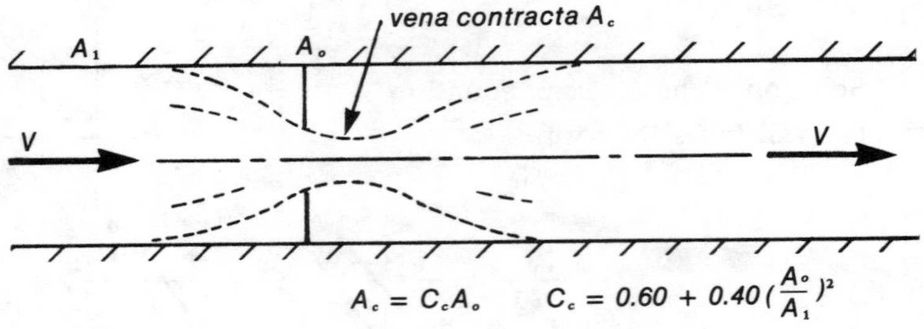

$$
\begin{aligned}
A_c &= C_c A_0 \\
&= [0.6 + 0.4\left(\frac{A_0}{A_1}\right)^2] A_0 = [0.6 + 0.4\left(\frac{1}{2}\right)^2] \frac{A_1}{2} = 0.35\, A_1 .
\end{aligned}
$$

Using the loss coefficient for an enlargement from Table 11.4, there results

$$
\begin{aligned}
h_L &= K \frac{V_c^2}{2g} \\
&= \left(1 - \frac{A_c}{A_1}\right)^2 \frac{V_c^2}{2g} = (1 - 0.35)^2 \frac{1}{0.35^2} \frac{V_1^2}{2g} = 3.45 \frac{V_1^2}{2g} .
\end{aligned}
$$

where the continuity equation $A_c V_c = A_1 V_1$ has been used. The loss coefficient for the orifice plate is thus

$$K = 3.4.$$

Note: Two-place accuracy is assumed since C_c is known to only two significant figures.

11.5 Open Channel Flow

If liquid flows down a slope in an open channel at a constant depth, the energy equation (11.4.3) takes the form

$$
-\cancelto{0}{\frac{\dot{W}_s}{\gamma Q}} = \cancelto{0}{\frac{V_2^2 - V_1^2}{2g}} + \cancelto{0}{\frac{p_2 - p_1}{\gamma}} + z_2 - z_1 + h_L \tag{11.5.1}
$$

which shows that the head loss is given by

$$
\begin{aligned}
h_L &= z_2 - z_1 \\
&= LS
\end{aligned} \tag{11.5.2}
$$

where L is the length of the channel between the two sections and S is the slope. Since we normally have small angles, we can use $S = \tan\theta = \sin\theta = \theta$ where θ is the angle that the channel makes with the horizontal.

The Chezy-Manning equation is used to relate the flow rate to the slope and the cross section; it is

$$Q = \frac{1.0}{n} AR^{2/3}S^{1/2} \tag{11.5.3}$$

where R is the hydraulic radius (see Eq. (11.4.9)), A is the cross-sectional area, and n is the Manning n, given in Table 11.5. The constant 1.0 must be replaced by 1.49 if English units are used.

TABLE 11.5. Average Values* of the Manning n

Wall Material	Manning n
Planed wood	.012
Unplaned wood	.013
Finished concrete	.012
Unfinished concrete	.014
Sewer Pipe	.013
Brick	.016
Cast iron, wrought iron	.015
Concrete pipe	.015
Riveted steel	.017
Earth, straight	.022
Corrugated metal flumes	.025
Rubble	.03
Earth with stones and weeds	.035
Mountain streams	.05

*If $R > 3$ m increase n by 15%.

EXAMPLE 11.17

A 2-m-dia concrete pipe transports water at a depth of 0.8 m. What is the flow rate if the slope is 0.001?

Solution. Calculate the geometric properties

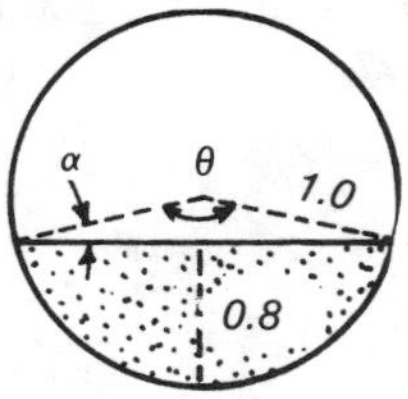

$$\alpha = \sin^{-1}\frac{0.2}{1.0} = 11.54^\circ.$$

$$\therefore \theta = 156.9^\circ$$

$$A = \pi \times 1^2 \times \frac{156.9}{360} - 0.2 \times \cos 11.54^\circ \times \frac{1}{2} \times 2 = 1.174 \text{ m}^2.$$

$$P = \pi \times 2 \times \frac{156.9}{360} = 2.738 \text{ m}.$$

Using a Manning n for concrete pipe of $n = 0.015$, we have

$$Q = \frac{1.0}{n} AR^{2/3}S^{1/2}$$

$$= \frac{1.0}{0.015} \times 1.174 \times \left(\frac{1.174}{2.738}\right)^{2/3} \times 0.001^{1/2} = 1.41 \text{ m}^3/\text{s}.$$

11.6 Compressible Flow

A gas flow with a Mach number below 0.3 (at standard conditions this means velocities less than about 100 m/s) can be treated as an incompressible flow, as in the previous articles. If the Mach number is greater than 0.3, the density variations must be accounted for. For such problems, we use the control volume equations as

continuity: $$\rho_1 A_1 V_1 = \rho_2 A_2 V_2 \tag{11.6.1}$$

momentum: $$\Sigma \vec{F} = \dot{m}\,(\vec{V}_2 - \vec{V}_1) \tag{11.6.2}$$

energy: $$\frac{\dot{Q} - \dot{W}_s}{\dot{m}} = \frac{V_2^2 - V_1^2}{2} + c_p\,(T_2 - T_1) \tag{11.6.3}$$

where

$$\dot{m} = \rho A V = \text{mass flux} \tag{11.6.4}$$

$$c_p = \text{constant pressure specific heat}$$

We often use the ideal gas relations (they become inaccurate at high pressure or low temperature)

$$p = \rho R T \qquad c_p = c_v + R \qquad k = c_p/c_v\,. \tag{11.6.5}$$

The energy equation, for an ideal gas, takes the form

$$\frac{\dot{Q} - \dot{W}_s}{\dot{m}} = \frac{V_2^2 - V_1^2}{2} + \frac{k}{k-1}\left(\frac{p_2}{\rho_2} - \frac{p_1}{\rho_1}\right). \tag{11.6.6}$$

We recall that the speed of sound and Mach number are given by

$$c = \sqrt{kRT} \qquad M = V/c. \tag{11.6.7}$$

Subsonic flow occurs whenever $M < 1$ and supersonic flow whenever $M > 1$. In subsonic flows, losses are quite small and isentropic flows can usually be assumed; thus, we can relate the properties by the isentropic relations

$$\frac{T_2}{T_1} = \left(\frac{p_2}{p_1}\right)^{\frac{k-1}{k}} \qquad \frac{p_2}{p_1} = \left(\frac{\rho_2}{\rho_1}\right)^k. \tag{11.6.8}$$

For air $k = 1.4$, $c_p = 1.00$ kJ/kg·K, $R = 0.287$ kJ/kg·K. The isentropic flow Table 11.6 can also be used.

For supersonic flows, shock waves are encountered; across a normal shock wave Table 11.7 can be used with isentropic flow assumed before and after the shock. If the entropy change is desired it is given by

$$\Delta s = c_p \ln \frac{T_2}{T_1} - R \ln \frac{p_2}{p_1}\,. \tag{11.6.9}$$

Supersonic flow behaves quite differently from subsonic flow; its velocity increases with increasing area and decreases with decreasing area. Hence, to obtain a supersonic flow from a reservoir, the flow must first accelerate through a converging section to a throat where $M = 1$; then an enlarging section will allow it to reach supersonic speed with $M > 1$. Supersonic flow cannot occur in a converging section only.

EXAMPLE 11.18

Air is flowing from a 20°C reservoir to the atmosphere through the converging-diverging nozzle shown. What reservoir pressure will locate a normal shock wave at the exit? Also, find V_c and the mass flux.

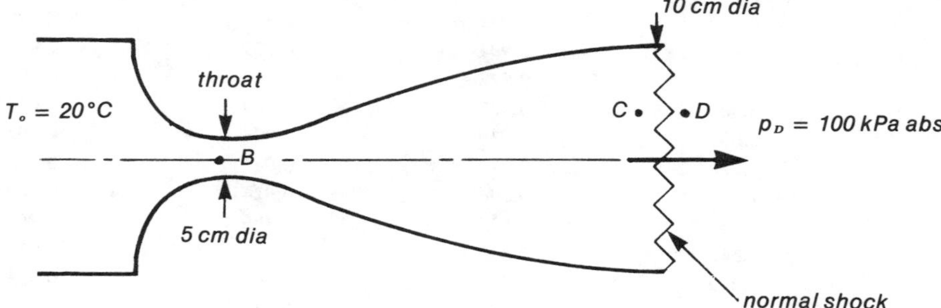

Solution. We know that at the throat the Mach number is unity. Such an area is the *critical area* and is designated A^* in the isentropic flow table. Between points B and C we have

$$\frac{A_C}{A_B} = \frac{A}{A^*} = \frac{\pi \times 5^2}{\pi \times 2.5^2} = 4.$$

Using Table 11.6 we find

$$M_c = 2.94.$$

Then from Table 11.7 we find, across the shock,

$$\frac{p_C}{p_D} = 9.92. \qquad \therefore p_c = 9.92 \times 100 = 992 \text{ kPa}.$$

The isentropic flow Table 11.6 then gives, at $M = 2.94$,

$$\frac{p_C}{p_o} = 0.0298. \qquad \therefore p_o = \frac{992}{0.0298} = 33\,300 \text{ kPa}.$$

To find V_c we must determine T_c. Using Table 11.6 at $M = 2.94$ we find

$$\frac{T_C}{T_o} = 0.3665. \qquad \therefore T_C = (273 + 20) \times 0.3665 = 107.4 \text{ K}.$$

The velocity is then found to be

$$V_C = M_C \sqrt{kRT_C}$$

$$= 2.94\sqrt{1.4 \times 287 \times 107.4} = 611 \text{ m/s}.$$

The mass flux is

$$\dot{m} = \rho_C A_C V_C$$

$$= \frac{p_C}{RT_C} A_C V_C = \frac{992\,000}{287 \times 107.4} \times \pi \times 0.05^2 \times 611 = 154 \text{ kg/s}.$$

TABLE 11.6 Isentropic Flow (air, $k = 1.4$)

M_1	A/A^*	p/p_0	ρ/ρ_0	T/T_0	V/V^*
0.00	∞	1.000	1.000	1.000	0.000
0.05	11.591	0.998	0.999	1.000	0.055
0.10	5.822	0.993	0.995	0.998	0.109
0.15	3.910	0.984	0.989	0.996	0.164
0.20	2.964	0.972	0.980	0.992	0.218
0.25	2.403	0.957	0.969	0.988	0.272
0.30	2.035	0.939	0.956	0.982	0.326
0.35	1.778	0.919	0.941	0.976	0.379
0.40	1.590	0.896	0.924	0.969	0.431
0.45	1.449	0.870	0.906	0.961	0.483
0.50	1.340	0.843	0.885	0.952	0.535
0.55	1.255	0.814	0.863	0.943	0.585
0.60	1.188	0.784	0.840	0.933	0.635
0.65	1.136	0.753	0.816	0.922	0.684
0.70	1.094	0.721	0.792	0.911	0.732
0.75	1.062	0.689	0.766	0.899	0.779
0.80	1.038	0.656	0.740	0.887	0.825
0.85	1.021	0.624	0.714	0.874	0.870
0.90	1.009	0.591	0.687	0.861	0.915
0.95	1.002	0.559	0.660	0.847	0.958
1.00	1.000	0.528	0.634	0.833	1.000
1.10	1.008	0.468	0.582	0.805	1.081
1.20	1.030	0.412	0.531	0.776	1.158
1.30	1.066	0.361	0.483	0.747	1.231
1.40	1.115	0.314	0.437	0.718	1.300
1.50	1.176	0.272	0.395	0.690	1.365
1.60	1.250	0.235	0.356	0.661	1.425
1.70	1.338	0.203	0.320	0.634	1.482
1.80	1.439	0.174	0.287	0.607	1.536
1.90	1.555	0.149	0.257	0.581	1.586
2.00	1.687	0.128	0.230	0.556	1.633
2.10	1.837	0.109	0.206	0.531	1.677
2.20	2.005	0.094	0.184	0.508	1.718
2.30	2.193	0.080	0.165	0.486	1.756
2.40	2.403	0.068	0.147	0.465	1.792
2.50	2.637	0.059	0.132	0.444	1.826
2.60	2.896	0.050	0.118	0.425	1.857
2.70	3.183	0.043	0.106	0.407	1.887
2.80	3.500	0.037	0.095	0.389	1.914
2.90	3.850	0.032	0.085	0.373	1.940
3.00	4.235	0.027	0.076	0.357	1.964
3.50	6.790	0.013	0.045	0.290	2.064
4.00	10.719	0.007	0.028	0.238	2.138
4.50	16.562	0.003	0.017	0.198	2.194
5.00	25.000	0.002	0.011	0.167	2.236
5.50	36.869	0.001	0.008	0.142	2.269
6.00	53.180	0.001	0.005	0.122	2.295
6.50	75.134	0.000	0.004	0.106	2.316
7.00	104.143	0.000	0.003	0.093	2.333
7.50	141.842	0.000	0.002	0.082	2.347
8.00	190.110	0.000	0.001	0.072	2.359
8.50	251.086	0.000	0.001	0.065	2.369
9.00	327.189	0.000	0.001	0.058	2.377
9.50	421.130	0.000	0.001	0.052	2.384
10.00	535.936	0.000	0.000	0.048	2.390
∞	∞	0.000	0.000	0.000	∞

TABLE 11.7 Normal Shock Wave (air, $k = 1.4$)

M_1	M_2	p_2/p_1	T_2/T_1	ρ_2/ρ_1	p_{02}/p_{01}
1.00	1.000	1.000	1.000	1.000	1.000
1.05	0.953	1.120	1.033	1.084	1.000
1.10	0.912	1.245	1.065	1.169	0.999
1.15	0.875	1.376	1.097	1.255	0.997
1.20	0.842	1.513	1.128	1.342	0.993
1.25	0.813	1.656	1.159	1.429	0.987
1.30	0.786	1.805	1.191	1.516	0.979
1.35	0.762	1.960	1.223	1.603	0.970
1.40	0.740	2.120	1.255	1.690	0.958
1.45	0.720	2.286	1.287	1.776	0.945
1.50	0.701	2.458	1.320	1.862	0.930
1.55	0.684	2.636	1.354	1.947	0.913
1.60	0.668	2.820	1.388	2.032	0.895
1.65	0.654	3.010	1.423	2.115	0.876
1.70	0.641	3.205	1.458	2.198	0.856
1.75	0.628	3.406	1.495	2.279	0.835
1.80	0.617	3.613	1.532	2.359	0.813
1.85	0.606	3.826	1.569	2.438	0.790
1.90	0.596	4.045	1.608	2.516	0.767
1.95	0.586	4.270	1.647	2.592	0.744
2.00	0.577	4.500	1.687	2.667	0.721
2.05	0.569	4.736	1.729	2.740	0.698
2.10	0.561	4.978	1.770	2.812	0.674
2.15	0.554	5.226	1.813	2.882	0.651
2.20	0.547	5.480	1.857	2.951	0.628
2.25	0.541	5.740	1.901	3.019	0.606
2.30	0.534	6.005	1.947	3.085	0.583
2.35	0.529	6.276	1.993	3.149	0.561
2.40	0.523	6.553	2.040	3.212	0.540
2.45	0.518	6.836	2.088	3.273	0.519
2.50	0.513	7.125	2.137	3.333	0.499
2.55	0.508	7.420	2.187	3.392	0.479
2.60	0.504	7.720	2.238	3.449	0.460
2.65	0.500	8.026	2.290	3.505	0.442
2.70	0.496	8.338	2.343	3.559	0.424
2.75	0.492	8.656	2.397	3.612	0.406
2.80	0.488	8.980	2.451	3.664	0.389
2.85	0.485	9.310	2.507	3.714	0.373
2.90	0.481	9.645	2.563	3.763	0.358
2.95	0.478	9.986	2.621	3.811	0.343
3.00	0.475	10.333	2.679	3.857	0.328
3.50	0.451	14.125	3.315	4.261	0.213
4.00	0.435	18.500	4.047	4.571	0.139
4.50	0.424	23.458	4.875	4.812	0.092
5.00	0.415	29.000	5.800	5.000	0.062
5.50	0.409	35.125	6.822	5.149	0.042
6.00	0.404	41.833	7.941	5.268	0.030
6.50	0.400	49.125	9.156	5.365	0.021
7.00	0.397	57.000	10.469	5.444	0.015
7.50	0.395	65.458	11.879	5.510	0.011
8.00	0.393	74.500	13.387	5.565	0.008
8.50	0.391	84.125	14.991	5.612	0.006
9.00	0.390	94.333	16.693	5.651	0.005
9.50	0.389	105.125	18.492	5.685	0.004
10.00	0.388	116.500	20.387	5.714	0.003
∞	0.378	∞	∞	6.000	0.000

11-22

Fluid Mechanics

Water

$\rho = 1000 \text{ kg/m}^3$
$= 1.94 \text{ slug/ft}^3$

$\gamma = 9800 \text{ N/m}^3$
$= 62.4 \text{ lb/ft}^3$

$\mu = 10^{-3} \text{ N·s/m}^2$
$= 2\times10^{-5} \text{ lb-sec/ft}^2$

$\nu = 10^{-6} \text{ m}^2\text{/s}$
$= 10^{-5} \text{ ft}^2\text{/sec}$

Air (standard conditions)

$\rho = 1.23 \text{ kg/m}^3$
$= 0.0024 \text{ slug/ft}^3$

$\gamma = 11.9 \text{ N/m}^3$
$= 0.077 \text{ lb/ft}^3$

$\mu = 2\times10^{-5} \text{ N·s/m}^2$
$= 4\times10^{-7} \text{ lb-sec/ft}^2$

$\nu = 1.6\times10^{-5} \text{ m}^2\text{/s}$
$= 1.6\times10^{-4} \text{ ft}^2\text{/sec}$

Statics

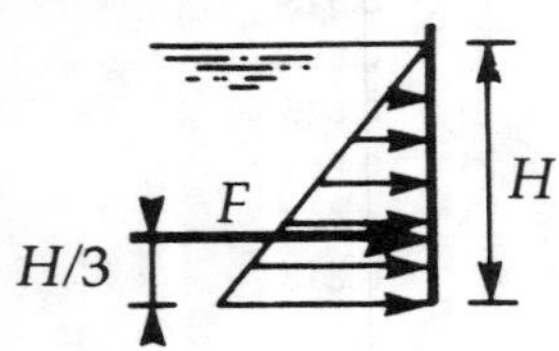

$p = \gamma h$

$F = \gamma \bar{h} A$

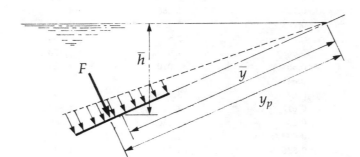

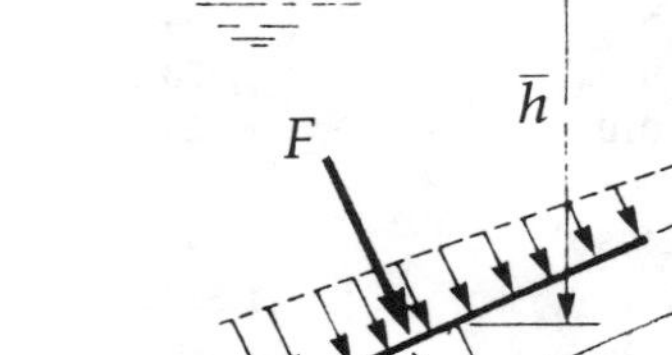

$F = \gamma \bar{h} A$

$y_P = \bar{y} + \frac{\bar{I}}{A\bar{y}}$

$\bar{I} = \frac{bh^3}{12}$

Pressure

$\Delta p = -\int \rho g dz$

$p_{atm} = 10^5 \text{ Pa}$
$= 100 \text{ kPa}$
$= 14.7 \text{ psi}$ $= 760$ *mm of mercury*
$= 2117 \text{ psf}$ $= 29.9$ *in of mercury*
$= 34$ *ft of water*

Bernoulli's equation

Assumptions:

1. *steady flow*
2. *no viscous effects*
3. *constant density*
4. *along a streamline*
5. *inertial reference frame*

$$\frac{V_1^2}{2g} + \frac{p_1}{\gamma} + z_1 = \frac{V_2^2}{2g} + \frac{p_2}{\gamma} + z_2$$

Dimensionless parameters

Reynolds No: $Re = \frac{Vl\rho}{\mu}$

Froude No: $Fr = \frac{V^2}{lg}$

Pressure coef: $C_P = \frac{\Delta p}{\frac{1}{2}\rho V^2}$

Mach No: $M = \frac{V}{c}$

Strouhal No. $St = \frac{l\omega}{V}$

Drag coef: $C_D = \frac{drag}{\frac{1}{2}\rho V^2 A}$

Control volume equations

continuity: $\rho_1 A_1 V_1 = \rho_2 A_2 V_2 = \dot{m}$ $\qquad A_1 V_1 = A_2 V_2 = Q$

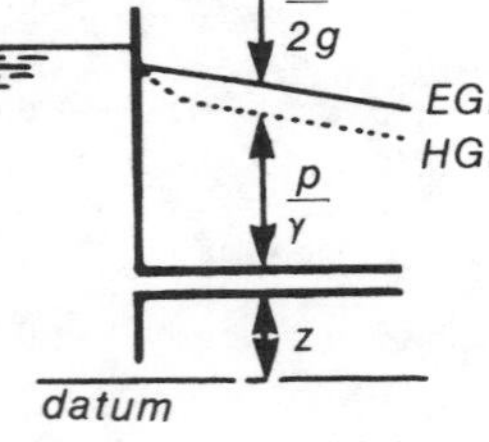

momentum: $\Sigma\vec{F} = \dot{m}\,(\vec{V}_2 - \vec{V}_1)$

energy: $-\frac{\dot{W}_s}{\rho Q g} = \frac{V_2^2 - V_1^2}{2g} + \frac{p_2 - p_1}{\gamma} + z_2 - z_1 + h_L$ $\qquad h_L = f\frac{L}{D}\frac{V^2}{2g}$

$(h_L)_{minor} = K\frac{V^2}{2g}$

open channel: $Q = \frac{1.0}{n} A\, R^{2/3} S^{1/2}$ *(metric)* $\qquad Q = \frac{1.49}{n} A\, R^{2/3} S^{1/2}$ *(english)*

$R = \frac{A}{P_{wetted}}$

Practice Problems (Metric Units)

(If only a few, selected problems are desired, choose those with a star.)

GENERAL

11.1 A fluid is a substance that

a) is essentially incompressible.
b) always moves when subjected to a shearing stress.
c) has a viscosity that always increases with temperature.
d) has a viscosity that always decreases with temperature.
e) expands until it fills its space.

11.2 Viscosity has dimensions of

a) FT^2/L b) F/TL^2 c) M/LT^2 d) M/LT e) ML/T

11.3 The viscosity of a fluid varies with

a) temperature.
b) pressure.
c) density.
d) temperature and pressure.
e) temperature, pressure, and density.

11.4 In an isothermal atmosphere the pressure

a) is constant with elevation.
b) decreases linearly with elevation.
c) cannot be related to elevation.
d) decreases near the surface but approaches a constant value.
e) decreases exponentially with elevation.

11.5 A torque of 1.6 N·m is needed to rotate the cylinder at 1000 rad/s. Estimate the viscosity (N·s/m²).

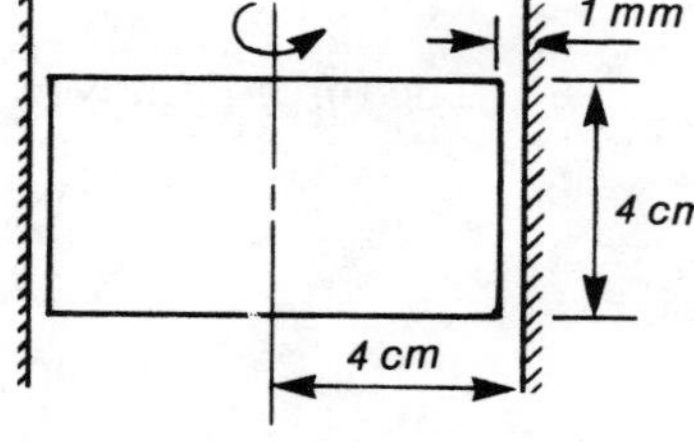

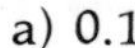

a) 0.1 b) 0.2 c) 0.3
d) 0.4 e) 0.5

11.6 A pressure of 500 kPa applied to 2 m³ of liquid results in a volume change of 0.004 m³. The bulk modulus, in MPa, is

a) 2.5 b) 25 c) 250 d) 2500 e) 2.5×10^6

11.7 Water at 20°C will rise, in a clean 1-mm-dia glass tube, a distance, in cm, of

a) 1 b) 2 c) 3 d) 4 e) 5

11.8 Water at 20°C flows in a piping system at a low velocity. At what pressure, in kPa abs, will cavitation result?

a) 35.6 b) 20.1 c) 10.6 d) 5.67 e) 2.45

11.9 A man is observed to strike an object and 1.2 s later the sound is heard. How far away, in meters, is the man?

a) 220 b) 370 c) 410 d) 520 e) 640

11.10 The viscosity of a fluid with specific gravity 1.3 is measured to be 0.0034 $N \cdot s/m^2$. Its kinematic viscosity, in m^2/s, is

a) 2.6×10^{-6} b) 4.4×10^{-6} c) 5.8×10^{-6} d) 7.2×10^{-6} e) 9.6×10^{-6}

FLUID STATICS

11.11 Fresh water 2 m deep flows over the top of 4 m of salt water ($S = 1.04$). The pressure at the bottom, in kPa, is

a) 60.4 b) 58.8 c) 55.2 d) 51.3 e) 47.9

11.12 What pressure, in kPa, is equivalent to 600 mm of Hg?

a) 100 b) 95.2 c) 80.0 d) 55.2 e) 51.3

11.13 What pressure, in MPa, must be maintained in a diving bell, at a depth of 1200 m, to keep out the ocean water ($S = 1.03$)?

a) 1.24 b) 5.16 c) 9.32 d) 12.1 e) 14.3

11.14 Predict the pressure, in kPa, at an elevation of 2000 m in an isothermal atmosphere assuming $T = 20°C$. Assume $P_{atm} = 100$ kPa.

a) 87 b) 82 c) 79 d) 71 e) 63

11.15 The force F, in Newtons, is

a) 25 b) 8.9 c) 2.5

d) 1.5 e)0.36

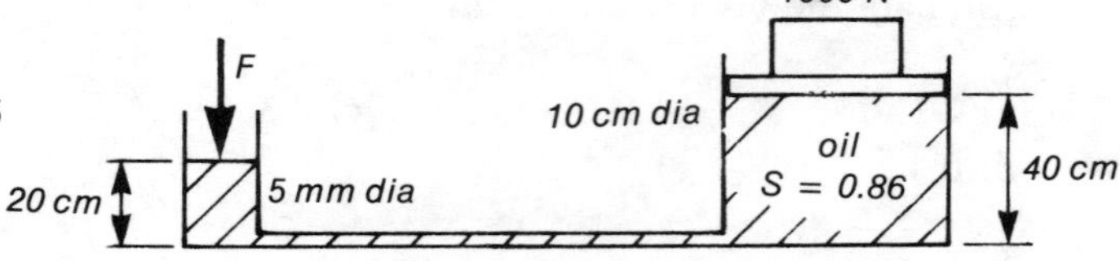

11.16 A U-tube manometer, attached to an air pipe, measures 20 cm of mercury. The pressure, in kPa, in the air pipe is

a) 26.7 b) 32.4 c) 38.6 d) 42.5 e) 51.3

11.17 The pressure p, in kPa, is

a) 51.3 b) 48.0 c) 45.2

d) 40.0 e) 37.0

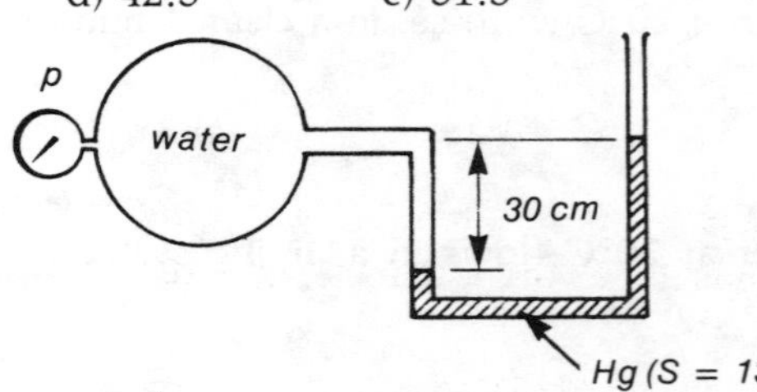

11.18 A 2-m-dia, 3-m-high, cylindrical water tank is pressurized such that the pressure at the top is 20 kPa. The force, in kN, acting on the bottom is

a) 195 b) 176 c) 155 d) 132 e) 106

11.19 The force, in kN, acting on one of the 1.5-m sides of an open cubical water tank (which is full) is

a) 18.2 b) 16.5 c) 15.3 d) 12.1 e) 10.2

11.20 The force P, in kN, for the 3-m-wide gate is

a) 55 b) 60 c) 65

d) 70 e) 75

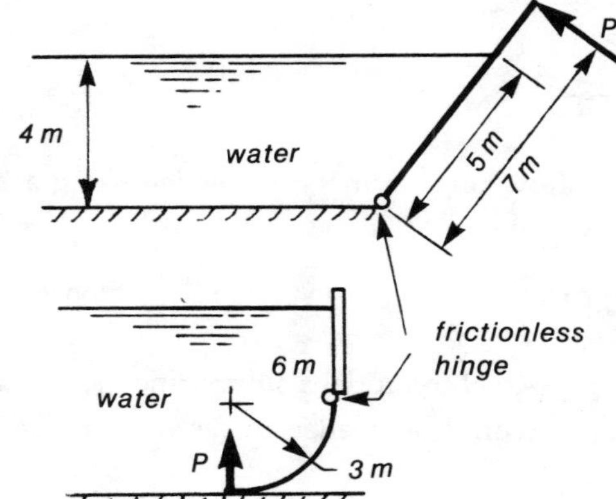

11.21 The force P, in kN, to just open the 4-m-wide gate is

a) 710 b) 762 c) 831

d) 983 e) 1220

*11.22 The force P, in kN, on the 5-m-wide gate is

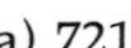

a) 721 b) 653 c) 602

d) 545 e) 498

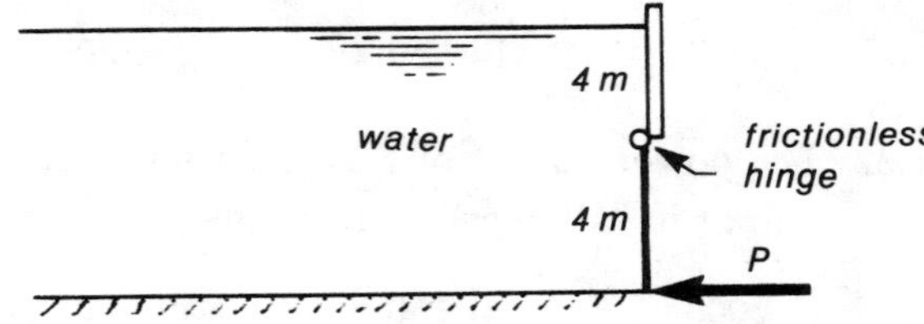

11.23 Four cars, with a mass of 1500 kg each, are loaded on a 6-m-wide, 12-m-long small car ferry. How far, in cm, will it sink in the water?

a) 15.2 b) 11.5 c) 10.2 d) 9.6 e) 8.3

11.24 An object weighs 100 N in air and 25 N when submerged in water. Its specific gravity is

a) 1.11 b) 1.22 c) 1.33 d) 1.44 e) 1.55

11.25 What pressure differential, in pascals, exists at the bottom of a 3-m vertical wall if the temperature inside is 20°C and outside it is −20°C? Assume equal pressures at the top.

a) 15 b) 12 c) 9 d) 6 e) 3

DIMENSIONLESS PARAMETERS

11.26 Arrange pressure p, flow rate Q, diameter D, and density ρ into a dimensionless group.

a) $pQ^2/\rho D^4$ b) $p/\rho Q^2 D^4$ c) $pD^4\rho/Q^2$ d) $pD^4/\rho Q^2$ e) $p/\rho Q^2$

*11.27 Combine surface tension σ, density ρ, diameter D, and velocity V into a dimensionless parameter.

a) $\sigma/\rho V^2 D$ b) $\sigma D/\rho V$ c) $\sigma\rho/VD$ d) $\sigma V/\rho D$ e) $\sigma D^2/\rho V$

11.28 The Reynolds number is a ratio of

a) velocity effects to viscous effects.
b) inertial forces to viscous forces.
c) mass flux to viscosity.
d) flow rate to kinematic viscosity.
e) mass flux to kinematic viscosity.

SIMILITUDE

*11.29 What flow rate, in m^3/s, is needed using a 20:1 scale model of a dam over which 4 m^3/s of water flows?

a) 0.010 b) 0.0068 c) 0.0047 d) 0.0022 e) 0.0015

11.30 It is proposed to model a submarine moving at 10 m/s by testing a 10:1 scale model. What velocity, in m/s, would be needed in the model study?

a) 1 b) 10 c) 40 d) 80 e) 100

11.31 The drag force on a 40:1 scale model of a ship is measured to be 10 N. What force, in kN, is expected on the ship?

a) 640 b) 520 c) 320 d) 160 e) 80

11.32 The power output of a 10:1 scale model of a hydro turbine is measured to be 20 W. The power output in kW expected from the prototype is

a) 200 b) 150 c) 100 d) 63 e) 2

CONTINUITY

11.33 The velocity in a 2-cm-dia pipe is 20 m/s. If the pipe enlarges to 5-cm dia the velocity, in m/s, will be

a) 8.0 b) 6.4 c) 5.2 d) 4.8 e) 3.2

11.34 A 2-cm-dia pipe transports water at 20 m/s. If it exits out 100 small 2-mm-dia holes, the exiting velocity, in m/s, will be

a) 120 b) 80 c) 40 d) 20 e) 10

11.35 Water flows through a 2-cm-dia pipe at 20 m/s. It then flows radially outward between two discs, 2 mm apart. When it reaches a radius of 40 cm its velocity, in m/s, will be

a) 5.0 b) 2.5 c) 2.25 d) 1.85 e) 1.25

BERNOULLI'S EQUATION

11.36 The pressure force, in Newtons, on the 15-cm-dia headlight of an automobile travelling at 25 m/s is

a) 10.4 b) 6.8 c) 5.6 d) 4.8 e) 3.2

11.37 The pressure inside a 4-cm-dia hose is 700 kPa. If the water exits through a 2-cm-dia nozzle what velocity, in m/s, can be expected inside the hose?

a) 20.4 b) 16.3 c) 12.4 d) 10.5 e) 9.7

11.38 Calculate V, in m/s.

a) 8 b) 7 c) 6

d) 5 e) 4

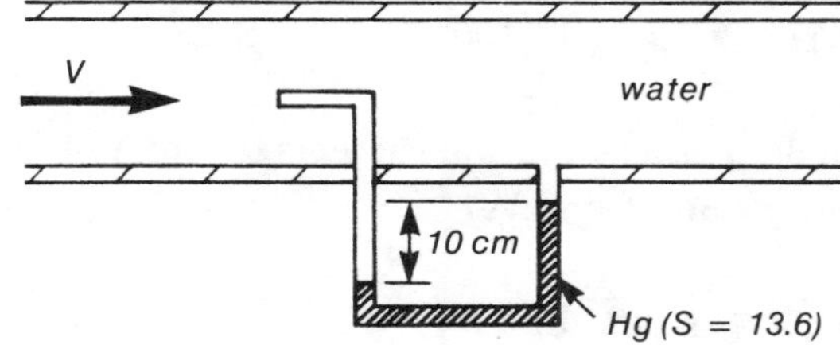

11.39 Water enters a turbine at 900 kPa with negligible velocity. What maximum speed, in m/s, can it reach before it enters the turbine rotor?

a) 52 b) 47 c) 45 d) 42 e) 38

MOMENTUM

11.40 If the density of the air is 1.2 kg/m³ find F, in Newtons.

a) 2.4 b) 3.6 c) 4.8

d) 7.6 e) 9.6

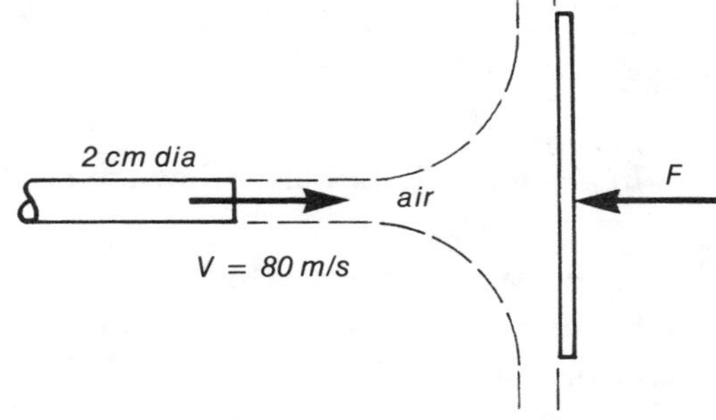

11.41 A rocket exits exhaust gases with $\rho = 0.5$ kg/m³ out a 50-cm-dia nozzle at a velocity of 1200 m/s. Estimate the thrust, in kN.

a) 420 b) 280 c) 140 d) 90 e) 40

11.42 A high-speed vehicle, travelling at 50 m/s, dips an 80-cm-wide scoop into water and deflects the water 180°. If it dips 5 cm deep what force, in kN, is exerted on the scoop?

a) 200 b) 150 c) 100 d) 50 e) 25

11.43 What force, in Newtons, acts on the nozzle?

a) 4020 b) 3230 c) 2420

d) 1830 e) 1610

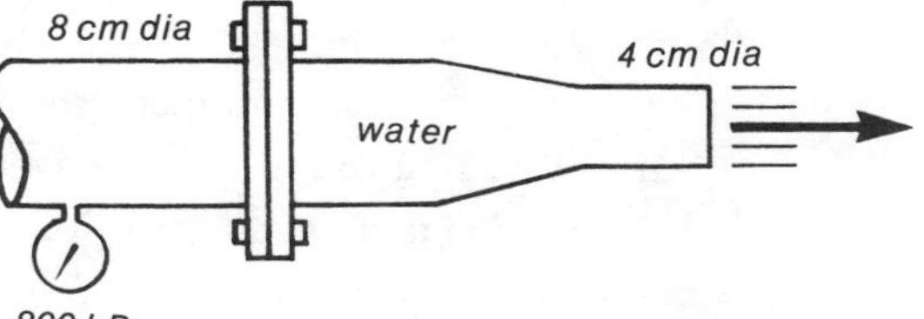

ENERGY

*11.44 The locus of elevations that water will rise in a series of pitot tubes is called

a) the hydraulic grade line.
b) the energy grade line.
c) the velocity head.
d) the pressure head.
e) the head loss.

*11.45 A pressure rise of 500 kPa is needed across a pump in a pipe transporting 0.2 m^3/s of water. If the pump is 85% efficient, the power needed, in kW, would be

a) 118 b) 100 c) 85 d) 65 e) 60

11.46 An 85% efficient turbine accepts 0.8 m^3/s of water at a pressure of 600 kPa. What is the maximum power output, in kW?

a) 820 b) 640 c) 560 d) 480 e) 410

11.47 If the turbine is 88% efficient, the power output, in kW, is

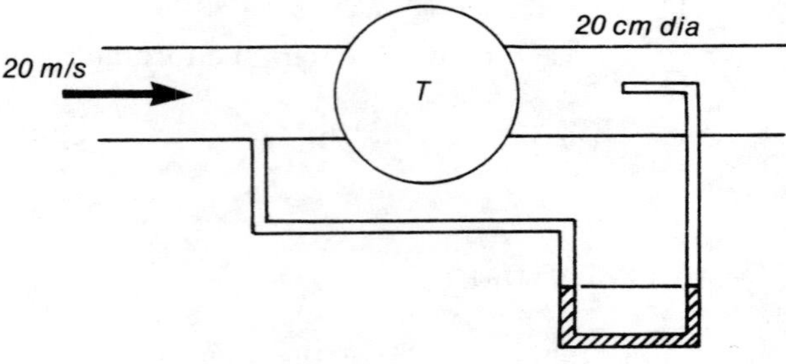

a) 111 b) 126 c) 135

d) 143 e) 176

LOSSES

*11.48 In a completely turbulent flow the head loss

a) increases with the velocity.
b) increases with the velocity squared.
c) decreases with wall roughness.
d) increases with diameter.
e) increases with flow rate.

*11.49 The shear stress in a turbulent pipe flow

a) varies parabolically with the radius.
b) is constant over the pipe radius.
c) varies according to the 1/7th power law.
d) is zero at the center and increases linearly to the wall.
e) is zero at the wall and increases linearly to the center.

*11.50 The velocity distribution in a turbulent flow in a pipe is often assumed to

a) vary parabolically.
b) be zero at the wall and increase linearly to the center.
c) vary according to the 1/7th power law.
d) be unpredictable and is thus not used.
e) be maximum at the wall and decrease linearly to the center.

*11.51 The Moody diagram is sketched. The friction factor for turbulent flow in a smooth pipe is given by curve

a) A b) B c) C

d) D e) E

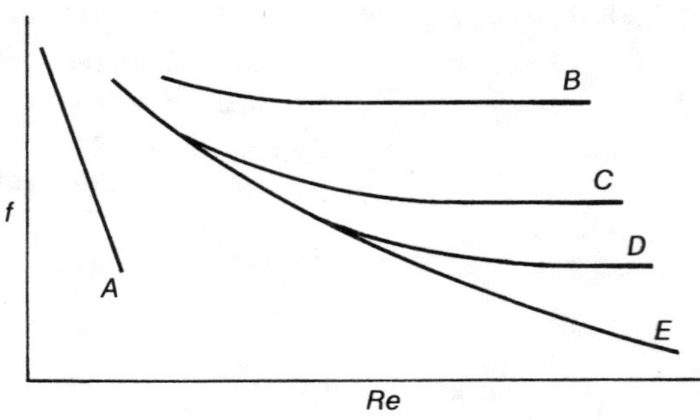

*11.52 For the Moody diagram given in Problem 11.51, the completely turbulent flow is best represented by curve

a) A b) B c) C d) D e) E

*11.53 The pressure gradient ($\Delta p/\Delta x$) in a developed turbulent flow in a horizontal pipe

a) is constant.
b) varies linearly with axial distance.
c) is zero.
d) decrease exponentially.
e) varies directly with the average velocity.

*11.54 The head loss in a pipe flow can be calculated using

a) the Bernoulli equation.
b) Darcy's law.
c) the Chezy-Manning equation.
d) the Momentum equation.
e) the Darcy-Weisbach equation.

*11.55 Minor losses in a piping system are

a) less than the friction factor losses, $f \dfrac{L}{D} \dfrac{V^2}{2g}$.
b) due to the viscous stresses.
c) assumed to vary linearly with the velocity.
d) found by using loss coefficients.
e) independent of the flow rate.

*11.56 In a turbulent flow in a pipe we know the

a) Reynolds number is greater than 10,000.
b) fluid particles move in straight lines.
c) head loss varies linearly with the flow rate.
d) shear stress varies linearly with radius.
e) viscous stresses dominate.

*11.57 Water flows through a 10-cm-dia, 100-m-long pipe connecting two reservoirs with an elevation difference of 40 m. The average velocity is 6 m/s. Neglecting minor losses, the friction factor is

a) 0.020 b) 0.022 c) 0.024 d) 0.026 e) 0.028

*11.58 Find the energy required, in kW, by the 85% efficient pump if $Q = 0.02\ m^3/s$.

a) 14 b) 20 c) 28

d) 35 e) 44

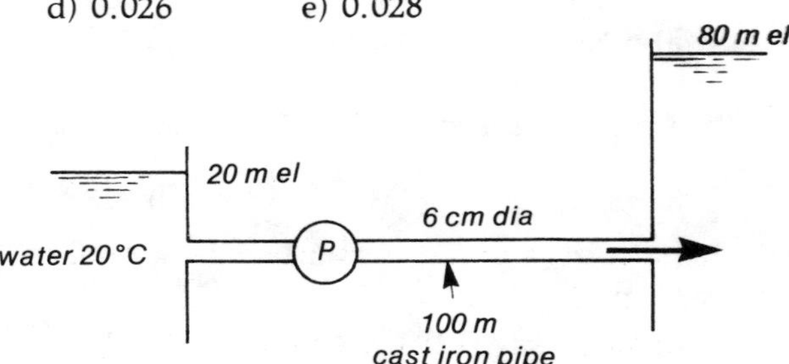

11.59 The pressure at section A in a 4-cm-dia, wrought-iron horizontal pipe is 510 kPa. A fully open globe valve, two elbows, and 50 meters of pipe connect section B. If $Q = 0.006$ m³/s of water, the pressure p_B, in kPa, is

a) 300 b) 250 c) 200 d) 250 e) 100

11.60 Air at 20°C and 100 kPa abs is transported through 500 m of smooth, horizontal, 15 cm × 40 cm rectangular duct with a flow rate of 0.3 m³/s. The pressure drop, in pascals, is

a) 300 b) 400 c) 500 d) 600 e) 700

11.61 Estimate the loss coefficient K in a sudden contraction $A_1/A_2 = 2$ by neglecting the losses up to the vena contracta A_c. Assume that $A_c/A_2 = 0.62 + 0.38\ (A_2/A_1)^3$ and $h_L = KV_2^2/2g$.

a) 0.40 b) 0.35 c) 0.30 d) 0.25 e) 0.20

11.62 An elbow exists in a 6-cm-dia galvanized iron pipe transporting 0.02 m³/s of water. Find the equivalent length of the elbow, in meters.

a) 6.3 b) 4.5 c) 3.6 d) 2.8 e) 2.2

OPEN CHANNEL FLOW

*11.63 The depth of water in a 3-m-wide, rectangular, finished concrete channel is 2 m. If the slope is 0.001, estimate the flow rate, in m³/s.

a) 14 b) 13 c) 12 d) 11 e) 10

11.64 At what depth, in meters, will 10 m³/s of water flow in a 4-m-wide, rectangular, brick channel if the slope is 0.001?

a) 1.3 b) 1.4 c) 1.5 d) 1.6 e) 1.7

11.65 A 2-m-dia, brick storm sewer transports 10 m³/s when it's nearly full. Estimate the slope of the sewer.

a) 0.0070 b) 0.0065 c) 0.0060 d) 0.0055 e) 0.0050

COMPRESSIBLE FLOW

*11.66 The pressure and temperature in a 10-cm-dia pipe are 500 kPa abs and 40°C, respectively. What is the mass flux, in kg/s, if the velocity is 100 m/s?

a) 6.63 b) 5.81 c) 5.36 d) 4.92 e) 4.37

*11.67 Air in a reservoir at 20°C and 500 kPa abs exits a hole with a velocity, in m/s, of

a) 353 b) 333 c) 313 d) 293 e) 273

11.68 A farmer uses 20°C nitrogen pressurized to 800 kPa abs. Estimate the temperature, in °C, in the nitrogen as it exits a short hose fitted to the tank.

a) −110 b) −90 c) −70 d) −50 e) −30

11.69 Estimate p_1, in kPa.

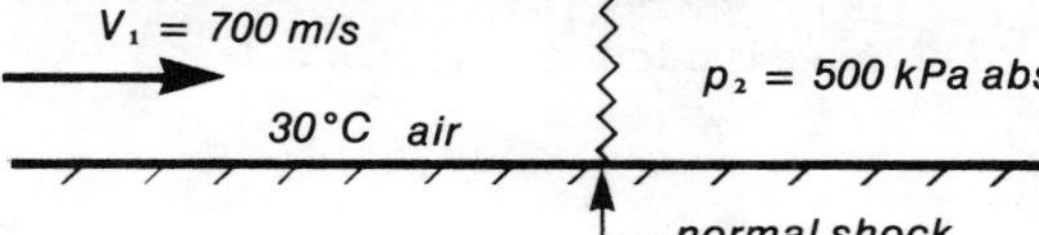

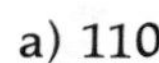

a) 110 b) 120 c) 130

d) 140 e) 150

11.70 Air leaves a reservoir and accelerates until a shock wave is encountered at a diameter of 10 cm. If the throat diameter is 6 cm, what is the Mach number before the shock wave?

a) 2.03 b) 2.19 c) 2.25 d) 2.40 e) 2.56

11.71 A supersonic aircraft flies at $M = 2$ at an elevation of 1000 m. How long, in seconds, after it passes overhead is the aircraft heard?

a) 2.3 b) 2.5 c) 2.7 d) 2.9 e) 3.1

11.72 Air at 20°C is to exit a nozzle from a reservoir. What maximum pressure, in kPa abs, can the reservoir have if compressibility effects can be neglected? $p_{atm} = 100$ kPa.

a) 115 b) 111 c) 109 d) 106 e) 104

(This page is intentionally blank.)

Practice Problems (English Units)

(If only a few, selected problems are desired, choose those with a star.)

GENERAL

*11.1 A fluid is a substance that

a) is essentially incompressible.
b) always moves when subjected to a shearing stress.
c) has a viscosity that always increases with temperature.
d) has a viscosity that always decreases with temperature.
e) expands until it fills its space.

11.2 Viscosity has dimensions of

a) FT^2/L b) F/TL^2 c) M/LT^2 d) M/LT e) ML/T

*11.3 The viscosity of a fluid varies with

a) temperature.
b) pressure.
c) density.
d) temperature and pressure.
e) temperature, pressure, and density.

11.4 In an isothermal atmosphere the pressure

a) is constant with elevation.
b) decreases linearly with elevation.
c) cannot be related to elevation.
d) decreases near the surface but approaches a constant value.
e) decreases exponentially with elevation.

11.5 A torque of 1.2 ft-lb is needed to rotate the cylinder at 1000 rad/sec. Estimate the viscosity (lb-sec/ft²).

a) 8.25×10^{-4} b) 7.16×10^{-4} c) 6.21×10^{-4}

d) 5.27×10^{-4} e) 4.93×10^{-4}

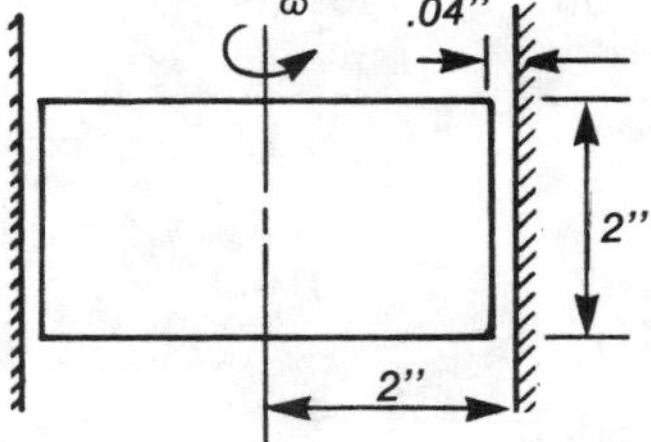

11.6 A pressure of 80 psi applied to 60 ft³ of liquid results in a volume change of 0.12 ft³. The bulk modulus, in psi, is

a) 20,000 b) 30,000 c) 40,000 d) 50,000 e) 60,000

11.7 Water at 70°F will rise, in a clean 0.04" dia glass tube, a distance, in inches, of

a) 1.21 b) 0.813 c) 0.577 d) 0.401 e) 0.311

11.8 Water at 70°F flows in a piping system at a low velocity. At what pressure, in psia, will cavitation result?

a) 0.79 b) 0.68 c) 0.51 d) 0.42 e) 0.34

11.9 A man is observed to strike an object and 1.2 sec later the sound is heard. How far away, in feet, is the man?

a) 1750 b) 1550 c) 1350 d) 1150 e) 950

11.10 The viscosity of a fluid with specific gravity 1.3 is measured to be 7.2×10^{-5} lb-sec/ft^2. Its kinematic viscosity, in ft^2/sec, is

a) 2.85×10^{-5} b) 1.67×10^{-5} c) 1.02×10^{-5} d) 9.21×10^{-4} e) 8.32×10^{-4}

FLUID STATICS

11.11 Fresh water 6 ft deep flows over the top of 12 ft of salt water ($S = 1.04$). The pressure at the bottom, in psi, is

a) 8 b) 9 c) 10 d) 11 e) 12

*11.12 What pressure, in psi, is equivalent to 28" of Hg?

a) 14.6 b) 14.4 c) 14.2 d) 14.0 e) 13.8

11.13 What pressure, in psi, must be maintained in a diving bell, at a depth of 4000 ft, to keep out ocean water ($S = 1.03$)?

a) 1480 b) 1540 c) 1660 d) 1780 e) 1820

11.14 Predict the pressure, in psi, at an elevation of 6000 ft in an isothermal atmosphere assuming $T = 70°$F. Assume $p_{atm} = 14.7$ psi.

a) 10.7 b) 11.2 c) 11.9 d) 12.3 e) 12.8

11.15 The force F, in lbs, is

a) 15.7 b) 13.5 c) 12.7

d) 11.7 e)10.2

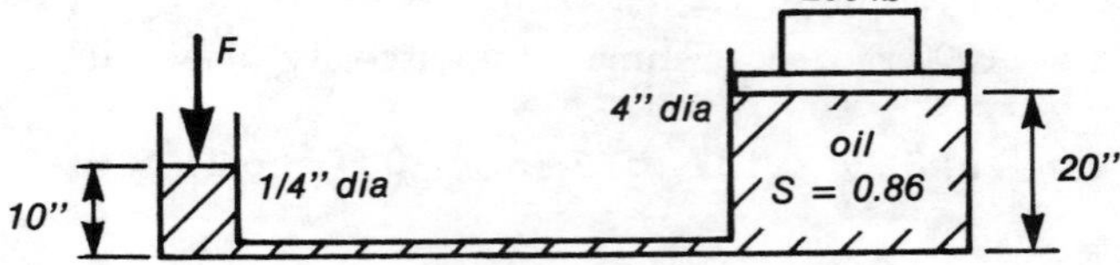

11.16 A U-tube manometer, attached to an air pipe, measures 10" of Hg. The pressure, in psi, in the air pipe is

a) 4.91 b) 4.42 c) 4.01 d) 3.81 e) 3.12

*11.17 The pressure p, in psi, is

a) 8.32 b) 7.51 c) 6.87

d) 6.21 e) 5.46

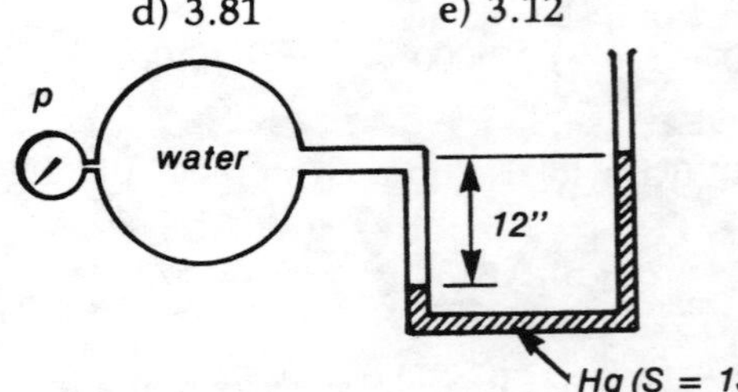

11.18 A 2-ft dia, 10-ft high, cylindrical water tank is pressurized such that the pressure at the top is 3 psi. The force, in lb, acting on the bottom is

a) 4250 b) 3960 c) 3320 d) 2780 e) 2210

11.19 The force, in lb, acting on one of the 5-ft sides of an open cubical water tank which is full is

a) 4200 b) 3900 c) 3600 d) 3300 e) 3000

11.20 The force P, in lb, for the 10-ft wide gate is

a) 17,000 b) 16,000 c) 15,000

d) 14,000 e) 13,000

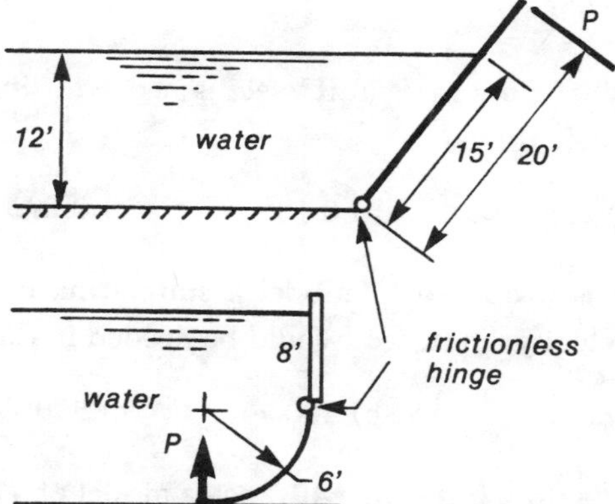

11.21 The force P, in lb, to just open the 12-ft wide gate is

a) 88,600 b) 82,500 c) 79,100

d) 73,600 e) 57,100

*11.22 The force P, in lb, on the 15-ft wide gate is

a) 90,000 b) 78,000 c) 62,000

d) 48,000 e) 32,000

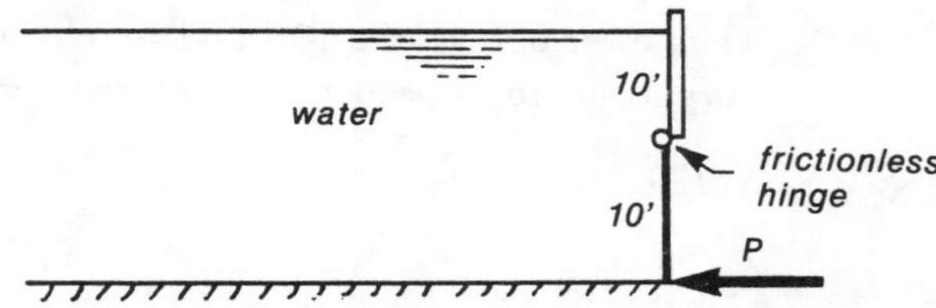

11.23 Four cars, with a mass of 3200 lb each, are loaded on a 20-ft-wide, 40-ft-long small car ferry. How far, in inches, will it sink in the water?

a) 7 b) 6 c) 5 d) 4 e) 3

11.24 An object weighs 25 lb in air and 6 lb when submerged in water. Its specific gravity is

a) 1.5 b) 1.4 c) 1.3 d) 1.2 e) 1.1

11.25 What pressure differential, in psf, exists at the bottom of a 10-ft-vertical wall if the temperature inside is 70°F and outside is −10°F? Assume equal pressures at the top.

a) 0.478 b) 0.329 c) 0.211 d) 0.133 e) 0.101

DIMENSIONLESS PARAMETERS

11.26 Arrange pressure p, flow rate Q, diameter D, and density ρ into a dimensionless group.

a) $pQ^2/\rho D^4$ b) $p/\rho Q^2 D^4$ c) $pD^4\rho/Q^2$ d) $pD^4/\rho Q^2$ e) $p/\rho Q^2$

*11.27 Combine surface tension σ, density ρ, diameter D, and velocity V into a dimensionless parameter.

a) $\sigma/\rho V^2 D$ b) $\sigma D/\rho V$ c) $\sigma\rho/VD$ d) $\sigma V/\rho D$ e) $\sigma D^2/\rho V$

11.28 The Reynolds number is a ratio of

a) velocity effects to viscous effects.
b) inertial forces to viscous forces.
c) mass flux to viscosity.
d) flow rate to kinematic viscosity.
e) mass flux to kinematic viscosity.

SIMILITUDE

*11.29 What flow rate, in ft³/sec, is needed using a 20:1 scale model of a dam over which 120 ft³/sec of water flows?

a) 0.20 b) 0.18 c) 0.092 d) 0.067 e) 0.052

11.30 It is proposed to model a submarine moving at 30 ft/sec by testing a 10:1 scale model. What velocity, in ft/sec, would be needed in the model study?

a) 300 b) 400 c) 500 d) 700 e) 900

11.31 The drag force on a 40:1 scale model of a ship is measured to be 2 lb. What force, in lb, is expected on the ship?

a) 128,000 b) 106,000 c) 92,000 d) 80,000 e) 60,000

11.32 The power output of a 10:1 scale model of a hydro turbine is measured to be 0.06 Hp. The power output, in Hp, expected from the prototype is

a) 130 b) 150 c) 170 d) 190 e) 210

CONTINUITY

11.33 The velocity in a 1″ dia pipe is 60 ft/sec. If the pipe enlarges to 2.5″ dia the velocity, in ft/sec, will be

a) 24.0 b) 20.0 c) 16.2 d) 12.4 e) 9.6

11.34 A 1″ dia pipe transports water at 60 ft/sec. If it exits out 100 small 0.1″ dia holes, the exiting velocity, in ft/sec, will be

a) 180 b) 120 c) 90 d) 60 e) 30

11.35 Water flows through a 1″ dia pipe at 60 ft/sec. It then flows radially outward between two discs, 0.1″ apart. When it reaches a radius of 20″ its velocity, in ft/sec, will be

a) 30 b) 22.5 c) 15.0 d) 7.50 e) 3.75

BERNOULLI'S EQUATION

11.36 The pressure force, in lb, on the 6″ dia headlight of an automobile travelling at 90 ft/sec is

a) 1.02 b) 1.83 c) 2.56 d) 3.75 e) 4.16

*11.37 The pressure inside a 2″ dia hose is 100 psi. If the water exits through a 1″ dia nozzle what velocity, in ft/sec, can be expected inside the hose?

a) 39.8 b) 37.6 c) 35.1 d) 33.7 e) 31.5

11.38 Calculate V, in ft/sec.

a) 19.7 b) 18.9 c) 17.4

d) 16.4 e) 15.2

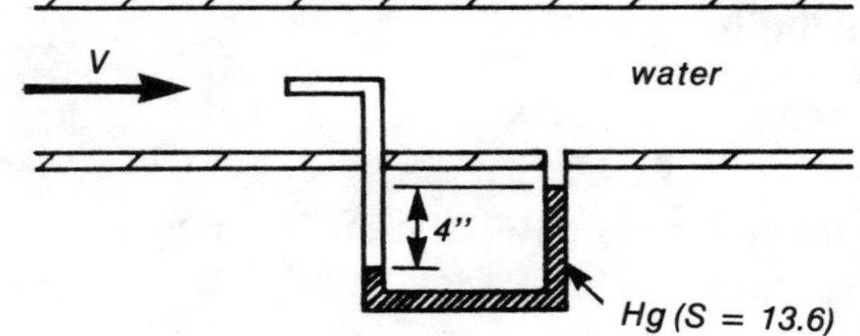

11.39 Water enters a turbine at 150 psi with negligible velocity. What maximum speed, in ft/sec, can it reach before it enters the turbine rotor?

a) 119 b) 131 c) 156 d) 168 e) 183

*11.40 If the density of the air is 0.0024 slug/ft³, find F, in lb.

a) 0.524 b) 0.711 c) 0.916

d) 1.17 e) 2.25

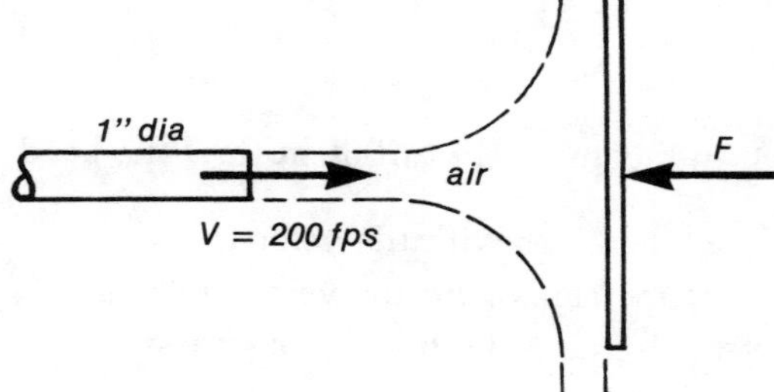

11.41 A rocket exits exhaust gases with $\rho = 0.001$ slug/ft³ out a 20″ dia nozzle at a velocity of 4000 ft/sec. Estimate the thrust, in lb.

a) 24,600 b) 30,100 c) 34,900 d) 36,200 e) 41,600

11.42 A high-speed vehicle, travelling at 150 ft/sec, dips a 30″ wide scoop into water and deflects the water 180°. If it dips 2″ deep what force, in lb, is exerted on the scoop?

a) 36,400 b) 32,100 c) 26,200 d) 22,100 e) 19,900

11.43 What force, in lb, acts on the nozzle?

a) 1700 b) 1600 c) 1500

d) 1400 e) 1300

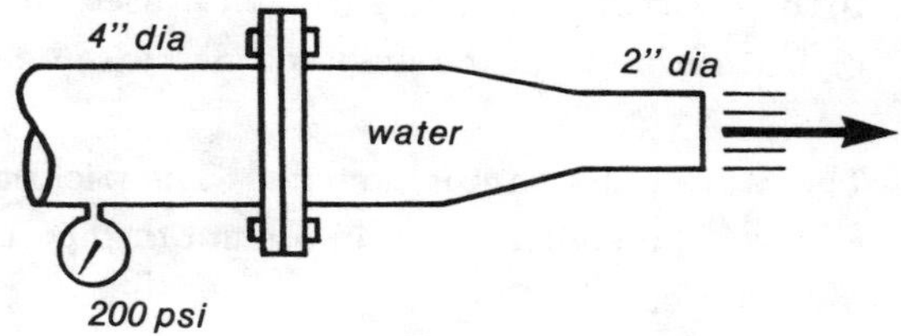

ENERGY

*11.44 The locus of elevations that water will rise in a series of pitot tubes is called

a) the hydraulic grade line.
b) the energy grade line.
c) the velocity head.
d) the pressure head.
e) the head loss.

*11.45 A pressure rise of 75 psi is needed across a pump in a pipe transporting 6 cfs of water. If the pump is 85% efficient, the power needed, in Hp, would be

a) 140 b) 130 c) 120 d) 110 e) 100

11.46 An 85% efficient turbine accepts 3 cfs of water at a pressure of 90 psi. What is the maximum power output, in Hp?

a) 100 b) 90 c) 80 d) 70 e) 60

11.47 If the turbine is 88% efficient, the power output, in Hp, is

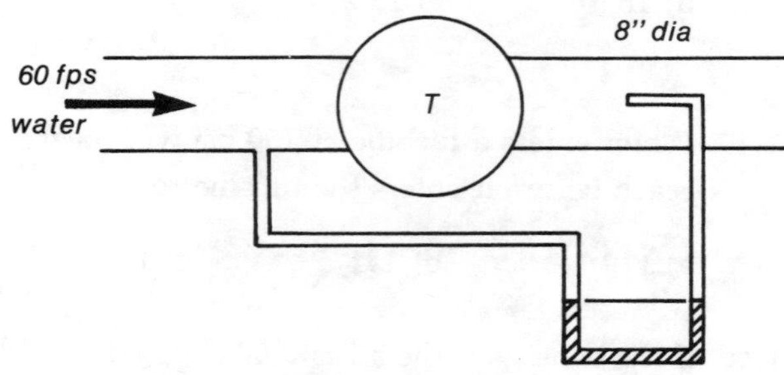

a) 60 b) 70 c) 80

d) 90 e) 120

LOSSES

*11.48 In a completely turbulent flow the head loss

a) increases with the velocity.
b) increases with the velocity squared.
c) decreases with wall roughness.
d) increases with diameter.
e) increases with flow rate.

*11.49 The shear stress in a turbulent pipe flow

a) varies parabolically with the radius.
b) is constant over the pipe radius.
c) varies according to the 1/7th power law.
d) is zero at the center and increases linearly to the wall.
e) is zero at the wall and increases linearly to the center.

*11.50 The velocity distribution in a turbulent flow in a pipe is often assumed to

a) vary parabolically.
b) be zero at the wall and increase linearly to the center.
c) vary according to the 1/7th power law.
d) be unpredictable and is thus not used.
e) be maximum at the wall and decrease linearly to the center.

*11.51 The Moody diagram is sketched. The friction factor for turbulent flow in a smooth pipe is given by curve

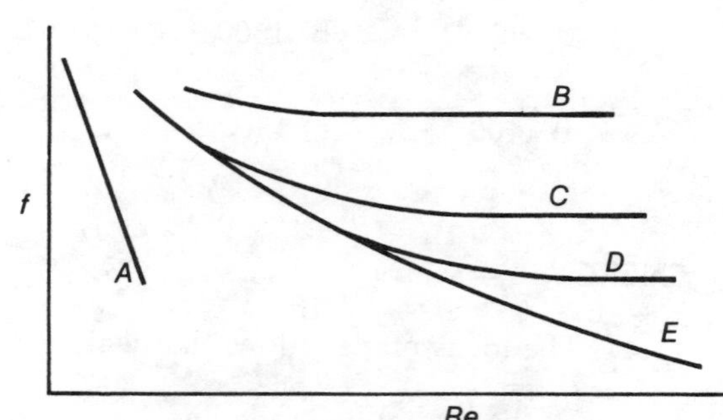

a) A b) B c) C

d) D e) E

*11.52 For the Moody diagram given in Problem 11.51, the completely turbulent flow is best represented by curve

a) A b) B c) C d) D e) E

*11.53 The pressure gradient ($\Delta p/\Delta x$) in a developed turbulent flow in a horizontal pipe

a) is constant.
b) varies linearly with axial distance.
c) is zero.
d) decreases exponentially.
e) varies directly with the average velocity.

*11.54 The head loss in a pipe flow can be calculated using

a) the Bernoulli equation.
b) Darcy's law.
c) the Chezy-Manning equation.
d) the momentum equation.
e) the Darcy-Weisbach equation.

*11.55 Minor losses in a piping system are

a) less than the friction factor losses, $f \frac{L}{D} \frac{V^2}{2g}$.
b) due to the viscous stresses.
c) assumed to vary linearly with the velocity.
d) found by using loss coefficients.
e) independent of the flow rate.

11.56 In a turbulent flow in a pipe we know the

a) Reynolds number is greater than 10,000.
b) fluid particles move in straight lines.
c) head loss varies linearly with flow rate.
d) shear stress varies linearly with radius.
e) viscous stresses dominate.

*11.57 Water flows through a 4" dia, 300-ft-long pipe connecting two reservoirs with an elevation difference of 120 ft. The average velocity is 20 fps. Neglecting minor losses, the friction factor is

a) 0.0257 b) 0.0215 c) 0.0197 d) 0.0193 e) 0.0182

*11.58 Find the power required, in Hp, by the 85% efficient pump if $Q = 0.6$ cfs.

a) 16.1 b) 14.9 c) 13.3
d) 11.2 e) 10.1

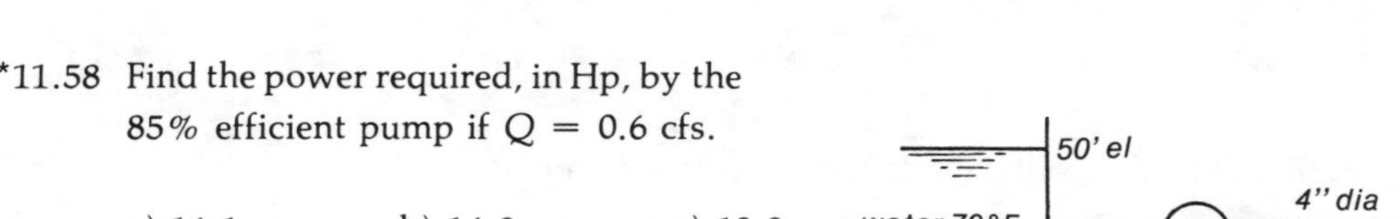

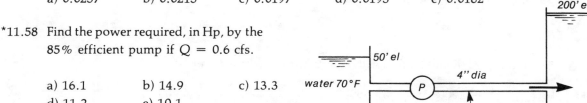

11.59 The pressure at section A in a 2" dia, wrought-iron, horizontal pipe is 70 psi. A fully open globe valve, two elbows, and 150 ft of pipe connect section B. If $Q = 0.2$ cfs of water, the pressure p_B, in psi, is

a) 35 b) 40 c) 45 d) 50 e) 55

11.60 Air at 70°F and 14.7 psia is transported through 1500 ft of smooth, horizontal, 40" x 15" rectangular duct with a flow rate of 40 cfs. The pressure drop, in psf, is

a) 2.67 b) 1.55 c) 0.96 d) 0.52 e) 0.17

11.61 Estimate the loss coefficient K in a sudden contraction $A_1/A_2 = 2$ by neglecting the losses up to the vena contracta A_c. Assume that $A_c/A_2 = 0.62 + 0.38\,(A_2/A_1)^3$ and $h_L = KV_2^2/2g$.

a) 0.28 b) 0.27 c) 0.26 d) 0.25 e) 0.24

11.62 An elbow exists in a 4" dia galvanized iron pipe transporting 0.6 cfs of water. Find the equivalent length of the elbow in ft.

a) 9.7 b) 10.3 c) 11.0 d) 12.1 e) 13.6

OPEN CHANNEL FLOW

*11.63 The depth of water in a 10-ft-wide, rectangular, finished concrete channel is 6 ft. If the slope is 0.001, estimate the flow rate, in cfs.

a) 460 b) 350 c) 290 d) 280 e) 270

11.64 At what depth, in ft, will 300 cfs of water flow in a 12-ft-wide, rectangular, brick channel if the slope is 0.001?

a) 7.5 b) 6.5 c) 5.5 d) 4.5 e) 3.5

11.65 A 6-ft-dia, brick storm sewer transports 100 cfs when it's nearly full. Estimate the slope of the sewer.

a) 0.0008 b) 0.002 c) 0.003 d) 0.004 e) 0.005

COMPRESSIBLE FLOW

*11.66 The pressure and temperature in a 4" dia pipe are 70 psia and 100°F, respectively. What is the mass flux, in slug/sec, if the velocity is 300 fps?

a) 2.43 b) 1.02 c) 0.763 d) 0.487 e) 0.275

*11.67 Air in a tank at 70°F and 75 psia exits a hole with a velocity, in fps, of

a) 1050 b) 1000 c) 960 d) 920 e) 900

11.68 A farmer uses 70°F nitrogen pressurized to 120 psia. Estimate the temperature, in °F, in the nitrogen as it exits a short hose fitted to the tank.

a) −170 b) −150 c) −130 d) −110 e) −90

11.69 Estimate p_1, in psia.

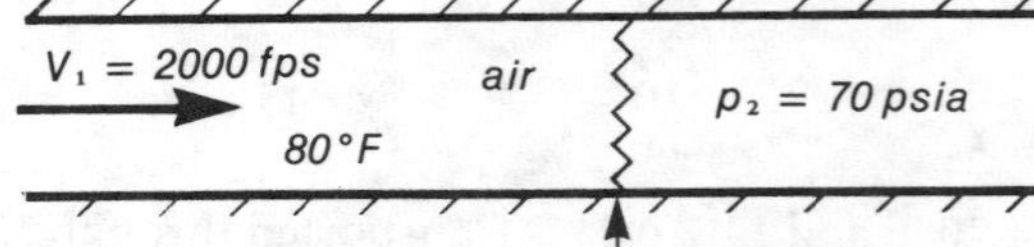

a) 20 b) 25 c) 30

d) 35 e) 40

11.70 Air leaves a reservoir and accelerates until a shock wave is encountered at a diameter of 4″. If the throat diameter is 3″, what is the Mach number before the shock wave?

a) 2.28 b) 2.22 c) 2.18 d) 2.12 e) 2.07

11.71 A supersonic aircraft flies at $M = 2$ at an elevation of 3000 ft. How long, in seconds, after it passes overhead is the aircraft heard? Assume $T = 70°F$.

a) 2.4 b) 2.3 c) 2.2 d) 2.1 e) 2.0

11.72 Air at 70°F is to exit a nozzle from a reservoir. What maximum pressure, in psia, can the reservoir have if compressibility effects can be neglected? $p_{atm} = 14.7$ psi.

a) 15.0 b) 15.2 c) 15.4 d) 15.6 e) 15.8

(This page is intentionally blank.)

12. Computer Science

Computing may be divided into two categories, analog and digital. Analog computing deals primarily with the solution and simulation of differential equations. Digital computing has much more general application possibilities. These two categories will be discussed in each of the following two articles. Emphasis will be placed on the major aspects only; many additional details would be required for a discussion of computing systems.

12.1 Analog Computing

Analog computing is primarily involved with the solutions of ordinary differential equations by simulation. Originally, an *analog* of the physical system to be studied was built using electrical or mechanical components. Another term for an analog computer is "differential analyzer." In addition, a number of programming languages for digital computers simulate the classical analog methods of solving differential equations.

12.1.1 Method of Solution

The functional basis of the analog computer is the "operational amplifier." It is a high negative gain amplifier with feedback and series input impedances. If both the series and feedback impedances are resistive, the operation is summing, or addition, as shown in Fig. 12.1a. The output voltage is given by

$$V_o = -R_f\left(\frac{V_1}{R_1} + \frac{V_2}{R_2} + \cdots + \frac{V_n}{R_n}\right) \tag{12.1.1}$$

If the series impedances are resistive and the feedback impedance is capacitive, the operation is integration as shown in Fig. 12.1b. A voltage, $V(0)$, placed on the capacitor determines the initial condition

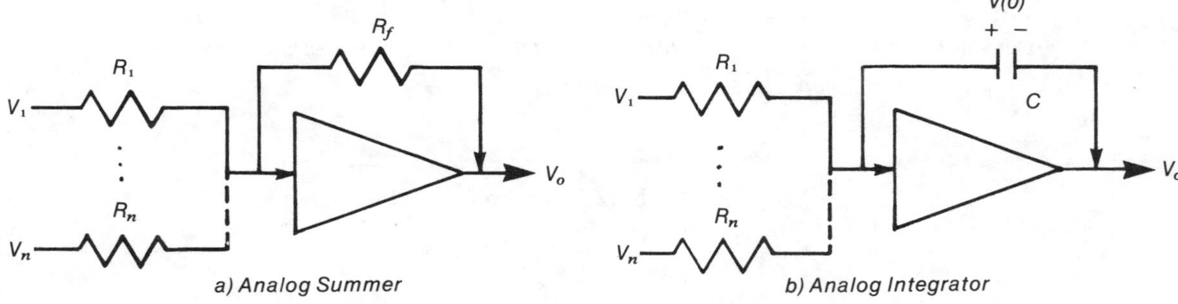

Figure 12.1 Analog computer elements.

of the integrator. The output voltage of the integrator is

$$V_o = \frac{-1}{C}\int_0^T \left(\frac{V_1}{R_1} + \frac{V_2}{R_2} + \cdots + \frac{V_n}{R_n}\right) dt - V(0) \qquad (12.1.2)$$

The general method of solution is to assume that the highest derivative of a variable is available and can be passed through a series of integrators to obtain the lower order derivatives and the variable itself. From the differential equation, the highest order derivative is the weighted sum of the lower order derivatives, the variable, and the driving function. Thus, a differential equation solution is one or more loops of operational amplifiers. Simultaneous differential equations may also be solved by adding additional loops of operational amplifiers.

In principle it is possible to solve a differential equation by differentiation rather than integration. In practice this is never done because differentiation causes "noise." An operational amplifier with capacitive series elements and a resistive feedback element acts as a differentiator.

It is understood that the ordinary differential equations to be solved are linear with constant coefficients. Nonlinear function elements are possible, but are difficult to model in practice; if the solution is to be done by a digital computer simulation, then nonlinear terms and variable coefficients pose no problem.

The driving function $f(t)$, which is the nonhomogeneous term in a differential equation, is usually obtained as a solution to another differential equation; however, it is often simply shown as $f(t)$ in an analog solution.

Operational amplifiers have input and output limitations. That is, their output voltage is limited to some maximum value, and, because of noise, the input voltage should be above some minimum level. Likewise, the constant coefficient is limited to the approximate range of 0.1 to 10.0 and is always negative, or "inverting." In commercial practice the coefficients are "round" values. Thus, a constant multiplier or "coefficient box" with a variable gain ($0 \leqslant g \leqslant 1$) may have to be used in order to obtain the desired value. Symbolic notation is shown in Fig. 12.2.

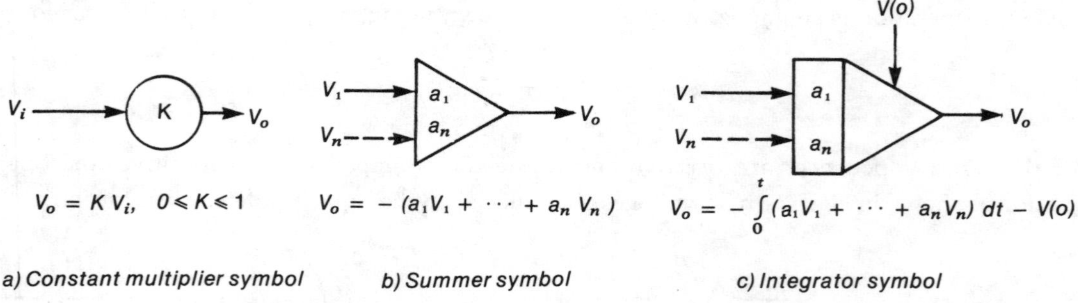

Figure 12.2. Symbolic notation.

It may be necessary to scale the differential equation being solved so that all amplifier voltages are within the linear operating range. With digital simulation, magnitude scaling can most often be ignored. It may also be desirable to change the time scale of the equations. If, for example, a certain physical system has time events occurring in the micro or millisecond range, it would be necessary to slow down the simulation so that the time behavior of the variables could be recorded with a pen and ink recorder. Likewise, it may be desirable in the simulation of a geological process to speed up time. Time scaling and magnitude scaling are separate operations.

12.1.2 Time Scaling

As an example, consider the third order differential equation

$$\dddot{x} + a_2\ddot{x} + a_1\dot{x} + a_0 x + f(t) = 0 \tag{12.1.3}$$

There is no loss of generality if the coefficient of the highest order derivative is unity. One possible operational amplifier solution is shown in Fig. 12.3.

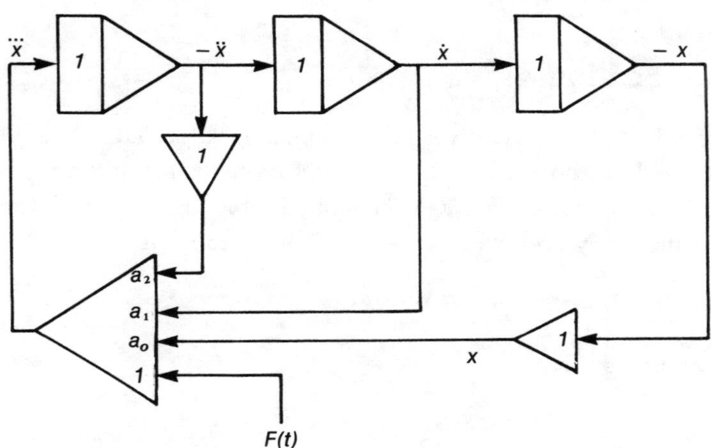

Figure 12.3. Analog solution to Eq. 12.1.3.

Time scaling can be defined as

$$T = h\,t \tag{12.1.4}$$

where T is the simulation, or computer time; t is real, or problem time; and h is the scaling factor. If $h > 1$, then computer time is slower than real time and hence the simulation runs slower. Conversely, if $h < 1$, then the simulation will be faster than real time. Substituting the change of time scale yields the simulation equation

$$h^3\dddot{x} + h^2 a_2\ddot{x} + h\,a_1\dot{x} + a_0 x + f(T/h) = 0 \tag{12.1.5}$$

An operational amplifier solution to this equation is displayed in Fig. 12.4.

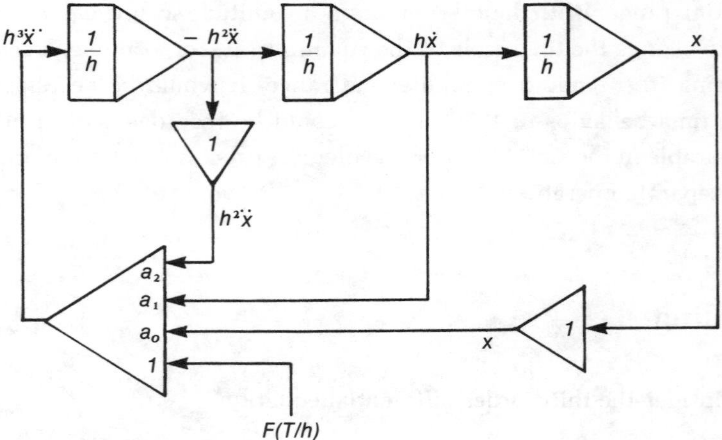

Figure 12.4 Analog solution to Eq. 12.1.4 with time scaling.

Note that time scaling may be achieved in the practical case by reducing the gain of each integrator by the factor h, the time scaling factor. The driving function is also changed in time scale. This general principle holds for all solutions. All simultaneous equations should be time scaled by the same factor.

12.1.3 Magnitude Scaling

The dependent variable and its derivatives must be magnitude scaled so the full linear range of each operational amplifier is used but not exceeded. It is therefore necessary to know the maximum expected values of all variables and derivatives. If x is a dependent variable, derivative, or forcing function and X is the simulation variable, then they are related by the scaling factor k as

$$X = k\,x \tag{12.1.6}$$

The scaling factor k is chosen so that

$$k \leqslant X_{max}/x_{max} \tag{12.1.7}$$

The scaling factor can be incorporated into each term in Eq. 12.1.3 by writing the equation as

$$(k_3\dddot{x})/k_3 + a_2(k_2\ddot{x})/k_2 + a_1(k_1\dot{x}_1x)/k_1 + a_0(k_0x)/k_0 + (f_mf(t))/f_m = 0 \tag{12.1.8}$$

Note that this procedure does not change the equation since each term is multiplied and divided by the scale factor; it simply scales the dependent variable and its derivatives. In simplifying the equation, the terms in the parentheses are kept intact. The reason that the simulation variables are not used will become apparent directly.

Consider the following example.

EXAMPLE 12.1

Using an analog computer simulation, solve the differential equation

$$2\ddot{x} + 18\,\dot{x} + 250\,x + 400\,e^{-0.8\,t} = 0$$

The initial conditions are $x(0) = 35$ and $\dot{x}(0) = 10$. The maximum values of the variable and derivatives presumed to be $x_{max} = 35$, $\dot{x}_{max} = 250$, and $\ddot{x}_{max} = 2500$.

Solution. Assuming that the maximum simulation voltage is 10 volts, then the scaling factors are determined using a maximum of 10; that is,

$$k_2 = 10/2500 = 0.004, \qquad k_1 = 10/250 = 0.04$$

$$k_0 = 10/35 = 0.2 \qquad f_m = 10/400 = 0.025$$

The simulation equation then becomes

$$(0.004\ \ddot{x}) + 0.9\ (0.04\ \dot{x}) + 2.5\ (0.2\ x) + 0.8\ e^{-0.8t} = 0$$

An operational amplifier solution to the equation is shown below and includes the integrator initial values.

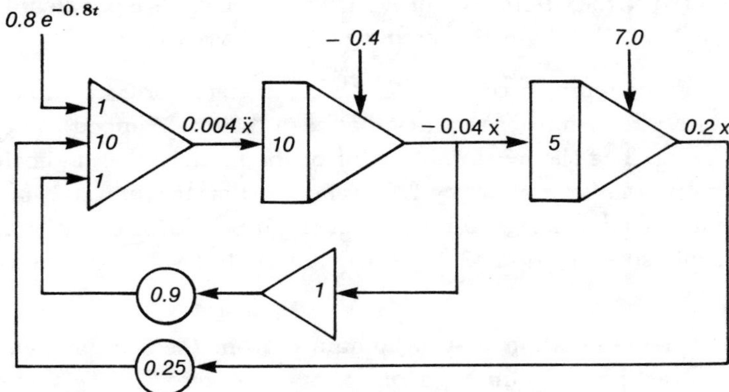

Note that by using the scaled problem variables rather than the simulation variables, the required operational amplifier coefficient values are immediately obvious. If it is desired to additionally time scale the solution by a factor of $h = 10$, then the driving function becomes $\exp(-0.08\ T)$ and the coefficient, or gain of the two integrators, is reduced by a factor of 10.

12.1.4 Operation Modes

Commercial analog computers have reset, hold, and compute modes. During the reset mode, initial condition voltages may be set into the integrator capacitors. The hold mode stops solution for closer examination. Afterwards, the solution may continue in the compute mode. Some analog computers can be automatically cycled between the reset and compute modes. This "repetitive operation" is most useful for adjusting parameters to obtain an "optimal" solution by observing a display of the solution on an oscilloscope.

12.2 Digital Computing

The range of subject material under this category is vast. Accordingly, the material presented here assumes a minimal level of knowledge. Textbooks on the subject can be consulted for more detailed information and thorough presentations.

12.2.1 Number Systems

Most digital computers represent numbers internally in base 2 (binary) rather than base 10 (decimal). A number may be represented as a polynomial in powers of its base, or radix. For clarification, consider the following example.

EXAMPLE 12.2

Evaluate the base 2 number 1101.01 using base 10.

Solution. $1101.1_2 = 1*2^3 + 1*2^2 + 0*2^1 + 1*2^0 + 0*2^{-1} + 1*2^{-2} = 13.25_{10}$

Of necessity, all numbers in a digital computer are of finite length and hence numerical precision must always be a consideration. It is convenient in handling long binary numbers to group the digits into groups of three or four. If grouped in threes, the numerical value for each group is between 0 and 7; if grouped in fours, the range is from 0 to 15. Thus, by grouping digits, a binary number may be represented either as a base 8 (octal) or base 16 (hexadecimal) number. In the case of hexadecimal numbers a convenient notation for digits larger than 9 is to use the letters "A" through "F" for the digits equivalent to 10 through 15. Thus, the integer 28 is 34 base 8 or 1C in base 16; the integer 3AC base 16 is 940 base 10.

Negative integers may be represented as either 1's complement or 2's complement notation. The 1's complement negation of a binary integer requires that each digit be "flipped" independently; that is, a 0 becomes a 1 and vice versa. The left-most digit or bit of the number then indicates its algebraic sign, a positive number by 0 and a negative number by 1. This implies that the integer 27 be represented by at least six bits as 011011. If interpreted as a signed integer, the 5-bit pattern 11011 is equivalent to −4 and not +27. The 1's complement representation of 0 is either 0..0 or 1..1 which are called plus zero and minus zero, respectively.

The 2's complement representation may be obtained from the 1's complement representation by adding 0..01 or by subtracting the number from a numeric zero. In 2's complement, zero is only represented by 0..0.

All arithmetic operations with signed numbers must be done with regard to the number of digits in the representation. If 1's complement is used, all carries (borrows) from the left-most position must be brought around and added (subtracted) into the right-most bit position. The end-around carry (borrow) is ignored with 2's complement arithmetic, as the following example illustrates:

EXAMPLE 12.3

Evaluate the base 2 number 1101.01 using base 10.

Solution. Conventional	1's Complement	2's Complement
+7	000111	000111
+ (−5)	+ 111010	+ 111011
+2	1 000001	1 000010 ignore carry
	+ 1 end-around carry	000010
	000010	

In all cases, the sign bit is treated as an ordinary numerical bit. The end-around carry (borrow) in 1's complement signals the need for the result correction resulting from the dual representation of zero. Numerical overflow during addition is signaled by two numbers of the same sign yielding a sum of the opposite sign. Numerical overflow during subtraction may occur only with oppositely signed numbers. It occurs when the sign of the difference is the opposite of the sign of the minuend (second number).

A collection of 8 bits, without regard to any numerical implications is referred to as a *byte*. Normally, a character in a text requires one byte. A half-byte, or 4 bits, is referred to as a *nibble*. When dealing with large binary numbers, the base 10 value 1024 = 2**10 is abbreviated by *K*. The value 1,048,576 = 2**20 is abbreviated by *M*. Thus, a computer memory that has 64K bytes actually has 65,536 bytes of capacity.

Real numbers are represented by a signed integer exponent and a signed mantissa similar to standard scientific notation. The representation format is computer dependent.

12.2.2 Combinational Logic

Boolean algebra deals with logical values, such as false or true, zero or one. The basic operations are "negation" or "NOT," usually indicated by an overscore; "union" or "OR," usually indicated by a plus sign; and "intersection" or "AND," usually indicated by a multiplication sign, a dot, or implied. These basic operations are supplemented by the "NAND" (Not AND), "NOR" (Not OR), and "XOR" (exclusive OR). These operations are defined in Fig. 12.5 with the standard schematic symbols shown. Other schematic symbols are also in common usage. All the operations may be extended to functions of more than two variables.

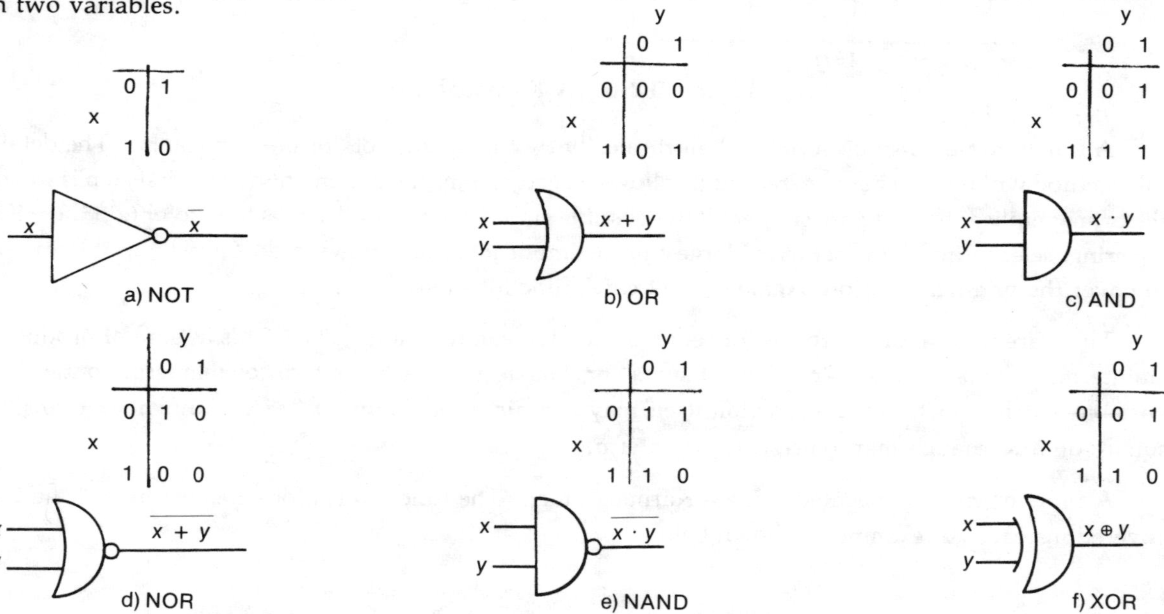

Figure 12.5. Schematic representations of the basic logic operations.

A given "black box" may not be classified as to its logical function without a knowledge of how its inputs and output(s) are mapped according to the Boolean values of "0" and "1." Consider a black box whose output is at 0 volts if one, two, or three of its three inputs are at 0 volts, and the output is at 5 volts only if all three inputs are at 5 volts. If 0 and 5 volts are equivalent to the Boolean 0 and 1, respectively, then this black box is a three-input AND gate. If, however, 0 and 5 volts are equivalenced to the Boolean 1 and 0, respectively, then this black box is a three-input OR gate.

The Karnaugh map is a tool for simplifying the functions of two, three, or four variables. But beyond four variables it rapidly loses its usefulness because of geometric considerations. The map simplifies functions with respect to NOT, OR, and AND operations. Because of input-output loading limitations (fan-in, fan-out) on the electrical circuitry of the gates, the most economical implementation of a Boolean function is not necessarily the simplest. This is particularly true if NAND, NOR, or XOR gates are to be used.

The Karnaugh map displays all 2***N* combinations of *N* variables such that vertically or horizontally adjacent squares differ by only one variable change. The left and right edges are considered adjacent and the top and bottom edges are adjacent. After plotting the function, simplification begins by trying to surround the 1's of the function with the largest possible "circles" consisting of 1, 2, 4, 8, or 16 squares. The minimum number of "circles" are selected such that all the 1's of the function are covered.

There are times when it is known that certain combinations of the variables cannot exist from the physical consideration of the problem. The combinations are called "don't care" functions and may be used in simplification. They are plotted on the Karnaugh map using an "X" rather than a "1." The name "don't care" arises because the use of an "X" is optional; they may be used as needed to make the largest possible circles and they do not have to be covered. Fig. 12.6 illustrates.

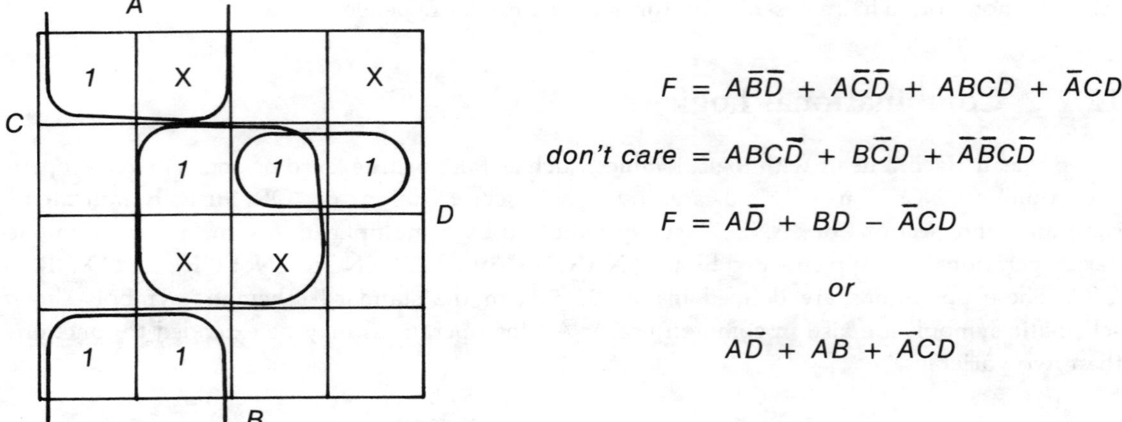

Figure 12.6. A Karnaugh map.

A function may also be simplified algebraically by the method of "prime implicants." The details of this method will not be shown here, but it follows directly from the map method. The first step is to obtain the prime implicants. They are simply all the possible circles. The second step is then to obtain the minimal covering: select a minimum set of the largest prime implicants (the ones with the fewest variables) in order to cover the original function. Again, "don't care" functions may be used.

There are two canonical forms for representing Boolean functions. The first is a "sum of products" or "minterm." The second is a "product of sums" or "maxterm." Each minterm or maxterm contains all N variables of the function. The function itself may contain a maximum of 2**N min or max terms. Each square on a Karnaugh map represents a minterm.

A function may be negated using a Karnaugh map. The function is plotted as before but the 0's are covered instead. An example is shown below.

EXAMPLE 12.4

Negate the function $A + \bar{B}C$ using a Karnaugh map.

Solution.

1	1		1
1	1		

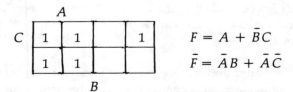

$F = A + \bar{B}C$

$\bar{F} = \bar{A}B + \bar{A}\bar{C}$

A function may be negated algebraically with DeMorgan's Law. All variables are complemented, with AND and OR interchanged. The trick is to properly parenthesize the expression so that the priority of AND over OR is explicit rather than implicit in all cases and then doing the interchanging and complementing. Some parentheses may be removed afterwards. Consider the following examples.

EXAMPLE 12.5

Negate the function $F = A + \bar{B}C$ with DeMorgan's Law.

Solution. We may write

$$F = A + \bar{B}C = A + (\bar{B} * C)$$

Upon complementing the variables and interchanging the AND-OR operators, the function becomes

$$\bar{F} = \bar{A} * (B + \bar{C}) = \bar{A}B + \bar{A}\bar{C}.$$

The dual of a function is obtained by interchanging the AND and OR operators. It is similar to negation but the variables are not complemented. Again, the proper parenthesization should be done beforehand.

12.2.3 Sequential Logic

The output in a combinational or memoryless logic circuit is a function of the current input only. If the output is a function of both the current input and past inputs (or outputs), then the circuit is sequential or has memory in it.

The behavior of a sequential circuit is described by a "state table." The state table describes the output(s) and the next state as functions of the input and present state. The "state" represents the memory. There may exist equivalent "machines" exhibiting identical behavior requiring a lesser number of states. If the outputs of a sequential circuit or machine are a function only of the state, the circuit is classified as a *Moore machine*. In another type called a *Mealy machine*, the outputs are a function of the state and the input. A state is represented by a state variable configuration. N state variables can implement up to $2**N$ states.

Sequential machines are either synchronous or asynchronous. *Synchronous* machines are defined only at discrete times controlled by an external clock. Their state variables must be implemented with flip-flops in order to hold the state variable values between clock pulses. The *asynchronous* machine is defined for all time and therefore does not need explicit memory for the state variables.

The asynchronous machine, although simpler in concept than a synchronous one, has two implementation restrictions necessary to ensure correct operation due to inherent circuit delays:

1. No more than one input variable may change at a time.
2. State variables must be assigned in such a way that no more than one state variable changes for any possible state changes.

Because of finite signal propagation times, it cannot be assumed that two variables which are supposed to change simultaneously will indeed do so or will always pass through the same intermediate configuration. The actual state transition may depend upon which path wins the "race." For example, the variable change 00 → 11 may actually be 00 → 01 → 11 or 00 → 10 → 11. Asynchronous machines will not be discussed further. There are no restrictions on the number of variable changes for synchronous machines. Either machine may have "don't care" entries.

Four types of flip-flops are in common usage. If X is the current flip-flop output, these four may be defined in either state table or equation form as shown in Fig. 12.7 below. The flip-flops also have a "clear" input, which when asserted forces the output to 0.

The general form of the state variable equation is

$$X_{next} = AX + B\bar{X} \tag{12.2.1}$$

where X is the state variable and A,B are not functions of X. For a given state variable application, the

D	X_{next}
0	0
1	1

$X_{next} = D$

D Flip-Flop

T	X_{next}
0	X
1	$\bar{X}$

$X_{next} = \bar{T}X + T\bar{X}$

T Flip-Flop

J K	X_{next}
0 0	X
0 1	0
1 0	1
1 1	$\bar{X}$

$X_{next} = \bar{K}X + J\bar{X}$

J-K Flip-Flop

S R	X_{next}
0 0	X
0 1	0
1 0	1
1 1	undefined

$X_{next} = S + \bar{R}X$, where $SR = 0$

S-R Flip-Flop

Figure 12.7. The four common flip-flops.

flip-flop input(s) must be found. This may be done by equating the desired flip-flop equation (Fig. 12.7) to the state variable equation (Eq. 12.2.1) and finding solutions for the input(s). The general solutions for each type are:

$$D = AX + B\bar{X} \qquad \text{D Flip-Flop} \tag{12.2.2}$$

$$T = \bar{A}X + B\bar{X} \qquad \text{T Flip-Flop} \tag{12.2.3}$$

$$J = B, \qquad K = \bar{A} \qquad \text{J-K Flip-Flop} \tag{12.2.4}$$

$$S = B\bar{X}, = B, \qquad R = \bar{A}X, = \bar{A} \text{ if } AB = 0 \qquad \text{S-R Flip-Flop} \tag{12.2.5}$$

The next example illustrates the above.

EXAMPLE 12.6

A synchronous Mealy machine with one input, one output, and six states as described below is to be realized. A is the initial state.

Present State	Input 0	Input 1
A	$C/0$	$C/0$
B	$A/0$	$A/0$
C	$D/0$	$E/0$
D	$B/0$	$F/0$
E	$F/0$	$F/0$
F	$A/0$	$A/1$

next state/output

Solution. This is a minimal state machine requiring at least three state variables. Although state variable assignment is arbitrary, the following rules of thumb seem to yield reduced cost logic:

1. Give adjacent assignments to a state and the state that follows it.
2. If two present states have the same next states, the present states should have adjacent assignments.
3. Use a minimum number of state variables.

These rules of thumb may be contradictory and compromises may have to be made in the assignment. Consider, for example, the following assignments:

$$A = 000,\ B = 100,\ C = 001,\ D = 110,\ E = 111,\ F = 010 \text{ and “don't care”} = 101{,}011$$

Note that with these assignments, the initial state may be forced by clearing the state variable flip-flops. By naming the state variables P, Q, and R, and the input variable I, the state and output equations become

$$P_{next} = P(Q\bar{R}\bar{I}) + \bar{P}(R)$$

$$Q_{next} = Q(R + PI) + \bar{Q}(R)$$

$$R_{next} = R(\bar{P}I) + \bar{R}(\bar{P}\bar{Q})$$

$$\text{output} = \bar{P}Q\bar{R}I$$

Note that the form of each state equation is that of Eq. 12.2.1.

Arbitrarily, let the P and Q state variables be implemented with J-K flip-flops and the R state variable be implemented by a T flip-flop. The flip-flop inputs become

$$J_P = R * \text{clock}, \qquad K_P = (\bar{Q} + R + I) * \text{clock}$$

$$J_Q = R * \text{clock}, \qquad K_Q = \bar{R}(\bar{P} + \bar{I}) * \text{clock}$$

$$T_R = (PR + \bar{I}R + \bar{P}\bar{R}\bar{Q}) * \text{clock}$$

Some flip-flops have a clock input built in and therefore do not require the explicit AND of the clock with all inputs. The resulting machine is shown below.

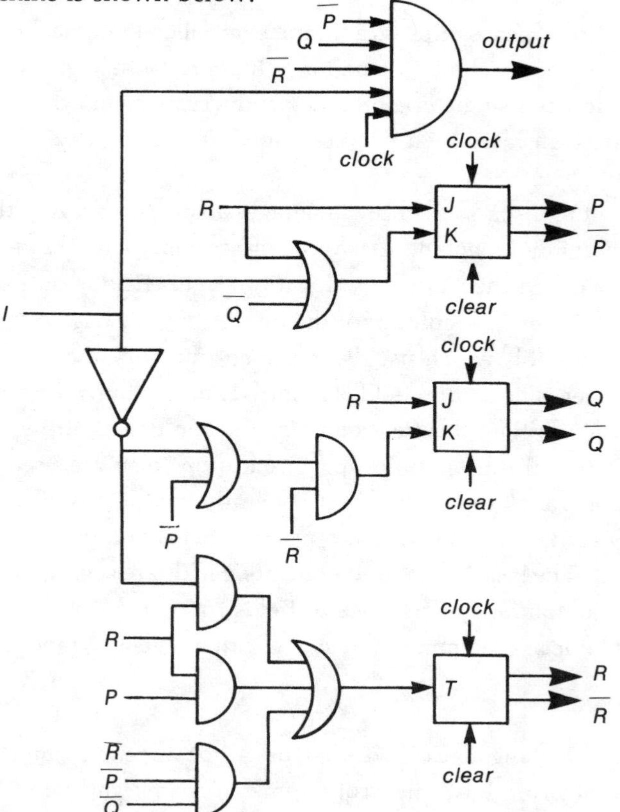

12.2.4 Operating Systems

An operating system is a program for interfacing between user and hardware. Its purpose is to provide an environment in which a user may execute programs conveniently and efficiently. The operating system may serve any number of users concurrently and independently.

A "batch" operating system never interacts between the user and the job once it is submitted. A measure of "goodness" is the average job turnaround time. A "time-sharing" or "interactive" system does allow a user to interact with the execution of the job. Interaction is accomplished by sharing the processor time and other resources by allocating each user a "time slice" in succession. The average response time to a user directive is the measure of goodness. A "real time" operating system is one that services external processes that have strict time constraints on response, such as for process control. Some interactive or real time operating systems process batch jobs in the "background" when there is no other "online" or external activity. Most operating systems are "multiprogrammed" wherein several user jobs reside within the computer system at any one time.

A typical operating system has text editing, language processing, library, and file storage capabilities. Text editing allows the user to enter, modify, and correct the program and data. Language processing consists of the "assemblers" and "compilers" for translating the user program written in a "source" language into the machine's instructions. The library facility, also called a "loader" or "linker," allows a user program to access utility programs. The file system permits the storage of user programs and data for future use. Files are stored on magnetic disks or magnetic tapes.

12.2.5 Programming

Assembly languages are a means of programming symbolically in machine language. Each line of code writing normally produces one machine instruction. Opposed to this are the high level languages such as FORTRAN or Pascal, wherein a single line of code produces many machine instructions. These so-called "high level" languages are translated by a process known as compiling.

It is extremely difficult to discuss assembly languages abstractly because they are very dependent upon the architecture of the individual machine. Each line of assembly code has four fields, some of which are optional. The first field is called the "label" or "location." This field is optional and if the line contains a label, it usually must start in the first column of the line. If the line has no label, column one of the line is left blank. Labels define symbols which usually represent memory location addresses. They are usually required to start with a letter. The second field is required and contains the operation. If the operation is to generate a machine instruction then this field contains the "op-code." If the operation is a directive to the assembler then this field contains a "pseudo-op." Pseudo-ops also are used to generate data items. The third field is called the "operand" or "address." It generally contains symbolic register names or memory locations for op-codes. For data generation pseudo-ops, this field contains the data; there are some operations which do not use this field. The last field is for comments. Fields are separated by one or more blanks or are assumed to start at certain columns in the case of fields following optional or not required fields. Most assemblers treat an entire line as a comment if it starts with a special symbol such as an asterisk or a "C" in column 1.

A feature that assembly languages have that does not appear in most high level languages is the "macro" definition. The macro allows the programmer to name and define parameterized sequences of code that may be referred to later in the program. This facility differs from a subroutine or procedure in that the body of the macro is inserted or assembled into the program at each place the macro name is invoked. If an assembler permits the use of macros it probably permits "conditional assembly." Conditional assembly allows the testing of parameter attributes and the subsequent conditional generation or assembly of code depending upon the test outcome. Strictly speaking, a conditional assembly facility is independent of the macro facility of an assembler. It is, however, of very limited utility outside the macro context.

High level languages are either compiled into native code for direct execution on the machine or into a pseudo-code for an interpretive execution. The advantage of an interpreter is that the programmer is allowed a much greater degree of interaction with the executing program than if it were allowed to execute directly on the machine itself. The chief disadvantage of an interpreter is that execution time is much slower than direct execution. Slow-down factors of 5 to 20 are common. FORTRAN is most often compiled into native code while BASIC is mostly interpreted. Pascal seems to be available both ways.

FORTRAN (FORmula TRANslator) was the first commercial high-level language and is the language still used most often by scientists and engineers. It is primarily a numerical computation language and one of the few which directly permit complex number arithmetic. It is easy to use and has a large library. BASIC (Beginner's All-purpose Symbolic Instruction Code) is a variant of FORTRAN. It does not require a differentiation between real and integer variables; nor are any variable declarations required. BASIC is a very easy language to use. Pascal (named after the French mathmatician Blaise Pascal) is a "modern" language. It requires that all variables be declared before their use and imposes a rigid structure upon the form of the program code and logic. The use of the "GOTO" is permitted but strongly discouraged by the availability of looping and control constructs. A program written in Pascal requires more preliminary design work before coding than would the same program written in either FORTRAN or BASIC. The net practical effect is that once the compiler-detectable errors are corrected, Pascal programs tend to be more execution-time error-free than either FORTRAN or BASIC programs. Current opinion among software engineers is that FORTRAN and BASIC are suitable for one-off programs, but for commercial quality, error free, large programs, a modern language such as Pascal is more suitable. Nevertheless, FORTRAN is the language of choice. The following paragraphs are a review of the main features of FORTRAN.

FORTRAN variable types may be INTEGER, REAL, DOUBLE PRECISION, COMPLEX, LOGICAL, and CHARACTER. Variables typed by default are INTEGER if the first letter is *I*, *J*, *K*, *L*, *M*, or *N* and are REAL if the first letter is any other letter. All array variables such as *X*(*I*,*J*), must be declared and their subscript bounds given. The default lower subscript is 1. Arrays may be declared in DIMENSION, COMMON, or type statements. With the EQUIVALENCE statement the same variable may have several names, types, and shapes. Variables are local to the program, subroutine, or function in which they are implicitly or explicitly declared, unless they have been declared in COMMON blocks, in which case they become shared. Variables within a COMMON block correspond by position and not by name. Variables can be initialized at compile time by the DATA statement.

Arithmetic expressions are evaluated following standard parenthesizing and operator conventions. The operators in order of priority from highest to lowest are exponentiation, multiplication and division, addition and subtraction, and negation. Expressions and subexpressions containing equal precedence operators are evaluated from left to right. The type of an expression or subexpression depends upon the types of the operands. If both operands are the same type, the result is that type. For mixed mode situations such as real/integer, the result is real. Mode conversion takes place also in assignment statements. Consider the following example using default variable typing.

EXAMPLE 12.7

What would the following FORTRAN statements print out?

```
      K=4/3
      L=10.0/4.0
      A=3/4
      B=4.0/10.0 + 3/4
      PRINT 6,K,L,A,B
6     FORMAT(1X,2I4,2F6.2)
```

Solution. Because of integer arithmetic truncation, the printout would be:

```
      1   2   .00   .40
```

Relational expressions can appear only with logical expressions. A relational expression is used to compare the values of two arithmetic or two character expressions and produces a logical result. The evaluation of a logical expression is determined by precedence among the logical operators unless modified by the use of parentheses. The logical operator precedence from highest to lowest is .NOT., .AND., .OR., and (.EQV. or .NEGV.). For example,

```
      (-4 .GT. 3.0) .OR. ((2 .NE. .3).AND.(1. .LE. 2.))
```

is evaluated as being '.TRUE.'.

The GOTO statement transfers control to the statement identified by the specific label. The computed GOTO statement transfers control to the statement label whose ordinal corresponds to the integer value of the control expression. Suppose we have the statements

```
      GOTO 15

      GOTO (19,4,17,100)   M/3 -2
```

The first statement always branches to statement label 15, while, if $M = 12$, the second will branch to label 4.

There are three types of IF statements: the arithmetical IF, the logical IF, and the block IF. The arithmetical IF has an arithmetical control expression and three statement labels. If the expression has a negative, zero, or positive value, control is transferred to the first, second, or third label, accordingly. The logical IF has a logical control expression and a statement to be executed if the expression is true. The statement is not executed if the control expression evaluates to false. The block IF statement is used with the ENDIF and, optionally, the ELSE statement. Good programming practice dictates, but doesn't require, that the statements within the block IF are to be indented. Examples of IF statements follow.

EXAMPLE 12.8

Write an example of each of the three types of IF statements.

Solution.

```
C  ARITHMETIC IF
      IF(I-1) 17,3,19

C  LOGICAL IF
      IF(I .EQ. -1) GOTO 3
      IF(I .GE. 0) J=J+K

C  BLOCK IF
      IF(K .NE. 15) THEN
        CALL YES(I,X)
      ELSE
        I=NO(X)
        A=3.5
      ENDIF
```

The DO statement is used to specify a loop that repeats a group of statements. It has a loop control variable whose initial, final, and increment values are specified. The increment defaults to 1 if not specified. Although not required, good programming practice dictates that each DO loop should terminate on its own CONTINUE statement and that its "body" be indented. The following example illustrates.

EXAMPLE 12.9

Write a DO loop within a DO loop using indentation.

Solution.

```
      DO  14  M  =  1,17
         DO 19 J  =  M,10,2
                A(M,J)=3*M-J
19       CONTINUE
         X(M)=0.2+M
14    CONTINUE
```

Input and output is done using READ, WRITE, and PRINT statements in conjunction with FORMAT statements. Unless unformatted input-output is to be used, each variable that is to be read in or written out must have a format descriptor. The format descriptor indicates how many columns the variable occupies or is to occupy. It also indicates the type of the variable. Integer variables should use an "Iw" descriptor and real variables should use either a "Fw.d" or "Ew.d" descriptor. The "w" indicates the field width in columns and the "d" indicates the number of digits after the decimal point. A partial example is shown in Ex. 12.7 above. Note: if the type specified in the format descriptor is different from the type specified in declaration statements, then disastrous results are very likely. Additionally, format descriptors permit the skipping of columns on input and the printing of characters or blanks on output.

Subroutines and functions have essentially the same structure as programs except for the word SUBROUTINE or FUNCTION appearing in their first line. Programs, subroutines, and functions are independent units and are compiled separately. They communicate only through parameter lists or by variables placed in common blocks. Subroutines are invoked by the CALL statement and functions are invoked by their name. It should be noted that a function invocation looks identical to an array reference. The fact that all arrays must be declared allows the compiler to know the difference. As with common blocks, the items in the parameter list of the invocation correspond to the variables in the subroutine/function definition only by position and not by name. The actual parameters in the invocating routine and the formal (dummy) parameters in the definition should be of the same or of a compatible type and shape (dimension). Likewise the type of a function definition should be the same as the type expected in the invoking subprogram.

The last statement of a program, subroutine, or function should be the END statement. In the case of a program it terminates execution. In a subroutine or function it serves as an implied RETURN statement. Termination may be done anywhere with the STOP statement.

12.2.6 Data Communications

Data may be transmitted either in serial, one bit at a time, or in parallel, several bits at a time. Within a piece of equipment, or between physically close units, parallel transmission is usually used. The connection is a multiconductor cable and the signals are usually sent "baseband." An additional signal, called either "strobe" or "clock," is necessary to indicate the presence of valid data. Over a relatively long distance, a single data path such as a telephone line or radio link is usually used. Consequently, data transmission is serial. Baseband is not used for distances because the low frequency limitations of such a data channel do not allow zero frequency or DC components. Instead, frequency or phase modulation of a carrier is employed.

With serial data transmission it is necessary to know the rate at which the individual bits are sent as there is no way to "strobe" the individual bits in a stream because, by the very definition of a serial channel, no marking pulses may be sent in addition to the data. It is necessary to have a locally generated clock at the receiving end. If it is possible to derive the beginning and end of each character from the bit stream itself, then the data transmission is termed "asynchronous." With "synchronous" transmission the separation of a bit stream into characters can be done only by counting bits from the start of the previous character.

A standard encoding is the 8 bits per character ASCII code. ASCII defines a 128-character set and thus only needs 7 bits. The extra bit (the leftmost bit) may be used as a parity bit for detecting an odd number of changed-bit errors or to indicate an alternate character set.

The convention in asynchronous transmission is to send a binary "1" (called a mark) when no data is being sent during an idle channel. Every character is prefixed by a 0 bit. This first bit is called a "start" bit and it signals the receiver local clock to start sampling the line for the succeeding bits. A terminal 1 or "stop" bit is appended to each character to ensure the idle condition. Consequently, each 8-bit character is actually sent as a 10-bit string. For a binary channel, bits per second is identical with the term "baud." Consequently, if characters are sent at 30 per second, the data rate is 300 baud. For historical reasons, data sent at 10 characters/second has a second stop bit appended, making an 11-bit string for each character, and is referred to as 110 baud.

The synchronous channel requires that a valid character be sent during any idle period in order to keep the receiver clock synchronized with the transmitter clock. It should be noted that no start-stop bits are used and hence there are 20% fewer bits than with the asynchronous channel. However, if synchronization is lost, resynchronization may take many character times to accomplish, rather than just one.

For transmission over the data channel, the baseband signal must modulate a carrier at the transmitter and demodulate it at the receiver. Two-way communication requires each end to have a modulator-demodulator or "modem." If data transmission is one-way only, it is said to be "simplex." If data can be transmitted in both directions over the channel simultaneously, the channel is said to be "full duplex." If data can be transmitted in only one direction at a time, then the channel is said to be "half duplex." The dial-up telephone network is limited to about 1200 baud. The local network is full duplex while the long distance network is half duplex.

(This page is intentionally blank.)

Practice Problems

ANALOG COMPUTING

12.1 An analog computer simulation is best suited to solve differential equations that are

a) nonlinear, constant coefficients.
b) linear, constant coefficients.
c) nonlinear, nonhomogeneous.
d) linear, variable coefficients.
e) nonlinear, variable coefficients.

PROBLEMS 12.2-12.4 RELATE TO THE INFORMATION THAT FOLLOWS.

An operational amplifier solution to the differential equation

$$25\ddot{x} + 50\dot{x} + 100x = 0$$

where $x_{max} = 5$, $\dot{x}_{max} = 10$, $\ddot{x}_{max} = 50$, $x(0) = 5$, and $\dot{x}(0) = 0$

assuming a maximum computer voltage of 10 V, is

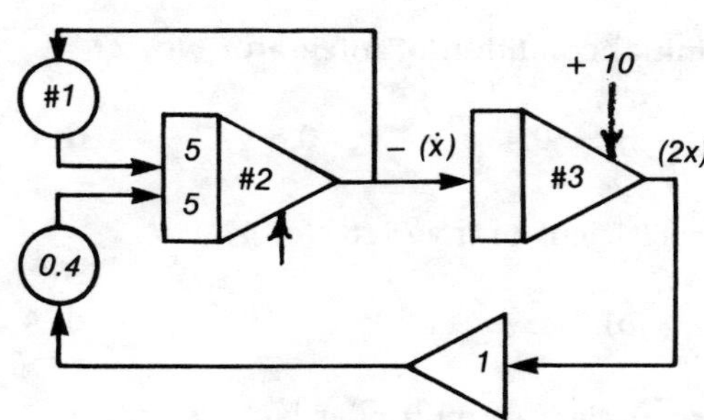

12.2 What is the value of the coefficient in multiplier No. 1?

a) 0.2 b) 1.0 c) 0.4 d) 5.0 e) 0.25

12.3 What is the value of the coefficient in integrator No. 3?

a) 1 b) 2 c) 4 d) 5 e) 10

12.4 The initial condition value of integrator No. 2

a) 0.0 b) 0.1 c) 1.0 d) −1.0 e) −0.1

PROBLEMS 12.5-12.9 RELATE TO THE INFORMATION THAT FOLLOWS.

The simultaneous differential equations

$$100\,\dot{x} + (x - y)/0.3 = 20$$

$$50\,\dot{y} + (y - x)/0.3 + y/0.2 = 0$$

where $x(0) = 6$, $y(0) = 4$, $x_{max} = y_{max} = 10$, $\dot{x}_{max} = 2$, $\dot{y}_{max} = 5$, have the operational amplifier solution for a maximum computer voltage of 1 V as shown.

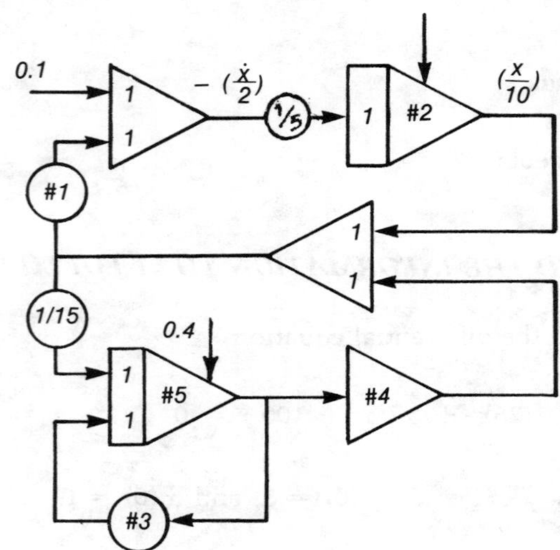

12.5 What is the value of coefficient in multiplier No. 1?

a) 1 b) 1/2 c) 1/4 d) 1/6 e) 1/10

12.6 What is the initial condition of integrator No. 2?

a) 0.4 b) −0.4 c) 0.6 d) 6.0 e) −6.0

12.7 What is the coefficient of inverter No. 4?

a) 1 b) 2 c) 3 d) 4 e) 5

12.8 What is the coefficient of multiplier No. 3?

a) 1/2 b) 1/5 c) 1/6 d) 1/10 e) 1/15

12.9 What is the scaling on the output of integrator No. 5?

a) y b) $2y$ c) $10y$ d) $y/10$ e) $y/15$

12.10 Repetitive operation in analog computers is most useful

a) in freezing a solution.
b) in setting initial condition voltages.
c) in repeating solutions to obtain the correct solution.
d) in adjusting parameters to obtain an optimal solution.
e) in obtaining simultaneous solutions.

NUMBER SYSTEMS

Convert the following numbers to their base 10 equivalents

12.11 101101.11 (base 2).

a) 42.25 b) 45.75 c) 48.68 d) 52.3 e) 55.5

12.12 24.6 (base 8)

a) 12.3 b) 14.75 c) 20.6 d) 20.75 e) 24.6

12.13 3FC.A (base 16)

a) 986.5 b)995.625 c) 1010.25 d) 1015.75 e) 1020.625

Convert the following numbers to 8-digit binary using 1's complement.

12.14 21 (base 10)

a) 00010001 b) 00010101 c) 10010101 d) 11101110 e) 11101010

12.15 −29 (base 10)

a) 00101001 b) 00011101 c) 11100010 d) 11100000 e) 11100011

12.16 −40 (base 10)

a) 11000000 b) 00101000 c) 11011000 d) 11010110 e) 11010111

Convert the following numbers to 8-digit binary using 2's complement.

12.17 21 (base 10).

a) 0001001 b) 00010101 c) 10101010 d) 11101110 e) 11101010

12.18 −29 (base 10).

a) 11100010 b) 11100011 c) 11100100 d) 00101000 e) 1100100

12.19 −40 (base 10).

a) 11011000 b) 11010111 c) 00100111 d) 00101000 e) 01111110

12.20 How many bytes does a 48K computer memory contain?

a) 47256 b) 48000 c) 48512 d) 49152 e) 51200

COMBINATIONAL LOGIC

12.21 A certain 3-input, 1-output device has the voltage characteristics displayed (numerical values are volts):

A	B	C	out
0	0	—	0
—	—	5	0
0	5	0	5
5	—	0	5

If the logical 0 is equivalent to 5 volts and the logical 1 is equivalent to 0 volts, the logical equation describing this device is

a) $A + C$ b) $A\bar{B}C$ c) $AB + \bar{C}$ d) $\bar{A}C + \bar{B}C$ e) $A + \bar{B} + C$

12.22 A minterm is

a) the minimum term in a Boolean function.
b) a prime implicant.
c) always smaller than a maxterm.
d) a circle on a Karnaugh map.
e) a square on a Karnaugh map.

Simplify the following Boolean functions.

12.23 $AB + A\bar{B}C + \bar{A}C$

a) $AB + BC$ b) $AC + B$ c) $AB + C$ d) A e) $A + B$

12.24 $F = \bar{B}\bar{C} + \bar{A}C + A\bar{B}C$, don't care $= ABC + \bar{A}B\bar{C}$

a) AC b) $A + C$ c) $AB + \bar{C}$ d) $\bar{A} + B$ e) $\bar{B} + C$

12.25 $F = \bar{A}BC + \bar{A}BD$, don't care $= (AB + \bar{A}\bar{B})(C + D)$

a) $BC + BD$ b) $BC + \bar{A}D$ c) $\bar{A}C + A\bar{D}$ d) $\bar{A}C + BD$ e) $\bar{A}\bar{B}CD$

Give the simplified complement of the functions.

12.26 $\bar{C}D + BD$

a) $B + CD$ b) $\bar{C}D + BD$ c) $\bar{D} + \bar{B}C$ d) $\bar{B} + CD$ e) $C + \bar{B}D$

12.27 $A\bar{B}C + A\bar{B}\bar{D} + BD$

a) $\bar{A}\bar{B} + B\bar{D} + \bar{B}\bar{C}D$
b) $AB + \bar{B}\bar{D} + \bar{C}D$
c) $A\bar{B}C + D$
d) $AD + BC$
e) $\bar{A}\bar{B}\bar{C}\bar{D}$

SEQUENTIAL LOGIC

Given the transition table of the synchronous Mealy machine and the state variable assignment...

PQ	state	MN, input 00	01	11	10
00	a	a/0	a/0	b/0	b/0
01	b	c/0	d/0	c/1	c/1
11	c	a/0	a/0	a/1	-/-
10	d	c/0	a/0	a/0	c/0

12.28... what is the output equation?

a) MQ b) $N\bar{P} + Q$ c) $M\bar{Q}$ d) $\bar{P}\bar{N} + PN$ e) $P + MQ$

12.29... if the P state variable is implemented with a J-K flip-flop, what is the "J" equation?

a) Q b) P c) $Q + N$ d) $\bar{Q} + N$ e) $\bar{Q}$

12.30... if the Q state variable is implemented with a S-R flip-flop, what is the "R" equation?

a) $\bar{P}N + \bar{P}M$
b) $(\bar{P}N + \bar{P}M)Q$
c) $(P + \bar{M}N)Q$
d) Q
e) P

OPERATING SYSTEMS

12.31 An operating system that allows the user to correct input data has

a) language processing.
b) multiprogramming.
c) text editing.
d) file storage.
e) real time capability.

12.32 FORTRAN variables may not be declared as

a) INTEGER
b) REAL
c) COMPLEX
d) TRIPLE PRECISION
e) CHARACTER

12.33 What is the value of the FORTRAN integer variable M? $M = 2 + 6*2/3**3$

a) 1 b) 2 c) 3 d) 4 e) 66

12.34 What number does the FORTRAN code segment print?

```
      DO 1 I=3,7,3
         J=I
1     CONTINUE
      PRINT 2,J
2     FORMAT(1X,I3)
```

a) 0 b) 3 c) 6 d) 7 e) 9

12.35 What does the following subroutine compute?

```
      SUBROUTINE MYSTERY(X,S)
      T=X
      S=X
      Z=3,4
1     CONTINUE
         T= -T*X*X**2/(Z*(Z-1.0))
         S1=S
         S=S+T
         Z=Z+2.0
      IF (S  .NE.  S1)  GOTO  1
      END
```

a) $X + X^3 + X^5 + X^7 + \cdots$
b) $X - X^3 + X^5 - X^7 + \cdots$
c) $X - X^3/3 + X^5/5 - X^7/7 + \cdots$
d) $X - 3X^3 + 5X^5 - 7X^7 + \cdots$
e) $X - X^3/3! + X^5/5! - X^7/7! + \cdots$

12.36 The subroutine of the previous question

a) will never terminate.
b) should have MYSTERY declared REAL.
c) should have variables *T* and *Z* declared.
d) will terminate.
e) needs a RETURN statement.

13. Systems Theory

Systems theory is concerned with the modeling, analysis and control of input-output processes with special focus on their time behaviors. The underlying structure describing time changes is assumed to be a set of differential or difference equations. If the variables of the process are studied at every point in time, then *differential equations* are used. Otherwise, as in digital systems, *difference equation* models are used to describe variables at discrete points in time.

Two fundamental properties a system can possess are: *linearity* and *time-invariance*. A *linear system* is one which exhibits the principle of superposition; meaning, given inputs u and v generate outputs y and z respectively, then input $(au + bv)$ generates output $(ay + bz)$, for any numbers a and b. A *time-invariant system* has the property that if input signal $u(t)$ generates output signal $y(t)$, then a time-shifted input signal $u(t + \tau)$ generates output signal $y(t + \tau)$ for any τ. Although it may be impossible to prove that a specific engineering process is linear and time-invariant, these properties do hold for certain operating regions of the important variables. Moreover, many practical systems are approximately linear and analyses based upon linear, time-invariant assumptions permit realistic comparisons with actual data, and subsequent control and design work.

13.1 Modeling and Analysis for Differential Equation Systems

There are several forms for describing systems where the variables are to be analyzed in continuous time:

1. the differential equations themselves rearranged into a *state model* form;

2. the *impulse response* giving the output variable for the case of an impulse input;

3. the *transfer function* expressing the ratio of the Laplace transform of the output to the Laplace transform of the input.

Consider a linear, time-invariant system with input signal $u(t)$ and output signal $y(t)$ as pictured in the block diagram of Fig. 13.1. The arrows are reminders of the direction of processing signals; $U(s)$ and $Y(s)$

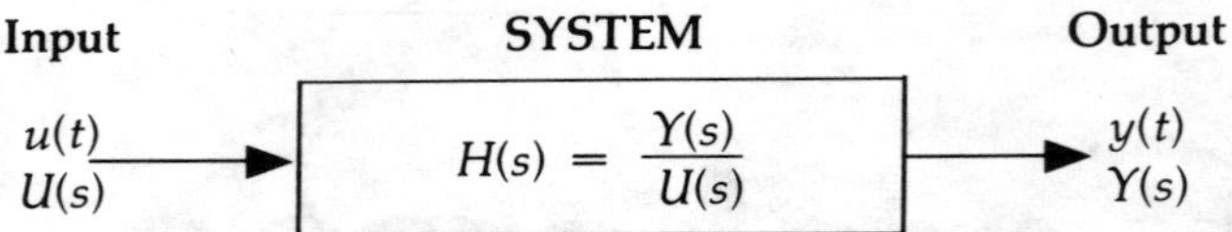

Figure 13.1. Block diagram of a linear, time-invariant system.

are the Laplace transforms of the input and output, respectively; and $H(s)$ is the system's transfer function. The inverse Laplace transform of $H(s)$ is the system's impulse response:

$$h(t) = \mathcal{L}^{-1}\{H(s)\}. \tag{13.1.1}$$

Using Tables 13.1 and 13.2, lets you solve a wide variety of continuous-time systems analysis problems.

TABLE 13.1 Laplace Transform Properties

$$\mathcal{L}[f(t)] = F(s) = \int_{0-}^{\infty} f(t)\, e^{-st}\, dt$$

$$\mathcal{L}[kf(t)] = kF(s),\ k \text{ a constant}$$

$$\mathcal{L}[f_1(t) + f_2(t)] = F_1(s) + F_2(s)$$

$\mathcal{L}[f_1(t)f_2(t)]$ does *not* equal $F_1(s)F_2(s)$

$$\mathcal{L}[f(t - T)] = e^{-sT}F(s),\ T \text{ a constant, provided } f(t) \text{ and } f(t - T) \text{ are both zero prior to } t = 0$$

$$\mathcal{L}[f(at)] = \frac{1}{a}F\left(\frac{s}{a}\right),\ a \text{ is a positive constant}$$

$$\mathcal{L}[e^{-at}f(t)] = F(s + a)$$

$$\mathcal{L}\left[\frac{df}{dt}\right] = sF(s) - f(0^-)$$

$$\mathcal{L}\left[\frac{d^2f}{dt^2}\right] = s^2F(s) - sf(0^-) - f'(0^-)$$

$$\mathcal{L}\left[\frac{d^nf}{dt^n}\right] = s^nF(s) - s^{n-1}f(0^-) - s^{n-2}f'(0^-) - \cdots - sf^{[n-2]}(0^-) - f^{[n-1]}(0^-)$$

$$\mathcal{L}\left[\int_{0-}^{t} f(t)dt\right] = \frac{F(s)}{s}$$

Initial value $f(0) = \lim_{s\to\infty} [s\,F(s)]$, provided limit exists.

Final value $f(\infty) = \lim_{s\to 0} [s\,F(s)]$, provided limit exists.

TABLE 13.2. LaPlace transform pairs

$f(t)$	$F(s)$
$\delta(t)$, unit impulse	1
$\underline{u}(t)$, unit step	$\dfrac{1}{s}$
$t\underline{u}(t)$	$\dfrac{1}{s^2}$
$t^n\underline{u}(t)$	$\dfrac{n!}{s^{n+1}}$
$e^{-at}\underline{u}(t)$, a a constant	$\dfrac{1}{s+a}$
$te^{-at}\underline{u}(t)$	$\dfrac{1}{(s+a)^2}$
$t^n e^{-at}\underline{u}(t)$	$\dfrac{n!}{(s+a)^{n+1}}$
$(\sin bt)\underline{u}(t)$, b a constant	$\dfrac{b}{s^2+b^2}$
$(\cos bt)\underline{u}(t)$	$\dfrac{s}{s^2+b^2}$
$(t \sin bt)\underline{u}(t)$	$\dfrac{2bs}{(s^2+b^2)^2}$
$(t \cos bt)\underline{u}(t)$	$\dfrac{s^2-b^2}{(s^2+b^2)^2}$
$(e^{-at} \sin bt)\underline{u}(t)$	$\dfrac{b}{(s+a)^2+b^2}$
$(e^{-at} \cos bt)\underline{u}(t)$	$\dfrac{(s+a)}{(s+a)^2+b^2}$
$2Me^{-at}\cos(bt+\theta)\underline{u}(t)$	$\dfrac{Me^{j\theta}}{s+a-jb} + \dfrac{Me^{-j\theta}}{s+a+jb} = \dfrac{\text{numerator}}{(s+a)^2+b^2}$

Note: In this chapter $j = \sqrt{-1}$ (i is reserved for current).

EXAMPLE 13.1

An R-L-C (resistor-inductor-capacitor) electrical network with output voltage $y(t)$ and driving supply voltage $u(t)$ is governed by the differential equation

$$\frac{d^2y}{dt^2} + 4\frac{dy}{dt} + 3y = u(t).$$

Find the system's impulse response. If the input is a step of 12 volts, determine the transient and steady-state components of the output.

Solution. First, we take the Laplace transform of the equation using zero initial conditions:

$$(s^2 + 4s + 3)\, Y(s) = U(s).$$

Then, the transfer function is

$$H(s) = \frac{Y(s)}{U(s)} = \frac{1}{(s+1)(s+3)} = \frac{0.5}{s+1} + \frac{(-0.5)}{s+3}$$

and the impulse response takes the form

$$h(t) = 0.5\, e^{-t} - 0.5\, e^{-3t}, \text{ for } t \geqslant 0.$$

With the step input, $u(t) = 12\, \underline{u}(t)$ and $U(s) = \frac{12}{s}$. Thus,

$$Y(s) = \frac{12}{s(s+1)(s+3)} = \frac{4}{s} + \frac{(-6)}{s+1} + \frac{2}{s+3}$$

yielding the output to the step input as

$$y(t) = 4 - 6\, e^{-t} + 2\, e^{-3t}, \text{ for } t \geqslant 0.$$

The first term is the steady-state response: $y(\infty) = 4 = \lim_{s \to 0} [sY(s)]$; the other two express the transients which fade to zero as t gets large.

EXAMPLE 13.2

A K-M-B (spring-mass-damper) mechanical accelerometer is designed so that the position $y(t)$ of the mass is related to the acceleration $u(t)$ of the vehicle in which it is mounted by the transfer function

$$H(s) = \frac{Y(s)}{U(s)} = \frac{1}{s^2 + (\frac{B}{M})\, s + \frac{K}{M}} .$$

Given an application where the mass $M = 0.05$ kg, find the spring and friction coefficients, K and B respectively, so that this system is critically damped with an undamped natural frequency of 100 rad/s.

Solution. To solve this problem, we compare the denominator of the transfer function with the standard form for second-order transient behavior:

$$s^2 + 2\xi\omega_n\, s + \omega_n^2$$

where ξ = the damping coefficient ($\xi = 1.0$ for critical damping) and ω_n = undamped natural frequency. The time behavior has the form

$$c\, e^{-\xi\omega_n t}\, [\cos \omega_n \sqrt{1 - \xi^2}\; t + \theta].$$

For the above data,

$$\omega_n^2 = 10^4 = \frac{K}{M} . \qquad \therefore K = 500.$$

Critical damping implies

$$\frac{B}{M} = 200 = \frac{B}{0.05} . \qquad \therefore B = 10.$$

EXAMPLE 13.3

A first-order linear system is modeled by the differential equation

$$\frac{dy}{dt} + 4\, y(t) = u(t).$$

Express the unit step and unit ramp responses in terms of the system eigenvalue and the system time constant.

Solution. We note that, in general, the system eigenvalues are the poles of the transfer function and that first-order transients have the denominator factor

$$s + \frac{1}{\tau}$$

where τ = time-constant = the time for the step response to rise to $1 - e^{-1}$ (that is, about 63%) of its steady-state value.

Here, the transfer function is

$$H(s) = \frac{1}{s + 4}$$

yielding time constant $\tau = 0.25$ and system eigenvalue $\lambda = -4$. For a unit step input,

$$Y(s) = H(s)\, U(s) = \left(\frac{1}{s + 4}\right)\left(\frac{1}{s}\right) = \frac{0.25}{s} + \frac{(-\,0.25)}{s + 4}$$

and the step response is

$$y(t) = 0.25\,[1 - e^{-4t}]\,.$$

Using $y(\infty)$ to denote the steady-state value (here 0.25), the general form of the step response can be written

$$y(t) = y(\infty)[1 - e^{-t/\tau}], \text{ for } t \geqslant 0.$$

For a unit ramp input, $u(t) = t\underline{u}(t)$ and $U(s) = 1/s^2$. Thus,

$$Y(s) = \left(\frac{1}{s + 4}\right)\left(\frac{1}{s^2}\right) = \frac{(-1/16)}{s} + \frac{0.25}{s^2} + \frac{1/16}{s + 4}$$

and the ramp response is

$$y(t) = -\frac{1}{16} + 0.25\, t + \frac{1}{16}\, e^{-4t}, \text{ for } t \geqslant 0.$$

We can use the above result to verify the general property of linear systems:

$$\frac{d}{dt}\,(\text{ramp response}) = \text{step response}.$$

EXAMPLE 13.4

For the system in Ex. 13.3, determine the frequency response to the sinusoidal waveform $u(t) = a \cos(\omega t + \theta)$ for arbitrary frequency ω. Generalize this result for any stable linear system with transfer function $H(s)$. Why is stability required?

Solution. We first note that "frequency response" refers to the steady-state sinusoidal response waveform that remains after the transient portions of the response have died out. The transient portions arise from the denominator factors in the transfer function. The roots of this denominator are the system eigenvalues. Thus, *stability*, which means that the transients die to zero as $t \to \infty$ (or, equivalently, that the system eigenvalues have negative real parts), is an essential ingredient for an analysis of frequency response.

The given sinusoidal waveform with frequency ω, amplitude a, and phase θ has Laplace transform:

$$U(s) = \frac{e^{j\theta}[a\ s]}{s^2 + \omega^2}\ .$$

Multiplying this by $H(s)$, we get $Y(s)$. After expanding in partial fractions — ignoring the terms which come from the denominator of $H(s)$ — and inverting the Laplace transforms, we obtain the sinusoidal steady-state response:

$$y(t) = a\ |H(j\omega)| \cos(\omega t + \theta + \arg H(j\omega)).$$

This equation shows how the output waveform has the same frequency as the input, has an amplitude gain (or attenuation) governed by the *gain factor* $|H(j\omega)|$, and has a phase shift relative to the input governed by the *phase angle factor arg* $H(j\omega)$, which is the angle of the complex number $H(j\omega)$ at frequency ω.

For the system of Ex. 13.3

$$H(j\omega) = \frac{1}{j\omega + 4} \to |H(j\omega)| = \frac{1}{\sqrt{16 + \omega^2}}$$

$$\arg H(j\omega) = -\tan^{-1}\left(\frac{\omega}{4}\right).$$

With,

$$u(t) = \cos(4t + 30°)$$

we get

$$y(t) = \frac{1}{\sqrt{32}} \cos(4t + 30° - 45°)$$

whereas, with

$$u(t) = \cos(40t + 30°),$$

we get

$$y(t) = \frac{1}{\sqrt{1616}} \cos(40t + 30° - \tan^{-1} 10).$$

Systems such as this last example, in which the gain factor $|H(j\omega)|$ becomes much smaller as ω progresses from low frequencies near zero to $\omega \to \infty$, are said to be *low pass filters.* When the gain factor is large for high frequencies and small for $\omega \approx 0$, we say the system is a *high pass filter.* If there is an intermediate range where $|H(j\omega)|$ is large, the system is said to be *band pass.* In all these cases a quantitative measure of the width of the frequency band for which $|H(j\omega)|$ exceeds 0.707 times its maximum value is defined to be the system's *bandwidth.*

EXAMPLE 13.5

For $H(s) = \dfrac{1}{s + 4}$, find the bandwidth BW of the low pass filter.

Solution. Considering $0 \leqslant \omega < \infty$, the maximum gain factor is obtained at $\omega = 0$:

$$|H(j0)| = 0.25.$$

Then

$$\frac{1}{\sqrt{16 + (\mathrm{BW})^2}} = (0.707)(0.25)$$

or

$$\mathrm{BW} = 4 \text{ rad/s}.$$

This gain factor and bandwidth of a low pass filter are displayed in the following figure.

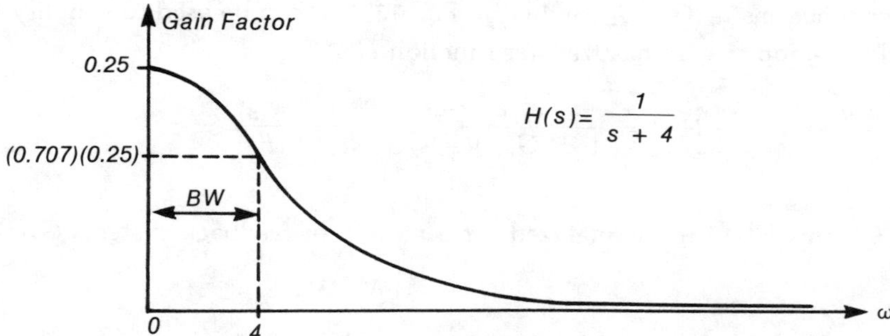

Note: We observe that for any first-order denominator factor $(s + \frac{1}{\tau})$ in $H(s)$, the bandwidth BW and time-constant τ are reciprocals. This illustrates the similarity of a design requirement for a "small time constant" or a "large bandwidth."

13.2 Feedback and Principles of Control Systems

A common problem in the analysis of systems is how to choose the input waveforms (i.e., control signals) or design parameters so that the output signal has a desirable form, as measured by some index or indices of performance. We find it convenient to designate that portion of the system we cannot change as the *plant* (see Fig. 13.2).

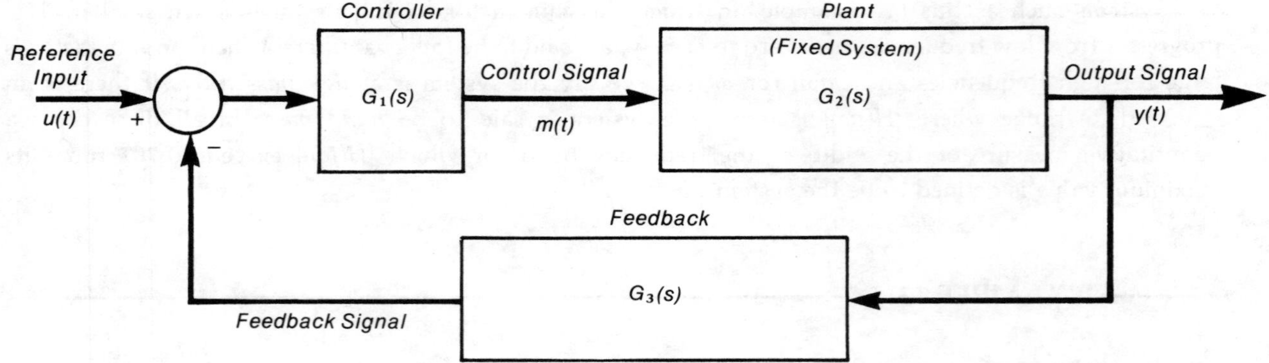

Figure 13.2. A feedback control system.

This plant is driven by the control signal $m(t)$ and its output is $y(t)$. The *control problem* is to choose $m(t)$ so that $y(t)$ behaves well (or optimally). Similarly, we can view $m(t)$ as the response of a system — called the *controller* — whose parameters we can choose, and whose input may be influenced by reference inputs and feedback signals (see Fig. 13.2).

The absence of a feedback signal is termed *open-loop control*; the opposite case is called *closed-loop control*. Obviously, the presence of feedback is important in that the control signal can adapt to actual changes in the plant model as exhibited by the output signal data. This feature gives a designer confidence in the system's reliability, especially when model parameters are not precisely known. Moreover, if the plant's performace is unsatisfactory (as in unstable systems, large or highly oscillatory transients, highly sensitive parameters, poor steady-state behavior, inadequate frequency response), then the addition of feedback may improve behavior.

When the controller, plant, and feedback systems are each modeled as a linear time-invariant system, appropriate transfer functions — G_1, G_2, and G_3 in Fig. 13.2 — can be used to simplify the descriptions. Then the overall closed-loop system has transfer function $H(s)$

$$H(s) = \frac{G_1(s)\,G_2(s)}{1 + G_1(s)G_2(s)G_3(s)} = \frac{Y(s)}{U(s)} \,. \tag{13.2.1}$$

For fixed plant $G_2(s)$, this $H(s)$ can be analyzed for a variety of feedback systems $G_3(s)$ and controllers $G_1(s)$.

One important case is unity feedback, $G_3(s) = 1$, and *PID* (proportional-integral-derivative) controller

$$G_1(s) = K_p + \frac{K_i}{s} + K_d s. \tag{13.2.2}$$

Here K_p = proportional control constant, K_i = integral control constant, and K_d = derivative control constant. Then the control problem is the choosing of parameters K_p, K_i, and K_d for good closed-loop system performance — e.g., stability, steady-state performance, quick but not unduly oscillatory transients, etc. If stability and transient behavior are central concerns, then $H(s)$ and the system eigenvalues — poles of $H(s)$ — need to be studied. The *root-locus diagram* is a plot of the system eigenvalues (characteristic roots) as some design parameter like K_p (or K_i or K_d) is varied. Two other frequently used techniques for investigating stability are the *Routh-Hurwitz Criterion* and the *Nyquist-Bode Method*.

EXAMPLE 13.6

Consider the position control system shown below with marginally unstable plant, unity feedback, and proportional controller gain K.

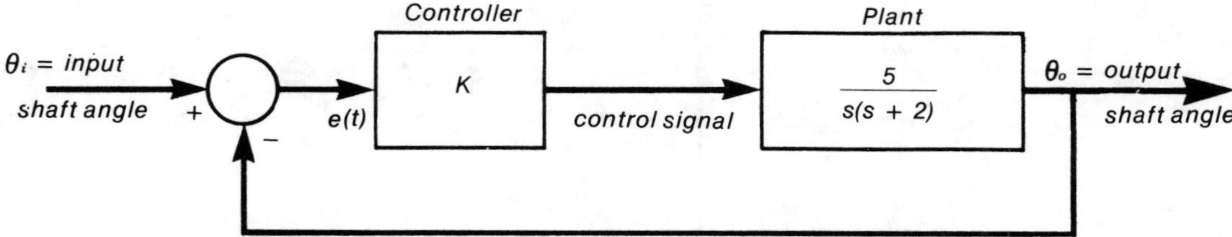

For what value of K will the closed-loop system be stable with an underdamped transient response having damping ratio $\xi = 0.5$? Find the undamped natural frequency ω_n and steady-state error $e(\infty) = \theta_i(\infty) - \theta_o(\infty)$ for unit ramp inputs corresponding to this value of K.

Solution. A control system with unity feedback where you want to make the output equal the input, i.e., to make the error $e = \theta_i - \theta_o$ be zero, is called a *servomechanism*. To analyze the servomechanism shown in the figure, we compute the closed-loop transfer function using Eq. 13.2.1

$$H(s) = \frac{5K}{s^2 + 2s + 5K} = \frac{\theta_o(s)}{\theta_i(s)} .$$

The denominator form $s^2 + 2\xi\omega_n s + \omega_n^2$ yields $\omega_n = \sqrt{5K}$ and $\xi\omega_n = 1$. Thus, $\omega_n = 2$ and $K = 0.8$ meets the specification of $\xi = 0.5$.

Steady-state error values can be obtained from $E(s)$ and the Laplace Final Value Theorem:

$$E(s) = \theta_i(s) - \theta_o(s) = [1 - H(s)]\,\theta_i(s)$$

$$e(\infty) = \lim_{s \to 0} s\,E(s) = \lim_{s \to 0} [1 - H(s)]\,s\,\theta_i(s).$$

Since $\theta_i(s) = \frac{1}{s^2}$ for unit ramp inputs,

$$e(\infty) = \lim_{s \to 0} \left[\frac{s + 2}{s^2 + 2s + 5K} \right] = \frac{2}{5K} = 0.5.$$

Note that the steady-state error for ramps generally improves (becomes smaller) as the gain K increases, whereas stability and transient performance usually deteriorate if the gain is made too large. For other inputs such as steps, a brief calculation in the example shows that $e(\infty)$ is 0 regardless of the gain K.

EXAMPLE 13.7

Obtain the root-locus diagram for the control system in the figure of Ex. 13.6 as gain K is varied through the range $0 \leqslant K < \infty$.

Solution. Noting that

$$\frac{\theta_o(s)}{E(s)} = \frac{5K}{s(s+2)}$$

we plot the two closed-loop system eigenvalues in the complex s-plane as shown in the root-locus diagram below.

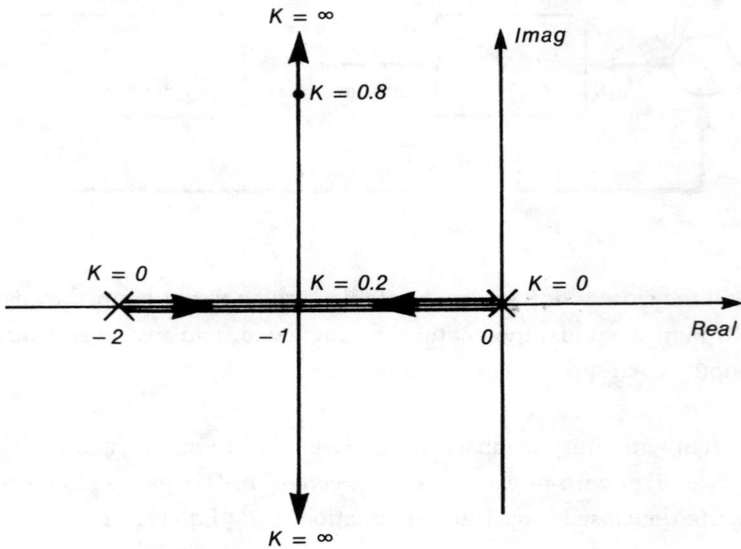

The extreme case $K = 0$ shows the marginal stability of the plant; $K = 0.2$ is critical damping with both eigenvalues at -1; $K > 0.2$ yields underdamped transients of the form $e^{-t}\,[\alpha \cos (\omega_n \sqrt{1-\xi^2}\; t + \beta)]$.

EXAMPLE 13.8

Consider the 4th-order control system with proportional plus integral controller shown below.

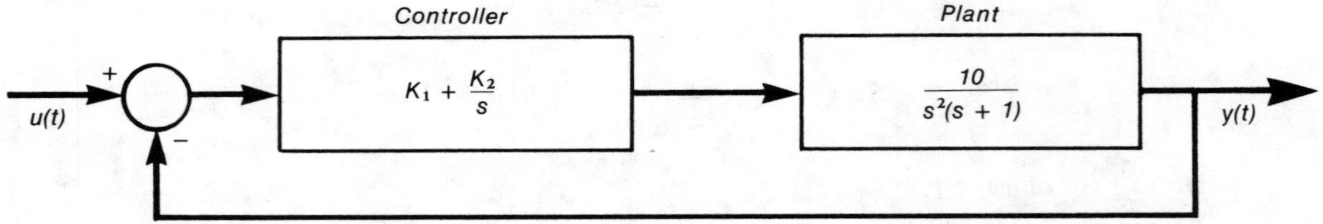

Use the Routh-Hurwitz Criterion to determine the number of closed-loop eigenvalues in the right-half s-plane when $K_1 = 1$ and $K_2 = 2$.

Solution. We first set the denominator of $H(s)$ equal to 0 to obtain the characteristic equation

$$1 + \left(1 + \frac{2}{s}\right) \frac{10}{s^2\,(s+1)} = 0.$$

Rearranging, we get

$$s^4 + s^3 + 10s + 20 = 0.$$

The Routh array is

$$\begin{array}{c||ccc} s^4 & 1 & 0 & 20 \\ s^3 & 1 & 10 & \\ s^2 & -10 & 20 & \\ s^1 & +12 & & \\ s^0 & +20 & & \end{array}$$

Here the array entries r_{ij} for the first two rows are obtained by inspection of the characteristic polynomial, $a_n s^n + a_{n-1} s^{n-1} + \cdots + a_o$, as follows: $r_{11} = a_n$, $r_{21} = a_{n-1}$, $r_{12} = a_{n-2}$, $r_{22} = a_{n-3}$, $r_{13} = a_{n-4}$, etc., until all the coefficients are used (including zeroes). Row 3 is obtained from rows 1 and 2 via the formulas

$$r_{31} = \frac{\begin{vmatrix} r_{11} & r_{12} \\ r_{21} & r_{22} \end{vmatrix}}{-\ r_{21}},$$

$$r_{32} = \frac{\begin{vmatrix} r_{11} & r_{13} \\ r_{21} & r_{23} \end{vmatrix}}{-\ r_{21}}, \text{ etc.}$$

The remaining entries are completed for row 3 (eventually zeroes are obtained). Then row 4 is obtained from rows 2 and 3 via the same pattern. (Special adjustments for completing the array can be made when a zero occurs in column 1.)

The Routh-Hurwitz Criterion states that the number of algebraic sign changes in the first column of this array equals the number of eigenvalues with positive real part. Thus, for these values of K_1 and K_2, 2 of the 4 eigenvalues are in the right-half s-plane and the system is unstable.

EXAMPLE 13.9

A position servo consists of a synchro, amplifier and a motor operating under load in the configuration shown in the figure.

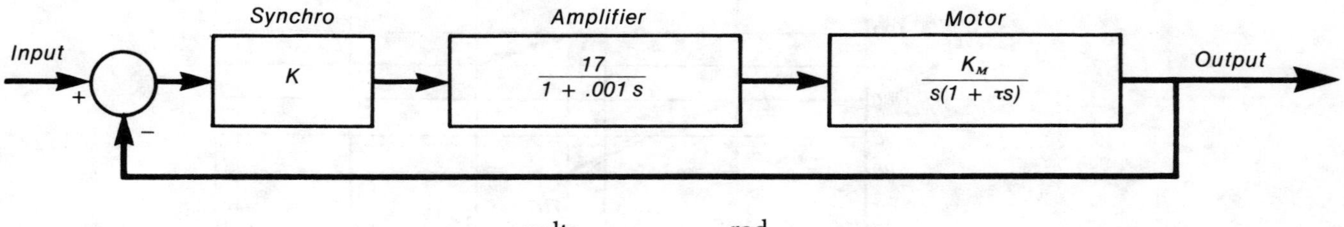

$$(K = 5\ \frac{\text{volts}}{\text{rad}},\ K_M = 12\ \frac{\text{rad}}{\text{volt}}).$$

For the motor in new condition unconnected to a load, the time constant τ is 0.039 s. Wear in the motor sleeve bearings and wear in the motor shaft couplings to the load increases the value of τ. Use the Routh-Hurwitz Criterion to determine the τ value for which the system becomes unstable.

Solution. The characteristic equation is

$$1 + \frac{5\ (17)(12)}{s(1 + .001\ s)(1 + \tau\ s)} = 0$$

Rearranging, we get

$$.001\ \tau s^3 + (.001 + \tau)\ s^2 + s + 1020 = 0$$

The Routh array is

$$\begin{array}{c||ll} s^3 & .001\ \tau & 1 \\ s^2 & .001 + \tau & 1020 \\ s^1 & \dfrac{.001 + \tau - 1.02\ \tau}{.001 + \tau} & \\ s^0 & 1020 & \end{array}$$

Stability requires all plus signs in column 1. Thus $0 < \tau < .05$ is the stable range. The τ value for which the system becomes unstable is .05.

EXAMPLE 13.10

The control system of Fig. 13.2 with proportional controller $G_1 = K$, plant $G_2 = \dfrac{100}{(s+1)^3}$, and unity feedback $G_3 = 1$ exhibits a frequency response for the open-loop as shown in the following Bode magnitude and angle plots:

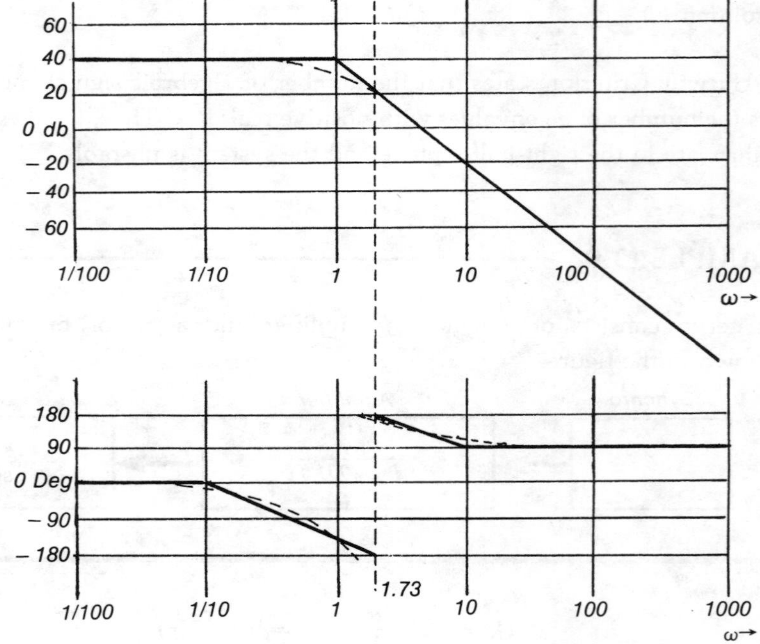

The vertical axes are the magnitude in decibels (i.e., $\log_{10} |G_2(j\omega)|$) and the phase angle of the plant. The design choice is to select a gain K from the set {0.001, 0.01, 0.1, and 1.0} for good stability margins.

Solution. The Nyquist method of analyzing stability for control systems focuses on the characteristic equation written in the form

$$1 + KG(s) = 0,$$

where K is the design parameter and $KG(s)$ is called the "open-loop transfer function." Here, $G(s) = G_2(s)$.

The theory considers a closed clockwise contour of complex s-values completely encircling the right-half plane. The corresponding values of $G(s)$ are plotted and the encirclements N around a "critical point" $-1/K$ are counted (clockwise encirclements viewed as positive). Then if $Z =$ the number of eigenvalues in the right-half plane — with $Z = 0$ for stability — and $P =$ the number of right-half plane poles of $G(s)$ which are known by inspection, then Nyquist's theorem states $N = Z - P$. Finally, look at the frequency response to envision the $G(s)$ plot because $G(j\omega)$ contains the all-essential information.

This example has $P = 0$ since the poles of $G(s)$ are at -1. Thus, for stability, N must be 0. The gains $K = 0.001$, 0.01, 0.1, and 1.0 correspond to critical points on the negative real axis at -1000, -100, -10, and -1, respectively. The frequency response curves indicate that a phase of $\pm 180°$ (negative real axis) occurs at $\omega = 1.73$ and $|G(j\omega)| \cong 22$ db $= 12.5$. This means that critical points -10 and -1 would be encircled, yielding an unstable feedback control system for the $K = 0.1$ and $K = 1$ settings of the proportional controller. The other two values of K yield stable closed-loop control.

The *gain margin* for the feedback system is the additional amplitude necessary to make $KG(s)$ unity at the phase crossover frequency. Thus at $K = 0.01$, the gain margin is 18 db; at $K = 0.001$, it is 38 db. While the latter provides a greater stability margin, the former choice of $K = 0.01$ provides adequate stability and would likely be preferred when steady-state considerations and speed of response are also important design objectives.

For example, the steady-state step response of this closed-loop system is

$$\frac{100\,K}{1 + 100\,K}$$

Thus, the steady-state difference between the input and output is 0.5 for $K = 0.01$ or 0.9 for $K = 0.001$. Often — as in this case — the larger K-values improve steady-state errors and speed of response while deteriorating stability margins somewhat.

13.3 State Model Representations

The scalar differential equations relating input signal $u(t)$ to output signal $y(t)$, and equivalently the transfer functions relating $U(s)$ to $Y(s)$, in the preceding two sections were commonly of order 2 or higher. In general, the *system order* n — which is a measure of the difficulty of analyzing the system — equals the order of the highest derivative in the scalar differential equation (or the degree of the denominator polynomial in the transfer function). Thus it also equals the number of initializing conditions required to completely specify the system and, for the case of linear time-invariant systems, is the number of system eigenvalues. Alternately, n may be viewed as the number of independent energy storage devices in the system.

To achieve a standard format for presenting the time-domain differential equations of a system — especially for cases where there may be multiple inputs and outputs or possibly nonlinear and time-varying behavior — it is convenient to define a *state vector* $x(t)$ of dimension n whose components $x_1(t)$, $x_2(t)$, $\cdots$, $x_n(t)$ [called *state variables*] completely describe the dynamic state of the system at time t. Then the scalar differential equations relating $u(t)$ and $y(t)$ can be rewritten in a vector-matrix form called a *state model* which consists of two equations:

1) a *state equation*, which is a vector differential equation of order *one* in the n-dimensional vector $x(t)$ related to input $u(t)$;
2) a *response equation*, which algebraically relates output $y(t)$ with state $x(t)$ and $u(t)$.

state equation $$\frac{d}{dt}x(t) = f(x(t), u(t), t) \tag{13.3.1}$$

response equation $$y(t) = g(x(t), u(t), t) \tag{13.3.2}$$

The linear, time-invariant state model becomes

state equation $$\frac{d}{dt}x(t) = A\,x(t) + B\,u(t) \tag{13.3.3}$$

response equation $$y(t) = C\,x(t) + D\,u(t). \tag{13.3.4}$$

Here A, B, C, D are constant matrices of dimensions $n \times n$, $n \times p$, $r \times n$, and $r \times p$, respectively. The state vector x is $n \times 1$, the input vector u is $p \times 1$, and the output vector y is $r \times 1$. Analytical solution of this linear state model can proceed using Laplace transforms of vector time functions. Of particular importance in explicit solutions is the exponential matrix e^{At} given by

$$e^{At} = \mathcal{L}^{-1}\left[(sI - A)^{-1}\right]. \tag{13.3.5}$$

The state model form is ideal for computer solution of the system for typical analysis problems such as: given $x(0)$ and $u(t)$, $0 \leqslant t \leqslant T$; find $x(t)$ and $y(t)$ for all t, $0 \leqslant t \leqslant T$. Moreover, especially with extensions to the nonlinear and time-varying cases where Laplace transform techniques lose their usefulness, standard forms like Eq. 13.3.1 and Eq. 13.3.2 are essential.

With a discrete-time system there is a general state model form analogous to Eq. 13.3.1 and Eq. 13.3.2 wherein the state equation becomes a first-order vector difference equation:

state equation $$x(T + 1) = f(x(T), u(T), T) \tag{13.3.6}$$

response equation $$y(T) = g(x(T), u(T), T) \tag{13.3.7}$$

The linear, time-invariant version of this state model — as well as scalar linear difference equations relating $u(T)$ and $y(T)$ and digital linear control systems — can be analyzed by techniques similar to those shown earlier in this chapter. The approach involves using the Z-transform of the discrete-time signals instead of the Laplace transform of the continuous-time waveforms. (These discrete-time techniques will *not* be discussed further here.)

EXAMPLE 13.11

Given the R-L-C electrical network of Ex. 13.1, find a state model representation of form Eq. 13.3.3 and Eq. 13.3.4, that is, define a state vector $x(t)$ and determine matrices A, B, C, D for the model.

Solution. We note the system order is 2 (there are 2 energy storage devices: the inductor and capacitor). A natural way to specify the dynamic state of this system at time t is to give $y(t)$ and $\frac{dy}{dt}(t)$. Letting $x_1 = y$ and $x_2 = \frac{dy}{dt}$ yields

$$\frac{d}{dt}\begin{bmatrix} x_1 \\ x_2 \end{bmatrix} = \begin{bmatrix} 0 & 1 \\ -3 & -4 \end{bmatrix}\begin{bmatrix} x_1 \\ x_2 \end{bmatrix} + \begin{bmatrix} 0 \\ 1 \end{bmatrix} u(t)$$

$$y(t) = [1 \quad 0]\begin{bmatrix} x_1 \\ x_2 \end{bmatrix} + [0]\, u(t).$$

The second state equation above is simply the differential equation of Ex. 13.1, namely

$$\frac{d^2y}{dt^2} = -4\frac{dy}{dt} - 3\,y + u = -4\,x_2 - 3\,x_1 + u.$$

EXAMPLE 13.12

Find the exponential matrix e^{At} for $A = \begin{bmatrix} 0 & 1 \\ -3 & -4 \end{bmatrix}$ and exhibit the solution to the state model in Example 13.11 given initial condition vector $x(o)$ and arbitrary input $u(t)$, $0 \leqslant t \leqslant T$.

Solution. The matrix of Eq. 13.3.5 is

$$sI - A = \begin{bmatrix} s & -1 \\ 3 & s+4 \end{bmatrix}$$

Thus,

$$(sI - A)^{-1} = \frac{\begin{bmatrix} s+4 & +1 \\ -3 & s \end{bmatrix}}{(s^2 + 4s + 3)} = \begin{bmatrix} \dfrac{s+4}{(s+1)(s+3)} & \dfrac{1}{(s+1)(s+3)} \\ \dfrac{-3}{(s+1)(s+3)} & \dfrac{s}{(s+1)(s+3)} \end{bmatrix}$$

Using partial fractions, Laplace transform tables, and Eq. 13.3.5, we obtain

$$e^{At} = \begin{bmatrix} (1.5\,e^{-t} - 0.5\,e^{-3t}) & (0.5\,e^{-t} - 0.5\,e^{-3t}) \\ (-1.5\,e^{-t} + 1.5\,e^{-3t}) & (-0.5\,e^{-t} + 1.5\,e^{-3t}) \end{bmatrix}$$

The solution to the state equation of Example 13.11 can now be expressed as

$$x(t) = e^{At}\,[x(0) + \int_0^t e^{-Av}\,Bu(v)dv]$$

where $B = \begin{bmatrix} 0 \\ 1 \end{bmatrix}$. We can verify this solution by checking at $t = 0$ noting that e^{At} at $t = 0$ is I, and by substituting for $\frac{dx}{dt}$ in the differential equation noting that

$$\frac{d}{dt}\,[e^{At}] = A\,e^{At}.$$

Knowledge of $x(t)$, $0 \leqslant t \leqslant T$, and matrices C and D yields output $y(t)$ in the response equation.

EXAMPLE 13.13

A state model for a linear, time-invariant continuous-time system has an exponential matrix

$$e^{At} = \begin{bmatrix} e^{-t} & 0 & 0 \\ 0 & (e^{-2t} - 2t\, e^{-2t}) & 4t\, e^{-2t} \\ 0 & -t\, e^{-2t} & (e^{-2t} + 2t\, e^{-2t}) \end{bmatrix}$$

Find the system matrix A and determine stability.

Solution. The general formula $A = \frac{d}{dt} e^{At}$ evaluated at $t = 0$ follows from the derivative behavior of the exponential matrix and $e^{A0} = I$. In this example, we obtain

$$A = \begin{bmatrix} -1 & 0 & 0 \\ 0 & -4 & 4 \\ 0 & -1 & 0 \end{bmatrix} .$$

Moreover, the characteristic equation

$$|\lambda I - A| = 0$$

can be rewritten

$$(\lambda + 1)(\lambda + 2)^2 = 0,$$

which yields the system eigenvalues -1, -2, -2. Stability follows, since these values are negative. We note that every entry in the e^{At} matrix has the form

$$c_1\, e^{-t} + c_2\, e^{-2t} + c_3\, t\, e^{-2t},$$

which corresponds naturally to the eigenvalue structure.

EXAMPLE 13.14

Consider the model of Eqs. 13.3.3 and 13.3.4 with A as in Example 13.13 and

$$B = \begin{bmatrix} 1 \\ 1 \\ 0 \end{bmatrix}, \quad C = \begin{bmatrix} 0 & 2 & 0 \\ 0 & 0 & 1 \end{bmatrix}, \quad D = 0.$$

Find whether or not all the system states can be controlled by the input signal $u(t)$; i.e., determine *system controllability*. Also find whether observation of the system output $y(t)$ and knowledge of $u(t)$ is adequate to compute the initial system states, i.e., determine *system observability*.

Solution. The state equations for this system are

$$\frac{d}{dt} x_1 = -x_1 + u$$

$$\frac{d}{dt} x_2 = -4\, x_2 + 4\, x_3 + u$$

$$\frac{d}{dt} x_3 = -x_2 \,.$$

At first inspection, we might presume that $u(t)$ cannot control state x_3 since the 3rd state equation does not contain u. That would be incorrect. To prove this by direct analysis of the time solutions for arbitrary bounded inputs is very tedious. Fortunately, we can use the Kalman Theorem as an alternative approach: the system is controllable if and only if matrix $M = [B, AB, \cdots, A^{n-1} B]$ has rank n.

Here $n = 3$ and

$$M = \begin{bmatrix} 1 & -1 & 1 \\ 1 & -4 & 12 \\ 0 & -1 & 4 \end{bmatrix} .$$

The determinant $|M|$ is nonzero, so rank $M = 3$ and the system is controllable. This means that an arbitrary initial state $x(0)$ can be steered to an arbitrary target $x(T)$ in finite T with a bounded control.

System observability holds if and only if matrix $Q = [C^T, A^T C^T, \cdots, (A^T)^{n-1} C^T]$ has rank n, where T denotes matrix transpose. For this example

$$Q = \begin{bmatrix} 0 & 0 & 0 & 0 & 0 & 0 \\ 2 & 0 & -8 & -1 & 24 & 4 \\ 0 & 1 & 8 & 0 & -32 & -4 \end{bmatrix} ,$$

so rank $Q = 2$ and the system is not observable. Inspection of the response equations $y_1(t) = 2\, x_2$ and $y_2(t) = x_3$ — together with the fact that the dynamics of $x_1(t)$ are uncoupled from the behavior of $x_2(t)$ and $x_3(t)$ in the state equations — provide an alternative verification of this system's lack of observability. Of course, in more complicated examples the matrix rank procedure is far superior than attempting to directly analyze the time equations.

(This page is intentionally blank.)

Practice Problems

13.1 What is the Laplace transform of a step function of magnitude a?

a) $\frac{1}{s + a}$ b) $s + a$ c) $\frac{a}{s}$ d) $\frac{a}{s^2}$ e) $\frac{a}{s + a}$

13.2 Which one of the following best defines a transfer function?

a) ratio of system response to system input function
b) ratio of system input function to system response
c) Laplace transform of system response minus the Laplace transform of the system input function
d) Laplace transform of the ratio of system response to system input function
e) ratio of the Laplace transform of the system response to the Laplace transform of the system input function

13.3 Analysis of control systems by Laplace transform techniques is *not* possible for which one of the following?

a) linear systems
b) time-invariant systems
c) feedback systems
d) discrete-time systems
e) unstable continuous-time systems

13.4 Which one of the following must have negative real parts for a stable system?

a) the system eigenvalues
b) the zeroes of the transfer function
c) the frequency response
d) the gain factor
e) the gain margin

13.5 Root locus diagrams exhibit which one of the following?

a) the response of a system to a step input
b) the frequency response of a system
c) the poles of the transfer function for a set of parameter values
d) the bandwidth of the system
e) none of the above

13.6 Well-designed first-order control systems have

a) a small negative time constant
b) a small bandwidth
c) a small negative eigenvalue
d) a large negative transfer function pole
e) a small positive eigenvalue

13.7 The unit step response of a linear time-invariant system can be obtained by

a) inverse Laplace transform of the transfer function
b) integration of the impulse response
c) differentiation of the impulse response
d) applying the Laplace final value theorem
e) using the Routh-Hurwitz Criterion

13.8 A unity feedback control system with plant $\frac{1}{s(s+3)}$ and proportional plus integral controller would require how many state variables for a state model description?

a) 1 b) 2 c) 3 d) 4 e) 5

13.9 Which one of the following can be extended to systems which are not linear?

a) Laplace transform
b) eigenvalue analysis
c) Nyquist Criterion
d) state variable analysis
e) Routh-Hurwitz Criterion

13.10 Which one of the following can be extended to systems which are time-varying?

a) state model representatives
b) transfer functions
c) Bode-Nyquist stability methods
d) Kalman controllability theorem
e) root locus design

QUESTIONS 13.11 — 13.15

Suppose an R-L-C electrical network with supply voltage $u(t)$ and output voltage $y(t)$ is described by

$$\frac{d^2y}{dt^2} + 6\frac{dy}{dt} + 25\,y(t) = 50\,u(t).$$

13.11 The transient behavior of this network would best be described as

a) a damped oscillation
b) an undamped oscillation
c) a critically damped oscillation
d) a decreasing exponential
e) an increasing exponential

13.12 If a step voltage of 10 volts is the input, what is the output steady-state voltage?

a) 5 b) 10 c) 15 d) 20 e) 25

13.13 The impulse response of this network would tend to which value as t approaches ∞?

a) 0 b) 5 c) 10 d) 20 e) ∞

13.14 The damping coefficient ξ for the impulse response belongs to which category?

a) $\xi < 0$ b) $\xi = 0$ c) $0 < \xi < 1$ d) $\xi = 1$ e) $\xi > 1$

13.15 If a sinusoidal input voltage with peak amplitude 150 volts and frequency 60 Hz is applied to this network, what is the frequency of the sinusoidal steady-state output?

a) the system's undamped natural frequency ω_n
b) the system's damped natural frequency $\omega_n\sqrt{1-\xi^2}$
c) the input frequency 60 Hz
d) somewhat less than 60 Hz
e) none of these

QUESTIONS 13.16 - 13.20

A model for a mass-damper system, with linear approximations, relates the applied force $u(t)$ to the output velocity $y(t)$ using the transfer function

$$H(s) = \frac{Y(s)}{U(s)} = \frac{6}{s+2}\,.$$

13.16 The system's time constant τ describing the time required for the step response to reach 63% of its steady-state value is

a) 2 b) 0.5 c) 6 d) 3 e) 1/6

13.17 Transient behavior of this system is

a) a damped oscillation
b) an undamped oscillation
c) a critically damped oscillation
d) a decreasing exponential
e) an increasing exponential

13.18 If this system is viewed as a processor of sinusoidal input signals to generate the frequency response, it would be a

a) low pass filter
b) high pass filter
c) band pass filter
d) band reject filter
e) unstable filter

13.19 What is the value for the system bandwidth?

a) 2 b) 0.5 c) 6 d) 3 e) 1/3

13.20 If a sinusoidal input force with frequency exactly at the system bandwidth is applied, what is the value of the gain factor?

a) 1 b) 2 c) 3 d) $3\sqrt{2}$ e) $3\sqrt{2}/2$

QUESTIONS 13.21 — 13.25

A chemical reactor with input flow $u(t)$ and output concentration $y(t)$ is modeled as a linear system with impulse response

$$h(t) = (4 + 8t)e^{-2t} \text{ for } t \geqslant 0.$$

13.21 The number of state variables needed to represent this system is

a) 1 b) 2 c) 3 d) 4 e) 8

13.22 Transients in this system are

a) undamped
b) critically damped
c) overdamped
d) unstable
e) oscillatory

13.23 The steady-state unit step response has magnitude

a) 0 b) 1 c) 2 d) 3 e) 4

13.24 The complete solution for the unit step response contains an exponential term of the form $c\,e^{rt}$ where the number r is

a) 1 b) -1 c) 2 d) -2 e) 4

13.25 In Problem 13.24 the number c is

a) 1 b) -1 c) 2 d) -2 e) -4

QUESTIONS 13.26 - 13.33

The block diagram for a numerically controlled lathe takes the form shown.

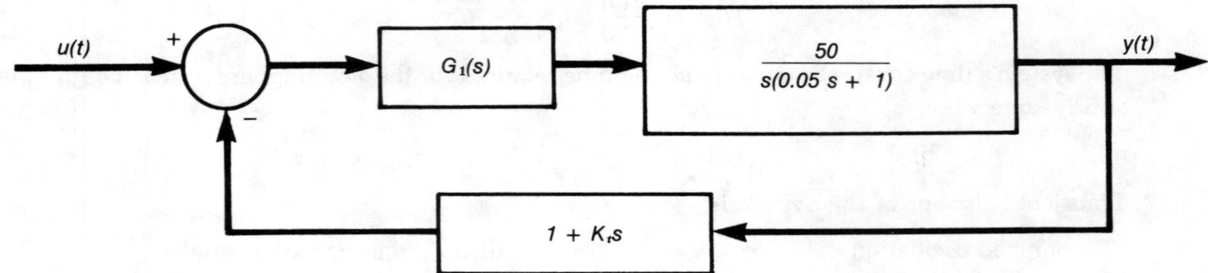

The signals $u(t)$ and $y(t)$ are the desired and actual shaft angles, respectively, G_1 is the controller, and the feedback includes tachometer measurements through the gain K_t.

13.26 With proportional control $G_1 = K$, this servomechanism behaves as a linear system of order n, where

a) $n = 0$ b) $n = 1$ c) $n = 2$ d) $n = 3$ e) $n = 4$

13.27 With $G_1 = K$, the steady-state error for a unit step input is

a) infinite
b) zero
c) dependent on K
d) dependent on K_t
e) none of these

13.28 With $G_1 = K$, the steady-state error for a unit ramp input is

a) infinite
b) zero
c) dependent on K and K_t
d) dependent on K_t but not on K
e) none of these

13.29 Assuming a positive K_t, the control system is unstable for proportional controller gain in which one of the following ranges?

a) $0 < K < 1$ b) $0 < K < 10$ c) $0 < K < 50$ d) $K > 50$ e) none of these

13.30 With $G_1 = K$ and $K_t = 0$, the effect of increasing the gain K for the step response is to

a) increase damping
b) increase oscillation
c) cause instability
d) increase system order
e) increase steady-state errors

13.31 With $G_1 = K$, a design requirement that the step response be slightly oscillatory with a rise time of 0.04 s for going between 10 and 90 percent of its final value implies that the system bandwidth should be in the range

a) 0-5 Hz b) 5-12 Hz c) 12-18 Hz d) 18-25 Hz e) 25-40 Hz

13.32 The presence of a properly adjusted tachometer with gain K_t causes the system to

a) be more oscillatory
b) decrease steady-state errors
c) have a much greater settling time
d) have a much greater rise time
e) have a much shorter rise time

13.33 Using a proportional plus integral controller, $G_1 = K + \frac{C}{s}$, causes the system to

a) have one fewer eigenvalue
b) improve steady-state performace
c) improve transient performance
d) decrease the system rise time
e) be less oscillatory

QUESTIONS 13.34 - 13.38

Consider a state model

$$\frac{d}{dt}x = \begin{bmatrix} -5 & -6 \\ 1 & 0 \end{bmatrix} x + \begin{bmatrix} 1 \\ 1 \end{bmatrix} u$$

$$y = \begin{bmatrix} 1 & 0 \end{bmatrix} x + \begin{bmatrix} 1 \end{bmatrix} u$$

describing the relationship between input signal $u(t)$ and output signal $y(t)$.

13.34 The number of eigenvalues for this linear system is

a) 0 b) 1 c) 2 d) 3 e) 4

13.35 The exponential matrix e^{At} which describes the transition between state $x(0)$ and state $x(t)$ has elements of the form $c_1e^{m_1t} + c_2e^{m_2t}$ where

a) $m_1 = 1$ and $m_2 = 5$
b) $m_1 = -1$ and $m_2 = -5$
c) $m_1 = 2$ and $m_2 = 3$
d) $m_1 = -2$ and $m_2 = -3$
e) $m_1 = 1$ and $m_2 = -5$

13.36 This linear system is

a) stable and time-varying
b) stable and time-invariant
c) unstable and time-varying
d) unstable and time-invariant
e) marginally stable

13.37 This linear system is

a) controllable and observable
b) controllable and unobservable
c) uncontrollable and observable
d) uncontrollable and unobservable
e) dynamically unstable

13.38 The impulse response of this system is best described as

a) exponentially decaying
b) exponentially increasing
c) highly oscillatory
d) damped but slightly oscillatory
e) an undamped sinusoid

13.39 A feedback control system of 4th order has characteristic equation

$$s^4 + s^3 + 6\,s^2 + 8\,s + 12 = 0.$$

The Routh-Hurwitz array contains the following entry in row 3, column 1.

a) 0 b) 1 c) -1 d) 2 e) -2

13.40 The system of Problem 13.39 has how many eigenvalues with positive real part?

a) 0 b) 1 c) 2 d) 3 e) 4

Solutions to Chapter 1

1.1 $100 = 10e^{2t}$. $\therefore \ln e^{2t} = \ln 10$. $\therefore 2t = 2.303$ $\therefore t = 1.151$ (c)

1.2 $\ln x = 3.2$. $\therefore e^{3.2} = x$. $\therefore x = 24.53$ (b)

1.3 $\ln_5 x = -1.8$. $\therefore 5^{-1.8} = x$. $\therefore x = 0.0552$ (c)

1.4 $x = \frac{-(-2) \pm \sqrt{2^2 - 4(3)(-2)}}{3\cdot 2} = 1.215$ (a)

1.5 $(4+x)^{1/2} = 4^{1/2} + \frac{1}{2}4^{-1/2}x + \frac{\frac{1}{2}(1-\frac{1}{2})}{2}4^{-3/2}x^2 + \cdots$
$= 2 + x/4 - x^2/64 + \cdots$ (e)

1.6 $\frac{2}{x(x^2-3x+2)} = \frac{A_1}{x} + \frac{A_2}{x-2} + \frac{A_3}{x-1}$
$= \frac{A_1(x^2-3x+2) + A_2(x^2-x) + A_3(x^2-2x)}{x(x^2-3x+2)}$.

$\therefore A_1 + A_2 + A_3 = 0$, $-3A_1 - A_2 - 2A_3 = 0$, $2A_1 = 2$ $\}$ $A_1 = 1$; $A_2 + A_3 = -1$, $-A_2 - 2A_3 = 3$ $\}$ $A_3 = -2$, $A_2 = 1$ (a)

1.7 $\frac{4}{x^2(x^2-4x+4)} = \frac{A_1}{x} + \frac{A_2}{x^2} + \frac{A_3}{x-2} + \frac{A_4}{(x-2)^2}$
$= \frac{A_1(x^3-4x^2+4x) + A_2(x^2-4x+4) + A_3(x^3-2x^2) + A_4x^2}{x^2(x-2)^2}$

$\therefore A_1 + A_3 = 0$, $-4A_1 + A_2 - 2A_3 + A_4 = 0$, $4A_1 - 4A_2 = 0$, $4A_2 = 4$. $\}$ $A_2 = 1$, $A_1 = 1$, $A_3 = -1$, $A_4 = 1$ (b)

1.8 at $t = 0$, population $= A$. $\therefore 2A = Ae^{0.4t}$. $\ln 2 = 0.4t$. $\therefore t = 1.733$ (d)

1.9 $\sin\theta = 0.7$ $\therefore \theta = 44.43°$. $\therefore \tan 44.43° = 0.980$ (a)

1.10 $\tan\theta = 5/7$. $\therefore \theta = 35.54°$ (e)

1.11 $\tan\theta \sec\theta (1 - \sin^2\theta)/\cos\theta = \tan\theta \frac{1}{\cos\theta} \cos^2\theta \frac{1}{\cos\theta} = \tan\theta$ (c)

1.12 $3^2 = 4^2 + 2^2 - 2\cdot 2\cdot 4\cos\theta$. $\therefore \cos\theta = 0.6875$. $\therefore \theta = 46.6°$
$\therefore$ rad $= 0.813$ (d)

1.13 $L^2 = 850^2 + 732^2 - 2\cdot 850\cdot 732 \cos 154°$ $\therefore L = 1542$ m (a)

1.14 $\cos 2\theta = \cos^2\theta - \sin^2\theta = 1 - \sin^2\theta - \sin^2\theta = 1 - 2\sin^2\theta$
$\therefore 2\sin^2\theta = 1 - \cos 2\theta$. (e)

1.15 $\theta = \pi(n-2)/n = \pi(8-2)/8$ radians.
$\frac{6\pi}{8} \times \frac{180}{\pi} = 135°$. (b)

1.16 Area = $\text{Area}_{top} + \text{Area}_{sides} = \pi R^2 + \pi DL$
$= \pi \times 7.5^2 + \pi \times 15 \times 10 = 648\ m^2$. $648 \div 10 = 65$ (a)

1.17 $y = mx + b$. $y = -2x + b$. $0 = -2(2) + b$. $\therefore b = 4$
$\therefore y = -2x + 4$. (a)

1.18 $y = mx + b$. $0 = 4m + b$ $-6 = b$. $\therefore m = 3/2$.
$\therefore y = 3x/2 - 6$ or $2y = 3x - 12$ (b)

1.19 $3x - 4y - 3 = 0$. $A = 3$, $B = -4$.
$d = \frac{|3\times 6 - 4\times 8 - 3|}{\sqrt{3^2 + (-4)^2}} = 3.4$ (d)

1.20 $B^2 - 4AC = 4^2 - 4\times 1\times 4 = 0$. $\therefore$ parabola (c)

1.21 $xy = \pm k^2 = 4$ (b)

1.22 $\frac{x^2}{25^2} + \frac{y^2}{50^2} = 1$. $\therefore 4x^2 + y^2 = 2500$. (c)

1.23 $x = r\cos\theta = 5 \times .866 = 4.33$ $y = r\sin\theta = 5 \times .5 = 2.5$
In spherical coordinates:
$r = \sqrt{4.33^2 + 2.5^2 + 12^2} = 13$ $\phi = \cos^{-1}\frac{z}{r} = \cos^{-1}\frac{12}{13} = 22.6°$
$\theta = \tan^{-1}\frac{y}{x} = \tan^{-1}\frac{2.5}{4.33}$. $\therefore \theta = 30°$. (b)

1.24 (a) is rectangular coord. (b) is spherical coord. (c)

1.25 $\frac{3-i}{1+i} = \frac{3-i}{1+i}\,\frac{1-i}{1-i} = \frac{3-1-4i}{1-(-1)} = \frac{1}{2}(2-4i) = 1-2i$ (a)

1.26 $1+i = re^{i\theta}$. $r = \sqrt{1^2+1^2} = \sqrt{2}$ $\theta = \tan^{-1}\frac{1}{1} = \pi/4$ rad.
$\therefore 1+i = \sqrt{2}\,e^{i\pi/4}$. $\therefore (1+i)^6 = (\sqrt{2})^6 e^{i\pi 3/2}$
$= 8\left(\cos\frac{3\pi}{2} + i\sin\frac{3\pi}{2}\right) = -8i$ (d)

1.27 $1+i = \sqrt{2}\,e^{i\pi/4}$. $\therefore (1+i)^{1/5} = 1.414^{1/5} e^{i\pi/20}$
$\therefore (1+i)^{1/5} = 1.072\left(\cos\frac{\pi}{20} + i\sin\frac{\pi}{20}\right) = 1.06 + .167i$. (b)

1.28 $(3+2i)(\cos 2t + i\sin 2t) + (3-2i)(\cos 2t - i\sin 2t)$
$= 6\cos 2t - 4i(i\sin 2t) = 6\cos 2t - 4\sin 2t$ (c)

1.29 $6(\cos 2.3 + i\sin 2.3) - 5(\cos 0.2 + i\sin 0.2)$
$= 6(-0.666 + 0.746i) - 5(0.98 + 0.199i)$
$= -8.90 + 3.48i$ (e)

1.30 $\begin{vmatrix} 3 & 2 & 1 \\ 0 & -1 & -1 \\ 2 & 0 & 2 \end{vmatrix} = -6 - 4 + 2 = -8$ (e)

1.31 $\begin{vmatrix} 1 & 0 & 1 & 1 \\ 2 & -1 & 0 & 1 \\ 0 & 0 & 2 & 0 \\ 3 & 2 & 1 & 1 \end{vmatrix} = 2\begin{vmatrix} 1 & 0 & 1 \\ 2 & -1 & 1 \\ 3 & 2 & 1 \end{vmatrix} = 2(-1+4+3-2) = 8$ (a)

1.32 $(-1)^3\begin{vmatrix} 2 & 1 \\ 0 & 2 \end{vmatrix} = -4$. (b)

1.33 $(-1)^7\begin{vmatrix} 1 & 0 & 1 \\ 2 & -1 & 0 \\ 3 & 2 & 1 \end{vmatrix} = -(-1+4+3) = -6$ (e)

1.34 $[a_{ij}]^+ = \begin{bmatrix} A_{11} & A_{21} \\ A_{12} & A_{22} \end{bmatrix} = \begin{bmatrix} 1 & 0 \\ -4 & 2 \end{bmatrix}$. (d)

1.35 $[a_{ij}] = \frac{[a_{ij}]^+}{|a_{ij}|} = \frac{\begin{bmatrix} 1 & -1 \\ -3 & 2 \end{bmatrix}}{-1} = \begin{bmatrix} -1 & 1 \\ 3 & -2 \end{bmatrix}$. (a)

1.36 $\begin{bmatrix} 2 & -1 \\ 3 & 2 \end{bmatrix}\begin{bmatrix} 2 \\ 1 \end{bmatrix} = \begin{bmatrix} 4-1 \\ 6+2 \end{bmatrix} = \begin{bmatrix} 3 \\ 8 \end{bmatrix}$ (b)

1.37 $\begin{bmatrix} 1 & 2 \\ 2 & 1 \end{bmatrix}\begin{bmatrix} -1 & 0 \\ 1 & 2 \end{bmatrix} = \begin{bmatrix} 1 & 4 \\ -1 & 2 \end{bmatrix}$ (a)

1.38 $[a_{ij}] = \begin{bmatrix} 3 & 2 & 0 \\ 1 & -1 & 1 \\ 4 & 0 & 2 \end{bmatrix}$. $[a_{ij}]^+ = \begin{bmatrix} -2 & -4 & 2 \\ 2 & 6 & -3 \\ 4 & 8 & -5 \end{bmatrix}$ $|a_{ij}| = -2$

$\therefore [a_{ij}]^{-1} = \begin{bmatrix} 1 & 2 & -1 \\ -1 & -3 & 3/2 \\ -2 & -4 & 5/2 \end{bmatrix}$. $[x_j] = [a_{ij}]^{-1}[r_i] = [a_{ij}]^{-1}\begin{bmatrix} -2 \\ 0 \\ 4 \end{bmatrix}$
$= \begin{bmatrix} -6 \\ 8 \\ 14 \end{bmatrix}$. (d)

1.39 $\frac{dy}{dx} = 6x^2 - 3 = 6(1)^2 - 3 = 3$ (a)

1.40 $\frac{dy}{dx} = \frac{1}{x} + e^x \cos x + e^x \sin x$ $(\cos 1 = \cos 57.3°)$

$= 1 + e \cos 1 + e \sin 1 = 4.76$ (c)

1.41 $\frac{dy}{dx} = 3x^2 - 3 = 0. \quad \therefore x^2 = 1, \quad \therefore x = \pm 1$

$\frac{d^2y}{dx^2} = 6x. \quad \therefore x = -1$ is a maximum. (d)

1.42 $y' = 3x^2 - 3 \quad y'' = 6x. \quad \therefore x = 0$ is inflection. (c)

1.43 $\lim_{x\to\infty} \frac{2x^2 - x}{x^2 + x} = \lim_{x\to\infty} \frac{4x-1}{2x+1} = \lim_{x\to\infty} \frac{4}{2} = 2.$ (a)

1.44 $\eta(x+h) = \eta + h\eta' + \frac{h^2}{2}\eta''.$ (c)

1.45 $e^x \sin x = (1 + x + \frac{x^2}{2})(x - \frac{x^3}{6})$

$= x + x^2 + \frac{x^3}{2} - \frac{x^3}{6} = x + x^2 + x^3/3.$ (b)

1.46 y=9, y=4, x. Area $= \int_4^9 x\,dy = \int_4^9 y^{1/2} dy = \frac{2}{3}(27 - 8) = 12\frac{2}{3}.$ (d)

1.47 (4,4) Area $= \int_0^4 (x_2 - x_1)\,dy = \int_0^4 (2y^{1/2} - \frac{y^2}{4})\,dy$

$= 2 \times \frac{2}{3} \times 8 - \frac{1}{12} \times 64 = 16/3$ (e)

1.48 $V = \int_0^2 2\pi y\, x\, dy = 2\pi \int_0^2 y^3 dy = 2\pi \times \frac{2^4}{4} = 8\pi.$ (c)

1.49 $\int_0^2 (e^x + \sin x)\,dx = e^x - \cos x \Big|_0^2$

$= e^2 - 1 - \cos 2 + 1 = 7.81$ (a)

1.50 $\int_0^1 e^x \sin x\, dx = e^x \sin x \Big|_0^1 - \int_0^1 e^x \cos x\, dx$

$u = \sin x \quad dv = e^x dx \quad u = \cos x \quad dv = e^x dx$

$du = \cos x\, dx \quad v = e^x \quad du = -\sin x\, dx \quad v = e^x$

$\therefore \int_0^1 e^x \sin x\, dx = e \sin 1 - \left[e^x \cos x \Big|_0^1 + \int_0^1 e^x \sin x\, dx \right].$

$\therefore 2\int_0^1 e^x \sin x\, dx = e \sin 1 - e \cos 1 + 1 \times 1 = 1.819.$

$\therefore \int_0^1 e^x \sin x\, dx = 0.909.$ (e)

1.51 $\int x \cos x\, dx = x \sin x - \int \sin x\, dx = x \sin x + \cos x + C$

$u = x \quad dv = \cos x\, dx$

$du = dx \quad v = \sin x$ (e)

1.52 linear and nonhomogeneous. The term (+2) makes it nonhomogeneous. $x^2 y'$ is linear. (b)

1.53 $\frac{dy}{dx} = -2xy. \quad \frac{dy}{y} = -2x\,dx. \quad \therefore \ln y = -x^2 + C.$ (a)

$\ln 2 = 0 + C. \quad \therefore C = \ln 2. \quad y(2) = \exp(-2^2 + \ln 2) = 0.0366$

1.54 $\frac{dy}{dx} = -2x. \quad dy = -2x\,dx. \quad \therefore y = -x^2 + C.$

$1 = 0 + C. \quad \therefore C = 1. \quad y(10) = -10^2 + 1 = -99$ (b)

1.55 $2m^2 + m + 50 = 0. \quad \therefore m = \frac{-1 \pm \sqrt{1 - 400}}{4} = -\frac{1}{4} \pm 4.99i.$

$\therefore y(t) = e^{-t/4}(A \cos 4.99t + B \sin 4.99t). \quad \therefore \omega = 4.99$ rad/s

$\therefore f = \frac{\omega}{2\pi} = \frac{4.99}{2\pi} = 0.794$ hertz.

1.56 $m^2 + 16 = 0. \quad \therefore m = \pm 4i. \quad \therefore y(t) = c_1 \cos 4t + c_2 \sin 4t$ (a)

1.57 $m^2 + 8m + 16 = 0 \quad (m+4)^2 = 0. \quad \therefore m = -4, -4.$

$\therefore y(t) = c_1 e^{-4t} + c_2 t e^{-4t}$ (e)

1.59 homogeneous: $m^2 + 16 = 0. \quad \therefore m = \pm 4i.$

$\therefore y_h(t) = c_1 \sin 4t + c_2 \cos 4t$

particular: $y_p = At \cos 4t.$ (this is resonance)

$\dot{y}_p = A \cos 4t - 4At \sin 4t.$ (a)

$\therefore -4A \sin 4t - 4A \sin 4t - 16At \cos 4t + 16At \cos 4t = 8 \sin 4t.$

$\therefore -8A = 8. \quad A = -1. \quad \therefore y = y_h + y_p = c_1 \sin 4t + c_2 \cos 4t - t \cos 4t$

1.58 $m^2 - 5m + 6 = 0. \quad m = \frac{5 \pm \sqrt{25 - 24}}{2} = 3, 2.$

$\therefore y_h = c_1 e^{3t} + c_2 e^{2t}$ Assume $y_p = Ae^t.$

then $Ae^t - 5Ae^t + 6Ae^t = 4e^t. \quad \therefore A = 2.$ (d)

1.60 $p(A_1 \text{ or } A_2) = p(A_1) + p(A_2) = \frac{5}{85} + \frac{30}{85} = 0.41$ (c)

1.61 $p(A_1 \text{ and } B_1) = p(A_1)\,p(B_1) = \frac{5}{85} \times \frac{30}{85} = 0.0208$

or 2/96.3 (b)

1.62 $p = \frac{30}{85} \times \frac{29}{84} \times \frac{28}{83} \times \frac{25}{82} \times \frac{24}{81} = .0037$ or 4/1077. (b)

1.63 $p(A_1 \text{ or } B_1) = p(A_1) + p(B_1) - p(A_1)\,p(B_1)$

$= \frac{1}{4} + \frac{1}{4} - \frac{1}{4} \times \frac{1}{4} = 7/16.$ (b)

1.64 The mode is the observation that occurs most frequently. It is 75. (b)

1.65 $\bar{x} = \frac{35 + 3\times45 + 6\times55 + 11\times65 + 13\times75 + 10\times85 + 2\times90}{1 + 3 + 6 + 11 + 13 + 10 + 2}$

$= \frac{3220}{46} = 70.0$ (d)

1.66 $\sigma = \sqrt{\frac{35^2 + 3\times45^2 + 6\times55^2 + 11\times65^2 + 13\times75^2 + 10\times85^2 + 2\times90^2 - 46\times70^2}{45}}$

$= \sqrt{\frac{8100}{45}} = 13.42$ (d)

Solutions to Chapter 2

1. By definition (c)

2. $E = \frac{hc}{\lambda} = \frac{6.63\times10^{-34}\cdot 3\times10^{9}}{5\times10^{-7}} = 3.98\times10^{-19}$ J

$3.98\times10^{-19}/1.6\times10^{-19} = 2.48$ eV (a)

3. Power $= WA = \sigma T^4 A = \sigma T^4 4\pi R^2$

$= 5.67\times10^{-8}\times 4000^4 \times 4\pi\times .005^2 = 4560$ W (d)

4. $P_{star} = W_{star} A_{star} = 600\, W_{sun} A_{sun}$.

$\sigma T^4_{star}\, 4\pi R^2_{star} = 600\, \sigma T^4_{sun}\, 4\pi R^2_{sun}$.

$\therefore R_{star}/R_{sun} = \sqrt{600}\; T^2_{sun}/T^2_{star} = 55$ (e)

5. (c)

6. $h\nu = E_{K_{max}} + \phi$. $E_{K_{min}} = 0$ when $h\nu = 0$. (a)

7. No. of Photons $= \frac{1.5\times10^{-18}}{6.63\times10^{-34}\times3\times10^{8}/6.625\times10^{-7}} = 5.$ (a)

8. $h\nu = \frac{hc}{\lambda} = \frac{1.24\times10^{4}}{8000} = 1.55$ eV (b)

9. $\frac{hc}{\lambda} = \frac{1.24\times10^{4}}{4000} = E_{K_{max}} + 2.1$. (e)

$\therefore E_{K_{max}} = 1.0 = \frac{1}{2}mV^2$. $\therefore V = \sqrt{\frac{2\times1.6\times10^{-19}}{9.11\times10^{-31}}} = 5.93\times10^{5}\ \frac{m}{s}$

10. No. of Photons $= \frac{3.31\times10^{-8}}{6.63\times10^{-34}\times3\times10^{8}/6\times10^{-7}}\times10^{-4}$

$\cong 10^{7}$. (e)

11. $2000 = \frac{M}{hc/\lambda}$. $\therefore M = 2000\,\frac{6.63\times10^{-34}\times3\times10^{8}}{5.89\times10^{-7}} = 6.75\times10^{-16}$ W (e)

12. $h\nu = E_{K_{max}} + \phi = 1.6 + 4.12 = 5.72$ eV.

$\therefore \nu = \frac{5.72}{6.63\times10^{-34}/1.6\times10^{-19}} = 1.38\times10^{15}$ Hz. (d)

13. $h\nu = \phi$ at threshold. $\frac{hc}{\lambda} = \frac{1.24\times10^{4}}{6525} = 1.9$ eV (a)

14. $hc/\lambda = E_{K_{max}} + \phi$. $\frac{1.24\times10^{4}}{3600} = E_{K_{max}} + 2.0$.

$\therefore E_{K_{max}} = 1.44$ eV, the stopping potential. (c)

15. $\frac{\text{Photon}}{\text{sec}} = \frac{0.1}{6.63\times10^{-34}\times3\times10^{8}/6.5\times10^{-7}} = 3.3\times10^{17}$. (c)

16. (c)

17. Least energy $\Rightarrow$ longest wave length. $\therefore n=3$ to $m=2$.

$\frac{1}{\lambda} = R_H\left(\frac{1}{m^2} - \frac{1}{n^2}\right) = 1.09737\left(\frac{1}{2^2} - \frac{1}{3^2}\right)\times10^{7}$. $\therefore \lambda = 6563$ Å (d)

18. (b)

19. $E_4 = -13.6/n^2 = -13.6/16 = -0.85$ eV (b)

20. First excited state: $n=2$. $E_2 = -13.6/4 = -3.4$ eV.

Emission: $E_4 - E_2 = -.85 + 3.4 = 2.55 = \frac{1.24\times10^{4}}{\lambda}$.

$\therefore \lambda = 4863$ Å (d)

21. $n=3 \rightarrow n=5$ Absorption. $E_3 = -\frac{13.6}{9} = -1.51$.

$E_5 = -13.6/25 = -0.54$. $\therefore \Delta E = -0.96$ eV. (b)

22. Second excited state: $n=3$. $\therefore$ Transition $n=3\rightarrow n=1$.

$E_3 - E_1 = -13.6\left(\frac{1}{9} - 1\right) = 12.09 = h\nu$.

$\therefore \nu = \frac{12.09}{6.63\times10^{-34}/1.6\times10^{-19}} = 2.92\times10^{15}$ Hz. (b)

23. 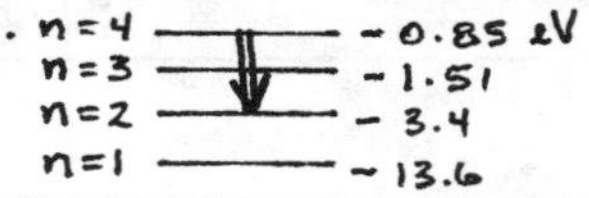 $n=4$ — -0.85 eV; $n=3$ — -1.51; $n=2$ — -3.4; $n=1$ — -13.6. See solution to Prob. 20.

$\therefore \lambda = 4863$ Å (d)

24. $E = -13.6\, Z^2/n^2 = -13.6\times4/1 = -54.4$ eV (e)

25. See Table 2.3 (a)

26. See Example 2.4 (c)

27. Defined by the physical system (d)

28. 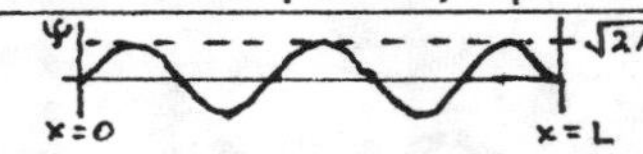 [sketch: ψ from $x=0$ to $x=L$, amplitude $\sqrt{2/L}$] (e)

29. 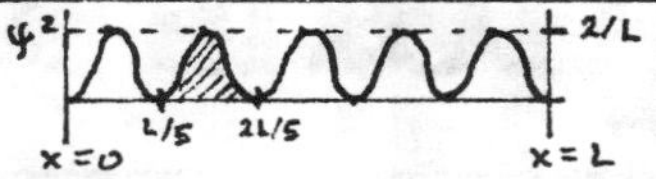 [sketch: ψ^2 from $x=0$ to $x=L$, max $2/L$, shaded between $L/5$ and $2L/5$] Probability "Area" $= 1/5$ (b)

30. 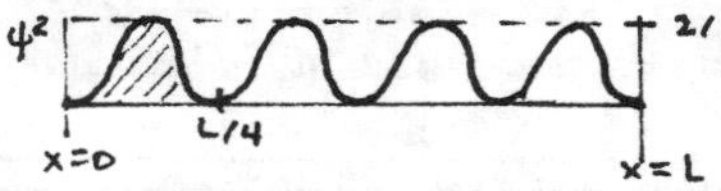 [sketch: ψ^2 from $x=0$ to $x=L$, max $2/L$, shaded from 0 to $L/4$] Probability "Area" $= 1/4$ $= 0.25$ (b)

31. (e)

32. See Prob. 29. Probability "Area" $= \frac{2}{5} = 0.4$ (c)

33. $E_n = \frac{n^2h^2}{8mL^2}$. $\therefore E_n \propto n^2$. (e)

34. $E_1 = \frac{n^2h^2}{8mL^2} = \frac{1^2\times(6.63\times10^{-34})^2}{8\times9.11\times10^{-31}\times(1\times10^{-10})^2} = 6.03\times10^{-18}$ J.

$6.03\times10^{-18}/1.6\times10^{-19} = 37.7$ eV. (d)

35. $E_3 - E_2 = \frac{h^2}{8mL^2}(3^2 - 2^2) = 5\,\frac{(6.63\times10^{-34})^2}{8\times9.11\times10^{-31}(1\times10^{-10})^2}/1.6\times10^{-19}$

$= 11.5$ eV (d)

36. Isotones (c)

37. $4\overline{)238}$ = 59: 20, 38, 36, 2. $4(59) + 2$. $\therefore n = 59$ (d)

38. $A = A_0 e^{-\lambda t}$. $5 = 20\, e^{-0.693\, t/6.98}$

$\therefore \ln 5/20 = -0.693\, t/6.98$. $\therefore t = 13.96$ days.

or $\frac{5}{20} = \frac{1}{4} = \left(\frac{1}{2}\right)^2$ ← this means 2 half-lives.

$\therefore 2\times6.98 = 13.96$ days (c)

39. Average life $= 1/\lambda$.

$$\frac{A}{A_0} = e^{-\lambda t} = e^{-\lambda \frac{1}{\lambda}} = e^{-1} = 0.368$$ (a)

40. $\frac{A}{A_0} = \frac{1}{10} = e^{-\lambda t} = e^{-\frac{.693}{T_{1/2}} \times 30}$.

$\therefore \ln 1/10 = -\frac{.693}{T_{1/2}} \times 30$. $\therefore T_{1/2} = 9.03$ min. (e)

41. $\frac{A}{A_0} = \frac{11.3}{15.3} = e^{-\frac{.693}{5730}t}$ $\therefore t = 2506$ yrs. (a)

42. $\frac{2}{200 \times 10^6 \times 1.6 \times 10^{-19}} = 6.25 \times 10^{10}$ fission/sec. (a)

43. ${}^{32}_{16}S + {}^{1}_{0}n \rightarrow {}^{29}_{16}S + {}^{4}_{2}He$. Charge 16 ≠ charge 18 (d)

44. ${}^{235}_{92}U + {}^{1}_{0}n \rightarrow {}^{143}_{57}La + {}^{87}_{35}Br + 6\,{}^{1}_{0}n$ (a)

45. ${}^{235}_{92}U + {}^{1}_{0}n \rightarrow {}^{97}_{40}Zr + {}^{134}_{52}Te + 5\,{}^{1}_{0}n$ (e)

46. ${}^{11}_{5}B + {}^{4}_{2}He \rightarrow {}^{1}_{0}n + {}^{14}_{7}N$

$$\begin{array}{r} 11.0083 \\ 4.0012 \\ \hline 15.0095 \\ -\ 1.0087 \\ \hline 14.0008 \end{array}$$ u for ${}^{14}_{7}N$

$\therefore 7 \times 14.0008 = 98.0056$ (d)

47. ${}^{14}_{7}N + {}^{1}_{0}n \rightarrow {}^{14}_{6}C + {}^{1}_{1}H$. Production of Radioactive C-14 nuclei (d)

Solutions to Chapter 3

3.1 If applied stress is greater than yield stress a permanant deformation occurs and the strain is irreversible. (c)

3.2 Off-set yield stress is defined as the stress corresponding to a specified plastic strain, usually 0.2%. (c)

3.3 When stress is calculated by using the original cross-sectional area, the stress is called the engineering stress. When necking occurs, cross-section decreases and the engineering stress decreases. It reaches a maximum value before the onset of necking. (d)

3.4 Higher the ductility smaller the cross-section at fracture, i.e., greater is the reduction in area. Reduction in area is zero for very brittle material. (c)

3.5 Yield point phenomenon is peculiar to carbon steels. Nonferrous alloys, such as aluminum alloy, do not show a yield point. (c)

3.6 In a crystalline solid slip is the primary mode of plastic deformation. Dislocation motion is necessary for slip. (b)

3.7 Slip produces a change in shape without producing a change in volume, like shearing a deck of cards. (a)

3.8 In pure aluminum as the amplitude of alternating stress decreases, the fatigue life monotonically increases. There is no threshhold stress and thus there is no endurance limit. (a)

3.9 The two elements in a binary alloy must be soluble in a solid state to produce a solid solution alloy. (c)

3.10 The eutectic temperature is invariant, i.e., the solidification of the eutectic composition occurs at a single temp. rather than over a range of temp. (b)

3.11 A liquid phase of a fixed composition reacts with a solid phase of another fixed composition to produce a new solid phase at the peritectic temperature. (c)

3.12 AB_3 is the only intermetallic compound in this diagram. It consists of 75 at.% B & 25 at.% A. (d)

3.13 The horizonal line corresponding to T_1 intersects the vertical, A+50 at.% B, within the L+γ region. (b)

3.14 According to Fig. 12a, 1147°C is the temperature the liquid solidified into (solid) and Fe_3C (solid). Thus, this is the eutectic temperature. (a)

3.15 The composition of alloy is A + 50 wt.% B. Composition of α is A + 25% B and that of β is A + 75% B. Thus, the lever rule can be applied as follows:

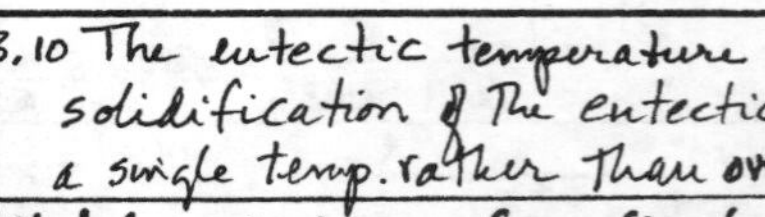

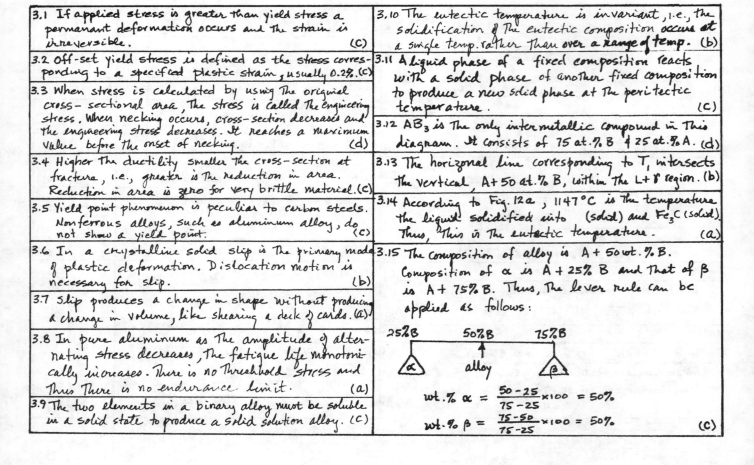

$$\text{wt.\% } \alpha = \frac{50-25}{75-25} \times 100 = 50\%$$

$$\text{wt.\% } \beta = \frac{75-50}{75-25} \times 100 = 50\%$$ (c)

Solutions to Chapter 4

4.1 A solution is defined as homogeneous. (a)

4.2 A element is a pure substance and is homogeneous. (d)

4.3 By definition. (d)

4.4 By definition. (d)

4.5 Conduction is a flow of electrons. (d)

4.6 The noble gases have closed shell, stable electronic configurations. (b)

4.7 Atomic radius increases down the periodic table, and decreases across (left to right). (b)

4.8 Metallic character increases down the table (a)

4.9 See the periodic table. (c)

4.10 By definition. (d)

4.11 The 13 electrons must be in the lowest energy orbitals available. Thus, at each quantum level ($n=1,2,3$) the s-orbitals, then the p-orbitals must be filled in order. (b)

4.12 The $5p_x^2$ configuration has two paired electrons which could be unpaired, i.e., $5p_x'$, $5p_y'$, $5p_z'$. (a)

4.13 Group VI is two electrons short of a closed shell configuration, therefore must have 2 electrons in a valence s-orbital, and 4 in the p-orbitals. (d)

4.14 Ba, $Z=56$, has a valence electron configuration $6s^2$. Sr, $Z=38$, has $5s^2$. (d)

4.15 By definition. (b)

4.16 If $l=0$, the orbital is an s-orbital. There can be only one s-orbital for a giver principle quantum no. Therefore, there can be only two electrons. (a)

4.17 There are 28 electrons in the $n=1,2,3$ energy levels (see Table 4.3). The configuration for the $n=4$ level is, therefore $4s^2, 4p^5$. If the 5 p-electrons are removed that leaves 30 electrons. (d)

4.18 Phosphorus, $Z=15$ (see Table 4.1), has electronic configuration $1s^2, 2s^2, 2p^6, 3s^2, 3p^3$ (see Table 4.3). (d)

4.19 See Table 4.1 for atomic number 15. (c)

4.20 The element must be P ($Z=15$). The 18 electrons are 3 in excess of the nuclear charge. (a)

4.21 Electrons are polarized toward the more electronegative atom. (d)

4.22 Alkane names always end in ane. (c)

4.23 Al and P are the closest in the periodic table. (b)

4.24 By definition. (c)

4.25 No net electron transfer. (e)

4.26 O is 2^-. Three oxygens is 6^-. The net charge is 2^-. ∴ S must be 4^+, i.e., $4-6=-2$ (d)

4.27 By definition. (d)

4.28 Molecular nitrogen is N_2. There are therefore 2 moles of N atoms in a mole of nitrogen molecules, and twice Avogadro's number of atoms. (c)

4.29 $50/27 = 1.85$ mol. (b)

4.30 Molecular wt. of $OF_2 = 16 + (2 \times 19) = 54$. $27g = 0.5$ mol. $0.5 \times 6 \times 10^{23} = 3 \times 10^{23}$. (a)

4.31 Mol wt. $Na_2SO_4 = (2 \times 23) + 32 + (4 \times 16) = 142$
$0.01 \times 142 = 1.42g$. (d)

4.32 Mol wt. $SO_2 = 32 + 2 \times 16 = 64$. $16g\ SO_2 = 16 \div 64 = \frac{1}{4}$ mol. One mol = 22.4 l at STP.
∴ 1/4 mol $= \frac{1}{4} \times 22.4 = 5.6$ l at STP. (c)

4.33 Mol wt $Al_2O_3 = 2 \times 27 + 3 \times 16 = 102$.
wt % Al $= (54/102) \times 100 = 53\%$ (d)

4.34 Compound contains $.475 \times 135 = 64.1$ g S per mol. ∴ there are $64.1/32 = 2.0$ mol S per mol compound. The remaining mol wt, $135 - 64 = 71$ is Cl (Atomic wt=35.5) There are ∴ 2 mol per mol compound. ∴ formula S_2Cl_2. (c)

4.35 Atomic ratio of C/H $= \frac{34.6}{12} \div \frac{3.8}{1} = 0.75$, i.e., 3:4 or C_3H_4. Malonic acid is $C_3H_4O_4$. To confirm, the atomic ratio of O/H $= \frac{61.5}{16} \div 3.8 = 1.0$. (d)

4.36 See Example 4.18. (c)

4.37 Solids do not enter into the K_{eq}. (b)

4.38 Mol wt CO = 28. 14g CO = 14/28 = .5 mol. Each mol of CH_4 gives 1 mol CO. ∴ $\frac{1}{2}$ mol $CH_4 \rightarrow \frac{1}{2}$ mol CO. (d)

4.39 $2Pb(NO_3)_2$ gives 4 N atoms, requiring $4NO_2$. (d)

4.40 There are 12 H atoms on the right requiring $6H_2O$ on the left (note that oxygen balances). (d)

4.41 An atom bonded only to itself has oxidation number = 0. In Mg_3N_2, Mg has oxidation no. 2^+ (an alkaline earth). ∴ to make a net 0, N must have 3^-. Any reduction in oxidation no. (0 to $^-3$) reduces the element. (d)

4.42 The balanced equation shows that 6 moles of HCl gives 2 moles of $AlCl_3$ or 3 moles gives 1 mole. (b)

4.43 Le Chatelier principle. (c)

4.44 An exothermic reaction has a favorable equilibrium constant, but the rate depends on the activation energy. (a)

4.45 Le Chatelier principle. (b)

Solutions to Chapter 5

5.1 Prestige. (d)

5.2 $1000 \times 1.06^3 = \$1191$. (d)

5.3 $150/1.08^2 = \$129$. (c)

5.4 $500\,(P/A)^{8}_{12} = 500 \times 7.536 = \$3768$. (b)

5.5 A: $300{,}000 + 35{,}000\,(P/A)^{8}_{30} - 50{,}000\,(P/F)^{8}_{30} = \$689{,}000$ (b)
B: $689{,}000\,(A/P)^{8}_{30}\,(P/A)^{8}_{\infty} = \$765{,}000$. (d)

5.6 $[1000(P/F)^{10}_{5} + 2000\,(P/F)^{10}_{10} + 3500(P/F)^{10}_{15}]\,(A/P)^{10}_{20} = \$262$. (b)

5.7 $54{,}000\,(A/P)^{8}_{30} = \$4800$ (d)

5.8 $10{,}000\,(F/P)^{6}_{4} - 3000 = \$9625$ (c)

5.9 $18{,}000\,(A/F)^{4}_{8} = \$1953$ (a)

5.10 $1000(P/A)^{10}_{\infty} + 1000(P/A)^{10}_{10} = \$16{,}145$ (a)

5.11 $20{,}000\,(P/A)^{8}_{6}\,(P/F)^{8}_{1} = \$85{,}600$ (a)

5.12 $e^{.10} - 1 = 0.10517$ or 10.517% (b)

5.13 $675 = 50 + 30(P/A)^{i}_{24}$. $\therefore (P/A)^{i}_{24} = 20.833$.
$\therefore$ by trial & error $i = 0.0116$. $\therefore 12i = 0.139$ or 13.9% (b)

5.14 $50{,}000 + [20{,}000 + 10{,}000\,(A/P)^{4}_{3}]\,(P/A)^{4}_{\infty} = \$640{,}000$. (b)

5.15 A: $-16{,}000\,(A/P)^{12}_{8} - 2000 + 2000\,(A/F)^{12}_{8} = -\$5058$
B: $-30{,}000(A/P)^{12}_{15} - 1000 + 5000(A/F)^{12}_{15} = -\$5270$ (a)

5.16 $P\,(A/P)^{10}_{20} = [30{,}000 + 10{,}000\,(P/F)^{10}_{4}]\,(A/P)^{10}_{8}$.
$\therefore P = \$58{,}760$. (d)

5.17 $120{,}000 + 9000\,(P/A)^{10}_{6} - 25{,}000\,(P/F)^{10}_{6} = \$145{,}000$. (b)

5.18 $[100{,}000\,(A/P)^{10}_{10} + 10{,}000]\,(P/A)^{10}_{\infty} = \$262{,}700$. (e)

5.19 A: $50{,}000 + 800\,(P/A)^{8}_{20} = \$57{,}900$.
B: $30{,}000 + 500(P/A)^{8}_{20} + [30{,}000 + 400(P/A)^{8}_{10}](P/F)^{8}_{10}$
$= \$50{,}000$. (d)

5.20 A: $-116 + 0.93 \times 206(P/A)^{10}_{8} = \$906$
B: $-60 + 0.89 \times 206(P/A)^{10}_{8} = \$918$ (b)

5.21 $[-25{,}000 + 83{,}000\,(P/F)^{10}_{3}](A/P)^{10}_{6} - 6000 + 13{,}000\,(A/F)^{10}_{6}$
$= \$4260$. (b)

5.22 $(20{,}000 + P)\,(A/P)^{8}_{9} - 300 - 5000\,(A/F)^{8}_{9}$
$= 20{,}000\,(A/P)^{8}_{6} - 5000\,(A/F)^{8}_{6}$. $\therefore P = \$7140$. (d)

5.23 $(23{,}000 - 15{,}000) + (23{,}000 - 32{,}500)(P/F)^{6}_{N} = 0$.
$(1.06)^{-N} = 0.84$. $\therefore N = 3$ yrs (b)

5.24 $-3500(A/P)^{8}_{5} + 12(200-50) = \$923$. (b)

5.25 Cost $= 6000 \times 18 + 85 \times 6000/X + (18X + 85) \times 0.10$.
Set the first derivative $= 0$.
$\frac{d\,Cost}{dx} = -85 \times 6000/x^2 + 1.8 = 0$. $\therefore x = 532$ (c)

5.26 $40{,}000X = 500{,}000\,(A/P)^{8}_{15} - 100{,}000(A/F)^{8}_{15} + 30{,}000X$.
$\therefore X = 5.47$. Thus use $x = 6$. (b)

5.27 A) $1200 + 40(20+X) = 90 \times 20 + 50X$. $\therefore X = 20$. (d)
B) $65 \times 50 - 65 \times 40 - 1200 = -\$550$. (b)

5.28 $3000\,(A/P)^{8}_{10} + (1+.01X)\,74.6/.9 = 1400(A/P)^{8}_{10}$
$\therefore X = 2200$ hr. B $\quad + (1+.01X)\,74.6/.8$. (d)

5.29 $10{,}000 - (10{,}000 - 1300)\,8/12 = \$4200$. (d)

5.30 $(90{,}000 - 18{,}000)/8 = \$9000$.
$90{,}000 - 5 \times 9000 = \$45{,}000$. (a)

5.31 $(90{,}000 - 18{,}000)\,8/36 = \$16{,}000$.
$(90{,}000 - 18{,}000)\,7/36 = \$14{,}000$.
$18{,}000 + (90{,}000 - 18{,}000)\,6/36 = \$30{,}000$. (c)

5.32 $90{,}000 \times 2/8 = \$22{,}500$
$(90{,}000 - 22{,}500)\,2/8 = \$16{,}875$
$(90{,}000 - 22{,}500 - 16{,}875 - 12{,}656 - 9492 - 7119) =$
$\$21{,}358$. (e)

5.33 $80{,}000\,(6/21) \times 0.5\,(P/A)^{10}_{6} - 80{,}000\,(1/21) \times .5(P/G)^{10}_{6}$
$-(80{,}000/6) \times 0.5\,(P/A)^{10}_{6} = \$2300$.
(a)

Solutions to Chapter 6

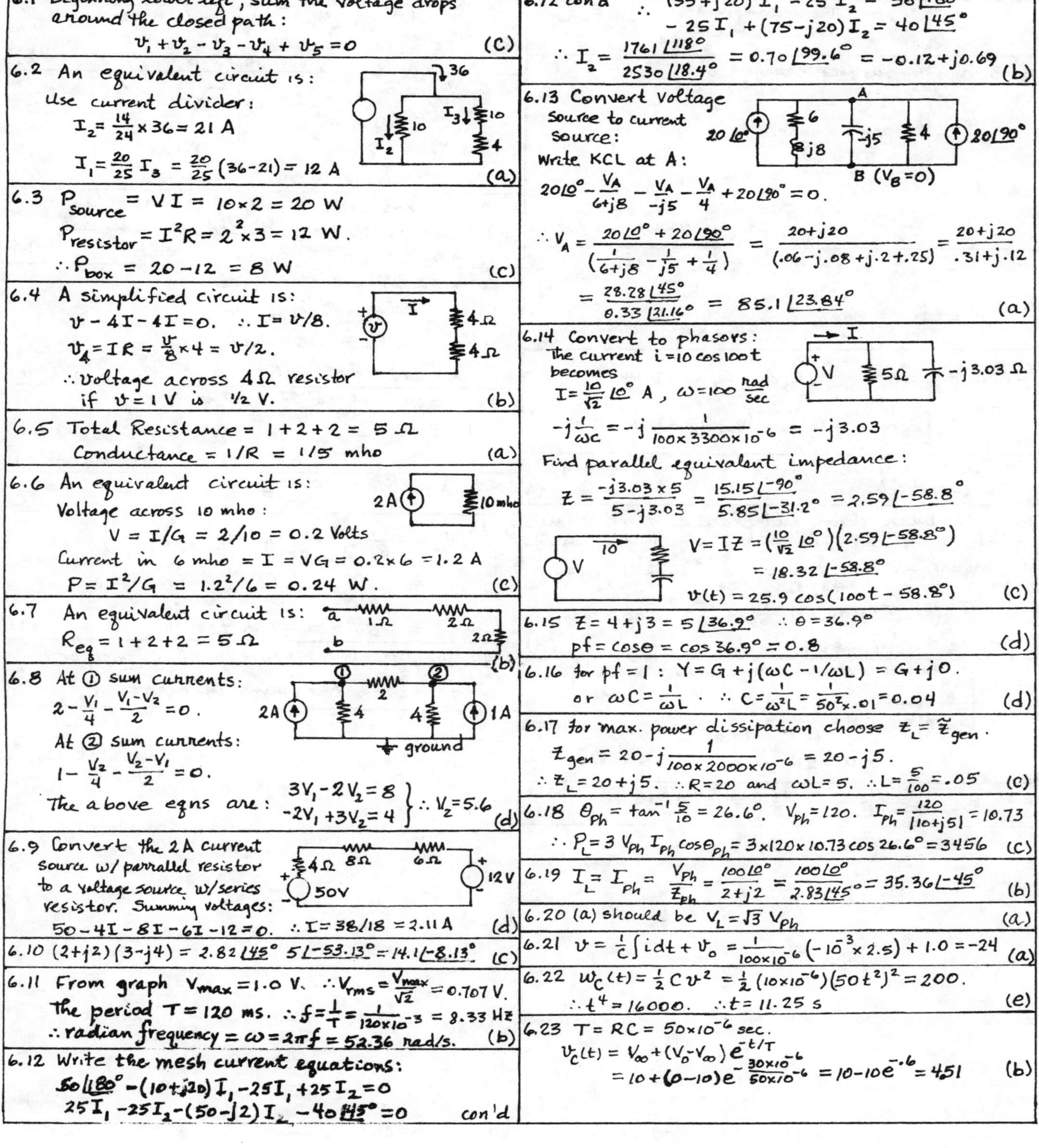

6.1 Beginning lower left, sum the voltage drops around the closed path:

$v_1 + v_2 - v_3 - v_4 + v_5 = 0$ (c)

6.2 An equivalent circuit is:

Use current divider:

$I_2 = \frac{14}{24} \times 36 = 21$ A

$I_1 = \frac{20}{25} I_3 = \frac{20}{25}(36-21) = 12$ A (a)

6.3 $P_{source} = VI = 10 \times 2 = 20$ W

$P_{resistor} = I^2R = 2^2 \times 3 = 12$ W.

$\therefore P_{box} = 20 - 12 = 8$ W (c)

6.4 A simplified circuit is:

$v - 4I - 4I = 0. \quad \therefore I = v/8.$

$v_4 = IR = \frac{v}{8} \times 4 = v/2.$

$\therefore$ voltage across 4 Ω resistor if $v = 1$ V is 1/2 V. (b)

6.5 Total Resistance = 1 + 2 + 2 = 5 Ω

Conductance = 1/R = 1/5 mho (a)

6.6 An equivalent circuit is:

Voltage across 10 mho:

$V = I/G = 2/10 = 0.2$ Volts

Current in 6 mho = $I = VG = 0.2 \times 6 = 1.2$ A

$P = I^2/G = 1.2^2/6 = 0.24$ W. (c)

6.7 An equivalent circuit is:

$R_{eq} = 1 + 2 + 2 = 5$ Ω (b)

6.8 At ① sum currents:

$2 - \frac{V_1}{4} - \frac{V_1 - V_2}{2} = 0.$

At ② sum currents:

$1 - \frac{V_2}{4} - \frac{V_2 - V_1}{2} = 0.$

The above eqns are: $\left.\begin{array}{l} 3V_1 - 2V_2 = 8 \\ -2V_1 + 3V_2 = 4 \end{array}\right\} \therefore V_2 = 5.6$ (d)

6.9 Convert the 2A current source w/ parralled resistor to a voltage source w/series resistor. Summing voltages:

$50 - 4I - 8I - 6I - 12 = 0. \quad \therefore I = 38/18 = 2.11$ A (d)

6.10 $(2+j2)(3-j4) = 2.82\angle 45^\circ \; 5\angle -53.13^\circ = 14.1\angle -8.13^\circ$ (c)

6.11 From graph $V_{max} = 1.0$ V. $\therefore V_{rms} = \frac{V_{max}}{\sqrt{2}} = 0.707$ V.

The period T = 120 ms. $\therefore f = \frac{1}{T} = \frac{1}{120 \times 10^{-3}} = 8.33$ Hz

$\therefore$ radian frequency $= \omega = 2\pi f = 52.36$ rad/s. (b)

6.12 Write the mesh current equations:

$50\angle 180^\circ - (10+j20)I_1 - 25I_1 + 25I_2 = 0$

$25I_1 - 25I_2 - (50-j2)I_2 - 40\angle 45^\circ = 0$ con'd

6.12 con'd $\therefore \begin{array}{l} (35+j20)I_1 - 25I_2 = 50\angle 180^\circ \\ -25I_1 + (75-j20)I_2 = 40\angle 45^\circ \end{array}$

$\therefore I_2 = \frac{1761\angle 118^\circ}{2530\angle 18.4^\circ} = 0.70\angle 99.6^\circ = -0.12 + j0.69$ (b)

6.13 Convert voltage source to current source:

Write KCL at A:

$20\angle 0^\circ - \frac{V_A}{6+j8} - \frac{V_A}{-j5} - \frac{V_A}{4} + 20\angle 90^\circ = 0.$

$\therefore V_A = \frac{20\angle 0^\circ + 20\angle 90^\circ}{\left(\frac{1}{6+j8} - \frac{1}{j5} + \frac{1}{4}\right)} = \frac{20+j20}{(.06-j.08+j.2+.25)} = \frac{20+j20}{.31+j.12}$

$= \frac{28.28\angle 45^\circ}{0.33\angle 21.16^\circ} = 85.1\angle 23.84^\circ$ (a)

6.14 Convert to phasors:

The current $i = 10\cos 100t$ becomes

$I = \frac{10}{\sqrt{2}}\angle 0^\circ$ A, $\omega = 100 \frac{\text{rad}}{\text{sec}}$

$-j\frac{1}{\omega C} = -j\frac{1}{100 \times 3300 \times 10^{-6}} = -j3.03$

Find parallel equivalent impedance:

$Z = \frac{-j3.03 \times 5}{5 - j3.03} = \frac{15.15\angle -90^\circ}{5.85\angle -31.2^\circ} = 2.59\angle -58.8^\circ$

$V = IZ = \left(\frac{10}{\sqrt{2}}\angle 0^\circ\right)(2.59\angle -58.8^\circ)$

$= 18.32\angle -58.8^\circ$

$v(t) = 25.9\cos(100t - 58.8^\circ)$ (c)

6.15 $Z = 4 + j3 = 5\angle 36.9^\circ \quad \therefore \theta = 36.9^\circ$

$pf = \cos\theta = \cos 36.9^\circ = 0.8$ (d)

6.16 for pf = 1 : $Y = G + j(\omega C - 1/\omega L) = G + j0.$

or $\omega C = \frac{1}{\omega L} \cdot \quad \therefore C = \frac{1}{\omega^2 L} = \frac{1}{50^2 \times .01} = 0.04$ (d)

6.17 for max. power dissipation choose $Z_L = \tilde{Z}_{gen}$.

$Z_{gen} = 20 - j\frac{1}{100 \times 2000 \times 10^{-6}} = 20 - j5.$

$\therefore Z_L = 20 + j5. \quad \therefore R = 20$ and $\omega L = 5. \quad \therefore L = \frac{5}{100} = .05$ (c)

6.18 $\theta_{ph} = \tan^{-1}\frac{5}{10} = 26.6^\circ. \quad V_{ph} = 120. \quad I_{ph} = \frac{120}{|10+j5|} = 10.73$

$\therefore P_L = 3 V_{ph} I_{ph} \cos\theta_{ph} = 3 \times 120 \times 10.73 \cos 26.6^\circ = 3456$ (c)

6.19 $I_L = I_{ph} = \frac{V_{ph}}{Z_{ph}} = \frac{100\angle 0^\circ}{2+j2} = \frac{100\angle 0^\circ}{2.83\angle 45^\circ} = 35.36\angle -45^\circ$ (b)

6.20 (a) should be $V_L = \sqrt{3}\, V_{ph}$ (a)

6.21 $v = \frac{1}{C}\int i\,dt + v_0 = \frac{1}{100 \times 10^{-6}}(-10^{-3} \times 2.5) + 1.0 = -24$ (a)

6.22 $w_C(t) = \frac{1}{2}Cv^2 = \frac{1}{2}(10 \times 10^{-6})(50t^2)^2 = 200.$

$\therefore t^4 = 16000. \quad \therefore t = 11.25$ s (e)

6.23 $T = RC = 50 \times 10^{-6}$ sec.

$v_C(t) = V_\infty + (V_0 - V_\infty)e^{-t/T}$

$= 10 + (0 - 10)e^{-\frac{30 \times 10^{-6}}{50 \times 10^{-6}}} = 10 - 10e^{-.6} = 4.51$ (b)

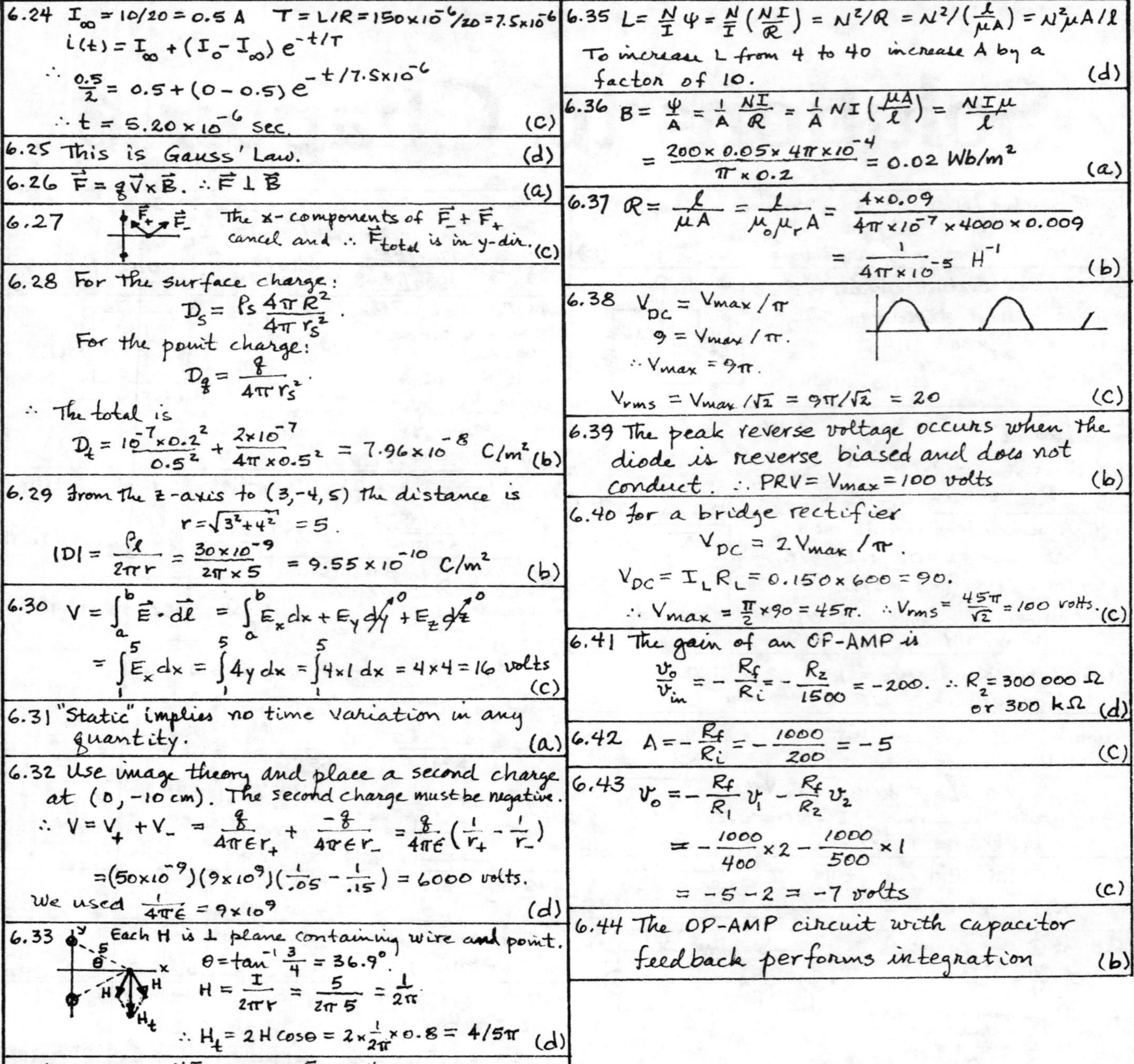

6.24 $I_\infty = 10/20 = 0.5$ A $T = L/R = 150\times10^{-6}/20 = 7.5\times10^{-6}$

$i(t) = I_\infty + (I_0 - I_\infty)e^{-t/T}$

$\therefore \frac{0.5}{2} = 0.5 + (0-0.5)e^{-t/7.5\times10^{-6}}$

$\therefore t = 5.20\times10^{-6}$ sec. (c)

6.25 This is Gauss' Law. (d)

6.26 $\vec{F} = q\vec{V}\times\vec{B}$. $\therefore \vec{F} \perp \vec{B}$ (a)

6.27 The x-components of $\vec{F}_- + \vec{F}_+$ cancel and $\therefore$ $\vec{F}_{total}$ is in y-dir. (c)

6.28 For the surface charge:

$$D_s = \rho_s \frac{4\pi R^2}{4\pi r_s^2}.$$

For the point charge:

$$D_q = \frac{q}{4\pi r_s^2}.$$

$\therefore$ The total is

$$D_t = \frac{10^{-7}\times0.2^2}{0.5^2} + \frac{2\times10^{-7}}{4\pi\times0.5^2} = 7.96\times10^{-8}\ \text{C/m}^2 \quad (b)$$

6.29 From the z-axis to (3,−4,5) the distance is

$$r=\sqrt{3^2+4^2}=5.$$

$$|D| = \frac{\rho_\ell}{2\pi r} = \frac{30\times10^{-9}}{2\pi\times5} = 9.55\times10^{-10}\ \text{C/m}^2 \quad (b)$$

6.30 $V = \int_a^b \vec{E}\cdot d\vec{\ell} = \int_a^b E_x\,dx + E_y\,dy + E_z\,dz$ (last two terms = 0)

$= \int_1^5 E_x\,dx = \int_1^5 4y\,dx = \int_1^5 4\times1\,dx = 4\times4 = 16$ volts (c)

6.31 "Static" implies no time variation in any quantity. (a)

6.32 Use image theory and place a second charge at (0, −10 cm). The second charge must be negative.

$\therefore V = V_+ + V_- = \frac{q}{4\pi\epsilon r_+} + \frac{-q}{4\pi\epsilon r_-} = \frac{q}{4\pi\epsilon}\left(\frac{1}{r_+} - \frac{1}{r_-}\right)$

$=(50\times10^{-9})(9\times10^{9})\left(\frac{1}{.05} - \frac{1}{.15}\right) = 6000$ volts.

We used $\frac{1}{4\pi\epsilon} = 9\times10^9$ (d)

6.33 Each H is ⊥ plane containing wire and point.

$\theta = \tan^{-1}\frac{3}{4} = 36.9°$.

$H = \frac{I}{2\pi r} = \frac{5}{2\pi 5} = \frac{1}{2\pi}$

$\therefore H_t = 2H\cos\theta = 2\times\frac{1}{2\pi}\times0.8 = 4/5\pi$ (d)

6.34 $H_{center} = \frac{NI}{\ell} = \frac{1000\times5}{0.5} = 10^4$ A/m. (a)

6.35 $L = \frac{N}{I}\Psi = \frac{N}{I}\left(\frac{NI}{\mathcal{R}}\right) = N^2/\mathcal{R} = N^2/\left(\frac{\ell}{\mu A}\right) = N^2\mu A/\ell$

To increase L from 4 to 40 increase A by a factor of 10. (d)

6.36 $B = \frac{\Psi}{A} = \frac{1}{A}\frac{NI}{\mathcal{R}} = \frac{1}{A}NI\left(\frac{\mu A}{\ell}\right) = \frac{NI\mu}{\ell}$

$= \frac{200\times0.05\times4\pi\times10^{-4}}{\pi\times0.2} = 0.02\ \text{Wb/m}^2$ (a)

6.37 $\mathcal{R} = \frac{\ell}{\mu A} = \frac{\ell}{\mu_o\mu_r A} = \frac{4\times0.09}{4\pi\times10^{-7}\times4000\times0.009}$

$= \frac{1}{4\pi\times10^{-5}}\ H^{-1}$ (b)

6.38 $V_{DC} = V_{max}/\pi$

$9 = V_{max}/\pi$.

$\therefore V_{max} = 9\pi$.

$V_{rms} = V_{max}/\sqrt{2} = 9\pi/\sqrt{2} = 20$ (c)

6.39 The peak reverse voltage occurs when the diode is reverse biased and does not conduct. $\therefore$ PRV $= V_{max} = 100$ volts (b)

6.40 for a bridge rectifier

$$V_{DC} = 2V_{max}/\pi.$$

$V_{DC} = I_L R_L = 0.150\times600 = 90.$

$\therefore V_{max} = \frac{\pi}{2}\times90 = 45\pi.$ $\therefore V_{rms} = \frac{45\pi}{\sqrt{2}} = 100$ volts. (c)

6.41 The gain of an OP-AMP is

$\frac{v_o}{v_{in}} = -\frac{R_f}{R_i} = -\frac{R_2}{1500} = -200.$ $\therefore R_2 = 300\,000\ \Omega$ or $300\ k\Omega$ (d)

6.42 $A = -\frac{R_f}{R_i} = -\frac{1000}{200} = -5$ (c)

6.43 $v_o = -\frac{R_f}{R_1}v_1 - \frac{R_f}{R_2}v_2$

$= -\frac{1000}{400}\times2 - \frac{1000}{500}\times1$

$= -5 - 2 = -7$ volts (c)

6.44 The OP-AMP circuit with capacitor feedback performs integration (b)

Solutions to Chapter 7
(Metric Units)

7.1 A tire. All other devices have fluid entering and/or leaving. (b)

7.2 Mass. All other quantities do not depend on the mass. (d)

7.3 Isobaric. The pressure will remain constant due to the inlets and outlets for air. (b)

7.4 All properties are uniform thru out the volume. (a)

7.5 A scientific law results from experimental observations. (e)

7.6 $\rho = p/RT = (-40+100)/.287\times(-40+273) = 0.897\ kg/m^3$. (e)

7.7 $v = v_f + x(v_g - v_f)$ $p = 0.2\ MPa$ TABLE 7.3.2
$0.5 = .001 + x(.8857 - .001)$
$\therefore x = 0.564$ (b)

7.8 If the state is beyond the Table 7.3.3 then use the equation of state.
$v = \frac{1}{\rho} = \frac{RT}{p} = \frac{0.4615\times1473}{4000} = 0.17\ m^3/kg$.
Use $T = 1200+273 = 1473\ K$, $p = 4000\ kPa$ (d)

7.9 $v = \frac{V}{m} = \frac{2}{20} = 0.1$, $p = 4\ MPa$.
From Table 7.3.3 interpolation gives
$T = \frac{0.1-0.09885}{0.1109-0.09885}\times100 + 600 = 610\ °C$. (b)

7.10 Using Table 7.3.2 we see that
$115°C < 120.2°C$. ∴ Compressed. (a)

7.11 $m = \frac{p_1V_1}{RT_1} = \frac{280\times.03}{.287\times263} = 0.1113\ kg$
$\therefore V_2 = \frac{mRT_2}{p_2} = \frac{.1113\times.287\times303}{310} = 0.0312\ m^3$ (c)

7.12 $\dot{Q} = \frac{1}{R}A\Delta T$ $A = 4\pi r^2$
$\frac{300}{1000} = \frac{1}{1.5}4\pi\times5^2(T-20)\frac{1}{3600}$. $\therefore T = 25.2°C$.
The 1000 converts J/s to kJ/s.
The 3600 converts hr to seconds. (e)

7.13 $W = F\times d = (10\times9.8)\times3 = 294\ N\cdot m$ input.
$\therefore Q = 294\ J$ out. (it is negative) (e)

7.14 $Q - W = E_a - E_b$
$20 - 5 = E_a - 10$. $\therefore E_a = 25\ kJ$. (d)

7.15 for $2\to3$ $Q-(-5) = 5$. $\therefore Q_{2-3} = 0$
$\therefore Q_{net} = 20 + 0 + 30 = 50$
$Q_{net} = W_{net} = 5 - 5 + W_{3-1} = 50$.
$\therefore W_{3-1} = 50\ kJ$. (a)

7.16 $Q = mc\Delta T$. Here $Q = \frac{1}{2}mV^2$, $m = V\rho$
$\frac{1}{2}\times2000\times25^2 = 10\,000\times10^{-6}\times1000\times4180\ \Delta T$
$\therefore \Delta T = 14.95°C$ (c)

7.17 $(mc\Delta T)_{ice} + (mc\Delta T)_{melted\ ice} + m\times h_{fusion} = (mc\Delta T)_{water}$
$10\times2.1\times10 + 10\times4.18(T-0) + 10\times320 = 100\times4.18(20-T)$
$\therefore T = 10.77°C$. (b)

7.18 The heat transfer exceeds the work done if $\Delta E > 0$. (b)

7.19 $W = mRT \ln p_1/p_2$
$= 2\times.287\times573 \ln 100/4000 = -1213\ kJ$ (a)

7.20 $\Delta U = m(u_2 - u_1)$ Table 7.3.3
$= 2(2725.3 - 2810.4) = -170.2\ kJ$ (b)

7.21 $Q = m\Delta h = m\,h_{fg}$
$= 2\times1940.8 = 3881.6\ kJ$ (e)

7.22 Assume p = const since doors would be opened. Cracks abound.
$Q = mc_p\Delta T$.
$\frac{400\times200}{3600} + \frac{30\times2000}{1000} = \frac{6000\times1.23\times1.00\ \Delta T}{20\times60}$.
$\therefore \Delta T = 13.37\ °C$ (e)

7.23 $Q = m\Delta h = mc_p\Delta T$
$522 = 1\times c_p\times(800-300)$ $\therefore c_p = 1.044\ \frac{kJ}{kg\cdot K}$ (d)

7.24 $Q = m\Delta h = mc_p\Delta T$
$= 10\times1.00\times(230-10) = 2200\ kJ$ (e)

7.25 $Q = m\Delta h$
$= 10[3072 - 42] = 30300\ kJ$ (c)

7.26 $W = p\Delta V = mp\Delta v$
$= 10(1.316 - .001)\times200 = 2630\ kJ$ (a)

7.27 $Q = m\Delta u$ $\quad u_1 = 851 + .5(2595 - 851) = 1723$ kJ/kg.

$T_2 = 500°C$ $\quad v_1 = .001 + .5(.1274 - .001) = .0642\ m^3/kg.$

$v_2 = .0642$

Use Table 7.3.3

$u_2 = \frac{.0642 - .0566}{.0864 - .0566}(3099 - 3082) + 3082 = 3086$

$\therefore Q = 2(3086 - 1723) = 2730$ kJ (a)

7.28 $\frac{p_1}{T_1} = \frac{p_2}{T_2}$. $\quad (p_2 + 100)273 = (100 + 100)343.$

$\therefore p_2 = 151$ kPa gage (d)

7.29 $Q = m\Delta u = mc_v\Delta T$ $\quad m = \frac{pV}{RT} = \frac{100 \times 10}{.287 \times 293} = 11.89$ kg

$= 11.89 \times .716 \times 60$

$= 511$ kJ (e)

7.30 $Q = m\Delta u$

$9000 = 10(u_2 - 42)$ $\quad \therefore u_2 = 942$ kJ/kg

$942 = 851 + x(2595 - 851)$ $\quad \therefore x = .0522$ (e)

7.31 $T_2 = T_1\left(\frac{p_2}{p_1}\right)^{k-1/k}$

$= 473\left(\frac{20}{400}\right)^{.4/1.4} = 201\ K$ or $-72°C$ (d)

7.32 $T_2 = T_1\left(\frac{v_1}{v_2}\right)^{k-1}$

$= 473\left(\frac{1}{3}\right)^{.4} = 305\ K$ or $31.8°C$ (a)

7.33 $s_2 = s_1 = 7.1685$. from Table 7.3.2,

$7.1685 = .649 + 7.502x_2$. $\therefore x_2 = .869$ (c)

7.34 $s_2 = s_1 = 7.1685$ from Table 7.3.3.

This is slightly less than $s = 7.171$ at $T = 200°C$ and $p = 0.4$ MPa. $\therefore T_2 = 199°C$. (b)

7.35 $T_2 = T_1\left(\frac{p_2}{p_1}\right)^{k-1/k} = 293\left(\frac{6000}{100}\right)^{.4/1.4} = 944\ K$

$W = -m\Delta u = -mc_v\Delta T$ $\quad$ let $m = 1$

$= -.716(944 - 293) = -466$ kJ/kg. (b)

7.36 The heat transfer must be zero (d)

7.37 $T_2 = T_1\left(\frac{p_2}{p_1}\right)^{n-1/n} = 283 \times 3^{.2/1.2} = 340\ K$ or $66.9°C$ (e)

7.38 $\rho_1 = \frac{1}{v_1} = \frac{1}{.06525} = 15.33$ $\quad$ Table 7.3.3

$\rho_2 = \frac{1}{v_2} = \frac{1}{7.649} = 0.1307$ $\quad$ Table 7.3.2

$\rho_1 A_1 V_1 = \rho_2 A_2 V_2$

$15.33\pi \times \frac{.2^2}{4} V_1 = .1307\pi \times \frac{.05^2}{4} V_2$. $\quad \frac{V_2}{V_1} = 1877$ (a)

7.39 $Q = h_2 - h_1$ $\quad h_1 = 251$ from Table 7.3.1

$= 3674 - 251 = 3423$ kJ/kg (e)

7.40 $\dot{Q} = \dot{m}_w c\Delta T = \dot{m}_s \Delta h$ $\quad \swarrow h_{fg}$ Table 7.3.2

$\dot{m} \times 4.18(30 - 20) = 10 \times 2358$ $\therefore \dot{m} = 564$ kg/s (b)

7.41 $s_2 = s_1 = 6.7698 = .649 + 7.502x_2$. $\therefore x_2 = .816$

$\therefore h_2 = 192 + .816 \times 2393 = 2144$

$w = h_1 - h_2 = 3213.5 - 2144 = 1069$ kJ/kg (c)

7.42 $\dot{W}_P = \dot{m}\frac{\Delta p}{\rho\eta} = 10\,\frac{6000 - 10}{1000 \times .75} = 79.9$ kW (e)

7.43 $\dot{W}_T = \dot{m}\frac{\Delta p}{\rho} = 60 \times 2 \times 2 \times 1000\,\frac{300}{1000} = 72000$ kW (e)

7.44 $T_2 = T_1\left(\frac{p_2}{p_1}\right)^{k-1/k} = 673\left(\frac{80}{2000}\right)^{.4/1.4} = 268\ K$

$V_2^2/2 = h_1 - h_2 = c_p(T_1 - T_2) = 1000(673 - 268)$

$\therefore V_2 = 900$ m/s. (e)

7.45 $h_1 = 3177$ $\quad s_{2'} = s_1 = 6.5415 = .832 + x_{2'}\, 7.0774$

$\therefore x_{2'} = .807$. $\quad \therefore h_{2'} = 251 + .807 \times 2358 = 2153$.

$W = 0.87 \times (3177 - 2153) = 891$ kJ/kg. (b)

7.46 $\eta_{max} = 1 - \frac{T_L}{T_H} = 1 - \frac{293}{393} = 0.254$

$\eta = \frac{4180}{10 \times 4.18(120 - 20)} = 1.00$. impossible (b)

7.47 $\Delta s_{net} > 0$ for all processes. (b)

7.48 Operates between const. temp. reservoirs. (d)

7.49 $\Delta S = Q/T = 90 \times 320/273 = 105.5$ kJ/K (d)

7.50 $40(320 + 4.18T_2) = 100 \times 4.18(20 - T_2)$. $\therefore T_2 = -7.6°C$.

Impossible. $\therefore T_2 = 0°C$. $\quad 320m = 100 \times 4.18 \times 20$. $\therefore m = 26.1$

$\Delta S = 26.1 \times 320/273 + 100 \times 4.18 \ln 273/293 = 1.04$ (e)

7.51 $\eta_{max} = 1 - \frac{T_L}{T_H} = 1 - 323/873 = 0.63$ (b)

7.52 $T_4 = P_4 v_4/R = 160 \times .5/.287 = 279\ K$

$W = Q\eta = 30(1 - 279/473) = 12.3$ kJ/kg (e)

7.53 $\eta = 1 - (400/40)^{.4} = 0.602$.

$\therefore Q = W/\eta = 100/.602 = 166$ kW. (a)

7.54 $r_c = 45/25 = 1.8$ $\quad r = 450/25 = 18$

$Q = W/\eta = 120/\left[1 - \frac{1.8^{1.4} - 1}{1.4(1.8 - 1)}\, 18^{-.4}\right] = 187$ kJ/s. (a)

7.55 $s_4 = s_3 = 7.1685 = .649 + 7.502x_4$. $\therefore x_4 = .869$

$h_4 = 192 + .869 \times 2393 = 2272$ $\quad h_3 = 3658$. $\quad h_1 = 192$

$\eta = \frac{W}{Q} = (3658 - 2272)/(3658 - 192) = 0.40$ (c)

7.56 $s_4 = s_3 = 7.168$. $\therefore$ ④ Superheat $\therefore h_4 = 2722$. $\quad h_3 = 3658$

after reheat $h_5 = 3704$. $\quad s_6 = s_5 = 8.778$

$\therefore h_6 = \frac{8.778 - 8.689}{8.9046 - 8.689}(2879.5 - 2783) + 2783 = 2823$

$\eta = \frac{W_{3-4} + W_{5-6}}{Q_{2-3} + Q_{4-5}} = \frac{3658 - 2722 + 3704 - 2823}{3658 - 192 + 3704 - 2722} = 0.408$ (c)

7.57 Decrease moisture content in the turbine (e)

7.58 $s_2 = s_1 = .7165$. Using $T_3 = 49.3°$, $P_3 = 1.2$ MPa (Table 7.4.2)

$\therefore h_2 = 218.4$ (interpolate). $\quad h_3 = 84.21 = h_4$. $\quad h_1 = 174.1$

$COP = \frac{Q_{4-1}}{W_{1-2}} = \frac{174.1 - 84.21}{218.4 - 174.1} = 2.03$. (b)

7.59 $h_4 = h_3 = 84.21 = h_f + x_4 h_{fg}$

$= 8.85 + 165.2x_4$. $\quad \therefore x_4 = 0.456$ (e)

7.60 $\dot{W}_{max} = \dot{m}(h_1 - T_0 s_1) - \dot{m}(h_2 - T_0 s_2)$

$= \dot{m}[(h_1 - h_2) + T_0(s_2 - s_1)]$

$= 10\left[1.00(100 - 20) + 293\left(1.00 \ln\frac{293}{373} - .287 \ln\frac{80}{6000}\right)\right]$

$= 3723$ kW (a)

Solutions to Chapter 7
(English Units)

7.1 A tire. All other devices have fluid entering an/or leaving. (b)

7.2 Mass. All other quantities do not depend on the mass. (d)

7.3 Isobaric. The pressure will remain constant due to the inlets and outlets for air. (b)

7.4 All properties are uniform throughout the volume (a)

7.5 A scientific law results from experimental observations. (e)

7.6 $\rho = p/RT = (-6+14.7)\times 144/(53.3\times 420)$ (e)
$= 0.056\ lbm/ft^3$. $(T=-40+460=420°R)$

7.7 $v = v_f + x(v_g - v_f)$ $p = 30$ psia Table 7.3.2E
$2 = .017 + x(13.75 - .017)$. $\therefore x = 0.144$ (b)

7.8 If the state is beyond Table 7.3.3E then use the equation of state $p = \rho RT$.
$v = \frac{1}{\rho} = \frac{RT}{p} = \frac{85.8\times 2660}{600\times 144} = 2.64\ ft^3/lbm$ (d)

7.9 $v = \frac{V}{m} = \frac{60}{40} = 1.5$, $p = 600$ psia.
From Table 7.3.3E interpolation gives
$T = \frac{1.5-1.411}{1.622-1.411}\times 200 + 1000 = 1084\ °F$ (b)

7.10 Using Table 7.3.2E we see that
$230°F < 233.9°F$. $\therefore$ Compressed liquid (a)

7.11 $m = \frac{p_1 V_1}{RT_1} = \frac{26\times 144\times 0.9}{53.3\times 474} = 0.1334\ lbm$.
$\therefore V_2 = \frac{mRT_2}{p_2} = \frac{.1334\times 53.3\times 550}{30\times 144} = 0.905\ ft^3$. (c)

7.12 $\dot{Q} = \frac{1}{R} A\Delta T$ $A = 4\pi r^2$
$\frac{300}{0.293} = \frac{1}{30}\times 4\pi\times 15^2\times (T-70)$ $\therefore T = 80.9°F$. (e)

7.13 $W = F\times d = 20\times 10 = 200$ ft-lb input.
$Q = -\frac{200}{778} = -0.257$ BTU. (heat out is neg.) (e)

7.14 $Q - W = E_a - E_b$.
$20 - 5 = E_a - 10$. $\therefore E_a = 25$ BTU (d)

7.15 for $2\to 3$ $Q - (-5) = 5$. $\therefore Q_{2-3} = 0$.
$\therefore Q_{net} = 20 + 0 + 30 = 50$.
$Q_{net} = W_{net} = 5 - 5 + W_{3-1} = 50$.
$\therefore W_{3-1} = 50$ BTU (a)

7.16 $Q = mc\Delta T$. Here $Q = \frac{1}{2}mV^2$, $m = V\rho$.
$\frac{1}{2}\times\frac{4000\times 90^2}{32.2\times 778} = \frac{600}{1728}\times 62.4\times 1.00\times\Delta T$. $\therefore \Delta T = 29.9°F$ (c)

7.17 $(mc\Delta T)_{ice} + (mc\Delta T)_{melted\ ice} + mh_{melt} = (mc\Delta T)_{water}$
$20\times 0.5\times(32-15) + 20\times 1.00(T-32) + 20\times 140 = 200\times 1.00(70-T)$ (b)
$\therefore T = 53.05°F$

7.18 The heat transfer exceeds the work done if $\Delta E > 0$ (b)

7.19 $W = mRT\ln p_1/p_2$
$= 4\times 53.3\times 1060\ln 15/600 = -834{,}000$ ft-lb. (a)

7.20 $\Delta U = m(u_2 - u_1)$ Table 7.3.3E
$= 4(1184.5 - 1218.4) = -135.6$ BTU (b)

7.21 $Q = m\Delta h = mh_{fg}$
$= 4\times 826.3 = 3305$ BTU (e)

7.22 Assume $p =$ const since doors would be opened. Cracks abound.
$Q = mc_p\Delta T$.
$200\times 400 + 3\times 20{,}000/0.293 = 200{,}000\times .076\times .24\ \Delta T$ (↙ V, ↙ ρ, ↙ c_p)
$\therefore \Delta T = 26°F$. (0.333, Δt) (e)

7.23 $Q = m\Delta h = mc_p\Delta T$ (d)
$520 = 2\times c_p(1000-100)$. $\therefore c_p = 0.289$ BTU/lbm-°F.

7.24 $Q = m\Delta h = mc_p\Delta T$
$= 20\times 0.24(400-50) = 1680$ BTU. (e)

7.25 $Q = m\Delta h$
$= 20[1332.1 - 8.02]$
$= 26{,}500$ BTU. (c)

7.26 $W = p\Delta V = mp\Delta v$
$= 20\times 60\times 144(10.425 - 0.017) = 1.800\times 10^6$ ft-lb (a)

7.27 $Q = m\Delta u$ $u_1 = 485 + .5(1117-485) = 801$
$v_1 = .020 + .5(.676 - .02) = .348$
$T_2 = 900°F$ and $v_2 = 0.348$. $\therefore$ Using Table 7.3.3E:
$u_2 = \frac{0.348-0.216}{0.353-0.216}\cdot(1277-1242) + 1242 = 1276$
$\therefore Q = 4(1276 - 801) = 1899$ BTU. (a)

7.28 $\frac{p_1}{T_1} = \frac{p_2}{T_2}$. $(p_2 + 14.7)\times 490 = (28+14.7)\times 620$.
$\therefore p_2 = 39.3$ psi gage (d)

7.29 $Q = m\Delta u = mc_v\Delta T$ $m = \frac{pV}{RT} = \frac{14.7\times 144\times 300}{53.3\times 530}$
$= 22.48\times .171(150-70)$ $= 22.48\ lbm$
$= 307.5$ BTU (e)

7.30 $Q = m\Delta u$

$9000 = 20(u_2 - 80)$. $\therefore u_2 = 458$ BTU/lbm

$458 = 353 + x_2(1114 - 353)$ $\therefore x_2 = 0.138$ (e)

7.31 $T_2 = T_1\left(\frac{p_2}{p_1}\right)^{\frac{k-1}{k}}$

$= 860\left(\frac{2}{60}\right)^{0.4/1.4} = 325\,°R$ or $-134\,°F$ (d)

7.32 $T_2 = T_1\left(\frac{v_1}{v_2}\right)^{k-1}$

$= 860\left(\frac{1}{3}\right)^{0.4} = 554\,°R$ or $94\,°F$ (a)

7.33 $s_2 = s_1 = 1.7632$, Table 7.3.3E. Using Table 7.3.2E

$1.7632 = .175 + x_2(1.7448)$. $\therefore x_2 = 0.910$ (c)

7.34 $s_2 = s_1 = 1.7632$. At 60 psia this is just less than s at 500°F. Interpolate and find

$T_2 = \frac{1.763 - 1.713}{1.768 - 1.713} \times 100 + 400 = 491\,°F$. (b)

7.35 $T_2 = T_1\left(\frac{p_2}{p_1}\right)^{\frac{k-1}{k}} = 530\left(\frac{400}{14.7}\right)^{0.2857} = 1362\,°R$.

$W = -m\Delta u = -mc_v\Delta T$. let $m = 1$.

$= 0.171(1362 - 530) = 142.3$ BTU/lbm

or $142.3 \times 778 = 110{,}700$ ft-lb/lbm (b)

7.36 The heat transfer must be zero. (d)

7.37 $T_2 = T_1\left(\frac{p_2}{p_1}\right)^{\frac{n-1}{n}} = 510 \times 3^{.2/1.2} = 612\,°R$

or $152\,°F$. (e)

7.38 $\rho_1 = \frac{1}{v_1} = \frac{1}{2.136} = 0.4682$ Table 7.3.3E

$\rho_2 = \frac{1}{v_2} = \frac{1}{173.75} = 0.005755$ Table 7.3.2E

$\rho_1 A_1 V_1 = \rho_2 A_2 V_2$

$.4682\,\pi \times 4^2 \times V_{in} = .005755\,\pi \times 1^2 \times V_{out}$. $\frac{V_{out}}{V_{in}} = 1302$ (a)

7.39 $Q = h_2 - h_1$ $h_1 = 88$ from Table 7.3.1E.

$= 1526.5 - 88 = 1438$ BTU/lbm. (e)

7.40 $\dot{Q} = \dot{m}_w C\Delta T = \dot{m}_s \Delta h$. h_{fg} Table 7.3.2E

$\dot{m}_w \times 1.00 \times (90 - 70) = 20 \times 1022.1$. $\therefore \dot{m}_w = 1022$ lbm/sec (b)

7.41 $s_2 = s_1 = 1.6751$ from Table 7.3.3E.

$= .175 + 1.7448\,x_2$. $\therefore x_2 = 0.860$.

$\therefore h_2 = 94 + .86 \times 1022 = 973$.

$W = h_1 - h_2 = 1368 - 973 = 395$ BTU/lbm.

$395 \times 778 = 3.07 \times 10^5$ ft-lb/lbm (c)

7.42 $\dot{W}_p = \dot{m}\frac{\Delta p}{\rho\eta} = 20\,\frac{398 \times 144}{62.4 \times 0.75} = 24{,}500$ ft-lb/sec. or 44.5 Hp (e)

7.43 $\dot{W}_T = \dot{m}\frac{\Delta p}{\rho} = (200 \times 6 \times 6 \times 62.4)\frac{50 \times 144}{62.4} = 5.18 \times 10^7\ \frac{\text{ft-lb}}{\text{sec}}$

or 94,300 Hp. (e)

7.44 $T_2 = T_1\left(\frac{p_2}{p_1}\right)^{\frac{k-1}{k}} = 1160\left(\frac{12}{100}\right)^{.2857} = 633\,°R$.

$V_2^2/2 = h_1 - h_2 = c_p(T_1 - T_2) = .24 \times 778(1160 - 633)$.

$\therefore V_2 = 444$ ft/sec (e)

7.45 $h_1 = 1362$ $s_{2'} = s_1 = 1.6397 = .175 + 1.7448\,x_{2'}$.

$\therefore x_{2'} = 0.840$. $\therefore h_{2'} = 94 + .84 \times 1022 = 952$. (b)

$W = 0.87(1362 - 952) = 357$ BTU/lbm or 2.78×10^5 ft-lb/lb

7.46 $\eta_{max} = 1 - \frac{T_L}{T_H} = 1 - \frac{530}{700} = 0.243$.

$\eta = \frac{2600 \times 550/778}{20 \times 1.00(240 - 70)} = 0.54$. $\therefore$ impossible (b)

7.47 $\Delta S_{net} > 0$ for all real processes (b)

7.48 Operates between constant temperature reservoirs. (d)

7.49 $\Delta S = Q/T = 200 \times 140/492 = 56.9$ BTU/°R. (d)

7.50 $80[140 + 1.00(T_2 - 32)] = 200 \times 1.00(60 - T_2)$ $\therefore T_2 = 2.9°F$.

That is impossible. $\therefore$ all the ice doesn't melt and $T_2 = 32°F$. $140\,m = 200 \times 1.00 \times (60 - 32)$. $\therefore m = 40$ lbm.

$\Delta S = \frac{40 \times 140}{492} + 200 \times 1.00 \ln\frac{492}{520} = 0.312$ BTU/°R. (e)

7.51 $\eta_{max} = 1 - \frac{T_L}{T_H} = 1 - \frac{560}{1460} = 0.616$ (b)

7.52 $T_4 = p_4 v_4/R = 24 \times 144 \times 8/53.3 = 518.7\,°R$

$W = Q\eta = 15 \times \left(1 - \frac{519}{860}\right) \times 778 = 4627$ ft-lb/lbm (e)

7.53 $\eta = 1 - (40/4)^{-.4} = 0.602$.

$\therefore Q = W/\eta = 130 \times \frac{0.746}{1.055}/0.602 = 153$ BTU/sec. (a)

7.54 $r_c = 5/3 = 1.67$ $r = 45/3 = 15$

$Q = W/\eta = \frac{160 \times 550}{778}\Big/\left[1 - \frac{1.67^{1.4} - 1}{1.4(1.67 - 1)}\,15^{-.4}\right] = 182$ BTU/sec (a)

7.55 $s_4 = s_3 = 1.7632 = .175 + 1.7448\,x_4$. $\therefore x_4 = 0.910$

$h_4 = 94 + .91 \times 1022 = 1024$. $h_3 = 1524$ $h_1 = 94$

$\eta = W/Q = (1524 - 1024)/(1524 - 94) = 0.3495$ (c)

7.56 $s_4 = s_3 = 1.7963$ $\therefore$ superheat $\therefore h_4 = 1208$ $h_3 = 1526$

after reheat $h_5 = 1534$ $s_6 = s_5 = 2.10$ (interpolate)

and $h_6 = \frac{2.10 - 2.051}{2.115 - 2.051}(1196 - 1150) + 1150 = 1185$

$\eta = \frac{W_{3-4} + W_{5-6}}{Q_{2-3} + Q_{4-5}} = \frac{1526 - 1208 + 1534 - 1185}{1526 - 70 + 1534 - 1208} = 0.374$ (c)

7.57 Decrease moisture content in the condenser (e)

7.58 $s_2 = s_1 = 0.171$. If $T_3 = 80\,°F$, $p_3 \cong 100$ psia (Table 7.4.1E)

$\therefore h_2 = 89.3$ (Table 7.4.2E). $h_3 = 26.4 = h_4$. $h_1 = 75.1$

$COP = \frac{Q_{4-1}}{W_{1-2}} = \frac{75.1 - 26.4}{89.3 - 75.1} = 3.43$ (b)

7.59 $h_4 = h_3 = 26.4 = h_f + h_{fg}\,x_4$

$= 4.24 + 70.87\,x_4$. $\therefore x_4 = 0.313$ (e)

7.60 $\dot{W}_{max} = \dot{m}(h_1 - T_0 s_1) - \dot{m}(h_2 - T_0 s_2)$

$= \dot{m}[(h_1 - h_2) + T_0(s_2 - s_1)]$

$= 20\left[.24(200 - 70) + 530\left(.24\ln\frac{530}{660} - \frac{.533}{778}\ln\frac{12}{400}\right)\right]$

$= 2660$ BTU/sec or 3760 Hp (a)

Solutions to Chapter 8

8.1 $\hat{i}_B = \frac{\hat{i}-2\hat{j}-2\hat{k}}{\sqrt{1+4+4}} = \frac{1}{3}(\hat{i}-2\hat{j}-2\hat{k})$

$\vec{A}\cdot\hat{i}_B = (15\hat{i}-9\hat{j}+15\hat{k})\cdot\frac{1}{3}(\hat{i}-2\hat{j}-2\hat{k})$

$= 5+6-10 = 1$ (a)

8.2 $\vec{A}+\vec{B}+\vec{C} = (2\hat{i}+5\hat{j})+(6\hat{i}-7\hat{k})+(2\hat{i}-6\hat{j}+10\hat{k})$

$= 10\hat{i} - \hat{j} + 3\hat{k}$.

magnitude $= \sqrt{10^2+1^2+3^2} = 10.49$ (c)

8.3 $\vec{M} = \vec{r}\times\vec{F} = (4\hat{i}-6\hat{j}+4\hat{k})\times(200\hat{i}+400\hat{j})$.

$M_y = 4\times 200 = 800$ since $\hat{k}\times\hat{i}=\hat{j}$. (e)

8.4 $\vec{M} = \vec{r}_1\times\vec{F}_1 + \vec{r}_2\times\vec{F}_2$

$= (2\hat{i}-4\hat{k})\times(50\hat{i}-40\hat{k}) + (-4\hat{i}+2\hat{j})\times(60\hat{j}+80\hat{k})$

$M_x = 2\times 80 = 160$ since $\hat{j}\times\hat{k}=\hat{i}$. (c)

8.5 Concurrent ⇒ all pass thru a point.
Coplanar ⇒ all in the same plane.
The forces are three-dimensional. (e)

8.6 $\Sigma\vec{F}=0$. $\therefore \vec{R}+141\hat{i}-141\hat{j}-200\hat{i}-100\hat{k}=0$.

$\therefore R = 59\hat{i}+141\hat{j}+100\hat{k}$. (b)

8.7 $\Sigma\vec{M}=0$. $\therefore \vec{M}_A + (4\hat{j}-3\hat{k})\times(-100\hat{k}) - 3\hat{k}\times(-200\hat{i})$

$+ 4\hat{i}\times(141\hat{i}-141\hat{j}) = 0$.

$\therefore \vec{M}_A = 400\hat{i} - 600\hat{j} + 564\hat{k}$. (c)

8.8 They must be concurrent, otherwise a resultant moment would occur. (b)

8.9 It is a two-force body. (a)

8.10 $\Sigma M_A = 0$. $F_B\times 8 = 400\times 4 + 400\times 6$.

$\therefore F_B = 500$ N. (b)

8.11 $M_A = 400\times 8 + 400\times 6 = 5600$ N·m. (a)

8.12 $\Sigma M_B = 0$.

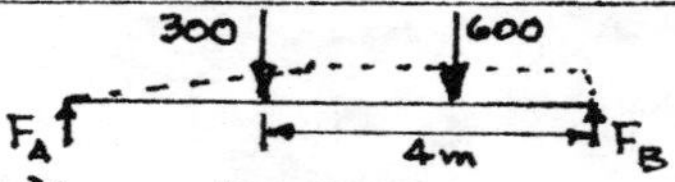

$6F_A = 4\times 300 + 600\times 3/2$. $\therefore F_A = 350$ N. (b)

8.13 $M_A = 0.6\times 100 - 141\times 0.6 + 141\times 0.8 = 88.2$ (c)

8.14 $M = 100\sin 45^\circ\times 4 = 282.8$ CW (a)

8.15 $\Sigma M_A = 0$. $\therefore 6\times 70.7 = 2\times .866F_1$. $\therefore F_1 = 245$.

$\Sigma F_x = 0$. $\therefore -70.7 - 245\times .5 + F_{Ax} = 0$. $F_{Ax} = 193$

$\Sigma F_y = 0$. $\therefore -70.7 - 245\times .866 + F_{Ay} = 0$. $F_{Ay} = 283$

$\therefore F_A = \sqrt{F_{Ax}^2+F_{Ay}^2} = \sqrt{193^2+283^2} = 343$. (c)

8.16 $\Sigma M_A = 0$. $\therefore -2F_B + 1.2\times 200 - 141.4\times 2 - 141.4\times 1.2 + 50 = 0$.

$\Sigma F_x = 0$. $\therefore F_{Ax} - 200 + 141.4 = 0$. $\therefore F_B = -81.2$

$\Sigma F_y = 0$. $\therefore F_{Ay} + 81.2 - 141.4 = 0$. $\therefore F_{Ax} = 58.6$

$\therefore F_A = \sqrt{58.6^2 + 60.2^2} = 84.0$ $F_{Ay} = 60.2$ (e)

8.17 $\Sigma M_A = 0$. $\therefore 500\,l + 200\times .866\,l - F_C\times 2l = 0$.

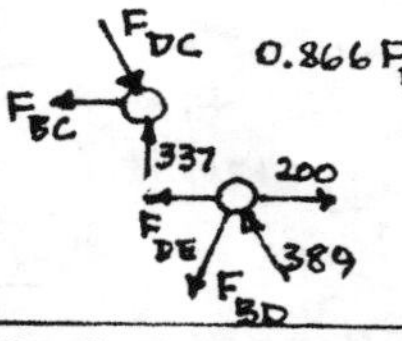

$0.866F_{DC} = 337$ $\therefore F_{DC} = 389$ $\therefore F_C = 337$

$.866\times 389 = .866F_{BD}$ $\therefore F_{BD} = 389$

$-F_{DE} + 200 - 389\times .5 - 389\times .5 = 0$. $\therefore F_{DE} = -189$. (e)

8.18 $\Sigma M_A = 0$. $\therefore 5\times 5000 = 10\times F_C$. $\therefore F_C = 2500$.

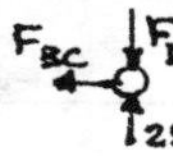

$F_{DC} = 2500$

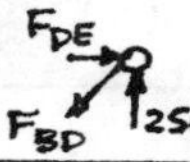

$.707F_{BD} = 2500$. $\therefore F_{BD} = 3536$

$.707\times 3536 = F_{DE}$ $\therefore F_{DE} = 2500$ (d)

8.19 $\Sigma M_A = 0$. $\therefore 4\times 2000 + 6\times 1000 = 8F_C$. $\therefore F_C = 1750$.

$.707F_{DC} = 1750$ $\therefore F_{DC} = 2475$

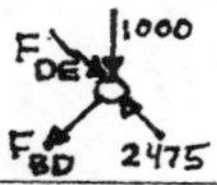

Sum forces in dir. of F_{DE}:

$F_{DE} - 2475 + 1000\times .707 = 0$.

$\therefore F_{DE} = 1768$. (b)

8.20 Sum forces in the dir. of F_{FB} at F. $\therefore F_{FB} = 0$. (a)

8.21 $\Sigma M_B = 0$. $\therefore 12F_F = 3\times 4000$. $\therefore F_F = 1000\downarrow$.

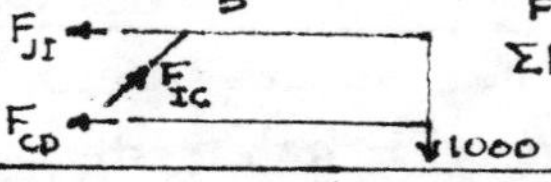

$\Sigma F_y = 0$. $\therefore 0.8F_{IC} = 1000$. $\therefore F_{IC} = 1250$ (c)

8.22 Cut vertically through link KA.

Then $F_{KA} = 5000$

Obviously, $F_{AL} = 0$. $\therefore F_{AB} = 3000$.

$\therefore F_{BC} = 3000$. (d)

8.23 At E we see that $F_{EC} = 0$. $\therefore$ At C $F_{FC} = 0$. (e)

8.24

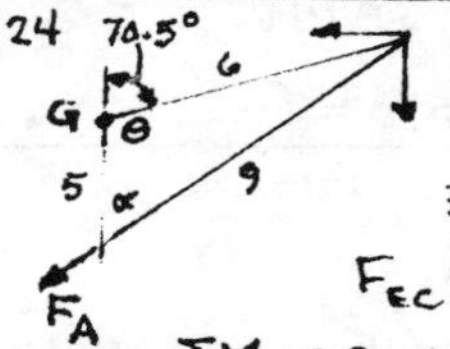

$9^2 = 6^2 + 5^2 - 2\times 5\times 6\cos\theta$ $\therefore \theta = 109.5^\circ$

$6^2 = 9^2 + 5^2 - 2\times 9\times 5\cos\alpha$ $\therefore \alpha = 38.9^\circ$

From pts E, C, F, B we see that

$F_{EC} = F_{FC} = F_{FB} = F_{GB} = 0$. Also, $F_A = F_{BC}$.

$\Sigma M_G = 0$. $\therefore 5\times F_A\sin 38.9^\circ + 5000\times 6\sin 70.5^\circ$

$= 5000\times 6\cos 70.5^\circ$.

$\therefore F_A = -5817$. (e)

8.25 Recognize that link BC is a two-force member.

$\Sigma M_A = 0$.

$\therefore .2\times 1000 + 100 = .08\times F_{BC}\times .8 + .2\times F_{BC}\times .6$

$\therefore F_{BC} = 1630$.

$A_x = 1630\times 0.8 = 1304$. $A_y = 1630\times .6 - 1000 = -22$.

$\therefore F_A = \sqrt{1304^2 + 22^2} = 1304$ N. (b)

8.26

$1800\times .8 = .6F_{BC}$ $F_{BC} = 2400$

$.3\times .6w = .6\times 2400$. $\therefore w = 8000$ (d)

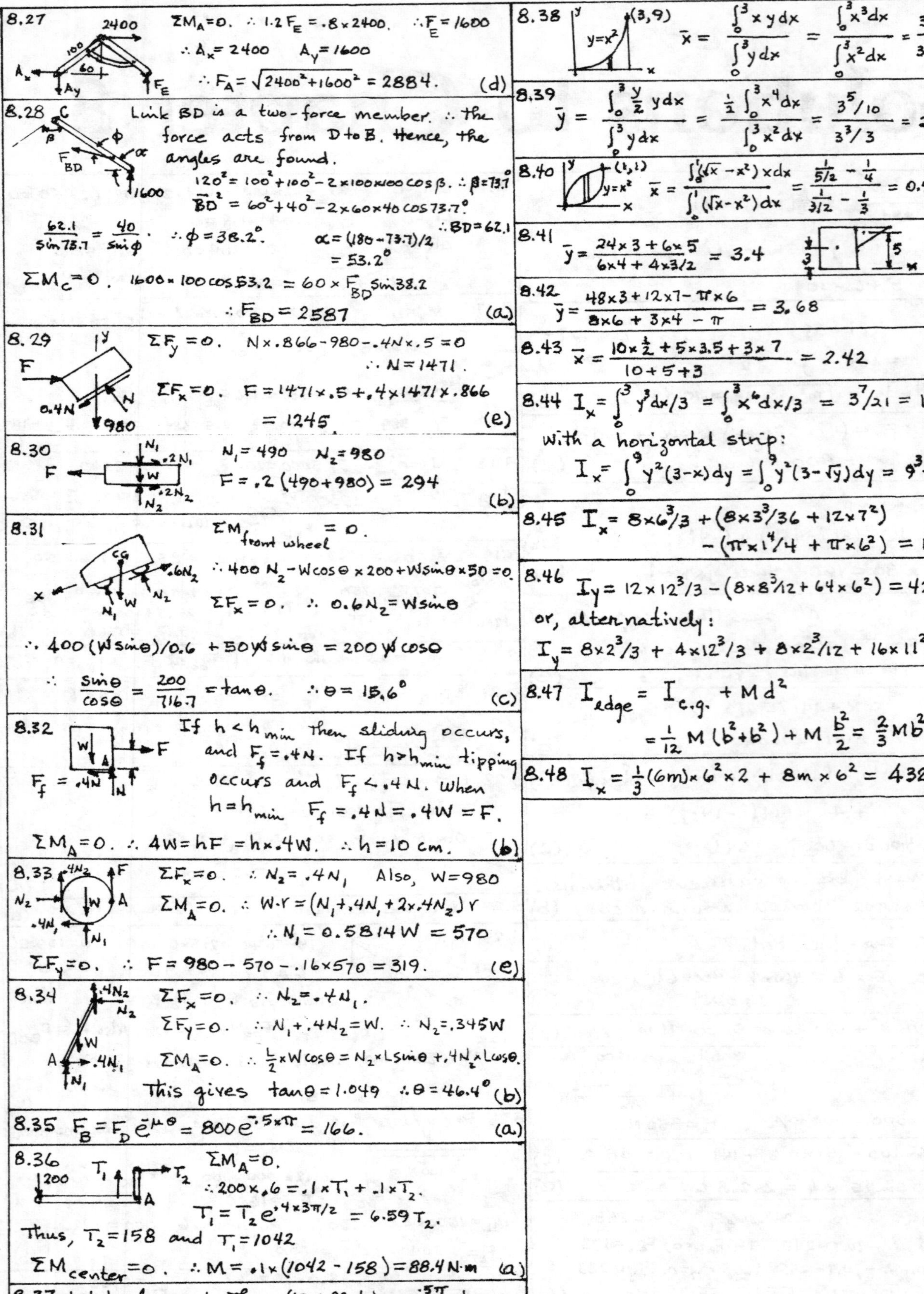

8.27 $\Sigma M_A = 0$. $\therefore 1.2F_E = .8\times 2400$. $\therefore F_E = 1600$

$\therefore A_x = 2400 \quad A_y = 1600$

$\therefore F_A = \sqrt{2400^2+1600^2} = 2884.$ (d)

8.28 Link BD is a two-force member. ∴ the force acts from D to B. Hence, the angles are found.

$120^2 = 100^2+100^2-2\times100\times100\cos\beta$. $\therefore \beta = 73.7°$

$\overline{BD}^2 = 60^2+40^2-2\times60\times40\cos 73.7°$. $\therefore BD = 62.1$

$\frac{62.1}{\sin 73.7} = \frac{40}{\sin\phi}$. $\therefore \phi = 38.2°$. $\quad \alpha = (180-73.7)/2 = 53.2°$

$\Sigma M_C = 0$. $1600\times 100\cos 53.2 = 60\times F_{BD}\sin 38.2$

$\therefore F_{BD} = 2587$ (a)

8.29 $\Sigma F_y = 0$. $N\times.866 - 980 - .4N\times.5 = 0$ $\therefore N = 1471$.

$\Sigma F_x = 0$. $F = 1471\times.5 + .4\times1471\times.866 = 1245$ (e)

8.30 $N_1 = 490 \quad N_2 = 980$

$F = .2(490+980) = 294$ (b)

8.31 $\Sigma M_{\text{front wheel}} = 0$.

$\therefore 400N_2 - W\cos\theta\times200 + W\sin\theta\times50 = 0$.

$\Sigma F_x = 0$. $\therefore 0.6N_2 = W\sin\theta$

$\therefore 400(W\sin\theta)/0.6 + 50W\sin\theta = 200W\cos\theta$

$\therefore \frac{\sin\theta}{\cos\theta} = \frac{200}{716.7} = \tan\theta$. $\therefore \theta = 15.6°$ (c)

8.32 If $h < h_{min}$ then sliding occurs, and $F_f = .4N$. If $h > h_{min}$ tipping occurs and $F_f < .4N$. When $h = h_{min}$ $F_f = .4N = .4W = F$.

$F_f = .4N$

$\Sigma M_A = 0$. $\therefore 4W = hF = h\times.4W$. $\therefore h = 10$ cm. (b)

8.33 $\Sigma F_x = 0$. $\therefore N_2 = .4N_1$ Also, $W = 980$

$\Sigma M_A = 0$. $\therefore W\cdot r = (N_1 + .4N_1 + 2\times.4N_2)r$

$\therefore N_1 = 0.5814W = 570$

$\Sigma F_y = 0$. $\therefore F = 980 - 570 - .16\times570 = 319$. (e)

8.34 $\Sigma F_x = 0$. $\therefore N_2 = .4N_1$.

$\Sigma F_y = 0$. $\therefore N_1 + .4N_2 = W$. $\therefore N_2 = .345W$

$\Sigma M_A = 0$. $\therefore \frac{L}{2}\times W\cos\theta = N_2\times L\sin\theta + .4N_2\times L\cos\theta$.

This gives $\tan\theta = 1.049$ $\therefore \theta = 46.4°$ (b)

8.35 $F_B = F_D e^{\mu\theta} = 800e^{.5\times\pi} = 166.$ (a)

8.36 $\Sigma M_A = 0$.

$\therefore 200\times.6 = .1\times T_1 + .1\times T_2$.

$T_1 = T_2 e^{.4\times3\pi/2} = 6.59T_2$.

Thus, $T_2 = 158$ and $T_1 = 1042$

$\Sigma M_{center} = 0$. $\therefore M = .1\times(1042-158) = 88.4$ N·m (a)

8.37 Let h = long end. Then, $(12-1.88-h)mg\,e^{.5\pi} = hmg$.

m = mass/unit length $\therefore h = 8.38$ m (d)

8.38 $\bar{x} = \frac{\int_0^3 xy\,dx}{\int_0^3 y\,dx} = \frac{\int_0^3 x^3 dx}{\int_0^3 x^2 dx} = \frac{3^4/4}{3^3/3} = 2.25$ (d)

8.39 $\bar{y} = \frac{\int_0^3 \frac{y}{2} y\,dx}{\int_0^3 y\,dx} = \frac{\frac{1}{2}\int_0^3 x^4 dx}{\int_0^3 x^2 dx} = \frac{3^5/10}{3^3/3} = 2.7$ (a)

8.40 $\bar{x} = \frac{\int_0^1(\sqrt{x}-x^2)x\,dx}{\int_0^1(\sqrt{x}-x^2)dx} = \frac{\frac{1}{5/2}-\frac{1}{4}}{\frac{1}{3/2}-\frac{1}{3}} = 0.45$ (c)

8.41 $\bar{y} = \frac{24\times3 + 6\times5}{6\times4 + 4\times3/2} = 3.4$ (b)

8.42 $\bar{y} = \frac{48\times3 + 12\times7 - \pi\times6}{8\times6 + 3\times4 - \pi} = 3.68$ (e)

8.43 $\bar{x} = \frac{10\times\frac{1}{2} + 5\times3.5 + 3\times7}{10+5+3} = 2.42$ (b)

8.44 $I_x = \int_0^3 y^3 dx/3 = \int_0^3 x^6 dx/3 = 3^7/21 = 104.1$ (b)

With a horizontal strip:

$I_x = \int_0^9 y^2(3-x)dy = \int_0^9 y^2(3-\sqrt{y})dy = 9^3 - \frac{9^{7/2}}{7/2} = 104.1$

8.45 $I_x = 8\times6^3/3 + (8\times3^3/36 + 12\times7^2) - (\pi\times1^4/4 + \pi\times6^2) = 1056$ (e)

8.46 $I_y = 12\times12^3/3 - (8\times8^3/12 + 64\times6^2) = 4267$

or, alternatively:

$I_y = 8\times2^3/3 + 4\times12^3/3 + 8\times2^3/12 + 16\times11^2 = 4267$ (a)

8.47 $I_{edge} = I_{c.g.} + Md^2$

$= \frac{1}{12}M(b^2+b^2) + M\frac{b^2}{2} = \frac{2}{3}Mb^2.$ (a)

8.48 $I_x = \frac{1}{3}(6m)\times6^2\times2 + 8m\times6^2 = 432m$ (e)

Solutions to Chapter 9

9.1 $v^2 = v_0^2 + 2as$
$0 = 20^2 - 2\times 5s \quad \therefore s = 40\text{ m}$ (d)

9.2 $a = 5t \quad t \le 2$
$v = \int a\,dt = \int_0^2 5t\,dt + \int_2^4 10\,dt$
$= 5\frac{2^2}{2} + 10(4-2) = 30\text{ m/s}$ (c)

9.3 $v = v_0 + at$
$0 = 40 - 9.8t \quad \therefore t = 4.08\text{ s.}$
$\therefore t_{total} = 2t = 8.16\text{ s}$ (c)

9.4 $\Delta s = v\Delta t = 25\times 0.3 = 7.5\text{ m}$
$v^2 = v_0^2 + 2as$
$0 = 25^2 - 2\times 6s \quad \therefore s = 52.1\text{ m}$
$\therefore s_{total} = 7.5 + 52.1 = 59.6\text{ m}$ (e)

9.5 $\theta = \alpha t^2/2 = 6\times 4^2/2 = 48\text{ rad.}$
$48/2\pi = 7.64\text{ rev.}$ (a)

9.6 $\omega = \omega_0 + \alpha t = 20 + 10t$
$\omega = v/L = 100/2 = 50$
$\therefore 20 + 10t = 50. \quad \therefore t = 3\text{ s.}$ (e)

9.7 $a = \frac{v^2}{r} = \frac{20^2}{40} = 10\text{ m/s}^2$ (c)

9.8 $\frac{v^2}{r} = g \quad \frac{v^2}{1.20} = 9.8 \quad \therefore v = 3.43\text{ m/s}$
$v = r\omega \quad \therefore \omega = 3.43/1.2 = 2.86\text{ rad/s}$ (e)

9.9 $H = v_0^2\sin^2\theta/2g = 100^2\times .707^2/2\times 9.8$
$= 255\text{ m}$ (c)

9.10 $y = v_0 t\sin\theta - gt^2/2$
$-10 = 100t\times 0.707 - 9.8t^2/2 \quad \therefore t = 14.6$ (a)

9.11 $x = v_0 t\cos\theta$
$= 100\times 14.6\times 0.707 = 1032\text{ m}$ (d)

9.12 The acceleration of the point of contact is v^2/r or $r\omega^2$. The acceleration of the center is 0. (c)

9.13 Motion is about the point of contact. Thus, $\omega = 10/0.2 = 50\text{ rad/s.}$
$v = r\omega = 0.4\times 50 = 20\text{ m/s.}$ (c)

9.14 $a = v^2/r = 20^2/0.4 = 1000\text{ m/s}^2$ (a)

9.15 $\vec{a}_A = \vec{a}_B + \vec{a}_{A/B}$
$= 60\hat{i} - 20\hat{j} - .4\times 20^2\hat{i} + .4\times 100\hat{j}$
$= -100\hat{i} + 20\hat{j}$ (c)

9.16 $v_B = 40\cos 45° = 28.3$
$= r\omega_{AB} = 1.0\,\omega_{AB}$
$\therefore \omega_{AB} = 28.3\text{ CCW}$ (d)

9.17 $v_{C/B} = 40\sin 45° = 28.3 = r_{BC}\omega_{BC}$
$= 1.0\,\omega_{BC}. \quad \therefore \omega_{BC} = 28.3\text{ CW}$ (e)

9.18 $\vec{A}_C = \vec{A}_B + \vec{A}_{C/B}$
$r_{AB}\omega_{AB}^2 = 1.0\times 28.3^2 = 800$
$r_{BC}\omega_{BC}^2 = 1.0\times 28.3^2 = 800$
Then, $r_{AB}\alpha_{AB} = 800. \quad \therefore \alpha_{AB} = 800\text{ CW}$ (a)

9.19 $\vec{a}_B = \vec{a}_A^{\,0} + \vec{a}_{B/A}$
$r\omega^2 = 1.0\times 28.3^2 = 800$
$r\alpha = 1.0\times 800 = 800$
$\therefore \vec{a}_B = -1130\hat{j}.$ (d)

9.20 $\vec{v}_A = \vec{v}_B + \vec{v}_{A/B}. \quad v_{A/B} = r\omega_{AB} \perp \overline{AB}.$
But, $\vec{v}_A$ and $\vec{v}_B$ are both horizontal. Thus, $v_{A/B} = 0$ and $\omega_{AB} = 0.$ (e)

9.21 $v_A = v_B = 20\times 0.1 = 0.04\,\omega_A.$
$\therefore \omega_A = 50. \quad r_B\omega_B^2 = .1\times 20^2 = 40$
$r_A\omega_A^2 = .04\times 50^2 = 100.$
$\therefore r_{AB}\alpha_{AB} = \frac{60}{0.8} = 75.$
But, $r_{AB} = 0.1. \quad \therefore \alpha_{AB} = 750\text{ CCW}$ (b)

9.22 $\vec{a} = -r\omega^2\hat{i} + 2\omega v\hat{j}$
$= -0.1\times 20^2\hat{i} + 2\times 20\times 1.0\hat{j}$
$= -40\hat{i} + 40\hat{j}$ (c)

9.23 $a_A = 2a_B$ (from small pulley above B)
$400\times 9.8 - 2T = 400\,\frac{a_A}{2}$ (body B)
$T - 0.2\times 500\times 9.8 = 500\,a_A$ (body A)
$\therefore a_A = 1.63\text{ m/s}^2$ (c)

9.24 $W\cos 60° - Wf\sin 60° = \frac{W}{g}a.$
$\therefore a = 9.8(0.866 - 0.3\times 0.5) = 7.02\text{ m/s}^2$
$s = at^2/2 = 7.02\times 10^2/2 = 351\text{ m}$ (a)

9.25 $W\sin\theta = \frac{W}{g}\frac{v^2}{r}\cos\theta$
$\tan\theta = \frac{25^2}{200\times 9.8}. \quad \therefore \theta = 17.7°$ (c)

9.26 $W = \frac{W}{g}\frac{v^2}{r}$.

$v = \sqrt{9.2 \times 6600000} = 7790$ m/s (a)

9.27 $F = k\frac{m_1 m_2}{r^2}$. $W = k\frac{m_e W/g}{r^2}$.

$\therefore m_e = r^2 g/k = 6400000^2 \times 9.8/6.67\times10^{-11}$

$= 6. \times 10^{24}$ kg (c)

9.28 $v^2 = v_0^2 + 2as$ $\quad -0.6W = \frac{W}{g}a$.

$0 = 25^2 - 2\times.6\times9.8s$ $\quad \therefore a = -0.6g$

$\therefore s = 53.2$ m (d)

9.29 $\Sigma M = 0$

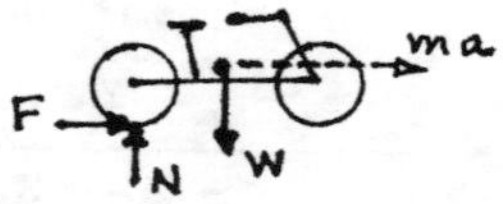

for maximum accel. the force on the front wheel $= 0$. Thus,

$30W = 80\,ma$. $\therefore a = \frac{30\times9.8}{80} = 3.68$ (e)

9.30 Before: $\Sigma \vec{F}_y = 0$.

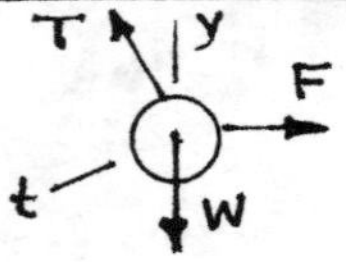

$T\cos30° = W$ $\quad \therefore T = \frac{W}{0.866}$

After: $F = 0$, $\Sigma F_n = ma_n^{\,0}$

$n \perp t$. $\therefore T = W\cos$ $= 0.866W$.

ratio $= \frac{\text{before}}{\text{after}} = \frac{1}{0.866\times0.866} = 1.33$ (b)

9.31 Take moments about the back wheels.

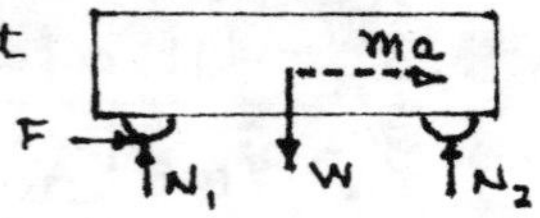

$4W - 8N_2 = 1.2\frac{W}{g}\times2$.

$\therefore N_2 = 0.469W = .469\times8000\times9.8$

$= 36\,800$ N (d)

9.32 Acceleration of block $= 0.2\alpha$. Thus, $T = 50\times0.2\alpha + 50\times9.8\times0.2$

$\Sigma M = I\alpha$. Use $I = mk^2$.

$(40\times9.8 - 40\alpha)\times0.1 - (10\alpha + 98)\times0.2 = 30\times.1^2\alpha$.

$\therefore \alpha = 7.26$ rad/s^2 (b)

9.33 $\Sigma M_o = I_o\alpha$

$mg\times\frac{l}{2} = \frac{1}{3}ml^2\alpha$. $\quad \therefore \alpha = \frac{3g}{2l}$.

$\Sigma F_y = m\bar{a}$ ($\bar{a}$ is acc. of mass center)

$mg - F_o = m\frac{l}{2}\alpha = m\frac{3g}{4}$. $\quad \therefore F_o = \frac{mg}{4}$. (c)

9.34 $W_{net} = \Delta E = \frac{1}{2}mv^2$.

$60\times10 - \frac{1}{2}\times100\times2^2 = \frac{1}{2}50v^2$

$\therefore v = 4$ m/s (b)

9.35 $W_{net} = \Delta KE = \frac{1}{2}mv^2 + \frac{1}{2}I\omega^2$. $\quad v = r\omega$

$100\times4 = \frac{1}{2}\times100(0.2\omega)^2 + \frac{1}{2}(\frac{1}{2}\times100\times.2^2)\omega^2$

$\therefore \omega = 11.55$ rad/s (d)

9.36 $W_{net} = \Delta KE = \frac{1}{2}mv^2$. $\quad \Delta x = 50 - 30 = 20$

$2\times9.8\times0.4 - \frac{1}{2}\times100\times.2^2 = \frac{1}{2}\times2v^2$

$\therefore v = 2.42$ m/s (d)

9.37 $W_{net} = \Delta KE = \frac{1}{2}mv^2 + \frac{1}{2}I\omega^2$

$\frac{1}{2}100(.4^2 - .1^2) - 10\times9.8(.25 - .20)$

$= \frac{1}{2}10(\frac{v}{2})^2 + \frac{1}{2}(10\times.5^2/12)(\frac{v}{.5})^2$

where $\bar{v} = v/2$ and $\omega = v/.5$. The above gives $v = 1.25$ m/s. (e)

9.38 $\Sigma M_c \Delta t = I_c\omega$ where C is the point of contact. $I_c = I_o + mr^2 = \frac{3}{2}mr^2$.

$100\times0.4\times4 = \frac{3}{2}\times100\times.2^2\omega$.

$\therefore \omega = 26.7$ rad/s. (b)

9.39 $\int F dt = m\Delta v$. $\quad F = 400t$.

$F_f = fN = 0.2\times100\times0.866\times9.8 = 170$

$400t = 100\times0.5\times9.8 + 170$. $\quad \therefore t = 1.65$ s when motion initiates.

$\int_{1.65}^{2.0} 400t\,dt - (170 + 50\times9.8)\times0.35 = 100v$.

$\therefore v = 0.245$ m/s (a)

9.40 $10m - 20m = v_A'm + v_B'm$

$\therefore v_A' + v_B' = -10$

$0.8 = \frac{v_B' - v_A'}{20 + 10}$

$\therefore v_B' - v_A' = 24$.

Simultaneous solution yields

$v_B' = 7$ m/s $\quad v_A' = -17$ m/s. (e)

9.41 $\Delta KE = \frac{1}{2}mv_A^2 + \frac{1}{2}mv_B^2$

$-\frac{1}{2}mv_A'^2 - \frac{1}{2}mv_B'^2$

$= \frac{2}{2}(10^2 + 20^2) - \frac{2}{2}(7^2 + 17^2)$

$= 162$ J. (c)

Solutions to Chapter 10

(English Units)

10.1 Isotropic (b)

10.2 Poisson's ratio (e)

10.3 $\sigma = E\Delta L/L = F/\pi r^2$

$\therefore F = E\Delta L\,\pi r^2/L = 30\times10^6\times .04\times\pi(\frac{1}{4})^2/48$

$= 4909$ lb. (d)

10.4 $\frac{F}{\pi r^2} = E\frac{\Delta L}{L}$. $\therefore \Delta L = \frac{FL}{\pi r^2 E} = \frac{3500\times100\times12}{\pi(1/4)^2\times30\times10^6}$

$= 0.71$ in. (d)

10.5 $\frac{F}{A} = E\frac{\Delta L}{L}$. $\therefore F = AE\Delta L/L$

$= \frac{1}{8}\times1\times30\times10^6\times\frac{1}{32}/36 = 3255$ (a)

10.6 $\frac{F}{A} = G\frac{\Delta L}{L}$. $\therefore G = \frac{FL}{A\Delta L} = \frac{5000\times6}{8\times\frac{1}{2}\times.0012} = 6.25\times10^6$ (b)

10.7 $\frac{F}{A} = E\epsilon$. $\therefore \epsilon = \frac{150{,}000}{\pi\times1^2\times30\times10^6} = 0.00159$

$\therefore \Delta d = \nu\epsilon d = 0.3\times.00159\times2 = .00095$.

$\therefore d_{after} = 2 - .00095 = 1.9990$ in. (a)

10.8 $\frac{F}{A} = E\epsilon$. $\therefore \epsilon = \frac{400{,}000}{\pi(5.923/2)^2\times10\times10^6} = .00145$

$\Delta d = \nu\epsilon d = 0.33\times.00145\times5.923 = .00283$

$\therefore d = d - \Delta d = 5.923 - .00283 = 5.920$ in. (c)

10.9 $\tau_{max} = \frac{1}{2}\sqrt{(\sigma_x-\sigma_y)^2+4\tau^2}$

$= \frac{1}{2}\sqrt{6000^2+4\times4000^2} = 5000$ psi. (d)

10.10 $\sigma_{max} = \frac{1}{2}(\sigma_x+\sigma_y)+\tau_{max} = 0+4000 = 4000$ psi. (a)

10.11 $\tau_{max} = \frac{1}{2}\sqrt{(3000+5000)^2+4\times3000^2} = 5000$ psi (c)

10.12 $\sigma = E\Delta L/L$

$16000 = 30\times10^6\,\Delta L/L$.

$\therefore \Delta L = 5.33\times10^{-4}L$.

$\Delta L_T = \alpha L\Delta T = 6.5\times10^{-6}\times50L = 3.25\times10^{-4}L$. (e)

$\therefore \Delta L_{final} = 2.08\times10^{-4}L$. $\therefore \sigma = 30\times10^6\times2.08\times10^{-4} = 6240$

10.13 $\Delta L = \alpha L\Delta T = 6.5\times10^{-6}\times1000\times12\times130 = 10.1''$ (a)

10.14 $\frac{\Delta L}{L} = \frac{\sigma}{E} = \alpha\Delta T$. $\therefore \frac{10\,000}{10\times10^6} = 12.8\times10^{-6}(80-T)$.

$\therefore T = 1.88°F$. (d)

10.15 It expands at a different rate. (d)

10.16 $\sigma = Mc/I$. $\therefore \sigma = \sigma_{max}$ at $c = c_{max}$. (b)

10.17 A triangle. V M (a)

10.18 $\Sigma M_{right\ end} = 0$. $\therefore 20F_{left} = 1000\times10+1000\times5$. $\therefore F = 750$

$M_A = 750\times10 = 7500$ ft-lb (a)

10.19 $M_A = 1000\times10+1000\times5 = 15{,}000$ ft-lb. (b)

10.20 $\delta_{max} = PL^3/48EI = 500\times240^3/48\times30\times10^6\times2^4/12 = 3.6''$ (d)

10.21 $\delta_{max} = \frac{PL^3}{48EI}+\frac{5wL^4}{384EI}$ $I = \frac{bh^3}{12} = \frac{4\times2^3}{12} = 2.67$ in^4

$= \frac{500\times(16\times12)^3}{48\times30\times10^6\times2.67}+\frac{5\times100/12\times(16\times12)^4}{384\times30\times10^6\times2.67} = 2.76''$ (e)

10.22 $\delta = \theta L_2 = \frac{PL^2}{16EI}\times L_2$

$\therefore 4 = \frac{P\times240^2\times120}{16\times30\times10^6\times\pi\times2^4/64}$ $\therefore P = 218$ lb. (a)

10.23 $10F_{right} = 5000\times5+1000\times14$. $\therefore F_{right} = 3900$ lb

$10F_{left} = 5000\times5-1000\times4$. $\therefore F_{left} = 2100$ lb

V: 2100, 4.2', 4, 1000

$M_{max} = 2100\times4.2/2 = 4410$ ft-lb
$=$ area under V-diagram

$\sigma = \frac{Mc}{I} = \frac{4410\times12\times5}{60.67} = 4361$ psi (a)

10.24 Compression occurs in bottom fibers over rt support. There $M = 4000$ ft-lb.

$\therefore \sigma = \frac{Mc}{I} = \frac{4000\times12\times5}{60.67} = 3956$ psi. (b)

10.25 $\tau_{max} = \frac{VQ}{Ib} = \frac{2900\times(5\times2.5)}{60.67\times1} = 597$ psi (e)

10.26 The max occurs on the N.A. with a sudden decrease when b goes from 1" to 8". Also, it's a parabolic distribution. (c)

10.27 $\sigma = \frac{Mc}{I}$. $\therefore \frac{I}{c} = \frac{M}{\sigma} = \frac{15{,}000\times12}{20000} = 9$ in^3. (c)

10.28 V: 250, −250, 1050

Using the area under the curve:
$M_{max} = 250\times8+800\times8/2 = 5200$

$\sigma_{max} = \frac{Mc}{I} = \frac{5200\times12\times1}{4\times2^3/12} = 23{,}400$ psi (b)

10.29 $\sigma_{max} = \frac{5200\times12\times2}{2\times4^3/12} = 11{,}700$ psi (e)

10.30 $V_{max} = 250$. $\tau_{max} = \frac{VQ}{Ib} = \frac{250\times2\times1/2}{2\times2^3/12\times2} = 93.8$ (e)

10.31 $\tau = \frac{Tc}{J} = \frac{2000\times1}{\pi\times2^4/32} = 1273$ psi (a)

10.32 $J = \pi(d_1^4-d_2^4)/32 = \pi(2^4-1.75^4)/32 = 0.650$ in^4

$\tau = \frac{2000\times1}{0.650} = 3077$ psi. (e)

10.33 $\tau = \frac{Tc}{J}$. $\therefore T = \frac{\tau J}{c} = \frac{20{,}000\times\pi\times4^4/32}{2} = 251{,}300$ in-lb

or $20{,}940$ ft-lb. (a)

10.34 $\theta = \frac{TL}{JG} = \frac{160\times24\times18}{(\pi\times(7/8)^4/32)\times12\times10^6} = 0.1001$ rad

or $5.73°$. (d)

10.35 $60 = \frac{L}{k} = \frac{L}{\sqrt{I/A}} = \frac{L}{\sqrt{4\times4^3/12/16}}$. $\therefore L = 69.3''$ or $5.77'$ (d)

10.36 Assume $P = 1000$ lb Using a factor of safety of 2.

$1000 = \frac{\pi^2\times10\times10^6\times\pi\times4^4/64}{4L^2}$. $\therefore L = 557''$ or $46.4'$. (c)

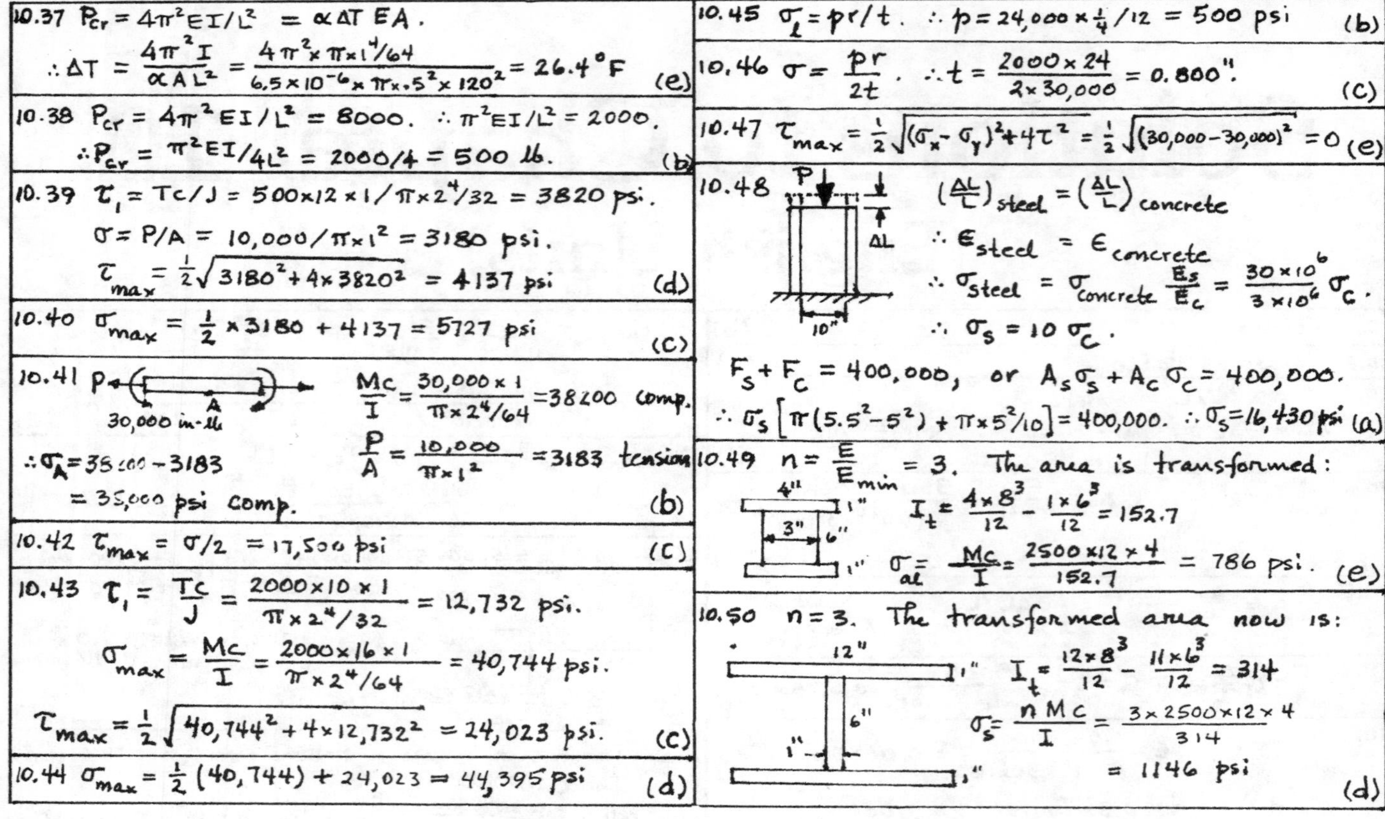

10.37 $P_{cr} = 4\pi^2 EI/L^2 = \alpha \Delta T\, EA$.
$\therefore \Delta T = \frac{4\pi^2 I}{\alpha A L^2} = \frac{4\pi^2 \times \pi \times 1^4/64}{6.5\times10^{-6}\times \pi \times .5^2 \times 120^2} = 26.4°F$ (e)

10.38 $P_{cr} = 4\pi^2 EI/L^2 = 8000$. $\therefore \pi^2 EI/L^2 = 2000$.
$\therefore P_{cr} = \pi^2 EI/4L^2 = 2000/4 = 500$ lb. (b)

10.39 $\tau_1 = Tc/J = 500\times12\times1/\pi\times2^4/32 = 3820$ psi.
$\sigma = P/A = 10,000/\pi\times1^2 = 3180$ psi.
$\tau_{max} = \frac{1}{2}\sqrt{3180^2+4\times3820^2} = 4137$ psi (d)

10.40 $\sigma_{max} = \frac{1}{2}\times3180 + 4137 = 5727$ psi (c)

10.41
$\frac{Mc}{I} = \frac{30,000\times1}{\pi\times2^4/64} = 38200$ comp.
$\frac{P}{A} = \frac{10,000}{\pi\times1^2} = 3183$ tension
$\therefore \sigma_A = 38200 - 3183$
$= 35,000$ psi comp. (b)

10.42 $\tau_{max} = \sigma/2 = 17,500$ psi (c)

10.43 $\tau_1 = \frac{Tc}{J} = \frac{2000\times10\times1}{\pi\times2^4/32} = 12,732$ psi.
$\sigma_{max} = \frac{Mc}{I} = \frac{2000\times16\times1}{\pi\times2^4/64} = 40,744$ psi.
$\tau_{max} = \frac{1}{2}\sqrt{40,744^2+4\times12,732^2} = 24,023$ psi. (c)

10.44 $\sigma_{max} = \frac{1}{2}(40,744) + 24,023 = 44,395$ psi (d)

10.45 $\sigma_t = pr/t$. $\therefore p = 24,000\times\frac{1}{4}/12 = 500$ psi (b)

10.46 $\sigma = \frac{pr}{2t}$. $\therefore t = \frac{2000\times24}{2\times30,000} = 0.800''$. (c)

10.47 $\tau_{max} = \frac{1}{2}\sqrt{(\sigma_x-\sigma_y)^2+4\tau^2} = \frac{1}{2}\sqrt{(30,000-30,000)^2} = 0$ (e)

10.48
$\left(\frac{\Delta L}{L}\right)_{steel} = \left(\frac{\Delta L}{L}\right)_{concrete}$
$\therefore \epsilon_{steel} = \epsilon_{concrete}$
$\therefore \sigma_{steel} = \sigma_{concrete}\frac{E_s}{E_c} = \frac{30\times10^6}{3\times10^6}\sigma_c$.
$\therefore \sigma_s = 10\sigma_c$.
$F_s + F_c = 400,000$, or $A_s\sigma_s + A_c\sigma_c = 400,000$.
$\therefore \sigma_s\left[\pi(5.5^2-5^2) + \pi\times5^2/10\right] = 400,000$. $\therefore \sigma_s = 16,430$ psi (a)

10.49 $n = \frac{E}{E_{min}} = 3$. The area is transformed:

$I_t = \frac{4\times8^3}{12} - \frac{1\times6^3}{12} = 152.7$
$\sigma_{al} = \frac{Mc}{I} = \frac{2500\times12\times4}{152.7} = 786$ psi. (e)

10.50 $n = 3$. The transformed area now is:

$I_t = \frac{12\times8^3}{12} - \frac{11\times6^3}{12} = 314$
$\sigma_s = \frac{nMc}{I} = \frac{3\times2500\times12\times4}{314}$
$= 1146$ psi (d)

Solutions to Chapter 10
(Metric Units)

10.1 Isotropic (b)

10.2 Poisson's ratio (e)

10.3 $\sigma = E\Delta L/L = F/\pi r^2$.
$\therefore F = E\Delta L\pi r^2/L = 210\times10^9\times.001\times\pi\times.01^2/1 = 66\,000$ N (d)

10.4 $\frac{F}{\pi r^2} = E\frac{\Delta L}{L}$. $\therefore \Delta L = \frac{14\,000\times30}{\pi\times.01^2\times210\times10^9} = 0.0064$ m (d)

10.5 $\frac{F}{A} = E\frac{\Delta L}{L}$. $\therefore F = AE\Delta L/L = .025\times.003\times210\times10^9\times.0008/1$
$= 12\,600$ N. (a)

10.6 $\frac{F}{A} = G\frac{\Delta L}{L}$ $\therefore G = \frac{FL}{A\Delta L} = \frac{20\,000\times.15}{.012\times.2\times.00003} = 41.7\times10^9$ Pa (b)

10.7 $F/A = E\epsilon$. $\therefore \epsilon = \frac{600\,000}{\pi\times.025^2\times210\times10^9} = 0.001455$
$\therefore \Delta d = \nu\epsilon d = 0.3\times.001455\times5 = .00218$ cm.
$\therefore d = d - \Delta d = 5 - .00218 = 4.9978$ cm (a)

10.8 $F/A = E\epsilon$. $\therefore \epsilon = \frac{1500\,000}{\pi\times.062^2\times70\times10^9} = 0.00190$.
$\Delta d = \nu\epsilon d_o = 0.33\times.00190\times12.015 = .0075$ cm.
$\therefore d = d_o - \Delta d = 12.015 - .0075 = 12.008$ cm (c)

10.9 $\tau_{max} = \frac{1}{2}\sqrt{(\sigma_x-\sigma_y)^2+4\tau^2}$
$= \frac{1}{2}\sqrt{60^2+4\times40^2} = 50$ MPa (d)

10.10 $\sigma_{max} = \frac{1}{2}(\sigma_x+\sigma_y) + \tau_{max} = 0 + 40 = 40$ MPa (a)

10.11 $\tau_{max} = \frac{1}{2}\sqrt{(30+50)^2+4\times30^2} = 50$ MPa (c)

10.12 L, ΔL, ΔL_T $\sigma = E\Delta L/L$
$100\times10^6 = 210\times10^9\times\Delta L/L$. $\therefore \Delta L = 4.76\times10^{-4}L$
$\Delta L_T = \alpha L\Delta T = 11.7\times10^{-6}\times30L = 3.51\times10^{-4}L$.
$\therefore \Delta L_{final} = 1.25\times10^{-4}L$. $\therefore \sigma = 210\times10^9\times1.25\times10^{-4} = 26.2\times10^6$ Pa (e)

10.13 $\Delta L = \alpha L\Delta T = 11.7\times10^{-6}\times300\times75 = 0.263$ m (a)

10.14 $\Delta L/L = \sigma/E = \alpha\Delta T$. $\frac{70\times10^6}{70\times10^9} = 23\times10^{-6}(30-T)$.
$\therefore T = -13.5°C$ (d)

10.15 It expands at a different rate (d)

10.16 $\sigma = Mc/I$. $\therefore \sigma_{max}$ occurs at $c = c_{max}$ (b)

10.17 A triangle. V M (a)

10.18 $\Sigma M_{\text{right end}} = 0. \quad \therefore 8F_{left} = 4000\times4 + 4000\times2. \quad \therefore F = 3000.$
$M_A = 3000\times4 = 12000$ N·m (a)

10.19 $M_A = 4000\times4 + 4000\times2 = 24\,000$ N·m. (b)

10.20 $\delta_{max} = PL^3/48EI = \dfrac{2000\times6^3}{48\times210\times10^9 .05^4/12} = 0.0823$ m (d)

10.21 $\delta_{max} = \dfrac{PL^3}{48EI} + \dfrac{5wL^4}{384EI}$ $\quad I = \dfrac{bh^3}{12} = \dfrac{.1\times.05^3}{12} = 1.04\times10^{-5}$ m^4
$= \dfrac{2000\times6^3}{48\times210\times10^9\times1.04\times10^{-5}} + \dfrac{1000\times5\times6^4}{384\times210\times10^9\times1.04\times10^{-5}} = .0118$ m (e)

10.22 $\delta = \theta\times4 = \dfrac{PL^2}{16EI}\times4$
$\theta.1 = \dfrac{P\times8^2\times4}{16\times210\times10^9\,\pi\times.05^4/64}. \quad \therefore P = 403$ N. (a)

10.23 $4F_{right} = 24000\times2 + 4000\times6. \quad \therefore F_{right} = 18\,000$ N.
$4F_{left} = 24000\times2 - 4000\times2. \quad \therefore F_{left} = 10\,000$ N.
V: 10000, 667 m, −14000, 4000. M_{max} = Area under V-diagram
$= 10000\times1.667/2$
$= 8330$ N·m.
$\sigma = \dfrac{Mc}{I} = \dfrac{8330\times.1}{885\times10^{-8}} = 94.1\times10^6$ Pa. (a)

10.24 Compression occurs in bottom fibers over rt. support. There $M = 4000\times2 = 8000$ N·m.
$\therefore \sigma = \dfrac{Mc}{I} = \dfrac{8000\times.1}{885\times10^{-8}} = 90.4\times10^6$ Pa (b)

10.25 $\tau_{max} = \dfrac{VQ}{Ib} = \dfrac{14\,000\,(.002\times.05)}{885\times10^{-8}\times.02} = 7.91\times10^6$ Pa. (e)

10.26 The max occurs on the N.A. with a sudden decrease when b goes from 2 cm to 16 cm. Also, it's a parabolic distribution. (c)

10.27 $\sigma = \dfrac{Mc}{I}. \quad \therefore \dfrac{I}{c} = \dfrac{M}{\sigma} = \dfrac{24\,000}{140\times10^6} = 171\times10^{-6}$ m^3 or 171 cm^3 (c)

10.28 V: 1000, −1000, 4000. Using the area under curve: $M_{max} = 1000\times3 + 3000\times3/2$
$= 7500$ N·m.
$\sigma_{max} = \dfrac{Mc}{I} = \dfrac{7500\times.025}{.1\times.05^3/12} = 180\times10^6$ Pa. (b)

10.29 $\sigma_{max} = \dfrac{Mc}{I} = \dfrac{7500\times.05}{.05\times.1^3/12} = 90\times10^6$ Pa. (e)

10.30 $V_{max} = 1000. \quad \tau_{max} = \dfrac{VQ}{Ib} = \dfrac{1000\,(.025\times.05\times.0125)}{(.05\times.05^3/12)\times.05}$
$= 600\times10^3$ Pa $= 0.6$ MPa (a)

10.31 $\tau = \dfrac{Tc}{J} = \dfrac{200\times.03}{\pi .06^4/32} = 4.72\times10^6$ Pa. (a)

10.32 $J = \pi(d_1^4 - d_2^4)/32 = \pi(.06^4 - .05^4)/32 = 65.9\times10^{-8}$ m^4
$\tau = \dfrac{Tc}{J} = \dfrac{200\times.03}{65.9\times10^{-8}} = 9.10\times10^6$ Pa (e)

10.33 $\tau = Tc/J \quad \therefore T = \tau J/c = \dfrac{140\times10^6\times\pi\times.1^4/32}{0.05}$
$= 27\,500$ N·m. (a)

10.34 $\theta = \dfrac{TL}{JG} = \dfrac{200\times.3\times.5}{83\times10^9\times\pi\times.01^4/32} = 0.368$ rad or $21.1°$ (d)

10.35 $60 = \dfrac{L}{k} = \dfrac{L}{\sqrt{I/A}} = \dfrac{L}{\sqrt{(.1\times.1^3/12)/0.01}}. \quad \therefore L = 1.73$ m (d)

10.36 Assume $P = 4000$ N using a factor of safety of 2.
$4000 = \dfrac{\pi^2\times70\times10^9\times\pi\times0.1^4/64}{4L^2}. \quad \therefore L = 14.55$ m. (c)

10.37 $P_{cr} = 4\pi^2EI/L^2 = \alpha\Delta TEA.$
$\therefore \Delta T = \dfrac{4\pi^2 I}{\alpha A L^2} = \dfrac{4\pi^2\times\pi\times.02^4/64}{11.7\times10^{-6}\times\pi\times.01^2\times4^2} = 5.27°C$ (e)

10.38 $P_{cr} = 4\pi^2EI/L^2 = 30\,000. \quad \therefore \pi^2EI/L^2 = 7500.$
$\therefore P_{cr} = \pi^2EI/4L^2 = 7500/4 = 1875$ N. (b)

10.39 $\tau_1 = Tc/J = 600\times.025/\dfrac{\pi\times.05^4}{32} = 24.45\times10^6$ Pa
$\sigma = P/A = 40\,000/\pi\times.025^2 = 20.37\times10^6$ Pa
$\therefore \tau_{max} = \frac{1}{2}\sqrt{20.37^2 + 4\times24.45^2}\times10^6 = 26.5\times10^6$ Pa (d)

10.40 $\sigma_{max} = (\frac{1}{2}\times20.37 + 26.5)\times10^6 = 26.7\times10^6$ Pa (c)

10.41 P, A, 3200 N·m. $\dfrac{Mc}{I} = \dfrac{3200\times.025}{\pi\times.05^4/64} = 261\times10^6$ Pa comp.
$\dfrac{P}{A} = \dfrac{40\,000}{\pi\times.025^2} = 20.4\times10^6$ Pa tension.
$\therefore \sigma_A = (261 - 20.4)\times10^6 = 241\times10^6$ Pa (b)

10.42 $\tau_{max} = \sigma/2 = 120\times10^6$ Pa $\quad VQ/Ib = 0$ (c)

10.43 $\tau_1 = \dfrac{Tc}{J} = \dfrac{8000\times.25\times.025}{\pi\times.05^4/32} = 81.5\times10^6$ Pa
$\sigma_{max} = \dfrac{Mc}{I} = \dfrac{8000\times.4\times.025}{\pi\times.05^4/64} = 261\times10^6$ Pa
$\tau_{max} = \frac{1}{2}\sqrt{261^2 + 4\times81.5^2}\times10^6 = 154\times10^6$ Pa (c)

10.44 $\sigma_{max} = (\frac{1}{2}\times261 + 154)\times10^6 = 284\times10^6$ Pa. (a)

10.45 $\sigma_\ell = pr/t. \quad \therefore p = 180\times10^6\times.005/.4 = 2250\times10^3$ Pa (b)

10.46 $\sigma = \dfrac{pr}{2t}. \quad \therefore t = \dfrac{8000\times10^3\times.6}{2\times200\times10^6} = 0.012$ m (c)

10.47 $\tau_{max} = \frac{1}{2}\sqrt{(\sigma_x - \sigma_y)^2 + 4\tau^2} = \frac{1}{2}\sqrt{(200-200)^2}\times10^6 = 0$ (e)

10.48 P, ΔL, .012 m, .25 m. $\left(\dfrac{\Delta L}{L}\right)_{steel} = \left(\dfrac{\Delta L}{L}\right)_{concrete}$
$\therefore \epsilon_{steel} = \epsilon_{concrete}$
$\therefore \sigma_{steel} = \dfrac{E_s}{E_c}\sigma_{concrete} = \dfrac{210\times10^9}{20\times10^9}\sigma_c$
$\therefore \sigma_c = \sigma_s/10$
$F_s + F_c = 2000000$ or $A_s\sigma_s + A_c\sigma_c = 2000\,000.$
$\therefore \sigma_s[\pi(.137^2 - .125^2) + \pi\times.125^2/10.5] = 2000\,000.$
$\therefore \sigma_s = 137\times10^6$ Pa. (a)

10.49 $n = \dfrac{E}{E_{min}} = 3.$ The area is transformed:
12 cm, 3 cm, 9 cm, 18 cm, 3 cm. $I_t = \dfrac{.12\times.24^3}{12} - \dfrac{.03\times.18^3}{12} = 1.237\times10^{-4}$ m^4
$\therefore (\sigma_{al})_{max} = \dfrac{Mc}{I} = \dfrac{2000\times2\times.12}{1.237\times10^{-4}}$
$= 3.88\times10^6$ Pa (e)

10.50 $n = \dfrac{E}{E_{min}} = 3.$ The area is transformed:
3 cm, 18 cm, 3 cm, 3 cm, 36 cm. $I_t = \dfrac{.36\times.24^3}{12} - \dfrac{.33\times.18^3}{12}$
$= 2.54\times10^{-4}$ m^4
$\sigma_s = \dfrac{nMc}{I} = \dfrac{3\times4000\times.12}{2.54\times10^{-4}}$
$= 5.67\times10^6$ Pa (c)

Solutions to Chapter 11

(Metric Units)

11.1 (a) is true of a liquid and low speed gas flows. (c) and (e) are true of gases. (d) is true for a liquid. (b) is correct. (b)

11.2 $\tau = \mu \frac{du}{dy}$. $\therefore \mu = \frac{\tau}{du/dy}$.

$[\mu] = \frac{F/L^2}{\frac{L}{T}/L} = \frac{FT}{L^2}$

$\frac{FT}{L^2} = \frac{ML}{T^2} \cdot \frac{T}{L^2} = \frac{M}{LT}$. (d)

11.3 Viscosity μ varies with temperature only. (a)

11.4 $dp = -\gamma dz = -\rho g dz$.

$p = \rho RT$ (air is an ideal gas)

$\therefore dp = -p \frac{g}{RT} dz$ or $\frac{dp}{p} = -\frac{g}{RT} dz$.

$\therefore \ln p = -Cz$ or $p = e^{-Cz}$. (e)

11.5 $T = \tau A r = \mu \frac{du}{dy} A r = \mu \frac{r\omega}{t} 2\pi r L r$

$1.6 = \mu \frac{0.04 \times 1000}{0.001} 2\pi \times 0.04 \times 0.04 \times 0.04$

$\therefore \mu = 0.1 \ N \cdot s/m^2$ (a)

11.6 $K = -V \frac{\Delta p}{\Delta V} = -2 \frac{500}{-0.004} = 250\ 000$

or 250 MPa. (c)

11.7 $\sigma \pi d = \gamma \pi r^2 L$

$0.0736 \times \pi \times 0.001$

$= 9800 \pi \times 0.0005^2 L$

$\therefore L = 0.03$ m or 3 cm. (c)

$\sigma\pi d$

$W = \gamma V$

11.8 Cavitation occurs when the pressure reaches the vapor pressure = 2.45 kPa (e) (see Table 11.2)

11.9 $L = V \Delta t = \sqrt{kRT}\ \Delta t$

$= \sqrt{1.4 \times 287 \times 288} \times 1.2 = 408$ m. (C)

assume standard temp = 15°C.

T must be in absolute. $\therefore T = 288$

11.10 $\nu = \frac{\mu}{\rho} = \frac{0.0034}{1.3 \times 1000} = 2.6 \times 10^{-6}$ (a)

11.11 $p = \gamma_1 \Delta h_1 + \gamma_2 \Delta h_2$ $\quad \gamma_2 = S\gamma_{H_2O}$

$= 9800 \times 2 + 1.04 \times 9800 \times 4$

$= 60\ 400$ Pa or 60.4 kPa (a)

11.12 $p = \gamma h$ $\quad \gamma_{Hg} = 13.6\gamma_{H_2O}$

$= 13.6 \times 9800 \times 0.6 = 80\ 000$ Pa (e)

11.13 $p = \gamma h$

$= 1.03 \times 9800 \times 1200 = 12.1 \times 10^6$ Pa

or 12.1 MPa (d)

11.14 $dp = -\gamma dz = -\rho g dz$ $\quad p = \rho RT$

$= -p \frac{g}{RT} dz$.

$\therefore \frac{dp}{p} = -\frac{g}{RT} dz$. $\quad \int_{100}^{p} \frac{dp}{p} = -\frac{g}{RT}\int_0^{2000} dz$.

$\therefore \ln \frac{p}{100} = -\frac{9.8}{287 \times 293} \times 2000$. $\quad \therefore p = 79.2$

(C)

11.15 $F = pA$.

$p = \frac{1000}{\pi \times 0.05^2} + 0.86 \times 9800 \times 0.2$

$= 129\ 000$ Pa.

$\therefore F = 129\ 000 \times \pi \times 0.0025^2 = 2.53$ N (C)

11.16 $p = \gamma h = (13.6 \times 9800) \times 0.2 = 26\ 700$ Pa

or 26.7 kPa.

(a)

11.17 $p + 9800 \times 0.3 = 13.6 \times 9800 \times 0.3$

$\therefore p = 37\,000$ Pa. (e)

11.18 $F = pA$

$= (20\,000 + 9800 \times 3)\pi \times 1^2$

$= 155\,000$ N. (c)

11.19 $F = \bar{p}A$

$= 9800 \times \frac{1.5}{2} \times 1.5^2 = 16\,500$ N. (b)

11.20 $7P = \frac{5}{3}F = \frac{5}{3}\gamma \bar{h} A$

$\therefore P = \frac{5}{21} \times 9800 \times 2 \times 15 = 70\,000$ N. (d)

11.21 All pressures on the curved section pass thru the center. Moments about the hinge give

$P = F_v = \gamma \times \text{Volume}$

$= 9800 \times \frac{9\pi}{4} \times 4 + 9800 \times 6 \times 3 \times 4$

$= 983\,000$ N. (d)

11.22 $l_p = l_c + \frac{\bar{I}}{l_c A} = 6 + \frac{5 \times 4^3/12}{6 \times 20} = 6.22$ m.

$4P = F \times 2.22 = \gamma \bar{h} A \times 2.22$.

$\therefore P = 9800 \times 6 \times 20 \times 2.22/4$

$= 653\,000$ N. (b)

11.23 $W = \gamma \Delta V$

$4 \times 1500 \times 9.8 = 9800 \times 6 \times 12 \times h$.

$\therefore h = 0.0833$ m. (e)

11.24 $25 = 100 - 9800V$. $\therefore V = 0.00765\ m^3$

$100 = 9800 S \times 0.00765$.

$\therefore S = 1.33$ (c)

11.25 $\Delta p = \Delta\gamma \times h$ $\quad p = \rho RT$

$= \left(\frac{1}{253} - \frac{1}{293}\right) \times \frac{100}{0.287} \times 3 \times 9.8$

$= 5.53$ Pa. (d)

11.26 $[p] = \frac{M}{LT^2}$ $[Q] = \frac{L^3}{T}$ $[D] = L$ $[\rho] = \frac{M}{L^3}$

$\frac{p}{\rho}\frac{D^4}{Q^2}$ (d)

11.27 $[\sigma] = \frac{M}{T^2}$ $[\rho] = \frac{M}{L^3}$ $[D] = L$ $[V] = \frac{L}{T}$

$\frac{\sigma}{\rho}\frac{1}{V^2}\frac{1}{D}$ (a)

11.28 inertial forces to viscous forces (b)

11.29 $(Fr)_m = (Fr)_p$ $\quad \frac{V_m^2}{l_m g} = \frac{V_p^2}{l_p g}$. $\therefore \frac{V_m^2}{V_p^2} = \frac{1}{20}$.

$Q^*_m = Q^*_p$

$\frac{Q_m}{V_m l_m^2} = \frac{Q_p}{l_p^2 V_p}$. $\therefore Q_m = 4 \times \frac{1}{20^2} \times \frac{1}{\sqrt{20}} = 0.0022$ (d)

11.30 $Re_m = Re_p$ $\quad \left(\frac{Vl}{\nu}\right)_m = \left(\frac{Vl}{\nu}\right)_p$. $\therefore \frac{V_m}{V_p} = \frac{l_p}{l_m} = 10$.

$\therefore V_m = 10V_p = 10 \times 10 = 100$ m/s. (e)

11.31 $(Fr)_m = (Fr)_p$ $\quad \frac{V_m^2}{l_m g} = \frac{V_p^2}{l_p g}$ $\quad \therefore \frac{V_m^2}{V_p^2} = \frac{1}{40}$

$(F_D^*)_m = (F_D^*)_p$

$\frac{(F_D)_m}{\rho_m V_m^2 l_m^2} = \frac{(F_D)_p}{\rho_p V_p^2 l_p^2}$. $\quad \therefore (F_D)_p = 10\frac{\rho_p}{\rho_m}\frac{V_p^2}{V_m^2}\frac{l_p^2}{l_m^2} = 10\frac{l_p^3}{l_m^3}$

$\therefore (F_D)_p = 10 \times 40^3 = 640\,000$ N. (a)

11.32 $(Fr)_m = (Fr)_p$. $\quad \frac{V_m^2}{l_m g} = \frac{V_p^2}{l_p g}$. $\therefore \frac{V_m^2}{V_p^2} = \frac{1}{10}$.

$\dot{W}^*_m = \dot{W}^*_p$.

$\frac{\dot{W}_m}{\rho_m V_m^3 l_m^2} = \frac{\dot{W}_p}{\rho_p V_p^3 l_p^2}$. $\quad \therefore \dot{W}_p = \frac{V_p^3}{V_m^3}\frac{l_p^2}{l_m^2}\dot{W}_m = 10^3\sqrt{10} \times 20$

$= 63\,250$ W. (d)

11.33 $V_2 = 20\pi \times 2^2/\pi \times 5^2 = 3.2$ m/s (e)

11.34 $V_2 = 20\pi \times 1^2/100 \times \pi \times 0.1^2 = 20$ m/s (d)

11.35 $V_2 \times 2\pi \times 40 \times 0.2 = 20 \times \pi \times 1^2$.

$\therefore V_2 = 1.25$ m/s (e)

11.36 $p = \rho V^2/2 = 1.23 \times 25^2/2 = 384$ Pa.

$F = pA = 384 \times \pi \times 0.075^2 = 6.79$ N. (b)

11.37 $V_2 A_2 = V_1 A_1$. $\therefore V_2 = V_1 \times 4^2/2^2 = 4V_1$.

$\frac{p_1}{\rho} + \frac{V_1^2}{2} = \frac{p_2}{\rho}\!\!\!\!\!\!\!\!\!\not\ \ {}^{0} + \frac{V_2^2}{2}$. $\quad \frac{700\,000}{1000} + \frac{V_1^2}{2} = \frac{16V_1^2}{2}$.

$\therefore V_1 = 9.66$. (e)

11.38 $p + \rho\frac{V^2}{2} + 9800 \times 0.1 = 13.6 \times 9800 \times 0.1 + p$.

$\therefore V^2 = 12.6 \times 9800 \times 0.1 \times 2/1000$.

$\therefore V = 4.97$ m/s (d)

11.39 Cavitation results if $p_2 = -100$ kPa.

$\frac{p_1}{\rho} + \frac{V_1^2}{2}\!\!\!\!\!\!\not\ \ {}^{0} = \frac{p_2}{\rho} + \frac{V_2^2}{2}$

$\frac{900\,000}{1000} = -\frac{100\,000}{1000} + \frac{V_2^2}{2}$. $\quad \therefore V_2 = 44.7$ m/s (c)

11.40 $F = \rho A V^2 = 1.2 \times \pi \times 0.01^2 \times 80^2 = 2.41$ (a)

11.41 $F = \rho A V^2 = 0.5 \times \pi \times 0.25^2 \times 1200^2 = 141\,000$ (c)

11.42 $-F = \rho A V(-V - V)$. $\quad \therefore F = 2\rho A V^2$.

$\therefore F = 2 \times 1000 \times 0.05 \times 0.8 \times 50^2 = 200\,000$ N. (a)

11.43 $p_1A_1 - F = \rho A_1V_1(V_2 - V_1)$

$V_2 = 4V_1 . \quad \frac{p_1}{\rho} + \frac{V_1^2}{2} = \frac{p_2 \nearrow^0}{\rho} + \frac{V_2^2}{2} = \frac{16V_1^2}{2}.$

$\therefore V_1 = \sqrt{\frac{800\,000}{1000} \times \frac{2}{15}} = 10.3 \qquad V_2 = 41.2$

$\therefore F = 800\,000 \times \pi \times 0.04^2 - 1000 \times \pi \times 0.04^2 \times 10.3 \times 30.9$
$= 2420 \text{ N}.$ (c)

11.44 The energy grade line. (b)

11.45 $\dot{W}_P = \gamma Q \frac{\Delta p}{\gamma} / 0.85$

$= 0.2 \times 500 / 0.85 = 117.6 \text{ kW}.$ (a)

11.46 $\dot{W}_T = \gamma Q \frac{\Delta p}{\gamma} \times 0.85$

$= 0.8 \times 600 \times 0.85 = 408 \text{ kW}$ (e)

11.47 $p_1 = p_2 + \frac{V_2^2}{2}\rho$ (manometer)

$-\frac{\dot{W}_T}{\gamma Q} = \frac{V_2^2}{2g} + \frac{p_2}{\gamma} - \frac{V_1^2}{2g} - \frac{p_1}{\gamma}.$ (100% efficient)

$\therefore \dot{W}_T = Q\frac{V_2^2}{2}\rho\eta = 20 \times \pi \times 0.1^2 \times \frac{20^2}{2} \times 1000 \times 0.88$

$= 111\,000 \text{ W}.$ (a)

11.48 varies with the velocity squared. (b)

11.49 increases linearly to the wall. (d)

11.50 vary as the 1/7 th power law. (c)

11.51 by curve E. (e)

11.52 by curve B. (b)

11.53 pressure varies linearly with distance. (a)

11.54 The Darcy-Weisbach eqn. (e)

11.55 found by using loss coefficients. (d)

11.56 shear stress varies linearly with radius. (d)

11.57 $h_L = f\frac{L}{d}\frac{V^2}{2g}. \quad \therefore f = 40\,\frac{0.1}{100}\,\frac{2 \times 9.8}{6^2} = 0.0218$ (b)

11.58 $V = Q/A = 0.02/\pi \times 0.03^2 = 7.07 \text{ m/s}.$

$Re = \frac{Vd}{\nu} = \frac{7.07 \times 0.06}{10^{-6}} = 4.2 \times 10^5 \quad \frac{e}{d} = \frac{.26}{60} = .0043$

from fig. 11.2 $f = 0.03$

$\dot{W}_P = \frac{\gamma Q}{\eta}\left(\frac{V_2^2}{2g}\nearrow^0 + \frac{p_2}{\gamma}\nearrow^0 + z_2 - \frac{V_1^2}{2g}\nearrow^0 - \frac{p_1}{\gamma}\nearrow^0 - z_1 + f\frac{L}{d}\frac{V^2}{2g} + K\frac{V^2}{2g}\right)$

$= \frac{9800 \times 0.02}{0.85}\left[80 - 20 + \left(0.03\,\frac{100}{.06} + \,:\, + .5\right)\frac{7.07^2}{2 \times 9.8}\right]$

$= 44\,000 \text{ W}.$ (e)

11.59 $V = Q/A = 0.006/\pi \times 0.02^2 = 4.77 \text{ m/s}$

$Re = \frac{4.77 \times 0.04}{10^{-6}} = 1.9 \times 10^5 \quad \frac{e}{d} = \frac{0.046}{40} = 0.0011$

from fig. 11.2 $f = 0.022$

$0 = \frac{p_B}{\gamma} - \frac{p_A}{\gamma} + f\frac{L}{d}\frac{V^2}{2g} + K\frac{V^2}{2g}$

$\therefore p_B = 510\,000 - \left(.022\,\frac{50}{.04} + 6.4 + 2 \times .9\right)\frac{4.77^2}{2 \times 9.8} \times 9800 = 104\,000 \text{ Pa}.$ (e)

11.60 $V = Q/A = 0.3/.15 \times .4 = 5 \text{ m/s} \quad R = \frac{40 \times 15}{110} = 5.45$

$Re = \frac{5 \times 4 \times .0545}{1.6 \times 10^{-5}} = 6.8 \times 10^4. \quad \therefore f = 0.02 \quad e/d = 0$

$\Delta p = f\frac{L}{4R}\frac{V^2}{2g}\gamma = .02\,\frac{500}{4 \times .0545}\,\frac{5^2}{2} \times 1.23 = 705 \text{ Pa}$ (e)

11.61 $\frac{A_c}{A_2} = .62 + .38 \times .5^3 = 0.668 \quad K_1 = (1 - .668)^2 = 0.11$

$0.11\frac{V_c^2}{2g} = K\frac{V_2^2}{2g}. \quad \therefore K = 0.11\left(\frac{A_2}{A_c}\right)^2 = .11 \times \frac{1}{.668^2} = 0.25$ (d)

11.62 $V = 0.02/\pi \times .03^2 = 7.07 \text{ m/s} \quad e/d = .15/60 = .0025$

$Re = \frac{7.07 \times .06}{10^{-6}} = 4.2 \times 10^5 \quad \therefore f = 0.024$

$L_e = Kd/f = 0.9 \times 0.06/0.025 = 2.16 \text{ m}$ (e)

11.63 $Q = \frac{1}{n}AR^{2/3}S^{1/2} = \frac{1}{.012} \times 6 \times .86^{2/3}.001^{1/2} = 14.3$

where $R = \frac{6}{3+4} = 0.86 \text{ m}$ (a)

11.64 $Q = \frac{1}{n}AR^{2/3}S^{1/2} = \frac{1}{.016}\,4h\left(\frac{4h}{4+2h}\right)^{2/3} \times .001^{1/2} = 10.$

trial-and-error : $h = 1.4 \text{ m}.$ (b)

11.65 $Q = \frac{1}{n}AR^{2/3}S^{1/2} = \frac{1}{.016}\,\pi \times 1^2 \times .5^{2/3}\,S^{1/2} = 10.$

$R = \frac{\pi \times 1^2}{\pi \times 2} = .5 \qquad \therefore S = 0.00654$ (b)

11.66 $\rho = p/RT = 500/.287 \times 313 = 5.57 \text{ kg/m}^3$

$\dot{m} = \rho AV = 5.57 \times \pi \times .05^2 \times 100 = 4.37 \text{ kg/s}$ (e)

11.67 $T_e = T_0\,(p_e/p_0)^{k-1/k} = 293 \times \left(\frac{100}{500}\right)^{.286} = 185 \text{ K}$

$V = Mc = 1 \times \sqrt{1.4 \times 287 \times 185} = 273 \text{ m/s}$ (e)

11.68 $T_e = T_0\,(p_e/p_0)^{k-1/k} = 293\left(\frac{100}{800}\right)^{.286} = 162 \text{ K}.$

$\therefore T_e = 162 - 273 = -111\,°\text{C}.$ (a)

11.69 $M_1 = \frac{V_1}{c_1} = \frac{700}{\sqrt{1.4 \times 287 \times 303}} = 2.01. \quad \therefore \frac{p_2}{p_1} = 4.54$ ↑ shock Table

$\therefore p_1 = p_2/4.54 = 500/4.54 = 110 \text{ kPa}$ (a)

11.70 $\frac{A}{A^*} = \frac{10^2}{6^2} = 2.78. \quad \therefore 2.5 < M < 2.6$ (Isentropic Flow Table) (e)

11.71 $\sin\phi = \frac{1}{M} = \frac{1}{2}. \quad \therefore \phi = 30°.$

$\tan 30° = \frac{1000}{L}.$

$\therefore L = 1732 \text{ m}.$

$\Delta t = \frac{L}{V} = \frac{1732}{2\sqrt{1.4 \times 287 \times 293}} = 2.52 \text{ s}$ (b)

V, φ, 1000 m, L

11.72 Assume $M_e = 0.3$, the maximum if ρ is assumed constant.

$\therefore V_e = M_ec_e = 0.3\sqrt{1.4 \times 287\,T_e} \quad \therefore V_e^2 = 36.2\,T_e$

energy: $0 = \frac{V_e^2 - V_0^2 \nearrow^0}{2} + c_p(T_e - T_0).$

$36.2\,T_e = 2 \times 1000(293 - T_e) \quad \therefore T_e = 287.8$

$p_0 = p_e(T_0/T_e)^{k/k-1} = 100\left(\frac{293}{287.8}\right)^{1.4/.4} = 106$ (d)

Solutions to Chapter 11
(English Units)

11.1 (a) is true for a liquid and low speed gas flows. (c) and (e) are true of gases. (d) is true for a liquid. (b) is correct. (b)

11.2 $\tau = \mu\, du/dy$. $\therefore \mu = \tau / du/dy$.
$[\mu] = \frac{F/L^2}{\frac{L}{T}/L} = \frac{FT}{L^2} = \frac{(ML/T^2)T}{L^2} = \frac{M}{LT}$ (d)

11.3 Viscosity μ varies with temperature only (a)

11.4 $dp = -\gamma dz = -\rho g dz$. $p = \rho RT$ (ideal gas)
$\therefore dp = -\frac{p}{RT} g dz$ or $\frac{dp}{p} = -\frac{g}{RT} dz$.
$\int \frac{dp}{p} = -\frac{g}{RT}\int dz$. $\therefore \ln p = -Cz$. $\therefore p = e^{-Cz}$. (e)

11.5 $T = \tau A r = \mu \frac{du}{dy} A r = \mu \frac{r\omega}{t} 2\pi r L\, r$.
$1.2 = \frac{2/12 \times 1000}{.04/12} \times 2\pi \times \frac{2}{12} \times \frac{2}{12} \times \frac{2}{12}\mu$. $\therefore \mu = 8.25 \times 10^{-4}$ (a)

11.6 $K = -V\frac{\Delta p}{\Delta V} = -60\frac{80}{0.12} = 40{,}000$ psi (c)

11.7 $\sigma 2\pi r = \gamma \pi r^2 L$ [sketch: $\sigma \pi d$, L, $W = \gamma V$]
$0.005 \times 2\pi \times \frac{.04}{12} = 62.4\pi\frac{.04^2}{144}L$
$\therefore L = 0.0481'$ or $0.577''$ (c)

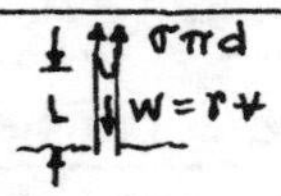

11.8 Cavitation occurs when the pressure reaches the vapor pressure = 0.34 psi abs. (see Table 11.1) (e)

11.9 $L = V\Delta t = \sqrt{kRT}\,\Delta t = \sqrt{1.4 \times 53.3 \times 32.2 \times 530} \times 1.2$
Assume $T = 70°F = 530°R$ $= 1354'$ (c)

11.10 $\nu = \frac{\mu}{\rho} = \frac{7.2 \times 10^{-5}}{1.3 \times 1.94} = 2.85 \times 10^{-5}$. ρ must be in slug/ft^3 (a)

11.11 $p = \gamma_1 \Delta h_1 + \gamma_2 \Delta h_2 = 62.4 \times 6 + (62.4 \times 1.04) \times 12$
$= 1153$ psf or 8.01 psi (a)

11.12 $p = \gamma h = (13.6 \times 62.4) \times 28/12 = 1980$ psf or 13.75 psi (c)

11.13 $p = \gamma h = (1.03 \times 62.4) \times 4000 = 257{,}000$ psf or 1785 psi (d)

11.14 $dp = -\gamma dz = -\rho g dz = -p\frac{g}{RT}dz$ using $p = \rho RT$
$\therefore \frac{dp}{p} = -\frac{g}{RT}dz$. $\int_{14.7}^{p} dp/p = -\frac{g}{RT}\int_0^{6000} dz$.
$\therefore \ln\frac{p}{14.7} = -\frac{32.2}{53.3 \times 32.2 \times 530} \times 6000$. $\therefore p = 11.89$ psia (c)

11.15 $F = pA$. $p = \frac{200}{\pi \times 2^2} + 0.86 \times 62.4 \times \frac{10}{12}/144 = 16.2$ psi
$\therefore F = 16.2 \times \pi \times (1/2)^2 = 12.74$ lb. (c)

11.16 $p = \gamma h = (13.6 \times 62.4) \times \frac{10}{12}/144 = 4.91$ psi (a)

11.17 $p + 62.4 \times 1 = 13.6 \times 62.4 \times 1$. $\therefore p = 786$ psf
or $p = 5.46$ psi (e)

11.18 $F = \bar{p}A = (3 \times 144 + 62.4 \times 10)\pi \times 1^2 = 3318$ lb. (c)

11.19 $F = \bar{p}A = 62.4 \times \frac{5}{2} \times 5^2 = 3900$ lb (b)

11.20 $20P = \frac{15}{3}(62.4 \times 6 \times 150)$. $\therefore P = 14{,}040$ lb. (d)

11.21 All pressures on the curved section pass thru the center. Moments about the hinge give
$P = F_V = \gamma\,\text{Volume} = 62.4 \times \pi 6^2/4 \times 12 + 62.4 \times 6 \times 8 \times 12$
$= 73{,}580$ lb. (d)

11.22 $\ell_p = \ell_c + \frac{I}{A\ell_c} = 15 + \frac{15 \times 10^3/12}{150 \times 15} = 15.56'$
$10P = 62.4 \times 15 \times 150 \times 5.56$ $\therefore P = 78{,}000$ lb. (b)

11.23 $W = \gamma V$. $4 \times 3200 = 62.4 \times 20 \times 40\, h$ $\therefore h = .256'$ or $3.08''$ (e)

11.24 $6 = 25 - 62.4V$. $\therefore V = 0.3045$ ft^3.
$25 = 62.4 S \times 0.3045$. $\therefore S = 1.316$ (c)

11.25 $\Delta p = \Delta\gamma \times h = \left(\frac{1}{450} - \frac{1}{530}\right)\frac{14.7 \times 144}{53.3} \times 10 = 0.1332$ psf.
We have used $p = \rho RT$ (d)

11.26 $[p] = \frac{M}{LT^2}$ $[Q] = \frac{L^3}{T}$ $[D] = L$ $[\rho] = \frac{M}{L^3}$.
$\frac{M}{LT^2} \cdot \frac{L^3}{M} \cdot \frac{T^2}{L^6} \cdot L^4 = p \cdot \frac{1}{\rho} \cdot \frac{1}{Q^2} \cdot D^4 = \frac{pD^4}{\rho Q^2}$ (d)

11.27 $[\sigma] = \frac{M}{T^2}$ $[\rho] = \frac{M}{L^3}$ $[D] = L$ $[V] = \frac{L}{T}$.
$\frac{M}{T^2} \cdot \frac{T^2}{L^2} \cdot \frac{L^3}{M} \cdot \frac{1}{L} = \sigma \cdot \frac{1}{V^2} \cdot \frac{1}{\rho} \cdot \frac{1}{D} = \frac{\sigma}{\rho D V^2}$ (a)

11.28 Inertial forces to viscous forces. (b)

11.29 $(Fr)_m = (Fr)_p$. $\therefore \frac{V_m^2}{\ell_m g} = \frac{V_p^2}{\ell_p g}$. $\therefore \frac{V_m^2}{V_p^2} = \frac{1}{20}$.
$Q_m^* = Q_p^*$ or $\frac{Q_m}{V_m \ell_m^2} = \frac{Q_p}{V_p \ell_p^2}$. $\therefore Q_m = 120 \times \frac{1}{20^2} \times \frac{1}{\sqrt{20}} = .0671$ cfs (d)

11.30 $Re_m = Re_p$. $\left(\frac{V\ell}{\nu}\right)_m = \left(\frac{V\ell}{\nu}\right)_p$. $\therefore \frac{V_m}{V_p} = \frac{\ell_p}{\ell_m} = 10$.
$\therefore V_m = 10V_p = 10 \times 30 = 300$ fps. (a)

11.31 $Fr_m = Fr_p$. $\left(\frac{V^2}{\ell g}\right)_m = \left(\frac{V^2}{\ell g}\right)_p$. $\therefore \frac{V_p^2}{V_m^2} = 40$.
$(F_D)_m^* = (F_D)_p^*$ or $\frac{(F_D)_m}{\rho_m V_m^2 \ell_m^2} = \frac{(F_D)_p}{\rho_p V_p^2 \ell_p^2}$. $\therefore (F_D)_p = 2\frac{\rho_p}{\rho_m}\frac{V_p^2}{V_m^2}\frac{\ell_p^2}{\ell_m^2} = 2\frac{\ell_p^3}{\ell_m^3}$.
$\therefore (F_D)_p = 2 \times 40^3 = 128{,}000$ lb. (a)

11.32 $Fr_m = Fr_p$. $\left(\frac{V^2}{\ell g}\right)_m = \left(\frac{V^2}{\ell g}\right)_p$. $\therefore \frac{V_p^2}{V_m^2} = 10$.
$\dot{W}_p^* = \dot{W}_m^*$. or $\frac{\dot{W}_m}{\rho_m V_m^3 \ell_m^2} = \frac{\dot{W}_p}{\rho_p V_p^3 \ell_p^2}$. $\therefore \dot{W}_p = .06\frac{V_p^3}{V_m^3}\frac{\ell_p^2}{\ell_m^2} = 189.7$ Hp (d)

11.33 $V_2 = 60\,\pi \times (\frac{1}{2})^2/\pi \times 1.25^2 = 9.6$ fps. (e)

11.34 $V_2 = 60\,\pi \times (\frac{1}{2})^2/100\pi \times .05^2 = 60$ fps (d)

11.35 $V_2 \times 0.1 \times 2\pi \times 20 = 60 \times \pi \times (\frac{1}{2})^2$. $\therefore V_2 = 3.75$ fps. (e)

11.36 $p = \rho V^2/2 = .0023 \times 90^2/2 = 9.32$ psf.
$F = pA = 9.32\,\pi \times 3^2/144 = 1.83$ lb. (b)

11.37 $V_2A_2 = V_1A_1$. $\therefore V_2 = V_1 \times 2^2/1^2 = 4V_1$.

$\frac{p_1}{\rho_1} + \frac{V_1^2}{2} = \frac{p_2^{\,0}}{\rho} + \frac{V_2^2}{2}$. $\frac{100\times144}{1.94} + \frac{V_1^2}{2} = \frac{16V_1^2}{2}$. $\therefore V_1 = 31.5$ (e)

11.38 $p + \rho\frac{V^2}{2} + 62.4\times\frac{4}{12} = p + 13.6\times62.4\times\frac{4}{12}$.

Using $\rho = 1.94\ slug/ft^3$ $V = 16.44$ fps (d)

11.39 Cavitation results if $p_2 = -14.7$ psi.

$p_1/\rho + V_1^2{}^{\,0}/2 = p_2/\rho + V_2^2/2$.

$150\times144/1.94 = -14.7\times144/1.94 + V_2^2/2$. $\therefore V_2 = 156$ (c)

11.40 $F = \rho AV^2 = .0024\times\pi\times\frac{.5^2}{144}\times200^2 = 0.524$ lb. (a)

11.41 $F = \rho AV^2 = .001\times\pi\times\frac{10^2}{144}\times4000^2 = 34{,}900$ lb. (c)

11.42. $-F = \rho AV(-V-V)$. $\therefore F = 2\rho AV^2$.

$\therefore F = 2\times1.94\times\frac{2\times30}{144}\times150^2 = 36{,}400$ lb. (a)

11.43 $V_2 = 4V_1$. $\frac{p_1}{\rho} + \frac{V_1^2}{2} = \frac{p_2^{\,0}}{\rho} + \frac{V_2^2}{2} = \frac{16V_1^2}{2}$. $\therefore V_1^2 = \frac{2p_1}{15\rho}$.

$\therefore V_1 = \sqrt{\frac{2\times200\times144}{15\times1.94}} = 44.5$. $\therefore V_2 = 178$.

$p_1A_1 - F = \rho A_1V_1(V_2 - V_1)$.

$\therefore F = 200\pi\times2^2 - 1.94\times\pi\times\frac{2^2}{144}\times44.5\times133.5 = 1508$. (c)

11.44 The energy grade line (b)

11.45 $\dot{W}_P = \gamma Q\frac{\Delta p}{\gamma}/0.85 = 6\times75\times144/.85 = 76{,}200$ ft-lb/sec or 139 Hp. (a)

11.46 $\dot{W}_T = \gamma Q\frac{\Delta p}{\gamma}\times0.85 = 3\times(90\times144)\times0.85 = 33{,}050$ $\frac{ft\text{-}lb}{sec}$ or 60.1 Hp (e)

11.47 manometer: $p_1 = p_2 + \rho V_2^2/2$.

$-\frac{\dot{W}_T}{\gamma Q} = \frac{V_2^2}{2g} + \frac{p_2}{\gamma} - \frac{V_1^2}{2g} - \frac{p_1}{\gamma}$ (100% efficient)

$\therefore \dot{W}_T = Q\frac{V_2^2}{2}\rho\eta = (60\times\pi\times\frac{4^2}{144})\frac{60^2}{2}\times1.94\times.88 = 64{,}400$ $\frac{ft\text{-}lb}{sec}$ or 117 Hp (a)

11.48 increases with the velocity squared. (b)

11.49 increases linearly to the wall. (d)

11.50 vary as the 1/7 th power law. (c)

11.51 by curve E. (e)

11.52 by curve B. (b)

11.53 pressure varies linearly with distance. (a)

11.54 The Darcy-Weisbach eqn. (e)

11.55 found by using loss coefficients. (d)

11.56 shear stress varies linearly w/ radius. (d)

11.57 $h_L = f\frac{L}{d}\frac{V^2}{2g}$. $\therefore f = 120\,\frac{4/12}{300}\,\frac{2\times32.2}{20^2} = .0215$ (b)

11.58 $V = Q/A = \frac{0.6}{\pi\times2^2/144} = 6.875$ fps.

$Re = \frac{Vd}{\nu} = 6.875\times\frac{4}{12}/10^{-5} = 2.3\times10^5$ $\frac{e}{d} = \frac{.00085}{4/12} = .0025$

from fig 11.2, $f = .02$

$\dot{W}_P = \frac{\gamma Q}{\eta}\left(\frac{V_2^2}{2g} + \frac{p_2}{\gamma} + z_2 - \frac{V_1^2}{2g} - \frac{p_1}{\gamma} - z_1 + f\frac{L}{d}\frac{V^2}{2g} + K\frac{V^2}{2g}\right)$

$= \frac{62.4\times.6}{.85}\left[200 - 50 + (.02\,\frac{300}{4/12} + 1 + .5)\frac{6.875^2}{64.4}\right] = 7310$ $\frac{ft\text{-}lb}{sec}$

or 13. Hp (c)

11.59 $V = Q/A = \frac{0.2}{\pi\times1^2/144} = 9.167$ fps

$Re = \frac{Vd}{\nu} = 9.167\times\frac{2}{12}/10^{-5} = 1.53\times10^5$. $\frac{e}{d} = \frac{.00015}{2/12} = .0009$

from fig 11.2 $f = .021$. $0 = \frac{p_B - p_A}{\gamma} + f\frac{L}{d}\frac{V^2}{2g} + K\frac{V^2}{2g}$.

$\therefore p_B = 70 - (.021\,\frac{150}{2/12} + 6.4 + 2\times.9)\frac{9.167^2}{2\times32.2}\times62.4/144 = 54.7$ psi (e)

11.60 $V = Q/A = 40/\frac{40}{12}\times\frac{15}{12} = 9.6$ fps. $R = \frac{40\times15}{110} = 5.455''$.

$Re = \frac{4\times9.6\times5.45/12}{1.6\times10^{-4}} = 1.10\times10^5$. With $\frac{e}{d} = 0$, $f = 0.0175$

$\gamma = p/RT = \frac{14.7\times144}{53.3\times530} = .075$

$\Delta p = f\frac{L}{4R}\frac{V^2}{2g}\gamma = .0175\,\frac{1500}{4\times5.45/12}\,\frac{9.6^2}{64.4}\times.075 = 1.55$ psf (b)

11.61 $A_c/A_2 = .62 + .38(.5)^3 = .6675$. $K_1 = (1 - .6675)^2 = .111$.

$.111V_c^2/2g = KV_2^2/2g$. $\therefore K = .111(A_2/A_c)^2 = .111/.6675^2 = .249$ (d)

11.62 $V = Q/A = .6/\pi\times2^2/144 = 6.875$ $e/d = .0005/\frac{1}{3} = .0015$

$Re = \frac{6.875\times4/12}{10^{-5}} = 2.3\times10^5$. $\therefore f = .022$.

$L_e = Kd/f = .9\times\frac{4}{12}/.022 = 13.6'$ (e)

11.63 $Q = \frac{1.49}{n}AR^{2/3}S^{1/2} = \frac{1.49}{.012}\,60\times2.73^{2/3}\times.001^{1/2} = 460$ cfs

where $R = 60/22 = 2.73'$ (a)

11.64 $Q = \frac{1.49}{n}AR^{2/3}S^{1/2} = \frac{1.49}{.016}\,12h\left(\frac{12h}{12+2h}\right)^{2/3}\times.001^{1/2} = 300$

Trial-and-error: $h = 4.52'$ (b)

11.65 $Q = \frac{1.49}{n}AR^{2/3}S^{1/2} = \frac{1.49}{.016}\,\pi\times3^2\times1.5^{2/3}\,S^{1/2} = 100$,

where $R = \frac{A}{P} = \frac{\pi r^2}{2\pi r} = \frac{r}{2} = 1.5'$. $\therefore S = .00084$ (b)

11.66 $\rho = p/RT = 70\times144/53.3\times32.2\times560 = .0105$ slug/ft³

$\dot{m} = \rho AV = .0105\times\frac{\pi\times2^2}{144}\times300 = 0.275$ slug/sec. (e)

11.67 $T_e = T_o(p_e/p_o)^{k-1/k} = 530(\frac{14.7}{75})^{1.4-1/1.4} = 339°R$

$V = Mc = 1\sqrt{1.4\times53.3\times32.2\times339} = 903$ fps. (e)

11.68 $T_e = T_o(p_e/p_o)^{k-1/k} = 530(\frac{14.7}{120})^{.2857} = 291°R$.

or $291 - 460 = -169°F$. (a)

11.69 $M_1 = \frac{V_1}{c_1} = \frac{2000}{\sqrt{1.4\times53.3\times32.2\times540}} = 1.76$ $\therefore \frac{p_2}{p_1} = 3.44$ (from Normal Shock Table)

$\therefore p_1 = p_2/3.44 = 70/3.44 = 20.3$ psia (a)

11.70 $\frac{A}{A^*} = \frac{4^2}{3^2} = 1.78$. $\therefore 2.0 < M_1 < 2.1$ (Isentropic flow Table) (e)

11.71 $\sin\phi = \frac{1}{M} = \frac{1}{2}$. $\therefore \phi = 30°$

$\tan 30° = \frac{3000}{L}$. $\therefore L = 5196'$

$\Delta t = \frac{L}{V} = \frac{5196}{2\sqrt{1.4\times53.3\times32.2\times530}} = 2.3$ sec. Assume $T = 70°$ (b)

11.72 Assume $M_e = 0.3$, the maximum if the density is assumed constant (i.e., $\rho_e = 0.97\rho_o$).

$V_e = M_ec_e = 0.3\sqrt{1.4\times53.3\times32.2\,T_e}$ $\therefore V_e^2 = 216T_e$.

energy: $0 = \frac{V_e^2 - V_o^2{}^{\,0}}{2} + c_p(T_e - T_o)$.

$\therefore \frac{216T_e}{2} = 0.24\times(778\times32.2)(530 - T_e)$. $\therefore T_e = 520.6$ (unit conversions $\frac{BTU}{lb} \to \frac{ft\text{-}lb}{slug}$)

$p_o = p_e(T_o/T_e)^{\frac{k}{k-1}} = 14.7(\frac{530}{520.6})^{\frac{1.4}{.4}} = 15.6$ psia

(d)

Solutions to Chapter 12

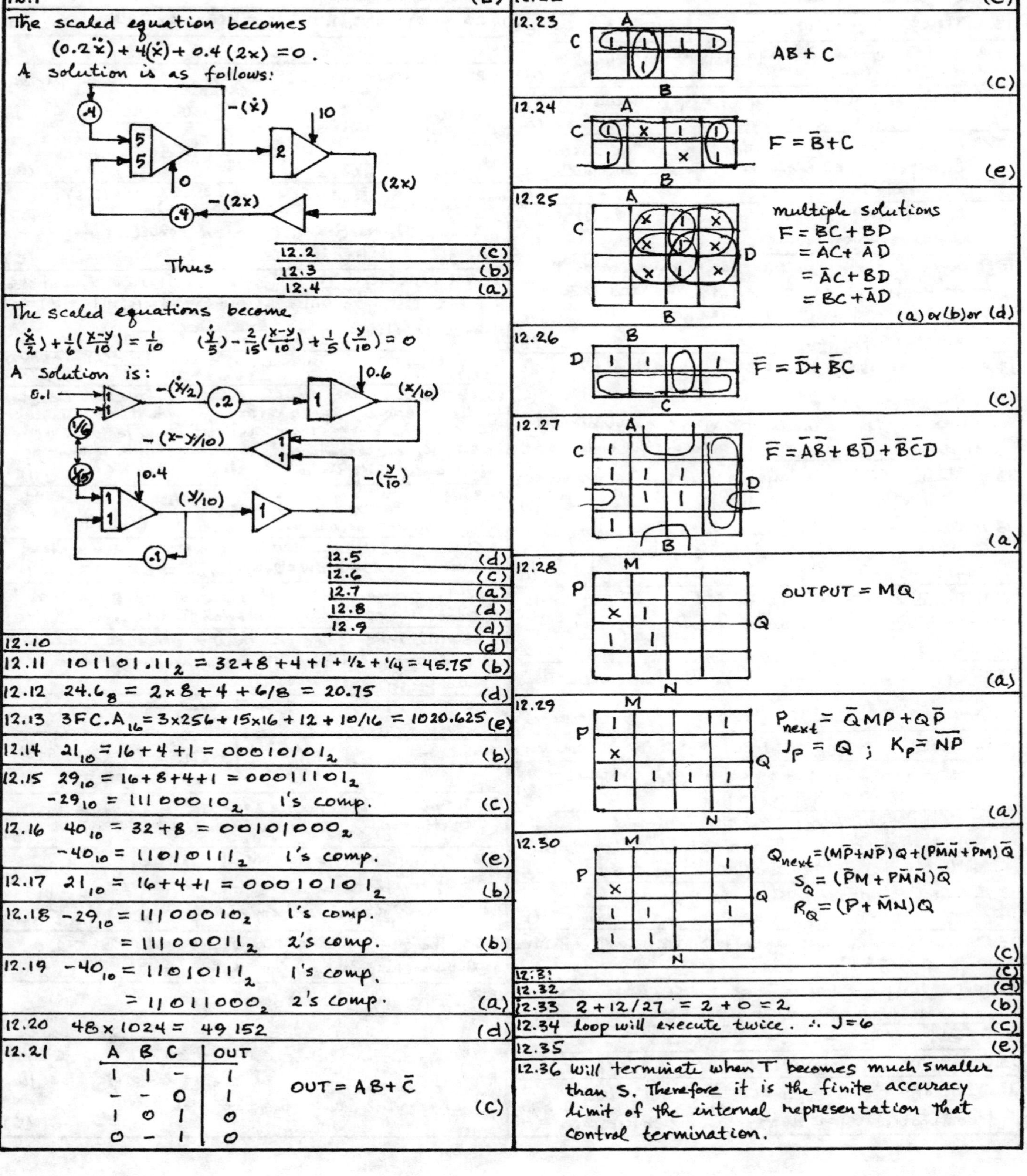

12.1 (b)

The scaled equation becomes

$(0.2\ddot{x}) + 4(\dot{x}) + 0.4(2x) = 0.$

A solution is as follows:

Thus

12.2 (c)

12.3 (b)

12.4 (a)

The scaled equations become

$(\frac{\dot{x}}{2}) + \frac{1}{6}(\frac{x-y}{10}) = \frac{1}{10}$ $\quad (\frac{\dot{y}}{5}) - \frac{2}{15}(\frac{x-y}{10}) + \frac{1}{5}(\frac{y}{10}) = 0$

A solution is:

12.5 (d)

12.6 (c)

12.7 (a)

12.8 (d)

12.9 (d)

12.10 (d)

12.11 $101101.11_2 = 32+8+4+1+1/2+1/4 = 45.75$ (b)

12.12 $24.6_8 = 2\times8+4+6/8 = 20.75$ (d)

12.13 $3FC.A_{16} = 3\times256+15\times16+12+10/16 = 1020.625$ (e)

12.14 $21_{10} = 16+4+1 = 00010101_2$ (b)

12.15 $29_{10} = 16+8+4+1 = 00011101_2$

$-29_{10} = 11100010_2$ 1's comp. (c)

12.16 $40_{10} = 32+8 = 00101000_2$

$-40_{10} = 11010111_2$ 1's comp. (e)

12.17 $21_{10} = 16+4+1 = 00010101_2$ (b)

12.18 $-29_{10} = 11100010_2$ 1's comp.

$= 11100011_2$ 2's comp. (b)

12.19 $-40_{10} = 11010111_2$ 1's comp.

$= 11011000_2$ 2's comp. (a)

12.20 $48\times1024 = 49\,152$ (d)

12.21

A	B	C	OUT
1	1	–	1
–	–	0	1
1	0	1	0
0	–	1	0

$OUT = AB + \bar{C}$ (c)

12.22 (e)

12.23 $AB + C$ (c)

12.24 $F = \bar{B} + C$ (e)

12.25 multiple solutions

$F = BC + BD$

$= \bar{A}C + \bar{A}D$

$= \bar{A}C + BD$

$= BC + \bar{A}D$

(a) or (b) or (d)

12.26 $\bar{F} = \bar{D} + \bar{B}C$ (c)

12.27 $\bar{F} = \bar{A}\bar{B} + B\bar{D} + \bar{B}\bar{C}D$ (a)

12.28 OUTPUT = MQ (a)

12.29 $P_{next} = \bar{Q}MP + Q\bar{P}$

$J_P = Q$; $K_P = \overline{NP}$ (a)

12.30 $Q_{next} = (M\bar{P} + N\bar{P})Q + (\bar{P}M\bar{N} + \bar{P}M)\bar{Q}$

$S_Q = (\bar{P}M + P\bar{M}\bar{N})\bar{Q}$

$R_Q = (P + \bar{M}N)Q$ (c)

12.31 (c)

12.32 (d)

12.33 $2 + 12/27 = 2 + 0 = 2$ (b)

12.34 loop will execute twice. ∴ J=6 (c)

12.35 (e)

12.36 Will terminate when T becomes much smaller than S. Therefore it is the finite accuracy limit of the internal representation that control termination.

Solutions to Chapter 13

13.1 $\mathcal{L}\{a\underline{u}(t)\} = a/s$ (c)

13.2 $H(s) = Y(s)/U(s)$ (e)

13.3 Z-transform is used for discrete-time systems. (d)

13.4 Stability if and only if eigenvalues have negative real part. (a)

13.5 A root locus shows the path of the closed-loop poles (i.e., eigenvalues) as a parameter is varied. (c)

13.6 A good first-order design has a large negative eigenvalue. (d)

13.7 Since a step input is the integration of the impulse function, linearity implies that a step response is the integration of the impulse response. (b)

13.8 Using Eqs 13.2.1 & 13.2.2 The characteristic eq. is $1 + \frac{1}{s(s+3)}[K_p + K_i/s] = 0$. Clearing the denominator yields a polynomial of degree 3 on the l.h.s. ∴ The system order is 3. (c)

13.9 State variable techniques are useful for linear and for nonlinear systems. (d)

13.10 State models with matrices A, B, C, D varying with t can be utilized for time-varying systems. (a)

13.11 The characteristic eqn $s^2 + 6s + 25 = 0$ yields a complex pair of eigenvalues with negative real part. (a)

13.12 $Y(s) = \frac{50}{s^2+6s+25}\left(\frac{10}{s}\right)$. The Final Value Thm gives $y(\infty) = \lim_{s\to 0} sY(s) = 500/25 = 20$. (d)

13.13 Stability means $h(t) \to 0$ as $t \to \infty$. (a)

13.14 $s^2 + 6s + 25 = s^2 + 2\zeta\omega_n s + \omega_n^2$ implies $\omega_n = 5$ and $\zeta = 0.6$. (c)

13.15 Stability implies that the transients damp out leaving an oscillatory steady-state output with the same frequency as the input. (c)

13.16 This 1st order system has $s+2 = s + 1/\tau$. Thus $\tau = 0.5$ (b)

13.17 The system is stable with transient of the form $Ce^{-t/\tau}$. (d)

13.18 $|H(j\omega)|$ has value 3 at $\omega = 0$ and decreases to 0 as ω increases thru positive values. (a)

13.19 $BW = 1/\tau = 2$. (a)

13.20 At $\omega = BW$, $|H(j\omega)| = H_{max}/\sqrt{2} = 3\sqrt{2}/2$ (e)

13.21 $H(s) = \frac{4}{s+2} + \frac{8}{(s+2)^2} = \frac{4s+16}{(s+2)^2}$. Thus the denominator degree and system order are 2. (b)

13.22 A repeated eigenvalue occurs at $s = -2$ (b)

13.23 $y(\infty) = \lim_{s\to 0}\left[H(s)\frac{1}{s}\right]s = H(0) = 16/4 = 4$ (e)

13.24 Eigenvalue is at -2. (d)

13.25 $Y(s) = \frac{4s+16}{s(s+2)^2} = \frac{a}{s} + \frac{b}{(s+2)^2} + \frac{c}{s+2}$ using partial where $c = \frac{d}{ds}\left[\frac{4s+16}{s}\right]_{s=-2} = \left.\frac{s(4)-(4s+16)}{s^2}\right|_{s=-2} = -4$. The step response is $y(t) = a + bte^{-2t} + ce^{-2t}$ where $a = H(0) = 4$. (e)

13.26 Using Eq. 13.2.1 $1 + K\frac{50}{s(.05s+1)}(1+K_t s) = 0$. Clearing The fraction yields a polynomial of degree 2 on The left hand side. (c)

13.27 $E(s) = U(s) - Y(s) = [1 - H(s)]U(s)$. With unit step input $U(s) = 1/s$. Also $e(\infty) = \lim_{s\to 0} sE(s) = 1 - H(0) = 0$ because $H(s) = \frac{50K}{s(.05s+1) + 50K(1+K_t s)}$ (b)

13.28 (c) $U(s) = 1/s^2$. Then $e(\infty) = \lim_{s\to 0}\frac{1-H(s)}{s} = \frac{1+50KK_t}{50K}$

13.29 The open-loop transfer function of the root locus has poles at $s=0$ and $s=-20$ corresponding to the eigenvalues for $K=0$. As $K\to\infty$ These eigenvalues move to The zero at $s = -1/K_t$ and asymtotically to $s = -\infty$, while remaining in the left-half s-plane. ∴ stable for $0 < K < \infty$. (e)

13.30 for $K_t = 0$ $G(s) = \frac{50}{s(.05s+1)}$. As K increases the transients become more oscillatory (still stable). (b)

13.31 Rise time t_r and BW in Hz for 2nd order systems are related by $BW = 2.4/2\pi t_r$. ∴ $BW = 9.5$ Hz. (b)

13.32 The primary effect of K_t designed properly is to reduce the oscillations & shorten The rise time. (e)

13.33 With integral control, there is one more eigenvalue & thus poorer transient response. However, the steady-state performance is enhanced. (b)

13.34 The A matrix is 2x2. ∴ The no. of eigenvalues and system order is 2. (c)

13.35 $|sI - A| = 0 = \begin{vmatrix} s+5 & 6 \\ -1 & s \end{vmatrix} = s^2 + 5s + 6 = (s+2)(s+3)$. ∴ $m_1 = -2$, $m_2 = -3$. (d)

13.36 Since A is constant and The eigenvalues are strictly negative, the system is stable & time invariant. (b)

13.37 Controllability matrix $M = [B \;\; AB] = \begin{bmatrix} 1 & -11 \\ 1 & 1 \end{bmatrix}$ Observability matrix $Q = [C^T \;\; A^TC^T] = \begin{bmatrix} 1 & -5 \\ 0 & -6 \end{bmatrix}$. Both have nonzero determinant and rank 2, which is n. (a)

13.38 The negative real eigenvalues mean that The impulse response decays exponentially. (a)

13.39 $r_{31} = \frac{\begin{vmatrix} 1 & 6 \\ 1 & 8 \end{vmatrix}}{-1} = \frac{8-6}{-1} = -2$. (e)

13.40 The first column entries are +1, +1, −2, +14, +12. Two changes in algebraic sign mean two eigenvalues with positive real part. (c)

Index

Index

Index

Please send a copy

of this book to

____

____

____

____

and bill the above person directly.

I just finished the exam

and you should make the following changes and/or additions to this book:

I think there are errors

on pages ______________. They should read as follows: